VOLUME ONE HUNDRED AND FORTY TWO

ADVANCES IN AGRONOMY

Quantifying and Managing Soil Functions in Earth's Critical Zone: Combining Experimentation and Mathematical Modelling

ADVANCES IN AGRONOMY

VOLUME ONE HUNDRED AND FORTY TWO

ADVANCES IN AGRONOMY

Quantifying and Managing Soil Functions in Earth's Critical Zone: Combining Experimentation and Mathematical Modelling

Edited by

STEVEN A. BANWART
Kroto Research Institute,
The University of Sheffield, Sheffield;
Integrating Chair in Soil/Agriculture/Water Research,
School of Earth and Environment, University of Leeds,
Leeds, United Kingdom

DONALD L. SPARKS
Delaware Environmental Institute,
Interdisciplinary Science and Engineering Laboratory,
University of Delaware, Newark, DE, United States

ELSEVIER

ACADEMIC PRESS
An imprint of Elsevier
elsevier.com

Academic Press is an imprint of Elsevier
50 Hampshire Street, 5th Floor, Cambridge, MA 02139, United States
525 B Street, Suite 1800, San Diego, CA 92101–4495, United States
The Boulevard, Langford Lane, Kidlington, Oxford OX5 1GB, United Kingdom
125 London Wall, London EC2Y 5AS, United Kingdom

First edition 2017

ISBN: 978-0-12-812222-8
ISSN: 0065-2113

Publisher: Zoe Kruze
Acquisition Editor: Alex White
Editorial Project Manager: Helene Kabes
Production Project Manager: Magesh Kumar Mahalingam
Cover Designer: Matthew Limbert

Typeset by SPi Global, India

CONTENTS

CONTRIBUTORS

M. Andrianaki
Geological Institute, ETH Zürich, Zürich, Switzerland

S.A. Banwart
Kroto Research Institute, The University of Sheffield, Sheffield; Integrating Chair in Soil/Agriculture/Water Research, School of Earth and Environment, The University of Leeds, Leeds, United Kingdom

S.M. Bernasconi
Geological Institute, ETH Zürich, Zürich, Switzerland

A. Blaud
Kroto Research Institute, The University of Sheffield, Sheffield, United Kingdom

W.E.H. Blum
Institute of Soil Research, University of Natural Resources and Life Sciences (BOKU), Vienna, Austria

F. Chabaux
Laboratory of Hydrology and Geochemistry of Strasbourg (LHyGeS), University of Strasbourg, Strasbourg, France

B. Davíðsdóttir
Environment and Natural Resources, University of Iceland, Gimli, Reykjavík, Iceland

P. de Ruiter
Wageningen University, Wageningen, The Netherlands

D.M. de Souza
Institute for Environment and Sustainability, Joint Research Centre—European Commission, Ispra, Italy

C. Duffy
The Pennsylvania State University, University Park, PA, United States

G.V. Giannakis
Technical University of Crete, School of Environmental Engineering, Chania, Greece

X.-Z. Han
National Observation Station of Hailun Agro-Ecology System, Key Laboratory of Mollisols Agroecology, Northeast Institute of Geography and Agroecology, Chinese Academy of Sciences, Harbin, PR China

J.Ö.G. Jónsson
Environment and Natural Resources, University of Iceland, Gimli, Reykjavík, Iceland

M. Kercheva
Institute of Soil Science, Agrotechnology and Plant Protection "N. Poushkarov", Sofia, Bulgaria

M. Kotronakis
Technical University of Crete, School of Environmental Engineering, Chania, Greece

P. Krám
Czech Geological Survey, Prague, Czech Republic

G.J. Lair
Institute of Soil Research, University of Natural Resources and Life Sciences (BOKU), Vienna, Austria

N. Li
National Observation Station of Hailun Agro-Ecology System, Key Laboratory of Mollisols Agroecology, Northeast Institute of Geography and Agroecology, Chinese Academy of Sciences, Harbin, PR China

K. Liu
National Observation Station of Hailun Agro-Ecology System, Key Laboratory of Mollisols Agroecology, Northeast Institute of Geography and Agroecology, Chinese Academy of Sciences, Harbin, PR China

Y.-L. Liu
National Engineering Laboratory for Efficient Utilization of Soil and Fertilizer, College of Resources and Environment, Shandong Agricultural University, Taian; National Observation Station of Hailun Agro-ecology System, Key Laboratory of Mollisols Agroecology, Northeast Institute of Geography and Agroecology, Chinese Academy of Sciences, Harbin, PR China

L. Lundin
Swedish University of Agricultural Sciences, Uppsala, Sweden

M. Menon
Kroto Research Institute, The University of Sheffield, Sheffield; School of Earth and Environment, The University of Leeds, Leeds, United Kingdom

K. Moirogiorgou
Technical University of Crete, School of Electronic & Computer Engineering, Chania, Greece

L. Montanarella
European Commission, Joint Research Centre, Sustainable Resources, Ispra (VA), Italy

D. Moraetis
Sultan Qaboos University, College of Science, Al Khoudh, Oman

N.P. Nikolaidis
Hydrogeochemical Engineering and Soil Remediation Laboratory; School of Environmental Engineering, Technical University of Crete, University Campus, Chania, Greece

M. Novak
Czech Geological Survey, Prague, Czech Republic

P. Panagos
European Commission, Joint Research Centre, Sustainable Resources, Ispra (VA), Italy

S.K. Panakoulia
School of Environmental Engineering, Technical University of Crete (TUC), Chania, Greece

N.V. Paranychianakis
School of Environmental Engineering, Technical University of Crete, University Campus, Chania, Greece

Y.-F. Qiao
National Observation Station of Hailun Agro-Ecology System, Key Laboratory of Mollisols Agroecology, Northeast Institute of Geography and Agroecology, Chinese Academy of Sciences, Harbin, PR China

K.V. Ragnarsdottir
School of Engineering and Natural Sciences, University of Iceland, Reykjavík, Iceland

D.A. Robinson
Centre for Ecology & Hydrology, Environment Centre Wales, Bangor, United Kingdom

S. Rousseva
Institute of Soil Science, Agrotechnology and Plant Protection "N. Poushkarov", Sofia, Bulgaria

E.C. Rowe
Centre for Ecology & Hydrology, ECW, Bangor, United Kingdom

J. Schiefer
University of Natural Resources and Life Sciences (BOKU), Vienna, Austria

T. Shishkov
Institute of Soil Science, Agrotechnology and Plant Protection "N. Poushkarov", Sofia, Bulgaria

F.E. Stamati
Technical University of Crete, School of Environmental Engineering, Chania, Greece

J. Valstar
Deltares Subsurface and Groundwater Systems, Utrecht, The Netherlands

B. van der Zaan
Deltares, Subsurface and Groundwater Systems, Princetonlaan, Utrecht, The Netherlands

P. van Gaans
Deltares, Utrecht, The Netherlands

Y. Wang
Key Laboratory of Agro-ecological Processes in Subtropical Regions, Changsha Research Station for Agricultural and Environmental Monitoring, Institute of Subtropical Agriculture, Chinese Academy of Sciences, Hunan, China

L. Weng
Wageningen University, Wageningen, The Netherlands

T. White
Earth and Environmental Systems Institute, The Pennsylvania State University, University Park, PA, United States

S.-H. Yao
National Engineering Laboratory for Improving Fertility of Arable Soils, Institute of Agricultural Resources and Regional Planning, Chinese Academy of Agricultural Sciences, Beijing, China

Y. Yigini
European Commission, Joint Research Centre, Sustainable Resources, Ispra (VA), Italy

M.-Y. You
National Observation Station of Hailun Agro-Ecology System, Key Laboratory of Mollisols Agroecology, Northeast Institute of Geography and Agroecology, Chinese Academy of Sciences, Harbin, PR China

Y.-R. Yuan
College of Geographical Science, Harbin Normal University, Harbin, PR China

B. Zhang
National Engineering Laboratory for Improving Fertility of Arable Soils, Institute of Agricultural Resources and Regional Planning, Chinese Academy of Agricultural Sciences, Beijing; National Observation Station of Hailun Agro-Ecology System, Key Laboratory of Mollisols Agroecology, Northeast Institute of Geography and Agroecology, Chinese Academy of Sciences, Harbin, PR China

W.-X. Zou
National Observation Station of Hailun Agro-Ecology System, Key Laboratory of Mollisols Agroecology, Northeast Institute of Geography and Agroecology, Chinese Academy of Sciences, Harbin, PR China

PREFACE

In March 2009 the UK Government's Chief Scientist warned of a perfect storm of converging global threats to natural resources. The risks are driven by growth in population, wealth and urbanization, and demands for food, fuel, and clean water during a period of increasing impacts from a changing climate. Since 2009, global population projections have increased, the projected requirements for productive land are estimated to be beyond Earth's environmental capacity, atmospheric greenhouse gas levels continue to rise, and the fifth report of the Intergovernmental Panel on Climate Change reports predicted declines in water availability and yields in key agricultural regions of the world. Global soils are on the frontline of these challenges and under tremendous pressure from intensifying land use.

This volume of *Advances in Agronomy* reports science advances and independent scientific evidence to help address the global soil challenge. The research combines two scientific frameworks: the integrated, mechanistic study of *Earth's Critical Zone*, the thin surface layer of the planet from "treetop to bedrock" that supplies most life-sustaining resources, and the concept of *soil functions* as biophysical flows and transformations of mass, energy, and biodiversity to deliver multiple environmental goods and services. The research presented focuses on the mechanistic links between the development of the internal pore structure of soil and the rates of soil processes that define soil functions.

The work develops new data sets and mathematical models to quantify the dynamics of soil structure and aggregation and applies the models and parameter sets to characterize soils from a range of Critical Zone Observatories that represent different stages in the development of soil functions. The stages include incipient soil development in an alpine glacial forefield, through productive land use for forestry and arable agriculture, to degraded soil under threat of desertification in the Mediterranean basin. Application of the new data tests how observation and simulation can be coupled to guide beneficial intervention in soils in order to protect and enhance soil functions. This collective work contributes along with other international efforts in critical zone science and interdisciplinary soils research to a research platform of sites, data sets, and models. Such platforms for research offer tremendous potential for far greater international collaboration and more rapid development of solutions to the urgent challenges facing Earth's soils.

Steven A. Banwart
Sheffield, United Kingdom
Leeds, United Kingdom

Donald L. Sparks
Newark, DE, United States

CHAPTER ONE

Soil Functions in Earth's Critical Zone: Key Results and Conclusions

S.A. Banwart*[,1,2], S.M. Bernasconi[†], W.E.H. Blum[‡], D.M. de Souza[§,3], F. Chabaux[¶], C. Duffy[||], M. Kercheva[#], P. Krám**, G.J. Lair[‡], L. Lundin[††], M. Menon*[,4], N.P. Nikolaidis[‡‡], M. Novak**, P. Panagos[§§], K.V. Ragnarsdottir[¶¶], D.A. Robinson[||||], S. Rousseva[#], P. de Ruiter[##,5], P. van Gaans***, L. Weng[##], T. White[†††], B. Zhang[‡‡‡,§§§]

*Kroto Research Institute, The University of Sheffield, Sheffield, United Kingdom
[†]Geological Institute, ETH Zürich, Zürich, Switzerland
[‡]Institute of Soil Research, University of Natural Resources and Life Sciences (BOKU), Vienna, Austria
[§]Institute for Environment and Sustainability, Joint Research Centre—European Commission, Ispra, Italy
[¶]Laboratory of Hydrology and Geochemistry of Strasbourg (LHyGeS), University of Strasbourg, Strasbourg, France
[||]The Pennsylvania State University, University Park, PA, United States
[#]Institute of Soil Science, Agrotechnology and Plant Protection "N. Poushkarov", Sofia, Bulgaria
**Czech Geological Survey, Prague, Czech Republic
[††]Swedish University of Agricultural Sciences, Uppsala, Sweden
[‡‡]Technical University of Crete, Chania, Greece
[§§]European Commission, Joint Research Centre, Sustainable Resources, Ispra (VA), Italy
[¶¶]School of Engineering and Natural Sciences, University of Iceland, Reykjavík, Iceland
[||||]Centre for Ecology & Hydrology, Environment Centre Wales, Bangor, United Kingdom
[##]Wageningen University, Wageningen, The Netherlands
***Deltares, Utrecht, The Netherlands
[†††]Earth and Environmental Systems Institute, The Pennsylvania State University, University Park, PA, United States
[‡‡‡]National Engineering Laboratory for Improving Fertility of Arable Soils, Institute of Agricultural Resources and Regional Planning, Chinese Academy of Agricultural Sciences, Beijing, China
[§§§]National Observation Station of Hailun Agro-Ecology System, Key Laboratory of Mollisols Agroecology, Northeast Institute of Geography and Agroecology, Chinese Academy of Sciences, Harbin, PR China
[2]Corresponding author: e-mail address: s.a.banwart@leeds.ac.uk

Contents

[1] Present address: School of Earth and Environment, Earth Surface Science Institute, University of Leeds, Leeds LS2 9JT, United Kingdom.

[3] Present address: Department of Energy and Land Technology, The Swedish University of Agricultural Sciences, 750 07 Uppsala, Sweden.

[4] Present address: Department of Geography, The University of Sheffield, Sheffield S10 2TN, United Kingdom.

[5] Present address: Institute for Biodiversity and Ecosystem Dynamics, University of Amsterdam, 1090 GE Amsterdam, The Netherlands.

Advances in Agronomy, Volume 142
ISSN 0065-2113
http://dx.doi.org/10.1016/bs.agron.2016.11.001

Abstract

This chapter summarizes the methods, results, and conclusions of a 5-year research project (SoilTrEC: Soil Transformations in European Catchments) on experimentation, process modeling, and computational simulation of soil functions and soil threats across a network of European, Chinese, and United States Critical Zone Observatories (CZOs). The study focused on the soil functions of *biomass production, carbon storage, water storage and transmission, water filtration, transformation of nutrients,* and *maintaining habitat and genetic diversity.*

The principal results demonstrate that soil functions can be quantified as biophysical flows and transformations of material and energy. The functions can be simulated with mathematical models of soil processes within the soil profile and at the critical zone interfaces with vegetation and atmosphere, surface waters and the below-ground vadose zone and groundwater. A new dynamic model for soil structure development, together with data sets from the CZOs, demonstrate both seasonal fluctuations in soil structure dynamics related to vegetation dynamics and soil carbon inputs, and long-term trends (decadal) in soil carbon storage and soil structure development.

Cross-site comparison for 20 soil profiles at seven field sites with variation in soil type, lithology, land cover, land use, and climate demonstrate that sites can be classified, using model parameter values for soil aggregation processes together with climatic conditions and soil physical properties, along a trajectory of soil structure development from incipient soil formation through productive land use to overly intensive land use with soil degradation.

A new modeling code, the Integrated Critical Zone model, was applied with parameter sets developed from the CZO site data to simulate the biophysical flows and transformations that quantify multiple soil functions. Process simulations coupled the new model for soil structure dynamics with existing modeling approaches for soil carbon dynamics, nutrient transformations, vegetation dynamics, hydrological flow and transport, and geochemical equilibria and mineral weathering reactions. Successful calibration, testing, and application of the model with data sets from horticulture plot manipulation experiments demonstrate the potential to apply modeling and simulation to the scoping and design of new practices and policy options to enhance soil functions and reduce soil threats worldwide.

1. INTRODUCTION

Increasing human population and wealth are placing unprecedented pressure on Earth's soil and water resources. Drivers for resource demand are a projected increase in human population to 9.7 billion by 2050 (Pison, 2013) with an associated quadrupling in the global economy (Manders et al., 2012) and projected doubling in demand for food (World Bank, 2008) and fuel (Manders et al., 2012) and a more than 50%

increase in demand for clean water (Leflaive et al., 2012). The resources will be required during a period of accelerating impacts from predicted changes in global climate with the Intergovernmental Panel on Climate Change (IPCC) fifth report projecting decreasing water availability in major global agricultural regions of Asia, Africa, and Latin America (IPCC, 2014) and with increasing global biodiversity decline driven by land-use intensification for agriculture and urbanization (Newbold et al., 2011). The United Nations (UN) Environment Programme (UNEP) report on Global Land Use summarizes evidence that by 2050 the demand for productive land will increase by 320–850 Mha which is estimated to be 10–45% greater than the environmental capacity for the sustainable use of Earth's land resources (UNEP, 2014).

Policy responses to these pressures on global soils define soil functions as environmental goods and services that are essential for economic growth and sustainable development and identify global soil threats that degrade soil functions (Banwart, 2011; EC, 2006a,b). European Commission policy (EC, 2012) defines soil as all unconsolidated material from the land surface to the bedrock and considers the role of water flow and transport to expose the deeper subsurface to contamination and other inputs from the land surface. This conceptualization of the soil system is coherent with scientific advances in the study of Earth's critical zone (CZ). The CZ is the thin surface layer that extends from the top of vegetation to the bottom of groundwater circulation and supplies humans with most life-sustaining resources (Anderson et al., 2004; Brantley et al., 2007; NRC, 2001). The study of Earth's CZ defines a new field of integrating environmental science that combines theory and observation methods from many fields of science. CZ science is largely congruent with other integrating sciences of the natural environment (Richter and Billings, 2015) but places a greater emphasis on understanding the vertical integration through the full depth of the CZ and on the mechanistic understanding of CZ processes across physical scales from nanometric to global and across temporal scales from subsecond to those of tectonic forcing. A major research challenge for the study of soil functions within CZ science is addressing the sharp vertical gradients in physical, chemical, and biological conditions within the CZ, from the oxic, rapid circulation of the atmosphere to the anoxic, slow-flowing circulation of aquifers, often encountered over depths on the order of only 5–10 m and including the full genetic and functional biodiversity of Earth's terrestrial surface.

Earth's soil layer is at the heart of the CZ, as a reactive layer that transmits mass, energy, and biodiversity and which transforms within the soil volume the input flows and generates output flows, both downward to the groundwater, upward to the vegetation and atmosphere, and laterally to surface waters (Fig. 1). Soil functions are defined as the sets of flows and transformations that describe a soil's capabilities. When they provide benefits for humans, soil functions align with the broad concept of environmental goods and services, most visibly articulated as the ecosystem services defined by the Millennium Ecosystem Assessment (MEA, 2005). The present study focuses on quantitative methods to understand the soil functions defined in Fig. 1, i.e., *biomass production, carbon and nitrogen storage* associated with climate regulation, *water storage and transmission, water filtration* and *attenuation of contaminants, transformation of nutrients* for soil fertility, and *maintaining habitat and genetic diversity.*

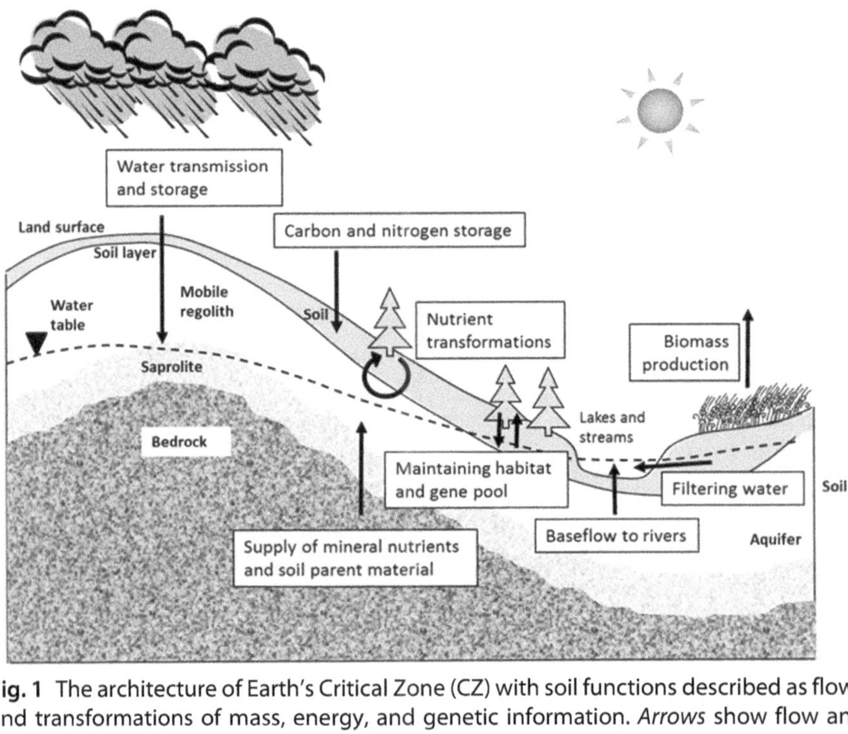

Fig. 1 The architecture of Earth's Critical Zone (CZ) with soil functions described as flows and transformations of mass, energy, and genetic information. *Arrows* show flow and transport pathways and classes of transformation processes that define soil functions which provide benefits to humans. *Text boxes* name the associated soil functions that are the focus of this study. *This figure is reproduced with permission from Banwart, S.A., Menon, M., Bernasconi, S.M., et al., 2012. Soil processes and functions across an International Network of Critical Zone Observatories: introduction to experimental methods and initial results. French Academy of Sciences symposium on land erosion and transformation, C. R. Geosci. 344 (11–12), 2012, 758–772.*

Within this framework, soil functions are degraded when the soil processes that maintain flows and transformations are altered, such that the benefits derived from these soil functions are decreased or lost. Soil threats are defined as human pressures from land use that are known to degrade soil functions and include *soil erosion, loss of soil organic matter, decline in soil biodiversity, soil acidification, mechanical compaction, sealing by infrastructure, salinization* from irrigation water evapotranspiration, and *industrial contamination* (EC, 2006a; FAO and ITPS, 2015). Recent compilations of scientific evidence for global soil decline include Amundson et al. (2015) who summarize current data and uncertainties on soil organic matter stocks and vulnerabilities for accelerated greenhouse gas emissions from soil, Banwart et al. (2014) who compiled evidence on soil organic matter decline and the potential for innovation in land management practices and policy to improve soil carbon stocks, and the UN Food and Agriculture Organization (FAO) report on the status of the World's soil resources (FAO and ITPS, 2015).

The aim of this study and outcomes reported in this volume are to provide independent scientific evidence and new methods of mathematical modeling of soil processes to quantify soil functions. The hypotheses and experimental design have been previously reported (Banwart et al., 2011) as have initial results of the research project (Banwart et al., 2012). Summarizing briefly, the hypotheses are that the intensity of human land use defines a life cycle of soil function, where soil functions develop from parent material, and support terrestrial ecosystems which humans utilize for agriculture, forestry, and other productive land uses. Furthermore, soil functions decline under increasingly overintensive land use, with the potential under conditions of extreme degradation to lose all soil functions through a complete loss of bulk soil by physical erosion of surface soils down to parent material.

A secondary hypothesis builds on the concepts of Graham et al. (2010) and Brantley (2010) that development of the subsurface porosity structure through combined weathering and fluid circulation gives rise to soil functions. This view of pedogenesis places greater emphasis on the definition of soil functions as the evolution of natural processes and their rates and physical extent. From this framework, the present study hypothesizes that the development of soil pore structure and soil aggregates strongly correlates with the development of soil functions (Banwart et al., 2011, 2012). The research design for this project thus focuses on quantifying through observation, and simulating through the development and application of new mathematical models, the dynamics of soil structure and processes in connection with the soil functions illustrated in Fig. 1.

The specific objectives of the full research project have been previously presented (Banwart, 2011; Banwart et al., 2012). The focus of the studies presented in this volume is to:

1. Develop a conceptual model for soil structure dynamics and translate it into mathematical models,
2. Obtain data on soil functions from CZOs across gradients of land-use intensity,
3. Study soil processes at field scale and gain data to test models for soil structure dynamics,
4. Develop a Integrated Critical Zone (ICZ) model that quantifies soil functions (Fig. 1) as flows and transformations of mass, energy, and biodiversity and their dependence on soil structure, and
5. Apply field observations from CZOs and results from computational simulation in order to quantify soil functions, assess the economic value of selected soil functions, and assess soil threats at the EU scale.

This chapter provides a summary of the key advances that were achieved and introduces the detailed experimental and modeling studies which comprised much of the project and are presented in subsequent chapters.

2. RESEARCH METHODS

The overarching experimental design is to quantify soil structure and process at research field sites that are CZOs located along a gradient of land-use intensity that defines a conceptual life cycle of soil functions (Banwart et al., 2011). Four European field sites were selected as CZOs that characterize key stages of land-use intensity of relevance for soil management and policy innovation (site descriptions provided in Banwart et al., 2011).

1. The Damma Glacier CZO, Switzerland, allows the study of incipient soil formation in the glacial forefield as the glacier retreats, exposing the underlying bedrock. A chronosequence on the order of centuries allows the earliest stages of soil formation to be observed.
2. The Lysina-Slavkov Forest CZO, Czech Republic, allows the study of soil processes during managed forest land use for intensive silvaculture.
3. The Fuchsenbigl-Marchfeld CZO, Austria, allows the study of soil processes during managed arable land use for production agriculture.
4. The Koiliaris River CZO, Crete, allows the study of highly degraded soils that have experienced millennia of intensive agricultural land use, including grazing, and are under additional threat from desertification due to modern climate change.

The research method was to select and characterize essential terrestrial variables across the four sites, and across different land cover and land use within the sites, at the physical scale of the soil profile. The selection of variables was prioritized in order to parameterize mathematical models of vegetation dynamics, soil structure dynamics, soil carbon dynamics, nutrient transformations, hydrological flow and reactive transport, mineral weathering, and a highly simplified model of the soil food web. Common soil variables are listed in Banwart et al. (2011), and additional variables are presented in the individual studies (Table 1) within this volume or otherwise cited from the literature.

Development of the conceptual model for soil functions as flows and transformations (Fig. 1) utilized concepts of mass flux balance by physical flow and transport processes across flux planes which defined physical interfaces of the CZ: atmosphere–vegetation, vegetation–soil, soil–vadose, vadose–groundwater (Fig. 2). Mass transformations were conceptualized within the plant–soil–water system as plant production of biomass, development of soil microbial biomass, aqueous chemical reactions, and mass transfer processes between soil particles and fluids. Translation to mathematical equations utilized theory of fluid flow and advective transport, diffusion, and dispersion; empirical rules for rates and stoichiometry of plant growth from known plant physiology and traits; thermodynamic and kinetic laws of mass action for geochemical reactions; and phenomenological zero- or first-order rate laws for soil carbon degradation, soil structure dynamics, and soil N and P transformations.

The terrestrial variables selected for experimental study were identified from the beginning of the study in order to support the parameterization of the process models. A new model for soil structure was developed, the Carbon Aggregation and Structure Turnover (CAST) model (Stamati et al., 2013), and applied in a number of studies reported in this volume. In brief, the model considered soil structure to be defined by three size classes of soil aggregates, particle size (d_p) $<$ 53 μm, 53 μm $< d_p <$ 250 μm, and $d_p >$ 250 μm, composed of mineral soil texture units (clay-, silt-, and sand-sized particles), living organisms, decomposing biomass, and fluids and solutes. As the mass fractions of soil aggregates change, new values for soil properties of bulk density, porosity, and saturated hydraulic conductivity are calculated. The model defines first-order rate laws for mass transfer of organic carbon and mineral texture units between aggregate size classes, and for mineralization of organic carbon within each size class, and calibrates the rate constants with site-specific data on organic matter inputs, soil texture, structure,

Table 1 Summary of Soil Characterization, Experimental, and Mathematical Modeling Studies

No.	Study Title	Summary Outcomes
1	Soil Water Characteristics of European SoilTrEC Critical Zone Observatories (Rousseva et al., 2017)	Soil sampling and characterization by horizon in soil profiles at the European CZO sites provide qualitative and quantitative data on soil texture, soil structure, water retention curves, particle size and porosity distribution, and permeability to evaluate soil state and provide parameter values for soil process models
2	Linkages between aggregate formation, porosity, and soil chemical properties (Regelink et al., 2015)	Results from physical characterization of dry- and wet-sieved aggregate size fractions quantified the distributions of micro- and mesoporosity as a function of aggregate mass fraction by size and demonstrate that formation of water stable aggregates correlated significantly with iron oxide content, suggesting that these oxides and their known role in forming organo–mineral complexes with soil organic matter may be important in soil structure development
3	Analyzing the impact of compaction of soil aggregates using X-ray microtomography and water flow simulations (Menon et al., 2015)	Soil aggregates from the Marchfeld CZO, Austria were subjected to compaction by controlled mechanical loading under laboratory conditions. Results from 3-D X-ray microtomography of initial and compacted soil were used to quantify soil structure and to parameterize Lattice-Boltzmann calculations of water mass and flow distribution. The results showed that compaction resulted in decreased water penetration in aggregate micropores and that water flow rate decreases linearly with increased bulk density
4	Reduced subsurface lateral flow in agroforestry system is balanced by increased water retention capacity: rainfall simulation and model validation (Wang et al., 2017)	Results from experimental rainfall simulations on soil planted with arable crops and trees are compared to elucidate the impact of plant roots to enhance vertical infiltration and reduce lateral flow, in order to provide mechanistic insights and quantitative data on soil water storage and transmission for agronomic benefit

5	Interception of subsurface lateral flow through enhanced vertical preferential flow in an agroforestry system observed using dye tracing and rainfall simulation experiments (Wang and Zhang, 2017)	Results from experimental rainfall simulations on soil planted with arable crops and trees, combined with flow balance calculations and digital imaging and image analysis of tracer transport with infiltrating water quantify vertical and lateral flow rates and provide evidence that soils with tree roots have enhanced vertical infiltration rates compared with soils in arable farming
6	Effects of dry- and wet sieving of soil on identification and interpretation of microbial community composition (Blaud et al., 2017)	Quantitative results on gene copy number for selected functional guilds of organisms from the soil microbiome show that wet sieving and dry sieving result in significant differences in apparent microbial diversity by aggregate size class and by land use illustrating, the importance of the soil microphysical structure as habitat that exerts selection pressure on the microbiome and that physical characterization methods can influence measurements of microbial biodiversity
7	An ecosystem approach to assess soil quality in organically and conventionally managed farms in Iceland and Austria (van Leeuwen et al., 2015)	Data on soil biodiversity combined with data on soil physical and chemical properties were used to compare soil functions on two pairs of farms, on in the subarctic and one in temperature central Europe, with each pair including a conventional farm and an organic farm. The food web structure in terms of trophic groups diversity was similar between the four sites but with generally greater microbial and nematode biomass and greater taxonomic diversity of microarthropods on organic farms
8	Soil mineralogy changes with different agricultural practices during 8-year soil development from the parent material of a Mollisol (Liu et al., 2017)	A manipulation experiment replaced the top 0.8 m of soil from the China black soil region with bare parent material from >2 m depth, as an analog to a degraded soil, and compared the role of different agricultural practices for arable farming to potentially restore soil functions on the parent material. Significant and rapid changes in clay mineralogy occurred during the 10-year experiment and were influenced significantly by the availability and cycling of potassium ion as an exchangeable cation mineral nutrient, and that the amount and mechanisms of organic matter association with mineral surfaces were substantially different for different clay mineral phases, and that there were substantial differences in clay mineral phases associated with different particle size classes

Continued

Table 1 Summary of Soil Characterization, Experimental, and Mathematical Modeling Studies—cont'd

No.	Study Title	Summary Outcomes
9	A coupled carbon, aggregation, and structure turnover (CAST) model for topsoils (Stamati et al., 2013)	Results are presented for a newly developed mathematical model of soil processes that describe the coupled dynamics of soil organic carbon and soil structure and quantify sequestration of soil organic matter. The study describes the underlying conceptual model and translation to a dynamic model of soil processes, the model code and equations, and includes parameter sets obtained from site-specific model calibration. The results demonstrate the soil processes responsible for the quantitative changes in soil carbon sequestration and/or loss during land-use change from arable agriculture to set-aside. Rapid degradation of particulate organic matter in large aggregates results in release of plant available forms of N explains the soil function of nutrient transformation, and slow degradation of humified organic matter bound to mineral surfaces in the clay–silt-sized mass fraction explains the soil function of carbon storage
10	Modeling soil aggregation at the early pedogenesis stage from the parent material of a Mollisol under different agricultural practices (Li et al., 2017)	Mathematical modeling was applied to a 2-year series of data on soil chemistry and structure under different practices of arable agriculture to restore degraded soils represented by parent material underlying Mollisols of the China black soil region. Rapid development of soil structure and larger aggregates was observed particularly for practices with the highest organic matter input. The results demonstrate the potential for arable farming practices to influence, through beneficial intervention, the soil carbon stocks and soil structure on the time scale of the annual cropping cycle
11	Quantifying the Incipient Development of Soil Structure and Functions within a Glacial Forefield Chronosequence (Andrianaki et al., 2017)	Results of mathematical modeling of soil function development along a chronosequence of soil formation in the Damma Glacier forefield demonstrated that dynamic modeling of soil organic carbon inputs and soil structure dynamics could reproduce the time series of soil function development along the chronosequence and that this development was sensitive to the time series input of biological productivity and associated inputs of particulate organic carbon to drive soil structure and soil function development

12	Factors controlling soil structure dynamics and carbon sequestration across different climatic and lithological conditions (Panakoulia et al., 2017)	Mathematical modeling results for soil structure development and soil organic carbon dynamics quantified model parameter values and soil carbon flux balances for soil profiles at 20 plots within seven different field sites representing a wide range of soil conditions and land use in the United States, Europe, and China. Model parameter values were combined with soil characterization data for the plots and were analyzed using principle component analysis which demonstrated the soils clustered discretely by sites along a hypothesized life cycle of soil function. Dominant factors in the clustering included carbon degradation rate constants, organic matter input, clay mineral content, microbial biomass, and humic material content of aggregates, evapotranspiration, and others
13	Integrated Critical Zone Model (1D-ICZ): A Tool for Dynamic Simulation of Soil Functions and Soil Structure (Giannakis et al., 2017)	Experimental time series data for a horticulture small plot experiment with mineral fertilizer addition to augment soil fertility allowed calibration of the Integrated Critical Zone Model and testing of the model calculation of state variable values against measured values. The study describes the soil processes included in the simulation methodology and includes soil carbon and soil structure dynamics, soil nutrient transformations, vegetation dynamics, a simplified soil food web model, hydrological flow and transport, and geochemical speciation and weathering kinetics. The study documents the mathematical structure and computational procedure for the model code, the parameter sets for the calibrated model, and demonstration of goodness of fit for state variables

Continued

Table 1 Summary of Soil Characterization, Experimental, and Mathematical Modeling Studies—cont'd

No.	Study Title	Summary Outcomes
14	Modeling the impact of carbon amendments on soil ecosystem functions using the 1D-ICZ model (Kotronakis et al., 2017)	Results from tomato horticulture plot manipulation experiments using different soil amendments with organic carbon provided data to apply the Integrated Critical Zone Model to assess the impact of different mineral fertilizer, compost, and manure additions on soil functions. Biomass production, C and N storage, nutrient transformation, and water transmission and filtration were evaluated. Soil structure, biomass production, and carbon storage improved most with the organic matter amendments, but the water filtration function was adversely affected with greater nitrate loss to drainage water compared to mineral fertilizer addition. The model application demonstrates the practical value in using computational simulation together with field observation to test soil management options, interpret field measurement, and scope and design potential innovations in soil management and agriculture
15	Valuation of Soil Ecosystem Services (Jónsson et al., 2017)	Biophysical flows and transformations that are quantified in field measurements and modeling studies of the Koiliaris CZO are translated into environmental service flows that quantitatively represent the soil functions of biomass production, water filtration, and carbon storage. Spatial variability in environmental service flows arising from variable soil type, land cover, and land use results in geospatial variation in economic value for soil functions that are mapped for the Koiliaris River watershed. Values are expressed in international dollars per hectare per year and include positive economic value for the soil functions of water filtration to remove nitrate from river discharge and biomass production from agriculture, and environmental costs for the soil function of carbon storage due to the ongoing loss of soil organic matter in the watershed

16	Modeling soil organic carbon stocks under future climate and land cover changes in Europe (Yigini and Panagos, 2016)	The first component of the model predicts current soil organic carbon content by using step-wise multiple linear regression and ordinary kriging, and the second component projects the soil organic carbon to the near future (2050) by using a set of environmental predictors. The regression–kriging as a base model is predicting current SOC stocks in Europe by using a set of covariates (Soil data, terrain, climate, and land cover) and dense SOC measurements coming from the LUCAS Soil Database. The base model delivers coefficients for each of the covariates to the future model. The overall model produced soil organic carbon maps which reflect the present and the future predictions (2050) based on climate and land cover projections
17	Modeling global soil organic carbon stocks in agricultural soils (Yigini et al., 2017)	The geostatistical model consists of two submodels, namely, donor and donee modules. The donor module uses statistics to quantify the relationships between soil organic carbon and environmental covariates. The multiple linear regression was used in the donor module by the selected covariates and dense SOC measurements coming from the LUCAS Soil Database. The donee model applies the statistical relationships of the first module at global scale by using as input the global databases for Climate (WorldClim), land Cover (GlobCover), Terrain (SRTM), and Soil (HWSD)

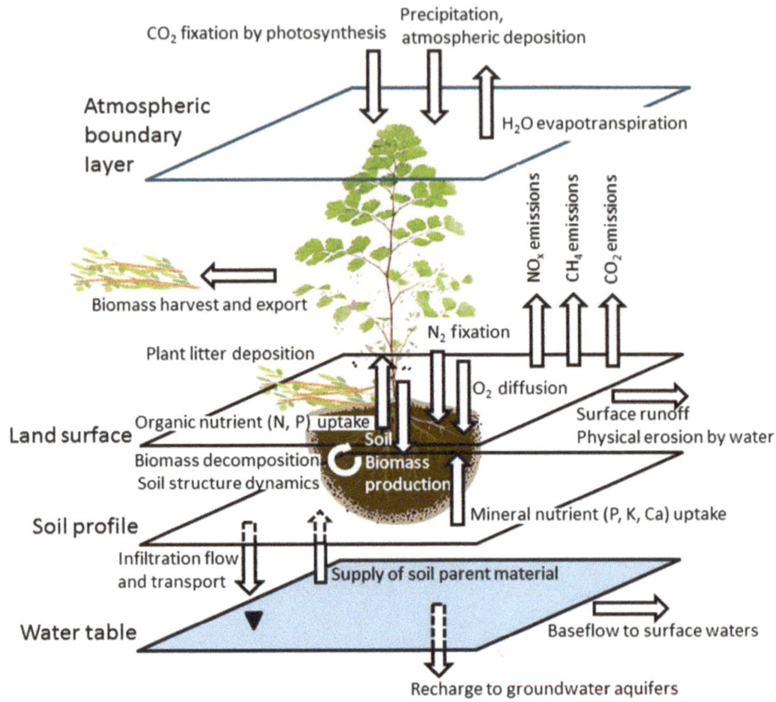

Fig. 2 Earth's critical zone soil layer and its functions. Translation of soil functions into mathematical descriptions of flows and transformations applies flux balance equations across hypothetical horizontal planes that are defined by the top of the vegetation canopy, the land surface, the bottom of the soil profile, the water table, and considers water mass exchange with surface waters. Mass transfer and transport equations apply principles of physics from continuum mechanics that describe fluid flow and mass transport by advection, diffusion, and dispersion in freshwaters and the atmosphere. Mass transformations are described by the thermodynamic law of mass action for reversible reactions and the kinetic law of mass action for irreversible reactions. Biodiversity is described by functional pools of organisms as producers, decomposers, and consumers by applying food web theory to this highly simplified network of functional groups.

and organic carbon content in aggregate size classes (Stamati et al., 2013). The code allows simulation of flows and transformations that define the soil functions of carbon storage and water transmission and storage.

A new mathematical model for soil processes, the ICZ model for computational simulation of multiple soil functions, in addition to those simulated with CAST, allowed calculation of process rates associated with water filtration, nutrient transformations, and biomass production. The ICZ model embeds the CAST model within a reactive transport model at soil profile scale which is driven by inputs from models of vegetation dynamics and soil hydrology. These processes are coupled with a highly simplified model of the soil food web which includes dynamic representation of biomass growth from producers (plant roots and their mycorrhizal fungi), heterotrophic biomass decomposers (bacteria and fungi), and grazing pressure exerted by consumers (fauna).

As described by Kotronakis et al. (2017), the ICZ model (Fig. 3) integrates process descriptions from the CAST model for structure and carbon dynamics (Stamati et al, 2013) and existing codes for the simulation of vegetation dynamics (PROSUM; Giannakis et al., 2017), 1-D water flow, reactive transport, and energy transfer (Hydrus 1-D; Šimůnek et al., 2009), geochemical

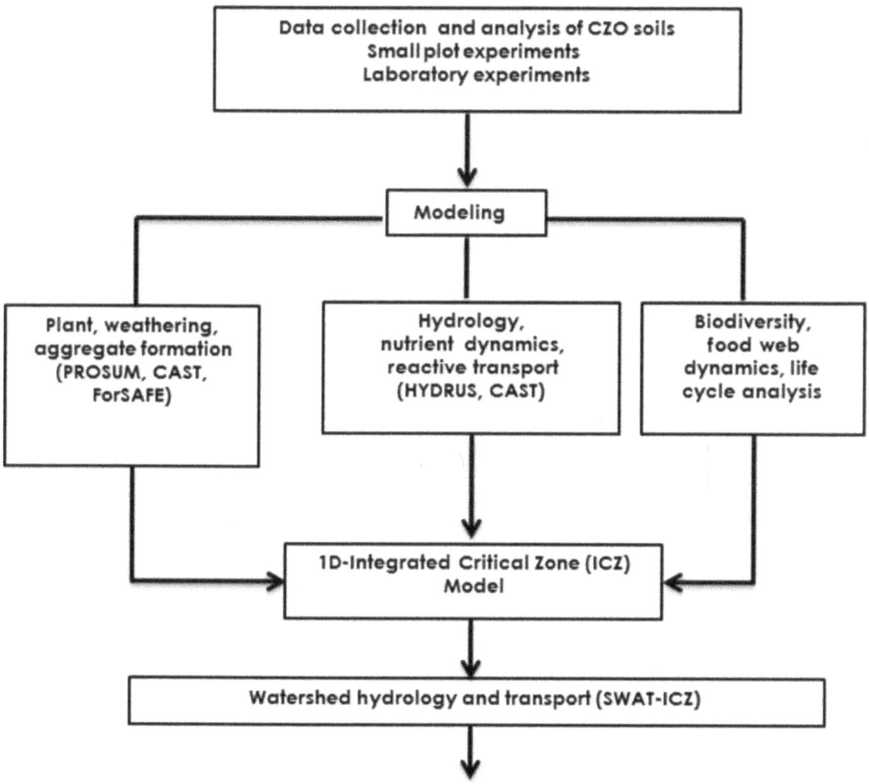

Fig. 3 A schematic diagram of the modeling approach applied with the Integrated Critical Zone (ICZ) model. Individual process models were applied using data from the CZOs sites in order to develop parameter sets and test model representations of site processes and rates. The individual modeling codes were integrated into the coupled 1-D ICZ model for simulation of multiple soil functions, and the 1-D ICZ model was also embedded as a module within the SWAT code for simulation of soil functions and their geospatial variation with soil properties and land cover and land use at catchment scale.

speciation based on the BRNS model (Aguilera et al., 2005), and weathering kinetics based on the SAFE component of the ForSAFE model (Wallman et al., 2005). The 1-D ICZ model has been added as a module within the Soil and Water Agriculture Tool (SWAT; Gassman et al., 2007) for 3-D dynamic simulation of soil functions at landscape scale (SWAT-ICZ).

3. OVERVIEW OF RESULTS

Conceptual site models of the four CZOs (Fig. 4A–D) incorporate information from existing site characterization and monitoring data (Panakoulia et al., 2017) and new soil characterization data from soil profile sampling across the CZOs (Rousseva et al., 2017). Mathematical models and parameter sets translated from these conceptual models using the CAST code quantified for soil profiles, the biophysical flows, and transformations that define the soil functions within and between CZOs (Panakoulia et al., 2017). Comparison of site characteristics and model results distinguish the state of soil functions between sites, exemplified by comparison of climate conditions, soil carbon flux balance, and organic matter decomposition rate constants for the four CZOs (Fig. 5).

Incipient soil formation at the Damma Glacier CZO correlates with low rates of biological productivity and organic matter input to the soil at the early stages (<20 years age) but demonstrates around half of the input organic carbon is stored in the soil each year. The conceptual model for the Damma Glacier CZO (Fig. 4A) demonstrates increasing mass of stored carbon in soil as an intensifying soil function at later stages of soil profile development within the chronosequence at the site. Comparison of the carbon flux balance for the Damma Glacier site with that of the Slavkov Forest CZO demonstrates a substantially greater input of organic matter to the forest soil resulting from the far greater biological productivity for the mature plantation forest, however, with a far smaller fraction of the organic carbon that is input to soil being stored, compared to the young soils of the Damma Glacier.

Comparison with the soil C flux results for the Marchfeld CZO (Fig. 5C) demonstrates nearly equal rates of organic carbon input to soil and mineralization, with a small loss of soil carbon calculated for the conditions of intensive mechanical agricultural practices on the arable farmland. The results for the Koiliaris CZO are substantially different, showing a significantly greater loss of soil organic C from mineralization, compared to inputs from biological productivity, indicating a substantial, ongoing loss of soil organic C. From these results, the calculated rates of soil C accumulation serve to quantify the soil function of carbon storage and indicate a degraded

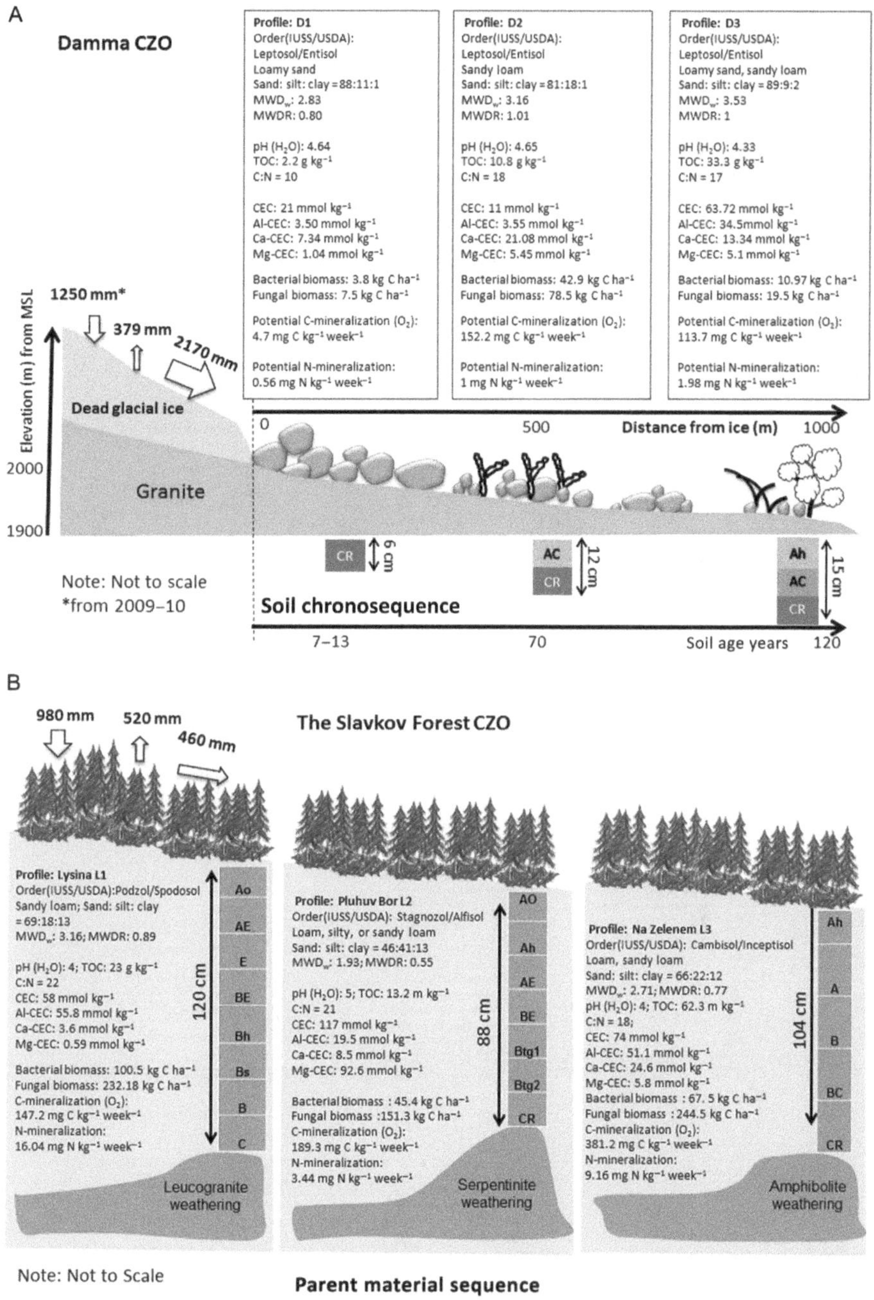

Fig. 4 Conceptual site models for the four European CZOs representing different stages of the life cycle for soil functions: (A) Damma Glacier CZO representing incipient soil formation from freshly exposed parent material; (B) Lysina-Slavkov Forest CZO representing productive plantation forestry; (C) Fuchsenbigl/Marchfeld CZO representing productive agriculture grassland, arable agriculture land and forest; and (D) the Koiliaris CZO representing degraded soils undergoing desertification under land-use pressures from long-term intensive agriculture.

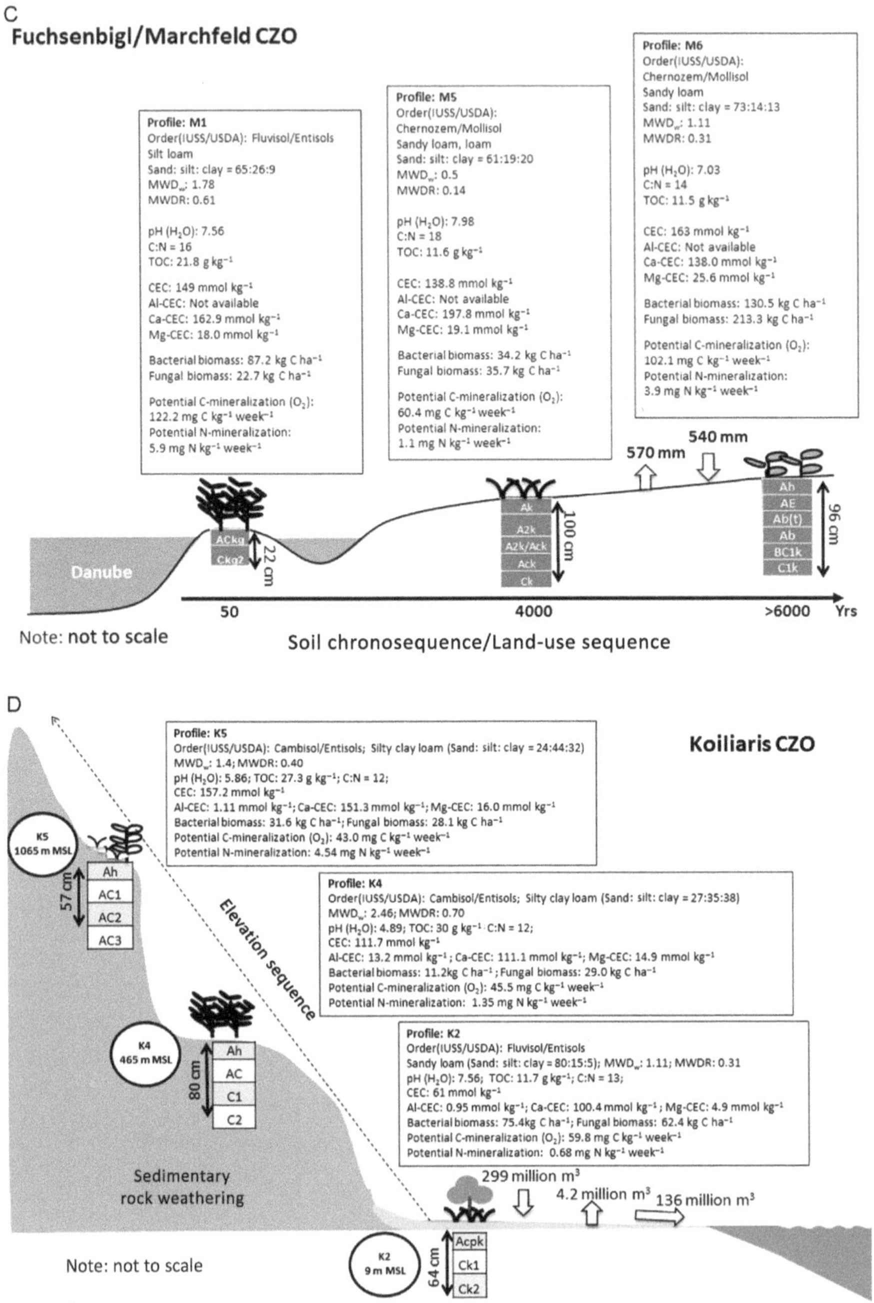
C
Fuchsenbigl/Marchfeld CZO
Profile: M1
Order(IUSS/USDA): Fluvisol/Entisols
Silt loam
Sand: silt: clay = 65:26:9
MWD$_w$: 1.78
MWDR: 0.61
pH (H$_2$O): 7.56
C:N = 16
TOC: 21.8 g kg^{-1}
CEC: 149 mmol kg^{-1}
Al-CEC: Not available
Ca-CEC: 162.9 mmol kg^{-1}
Mg-CEC: 18.0 mmol kg^{-1}
Bacterial biomass: 87.2 kg C ha^{-1}
Fungal biomass: 22.7 kg C ha^{-1}
Potential C-mineralization (O$_2$): 122.2 mg C kg^{-1} week^{-1}
Potential N-mineralization: 5.9 mg N kg^{-1} week^{-1}
Profile: M5
Order(IUSS/USDA): Chernozem/Mollisol
Sandy loam, loam
Sand: silt: clay = 61:19:20
MWD$_w$: 0.5
MWDR: 0.14
pH (H$_2$O): 7.98
C:N = 18
TOC: 11.6 g kg^{-1}
CEC: 138.8 mmol kg^{-1}
Al-CEC: Not available
Ca-CEC: 197.8 mmol kg^{-1}
Mg-CEC: 19.1 mmol kg^{-1}
Bacterial biomass: 34.2 kg C ha^{-1}
Fungal biomass: 35.7 kg C ha^{-1}
Potential C-mineralization (O$_2$): 60.4 mg C kg^{-1} week^{-1}
Potential N-mineralization: 1.1 mg N kg^{-1} week^{-1}
Profile: M6
Order(IUSS/USDA): Chernozem/Mollisol
Sandy loam
Sand: silt: clay = 73:14:13
MWD$_w$: 1.11
MWDR: 0.31
pH (H$_2$O): 7.03
C:N = 14
TOC: 11.5 g kg^{-1}
CEC: 163 mmol kg^{-1}
Al-CEC: Not available
Ca-CEC: 138.0 mmol kg^{-1}
Mg-CEC: 25.6 mmol kg^{-1}
Bacterial biomass: 130.5 kg C ha^{-1}
Fungal biomass: 213.3 kg C ha^{-1}
Potential C-mineralization (O$_2$): 102.1 mg C kg^{-1} week^{-1}
Potential N-mineralization: 3.9 mg N kg^{-1} week^{-1}
570 mm
540 mm
Danube
ACkg
Ckg2
22 cm
Ak
A2k
A2k/Ack
Ack
Ck
100 cm
Ah
AE
Ab(t)
Ab
BC1k
C1k
96 cm
50
4000
>6000
Yrs
Note: not to scale
Soil chronosequence/Land-use sequence
D
Koiliaris CZO
Profile: K5
Order(IUSS/USDA): Cambisol/Entisols; Silty clay loam (Sand: silt: clay = 24:44:32)
MWD$_w$: 1.4; MWDR: 0.40
pH (H$_2$O): 5.86; TOC: 27.3 g kg^{-1}; C:N = 12;
CEC: 157.2 mmol kg^{-1}
Al-CEC: 1.11 mmol kg^{-1}; Ca-CEC: 151.3 mmol kg^{-1}; Mg-CEC: 16.0 mmol kg^{-1}
Bacterial biomass: 31.6 kg C ha^{-1}; Fungal biomass: 28.1 kg C ha^{-1}
Potential C-mineralization (O$_2$): 43.0 mg C kg^{-1} week^{-1}
Potential N-mineralization: 4.54 mg N kg^{-1} week^{-1}
Profile: K4
Order(IUSS/USDA): Cambisol/Entisols; Silty clay loam (Sand: silt: clay = 27:35:38)
MWD$_w$: 2.46; MWDR: 0.70
pH (H$_2$O): 4.89; TOC: 30 g kg^{-1} C:N = 12;
CEC: 111.7 mmol kg^{-1}
Al-CEC: 13.2 mmol kg^{-1}; Ca-CEC: 111.1 mmol kg^{-1}; Mg-CEC: 14.9 mmol kg^{-1}
Bacterial biomass: 11.2kg C ha^{-1}; Fungal biomass: 29.0 kg C ha^{-1}
Potential C-mineralization (O$_2$): 45.5 mg C kg^{-1} week^{-1}
Potential N-mineralization: 1.35 mg N kg^{-1} week^{-1}
Profile: K2
Order(IUSS/USDA): Fluvisol/Entisols
Sandy loam (Sand: silt: clay = 80:15:5); MWD$_w$: 1.11; MWDR: 0.31
pH (H$_2$O): 7.56; TOC: 11.7 g kg^{-1}; C:N = 13;
CEC: 61 mmol kg^{-1}
Al-CEC: 0.95 mmol kg^{-1}; Ca-CEC: 100.4 mmol kg^{-1}; Mg-CEC: 4.9 mmol kg^{-1}
Bacterial biomass: 75.4kg C ha^{-1}; Fungal biomass: 62.4 kg C ha^{-1}
Potential C-mineralization (O$_2$): 59.8 mg C kg^{-1} week^{-1}
Potential N-mineralization: 0.68 mg N kg^{-1} week^{-1}
K5
1065 m MSL
Ah
AC1
AC2
AC3
57 cm
Elevation sequence
K4
465 m MSL
Ah
AC
C1
C2
80 cm
Sedimentary rock weathering
299 million m^3
4.2 million m^3
136 million m^3
K2
9 m MSL
Acpk
Ck1
Ck2
64 cm
Note: not to scale

Fig. 4—Cont'd

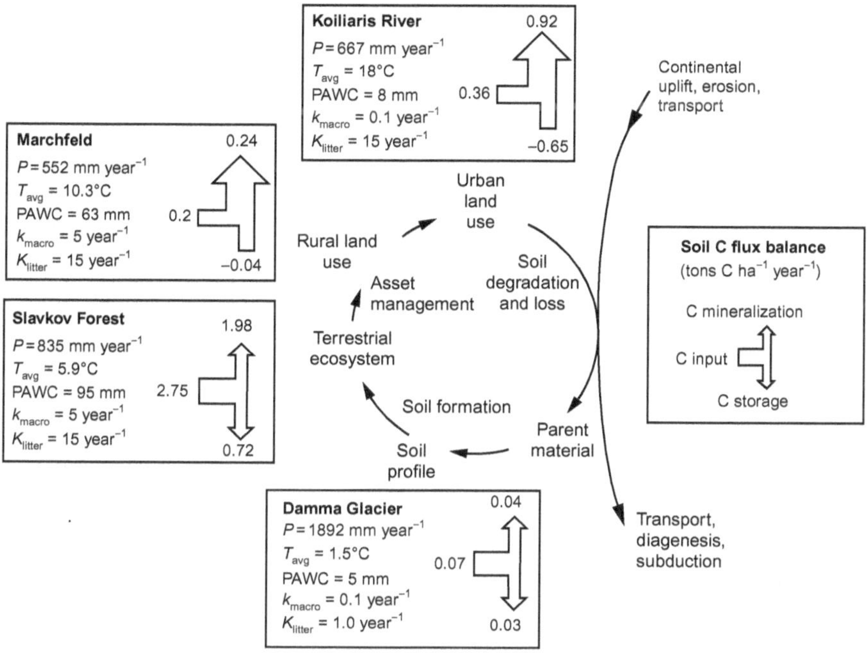

Fig. 5 A summary of climatic conditions (temperature, °C; mean annual precipitation, mm), carbon flows (tons C ha^{-1} $year^{-1}$), plant available water capacity (PAWC, mm), and decomposition rate constants in soil for fresh particulate organic matter (k_{litter}, $year^{-1}$) at the four European Critical Zone Observatories. The Slavkov Forest CZO results are for the Lysina site; the Marchfeld results are for the M4 site, intensively farmed arable land; and the Koiliaris CZO results are for site K2, intensively managed horticulture land near sea level. The sites are placed along a gradient of climate from the alpine through temperate to Mediterranean, and along a gradient of increasing intensively managed land. The Koiliaris CZO has been under long-term (centuries) intensive land use for horticulture and livestock grazing. This conceptual model of a life cycle of soil functions is represented by the *circle* within the center of the figure, which illustrates the following stages of soil function development following exposure or deposition of regolith at the land surface: soil profile development, terrestrial ecosystem development, human development of land use, intensification of land use, and potential loss of soil functions with physical erosion of the soil profile to again expose fresh regolith.

soil function under arable land use, however, with indications of relatively low net rates of soil C loss for the Marchfeld CZO. The results presented for the Koiliaris CZO represent intense arable agriculture and horticulture and demonstrate a severely degraded soil function with a loss of carbon storage function and an ongoing loss of soil organic C through mineralization.

Results of the ICZ model calibration for the Koiliaris CZO horticulture experiments, which are reported in detail by Giannakis et al. (2017), demonstrate the dynamics of soil structure development and the associated dynamics of organic C content within aggregate size fractions (Fig. 6A and B). For the simulated 10-year period, corresponding to conventional horticulture practices utilizing mineral fertilizer to supplement soil fertility, there are intraannual fluctuations in soil organic carbon content and soil structure due to organic C inputs and removal during the cropping season, and a long-term decline in soil organic carbon (on the order of 10%) primarily associated to loss of carbon from macroaggregates ($d_p > 250$ μm).

Incorporation of the CAST model code into the ICZ model allows bulk soil properties to be simulated dynamically at soil profile scale. The detailed results reported by Kotronakis et al. (2017) show how bulk density, porosity, water holding capacity, and saturated hydraulic conductivity are calculated from changes in soil structure. These dynamic properties of soil are in turn utilized by the modeling code to quantify water and gas content, plant available water and uptake, infiltration flow, and vertical advective transport velocity. From these calculations, the biophysical flows and transformations are quantified and define the multiple soil functions presented earlier.

The detailed studies (summarized in Table 1) published previously or reported in this volume illustrate this methodology to implement soil characterization and field experimental data (Blaud et al., 2017; Liu et al., 2017; Regelink et al., 2015; Rousseva et al., 2017; Wang and Zhang, 2017; Wang et al., 2017) in mathematical modeling studies to quantify changes in soil carbon and structure and the relation of soil structure changes to changes in soil functions (Andrianaki et al., 2017; Li et al., 2017; Menon et al., 2015; Stamati et al., 2013); synthesis of modeling results for soil C sequestration, soil aggregation, and soil structure development across multiple sites in the United States, Europe, and China (Panakoulia et al., 2017); application of parameter sets and model simulations to evaluate biophysical flows and transformations that define multiple soil functions (Giannakis et al., 2017; Kotronakis et al., 2017); and their translation into environmental service flows for economic valuation (Jónsson et al., 2017) and upscaling using the example of soil organic matter stocks to continental (Yigini and Panogos, 2016) and global (Yigini et al., 2017) scales to guide policy innovation.

4. DISCUSSION

This series of published results shows that the dynamics of soil structure is driven by temporal trends and fluctuations in soil organic matter input

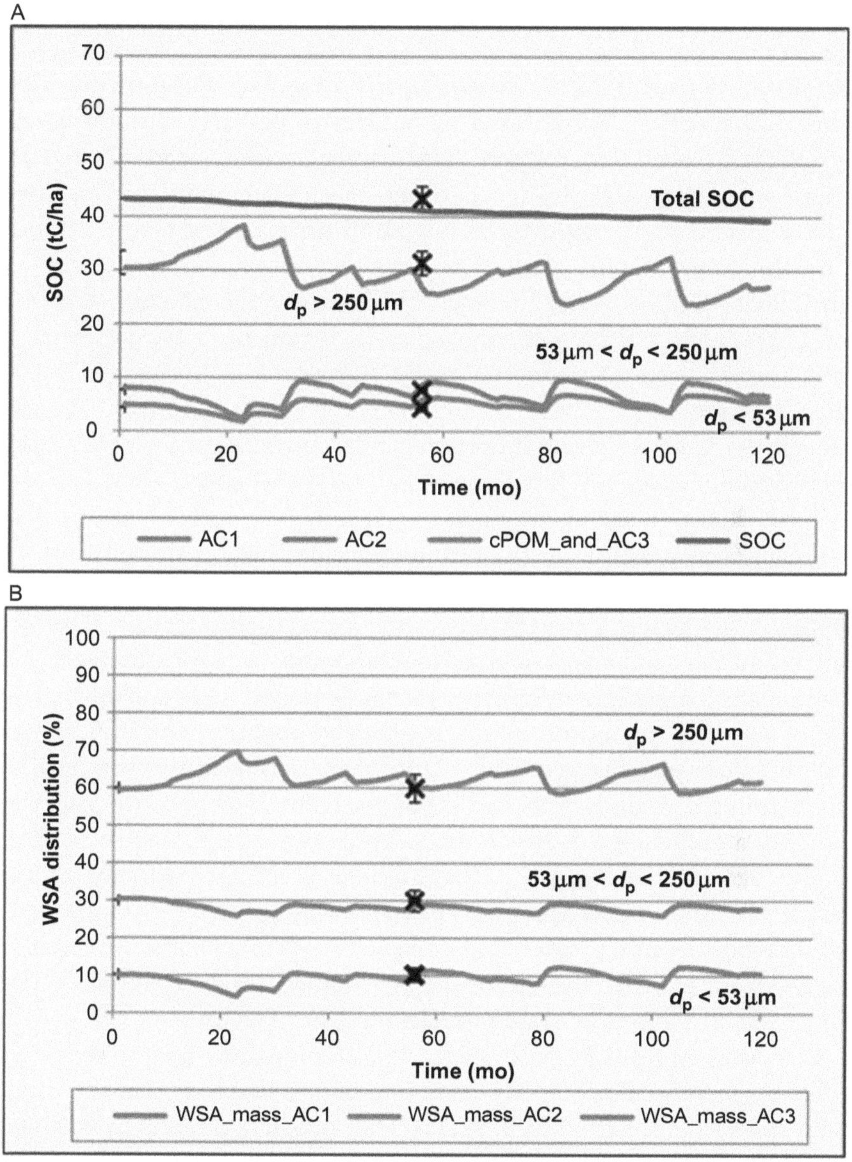

Fig. 6 Results of model calibration for the Integrated Critical Zone model applied to horticulture experiments at the Koiliaris CZO for (A) soil organic carbon (SOC, tonnes C/ha) and (B) water stable aggregate mass fraction (%) over a 10-year period. The experimental treatment is application of mineral fertilizer. Parameter values for mathematical rate expressions that calculate carbon degradation and aggregate formation and loss are constrained by data obtained at the start of the experiment during year 5 of the simulation period. *The figure is reproduced with permission from Giannakis, G.V., Nikolaidis, N.P., Valstar, J., Rowe, E.C., Moirogiorgou, K., Kotronakis, M., Paranychianakis, N.V., Rousseva, S., Stamati, F.E., Banwart, S.A., 2017. Integrated Critical Zone Model (1D-ICZ): a tool for dynamic simulation of soil functions and soil structure. Adv. Agron. 142, 277–314.*

to soil including that from dynamic vegetation production and land-use practices such as tillage, fertilization, and irrigation. The detailed results demonstrate that the simulated biophysical flows and transformations that define soil functions are affected by, and in some cases sensitive to, the dynamic behavior of soil structure. The mathematical models described here quantify the mechanistic linkages between physical, chemical, and biological processes that determine the flows and transformations and define soil functions. Bulk soil physical properties, particularly pore size distribution and permeability, impact water holding capacity and drainage rates (Rousseva et al., 2017). Increasing soil aggregation increases both the potential for water storage within the micropores of aggregating soil (Regelink et al., 2015; Rousseva et al., 2017) and the potential for improved drainage through the larger resulting interaggregate pores (Kotronakis et al., 2017). Therefore, enhanced soil functions of water storage and transmission correlate with increased soil aggregation.

Changes in porosity distribution and permeability that arise from soil aggregation also influence O_2 ingress for root respiration, water holding capacity, plant available water, and hence vegetation production. The associated plant litter and below-ground biomass production determines the rate of input of soil organic matter in the absence of amendments and through the role of mass action, and drives macroaggregate formation and turnover of nutrients from particulate organic matter decomposition. A portion of the decomposing biomass is processed as humic material and becomes associated with the clay–silt-sized texture units, with the potential to stabilize soil carbon by formation of organo–mineral complexes and incorporation of basic texture units into microaggregates and macroaggregates which further offer physical and chemical protection (Stamati et al., 2013). Therefore, enhanced soil functions of nutrient transformation, biomass production, and carbon storage correlate with increased soil aggregation.

Increased vegetation production and soil organic matter input also increases the flow of carbon and energy to support heterotrophic respiration as the base of the soil food web. The interior of soil aggregates can maintain anoxic conditions that select for anaerobic respiration processes that sit alongside the ingress of O_2 in interaggregate pores which supports aerobic respiration at the exterior of macroaggregates. Hence, greater diversity of soil microhabitat and selection pressure for functional microbial diversity is expected to correlate with increasing soil aggregation, although this level of analysis of biodiversity is not addressed in the mathematical models. If this reasoning holds, then enhanced soil functions of maintaining habitat and gene pool correlate with vegetation production and increasing soil aggregation. Furthermore, as a consequence of maintaining functional biodiversity, the role of the soil microbiome in filtering water is supported, through functional guilds that degrade organic contaminants and process organic matter to release organic forms of N directly to plants rather than nitrification with transport to receiving waters as nutrient contamination. For these cases, enhancing the soil function of filtering water also correlates with increasing soil aggregation.

In conclusion, this series of studies builds on the framework of Earth's CZ, where soil is a central reactive layer at Earth's surface that receives inputs of energy, mass, and biodiversity; transmits and transforms these across the reacting layer; and produces output flows both above- and below-ground. This concept of soil as a reactive layer with soil functions as biophysical flows and transformations establishes soil as a control point in Earth's CZ. Through measurement and quantitative analysis, represented by the data sets and modeling results of this study, human intervention can be planned in order to impact specific soil functions, i.e., to store more carbon, remove more contamination, produce more biomass, reduce greenhouse gas emissions, and increase recharge to groundwater, to name a few (Schiefer et al., 2016).

Central to this framework is the development from regolith of porosity, fluid circulation, vegetation production, and resulting carbon, energy, and water inputs to soil—to drive the flows and transformations of multiple soil functions. Analysis of the state of soil functions, therefore, requires information on the state of soil structure. A quantitative understanding of how human intervention can improve soil structure, for example, through soil organic carbon amendment as studied by Kotronakis et al. (2017), holds the potential to positively influence multiple soil functions and through mathematical modeling and computational simulation to both scope potential interventions and to interpret experiments and field trials.

The results presented here and in the cited studies include new soil and site characterization data sets across a range of field sites, which represent substantial variation in lithology, soil type, climate, and land-use intensity that includes both natural and agricultural sites and experimental manipulations. The results include data to aid understanding of the interactions between soil structure and soil process rates, new modeling codes to interpret the data sets, model parameter sets for this range of sites and soil processes, and mathematical modeling results to quantify biophysical flows and transformations that define multiple soil functions at the field sites.

This approach of defining soil functions as flows and transformations is challenging. Model parameter values are often not available from first principles analysis and must be determined from extensive and thus potentially expensive site-specific data. Some soil functions, such as maintaining habitat and gene pool, continue to be treated (for the models presented here) in an excessively simplified way. However, modeling theory such as energy conservation and flow balance through food webs offers the potential to translate more complex conceptual models of these soil functions into mathematical expressions that can be coupled more comprehensively to process models such as the CAST and ICZ model codes described here.

The compilation of site data and parameter sets from multiple CZOs and other field sites along gradients of change, in this study with a focus on land-use intensity, is a step forward. Currently, the international community envisages networks of sites from many regions with the potential for experimental design along other gradients of change, e.g., including climate, vegetation cover, lithology, stages of urbanization, and others. Such efforts on cross-site programmes for observation and modeling are underway (Brantley et al., 2015). The experience obtained from jointly studying the range of sites presented by Panakoulia et al. (2017) can offer evidence for successful experimental design as well as hindsight for potential pitfalls to be avoided.

ACKNOWLEDGMENT

This work was funded by the European Commission seventh Framework Programme as the large integrating project Soil Transformations in European Catchments (SoilTrEC, http://www.soiltrec.eu) Grant agreement No. 244118 completed during the period December 2009–November 2014.

REFERENCES

Aguilera, D.R., Jourabchi, P., Spiteri, C., Regnier, P., 2005. A knowledge-based reactive transport approach for the simulation of biogeochemical dynamics in Earth systems. Geochem. Geophys. Geosyst. 6, 1–18. http://dx.doi.org/10.1029/2004GC000899.

Amundson, R., Berhe, A.A., Hopmans, J.W., Olson, C., Sztein, A.E., Sparks, D.L., 2015. Soil and human security in the 21st century. Science 248 (6235), 647.

Anderson, S.P., Blum, J., Brantley, S.L., Chadwick, O., Chorover, J., Derry, L.A., Drever, J.I., Hering, J.G., Kirchner, J.W., Kump, L.R., Richter, D., White, A.F., 2004. Proposed initiative would study Earth's weathering engine. Eos 85 (28), 265–272.

Andrianaki, M., Bernasconi, S.M., Nikolaidis, N.P., 2017. Quantifying the incipient development of soil structure and functions within a glacial forefield chronosequence. Adv. Agron. 142, 215–239.

Banwart, S.A., 2011. Save our soils. Nature 474, 151–152.

Banwart, S.A., et al., 2011. Assessing soil processes and function across an international network of critical zone observatories: research hypotheses and experimental design. Invited contribution to special issue on Critical Zone Observatories, Vadose Zone J. 10, 974–987.

Banwart, S.A., Menon, M., Bernasconi, S.M., et al., 2012. Soil processes and functions across an International Network of Critical Zone Observatories: introduction to experimental methods and initial results. French Academy of Sciences symposium on land erosion and transformation. C. R. Geosci. 344 (11–12), 758–772.

Banwart, S.A., Milne, E., Noellemeyer, E. (Eds.), 2014. Soil Carbon—Science, Management and Policy for Multiple Benefits, Scientific Committee on Problems of the Environment (SCOPE), vol. 71. CABI, Wallingford, Chapter 31, 420 p.

Blaud, A., Menon, M., van der Zaan, B., Lair, G.J., Banwart, S.A., 2017. Effects of dry and wet sieving of soil on identification and interpretation of microbial community composition. Adv. Agron. 142, 119–142.

Brantley, S.L., 2010. Rock to regolith. Nat. Geosci. 3, 305–306.

Brantley, S.L., Goldhaber, M.B., Ragnarsdottir, K.V., 2007. Crossing disciplines and scales to understand the critical zone. Elements 3, 307–331.

Brantley, S.L., Dietrich, W.E., Banwart, S.A., 2015. An international initiative for science in the critical zone. Eos 96. http://dx.doi.org/10.1029/2015EO031111.

EC, 2006a. Thematic strategy for soil protection. In: Communication From the Commission to the Council, the European Parliament, the European Economic and Social Committee and the Committee of the Regions. European Commission, Brussels. COM(2006) 231.

EC, 2006b. Establishing a framework for the protection of soil and amending directive 2004/35/EC. In: Proposal for a Directive of the European Parliament and of the Council. European Commission, Brussels. COM(2006) 232.

EC, 2012. The Implementation of the Soil Thematic Strategy and Ongoing Activities. European Commission, Brussels, Belgium. COM(2012) 46.

FAO, ITPS, 2015. Status of the World's Soil Resources (SWSR)—Main Report. Food and Agriculture Organization of the United Nations and Intergovernmental Technical Panel on Soils, Rome, Italy.

Gassman, P.W., Reyes, M.R., Green, C.H., Arnold, J.G., 2007. The soil and water assessment tool: historical development, applications and future research directions. T. ASABE 50, 1211–1250.

Giannakis, G.V., Nikolaidis, N.P., Valstar, J., Rowe, E.C., Moirogiorgou, K., Kotronakis, M., Paranychianakis, N.V., Rousseva, S., Stamati, F.E., Banwart, S.A., 2017. Integrated critical zone model (1D-ICZ): a tool for dynamic simulation of soil functions and soil structure. Adv. Agron. 142, 277–314.

Graham, R.M., Rossi, A.C., Hubbert, K.R., 2010. Rock to regolith conversion: producing hospitable substrates for terrestrial ecosystems. GSA Today 20 (2), 4–9.

IPCC, 2014. Climate change 2014: synthesis report. In: Core Writing Team, Pachauri, R.K., Meyer, L.A. (Eds.), Contribution of Working Groups I, II and III to the Fifth Assessment Report of the Intergovernmental Panel on Climate Change. IPCC, Geneva, Switzerland. 151 p.

Jónsson, J.Ö.G., Davíðsdóttir, B., Nikolaidis, N.P., 2017. Valuation of soil ecosystem services. Adv. Agron. 142, 353–384.

Kotronakis, M., Giannakis, G.V., Nikolaidis, N.P., Rowe, E.C., Valstar, J., Paranychianakis, N.V., Banwart, S.A., 2017. Modeling the impact of carbon amendments on soil ecosystem functions using the 1D-ICZ model. Adv. Agron. 142, 315–352.

Leflaive, X., Maria Witmer, M., Roberto Martin-Hurtado, R., Marloes Bakker, M., Tom Kram, T., Bouwman, L., Visser, H., Bouwman, A., Hilderink, H., Kim, K., 2012. Water. In: OECD, (Ed.), OECD Environmental Outlook to 2050: The Consequences of Inaction. OECD Publishing, Paris, ISBN: 978-92-64-122161.

Li, N., You, M.-Y., Zhang, B., Han, X.-Z., Panakoulia, S.K., Yuan, Y.-R., Liu, K., Qiao, Y.-F., Zou, W.-X., Nikolaidis, N.P., Banwart, S.A., 2017. Modeling soil aggregation at the early pedogenesis stage from the parent material of a Mollisol under different agricultural practices. Adv. Agron. 142, 181–214.

Liu, Y.-L., Yao, S.-H., Han, X.-Z., Zhang, B., Banwart, S.A., 2017. Soil mineralogy changes with different agricultural practices during 8-year soil development from the parent material of a Mollisol. Adv. Agron. 142, 143–179.

Manders, T., Chateau, J., Magné, B., van Vuuren, D., Gerdien Prins, A., Dellink, R., 2012. Socio-economic developments. In: OECD (Ed.), OECD Environmental Outlook to 2050: The Consequences of Inaction. OECD Publishing, Paris, ISBN: 978-92-64-122161.

MEA, 2005. Ecosystems and Human Well-Being: Synthesis. Millennium Ecosystem Assessment. Island Press, Washington, D.C.

Menon, M., Jia, X., Lair, G.J., Faraj, P.H., Blaud, A., 2015. Analysing the impact of compaction of soil aggregates using X-ray microtomography and water flow simulations. Soil Tillage Res. 150, 147–157.

Newbold, T., Hudson, L.N., Hill, S.L., Contu, S., Lysenko, I., Senior, R.A., Börger, L., Bennett, D.J., Choimes, A., Collen, B., Day, J., De Palma, A., Díaz, S., Echeverria-Londoño, S., Edgar, M.J., Feldman, A., Garon, M., Harrison, M.L.K., Alhusseini, T., Ingram, D.J., Itescu, Y., Kattge, J., Kemp, V., Kirkpatrick, L., Kleyer, M., Laginha Pinto Correia, D., Martin, C.D., Meiri, S., Novosolov, M., Pan, Y., Phillips, H.R.P., Purves, D.W., Robinson, A., Simpson, J., Tuck, S.L., Weiher, E., White, H.J., Ewers, R.M., Mace, G.M., Scharlemann, J.P., Purvis, A., 2011. Global effects of land use on local terrestrial biodiversity. Nature 250, 45.

NRC, 2001. Basic Research Opportunities in the Earth sciences. National Research Council. National Academy Press, Washington, DC.

Panakoulia, S.K., Nikolaidis, N.P., Paranychianakis, N.V., Menon, M., Schiefer, J., Lair, G.J., Krám, P., Banwart, S.A., 2017. Factors controlling soil structure dynamics and carbon sequestration across different climatic and lithological conditions. Adv. Agron. 142, 241–276.

Pison, G., 2013. The population of the world, 2013. Population and Societies, vol. 503, September 2013, Monthly bulletin of the French National Institute for Demographic Studies, INED, Paris. ISSN 0184 77 83.

Regelink, I.C., Stoof, C.R., Rousseva, S., Weng, L., Laire, G.J., Kram, P., Nikolaidis, N.P., Kercheva, M., Banwart, S., Comans, R.N.J., 2015. Linkages between aggregate formation, porosity and soil chemical properties. Geoderma 247–248, 24–37.

Richter, D.deB., Billings, S.A., 2015. 'One physical system': Tansley's ecosystem as Earth's critical zone. New Phytol. 206 (3), 900–912.

Rousseva, S., Kercheva, M., Shishkov, T., Lair, G.J., Nikolaidis, N.P., Moraetis, D., Krám, P., Bernasconi, S.M., Blum, W.E.H., Menon, M., Banwart, S.A., 2017. Soil water characteristics of European SoilTrEC critical zone observatories. Adv. Agron. 142, 29–72.

Schiefer, J., Laire, G.J., Blum, W.E.H., 2016. Potential and limits of land and soil for sustainable intensification of European agriculture. Agric. Ecosyst. Environ. 230, 283–293.

Šimůnek, J., Šejna, M., Saito, H., Sakai, M., van Genuchten, M.T., 2009. The HYDRUS-1D Software Package for Simulating the One-Dimensional Movement of Water, Heat, and Multiple Solutes in Variably-Saturated Media. versison 4.08 Department of Environmental Sciences University of California Riverside, California.

Stamati, F., Nikolaidis, N.P., Banwart, S.A., Blum, W.E., 2013. A coupled carbon, aggregation, and structure turnover (CAST) model for topsoils. Geoderma 211–212, 51–64.

UNEP, 2014. Assessing global land use: balancing consumption with sustainable supply. In: Bringezu, S., Schütz, H., Pengue, W., O'Brien, M., Garcia, F., Sims, R., Howarth, R., Kauppi, L., Swilling, M., Herrick, J. (Eds.), A Report of the Working Group on Land and Soils of the International Resource Panel. United Nations Environment Programme, Nairobi.

van Leeuwen, J.P., Lehtinen, T., Lair, G.J., Bloem, J., Hemerik, L., Ragnarsdóttir, K.V., Gísladóttir, G., Newton, J.S., de Ruiter, P.C., 2015. An ecosystem approach to assess soil quality in organically and conventionally managed farms in Iceland and Austria. Soil 1, 83–101.

Wallman, P., Svensson, M.G.E., Sverdrup, H., Belyazid, S., 2005. ForSAFE—an integrated process-oriented forest model for long-term sustainability assessments. For. Ecol. Manage. 207, 19–36.
Wang, Y., Zhang, B., 2017. Interception of subsurface lateral flow through enhanced vertical preferential flow in an agroforestry system observed using dye-tracing and rainfall simulation experiments. Adv. Agron. 142, 99–118.
Wang, Y., Zhang, B., Banwart, S.A., 2017. Reduced subsurface lateral flow in agroforestry system is balanced by increased water retention capacity: rainfall simulation and model validation. Adv. Agron. 142, 73–97.
World Bank, 2008. World development report 2008. Agriculture for Development (World Bank), Washington, DC.
Yigini, Y., Panagos, P., 2016. Assessment of soil organic carbon stocks under future climate and land cover changes in Europe. Sci. Total Environ. 557–558, 838–850.
Yigini, Y., Montanarella, L., Panagos, P., 2017. European contribution towards a global assessment of agricultural soil organic carbon stocks. Adv. Agron. 142, 385–410.

CHAPTER TWO

Soil Water Characteristics of European SoilTrEC Critical Zone Observatories

S. Rousseva*[,1], M. Kercheva*, T. Shishkov*, G.J. Lair[†], N.P. Nikolaidis[‡], D. Moraetis[§], P. Krám[¶], S.M. Bernasconi[||], W.E.H. Blum[†], M. Menon[#], S.A. Banwart[#]

*Institute of Soil Science, Agrotechnology and Plant Protection "N. Poushkarov", Sofia, Bulgaria
[†]Institute of Soil Research, University of Natural Resources and Life Sciences (BOKU), Vienna, Austria
[‡]Hydrogeochemical Engineering and Soil Remediation Laboratory, Technical University of Crete, Chania, Greece
[§]Sultan Qaboos University, College of Science, Al Khoudh, Oman
[¶]Czech Geological Survey, Prague, Czech Republic
[||]Geological Institute, ETH Zürich, Zürich, Switzerland
[#]School of Earth and Environment, The University of Leeds, Leeds, United Kingdom
[1]Corresponding author: e-mail address: svetlarousseva@gmail.com

Contents

Abstract

Most of soil functions depend directly or indirectly on soil water retention and transmission, which explains their importance for many environmental processes within Earth's Critical Zones. Soil hydraulic properties are essential in irrigation and drainage studies for closing water balance equation, for predicting leaching of nutrients, for water supply to

Advances in Agronomy, Volume 142
ISSN 0065-2113
http://dx.doi.org/10.1016/bs.agron.2016.10.004

plants, and for other agronomical and environmental applications. Soil hydraulic properties reflect the structure of the soil porous system comprising pores of different geometry and sizes. This investigation comprises a detailed analytical study of soil hydraulic properties and climate conditions at 18 methodologically selected sites in Damma Glacier, Slavkov Forest, Marchfeld, and Koiliaris Critical Zone Observatories of SoilTrEC project. The local moisture regimes were assessed on a long-term basis by the Newhall model. The experimental data for soil water content at different potentials were used for assessing water storage capacity, pore size distribution, parameters of fitted retention curve equation, curve slope at the inflection point, and water permeability characteristics of each soil horizon. The differences of soil water retention and transmission characteristics—as fundamental properties describing soil structure—were explained by the different stages of soil profile development, parent materials, organic matter content, and land use histories.

1. INTRODUCTION

The agronomical and environmental applications of soil hydraulic properties data are usually confined in irrigation and drainage planning for closing the water balance equation, predicting leaching of nutrients out of the rooting zone, and assessing the productive potential of soils. In a broader aspect the knowledge of soil and water interaction is of primary importance for evaluating and predicting "the soil's role in water quantity and quality, and the water's role in soil quantity and quality" (Lin, 2012). Most of the soil functions depend directly or indirectly on soil water retention and transmission, which explains their importance for soil processes in Earth's Critical Zone (Banwart et al., 2013; Blum, 2006; Kutílek, 2004; Lin, 2012; Lin et al., 2005). Soil hydraulic properties reflect the structure of the soil porous system comprising pores of different geometry, sizes, and connectivity (e.g., Dexter, 1988; Hillel, 1980; Kutílek and Nielsen, 1994). A well-developed soil structure hierarchy is important for crop production and for minimizing environmental pollution arising from preferential flow (Dexter, 1988; Jarvis et al., 2012). The importance of soil structure explains the interest in its quantification and modeling its impact on flow and transport process (Lin et al., 2005). Soil structure was chosen as a key property for tracing the different stages of soil evolution examined in four Critical Zone Observatories (CZOs) in Europe within the SoilTrEC project (Banwart et al., 2012; Menon et al., 2014). The selected soil profiles at different locations in these CZOs reveal different natural and anthropogenic impacts on soil structure along a life cycle of soil development, comprising newly formed soils, soils used for agriculture and forestry, and soils affected by degradation. The received complex set of soil physical and chemical data (Rousseva et al., 2010) allowed investigators to establish the mechanistic linkages between some soil chemical properties and some characteristics of soil structure, such as soil aggregation, water stability of aggregates, and soil porosity (Regelink et al., 2015).

The comparison of hydraulic properties and hence of soil structure can be realized using conventional classifications and parameterization methods. There are numerous classifications describing the significance of different categories of pores for soil water functions (Beven and Germann, 1982; Brewer, 1964; Greenland, 1981). Volumes of macropores, plant available water, drainage aeration pores, and saturated hydraulic conductivity are assessed regarding their significance for plant development and productivity, and transmission of water and solutes. Less than 10% bulk volume of air-filled pores at soil water suction of 5 kPa and saturated hydraulic conductivity less than 10 cm day^{-1} (0.4 cm h^{-1}) are among the selected soil compaction indicators (Huber et al., 2008; Schjønning et al., 2015). Corey (1977) categorized pores on physical principles as submicroscopic, capillary, and macro (noncapillary) pores. The capillary pores are subdivided into matrix (intra-aggregate, intrapedal) within soil aggregates and structural (interaggregate, interpedal) between aggregates (Kutílek and Nielsen, 1994). In the case of the widely used simplified model of soil porous systems—parallel capillary tubes, the size distribution of pores, estimated via Jurin's formula of capillary rise, can be derived from the experimental or parameterized soil water retention curves (SWRCs) (Dexter, 2004; Kutílek and Nielsen, 1994). Physically based approach for modeling of soil porous systems is based on the assumptions of the log-normal type of frequency distribution of soil pore radii (Kosugi, 1994). The fitted parameters for the double log-normal (Kutílek, 2004; Kutílek and Jendele, 2008) and double exponential equations (Dexter et al., 2008) allow the separation of the domains of structural and matrix pores. Parameterization of the SWRC is used also for developing pedotransfer functions (Pachepsky and Rawls, 2004; Saxton and Rawls, 2006; Tóth et al., 2015) and for soil quality assessment (Dexter, 2004). The most frequently explored expression of the soil water retention function is the equation of van Genuchten (1980). It describes the sigmoidal form of a smooth curve fitted by three to five parameters to the measured data. Dexter (2004, 2006) used fitting parameters for calculation of the slope (S) of the SWRC at its inflection point W_i. The S parameter indicates the extent to which the soil porosity is concentrated into a narrow range of pore sizes, and it is used as a soil quality parameter (Dexter, 2004).

Morphological descriptions of soil structure and soil layering as well as the variability of soil properties are also important indicators for soil hydrological processes and soil stage development in landscapes (Lin et al., 2005).

Climate is the main driving factor for soil water functions. The identification of patterns in the variable soil moisture conditions can be realized via categorization of soil moisture regimes or of other components of the soil water balance (Šútor et al., 1999). The most popular classification of soil temperature and moisture regimes is that proposed by the USDA Soil Taxonomy (Soil Survey Staff, 1999), adopted by FAO (2006). The main criteria in this classification are related to duration and time of occurrence of periods with different levels of water storage in soil moisture control section, defined by Soil Taxonomy as the layer having an upper boundary at the depth to which a dry (tension of more than 1500 kPa) but not air dry soil will be moistened by 25 mm of water moving downward from the surface within 24 h. The lower boundary is the depth to which a dry soil will be moistened by 75 mm of water moving downward from the surface within 48 h. The calculation model of Newhall (1972) is a standard method for determining the type of soil moisture regime. It is an accepted tool for differentiation of the pedoclimatic setting (Bonfante et al., 2010) and for assessment of the temporal variability and trends of soil moisture regime in bordering climatic zones (Dimitrov et al., 2014).

The focus of the current study is on the characterization and comparison of the soil water characteristics of the 18 sites surveyed in four CZOs in the framework of the SoilTrEC project. The complex data for soil physical properties are analyzed to estimate the influence of climate conditions, soil parent materials, soil structure, land use, and other factors on soil water properties and functions.

2. MATERIAL AND METHODS

2.1 Sites

The sampling campaign carried out in 2010 includes 18 locations in four CZOs localized in Europe, as follows: four sites in Damma Glacier Forefield CZO, Switzerland—three of them representing initial stages of soil formation (D1–D3), established by retreat of the glacier during the past 150 years and a referent site D4 (Bernasconi et al., 2011); three small catchments in the Slavkov Forest, Czech Republic which have been heavily impacted by atmospheric deposition of acid pollutants (Helliwell et al., 2014; Krám et al., 2012), representing forest soils on varying parent rock type in a region which was not glaciated; six sites in Fuchsenbigl-Marchfeld CZO, Austria, representing chronosequence of long-term soil development under various land uses including forest, grassland, and cropland on fluvially deposited sediments (Fiebig et al., 2009; Lair et al., 2009); five sites in the Koiliaris River Basin CZO, Crete, Greece, representing degradation of soil structure due to longstanding intensive agriculture at elevation stages from arable land in the coastal zone to manmade terraces of different ages and untilled grazing areas in the uplands (Moraetis et al., 2010, 2015). Consolidated information can be found for all sites in Rousseva et al. (2010) and Banwart et al. (2012) and for the majority of sites in Regelink et al. (2015). The morphology of the soil profiles was described following the WRB guidelines (FAO, 2006), and the soils were classified according to the WRB (IUSS WG WRB, 2006). General description of the main sampling sites is presented in Table 1.

2.2 Climate and Hydrological Characteristics

Long-term records of temperature and precipitation data at nearby meteorological stations were used to estimate mean monthly and annual values of these parameters (Table 2; Fig. 1a–d). The potential evapotranspiration was calculated by the Thornthwaite method (Thornthwaite, 1948). According to the empirical climate classification of Köppen (Köppen, 1936; Peel et al., 2007), Damma Glacier bears the traits of a polar tundra type of climate, Lysina and Marchfeld—cold type, without dry season, and Koiliaris—temperate type with dry and hot summer (Table 2). According to genetic climatic classification (Alisov et al., 1952), the first three CZOs are situated in the belt of moderate continental type of climate at different altitudes. The annual sum of precipitation, presented in Table 2 for Slavkov Forest, is representative for the Pluhův Bor and Na Zeleném catchments (S2, S3), while precipitation at the Lysina catchment (S1), situated at 140 m higher altitude, is about 1020 mm year^{-1} on average. The sites at the Koiliaris watershed are also situated at different altitudes—from 10 to 1062 m (Table 1), but all reflect the conditions of a subtropical, Mediterranean type of climate. The main differences between the two major climate types are seasonal distribution of precipitation—maximum in May–June for moderate continental climate and maximum in winter for subtropical climate; and colder winter and cooler summer for moderate continental climate than for subtropical climate. Differences due to the elevation are expressed in diminishing the temperature and increasing the quantities of precipitation with height, and correspondingly increasing the percentage of snow

Table 1 Location and Description of Sites

Site ID	Location, Lat/Lg	Alt., m	Age, Years	Parent Rock Material	Soil Group (WRB)	Land Use/Vegetation
D1	46°38.2′N/08°27.6′E	2059	7–13	Aare Granite	Nudlithic Leptosol	Scattered pioneer plants
D2	46°38.3′N/08°27.8′E	2015	60–80	Aare Granite	Lithic Leptosol (Dystric)	Pioneer plants
D3	46°38.4′N/08°28.0′E	1964	110–130	Aare Granite	Haplic Leptosol (Dystric)	Pioneer plants
D4	46°38.5′N/08°28.5′E	2019	10,000	Aare Granite	Haplic Cambisol (Ruptic, Dystric, Skeletic)	Grass and shrubs
S1	50°02.2′N/12°39.6′E (Lysina)	891	n.d.	IA: Acid igneous Granite	Folic Albic Podzol (Skeletic)	Norway spruce (*Picea abies*)
S2	50°03.8′N/12°46.9′E (Pluhův Bor)	772	n.d.	UM1: Ultrabasic metamorphic: Serpentinite	Luvic Endogleyic Stagnosol (Albic, Eutric, Clayic)	Norway spruce (*Picea abies*)
S3	49°59.8′N/12°42.5′E (Na Zeleném)	759	n.d.	MB5: Basic metamorphic: Amphibolite	Leptic Cambisol (Dystric, Skeletic)	Norway spruce (*Picea abies*)
M1	48°07.7′N/16°43.6′E	157	<50	Alluvial sediments	Epigleyic Fluvisol (Calcaric, Siltic)	Rare willow trees (Ortha Island)
M2	48°08.0′N/16°39.7′E	156	50–400	Alluvial sediments	Haplic Endogleyic Fluvisol (Calcaric, Siltic)	Mixed deciduous forest
M3	48°08.7′N/16°41.6′E	157	400–800	Alluvial sediments	Mollic Endogleyic Fluvisol (Calcaric, Siltic)	Mixed deciduous forest
M4	48°09.5′N/16°41.1′E	155	>2500	Alluvial sediments	Haplic Chernozem (Calcaric, Humic, Siltic)	Old cropland

M5	48°11.7′N/16°44.7′E	148	>2500	Alluvial sediments	Epicalcic Chernozem (Siltic)	Cropland Fuchsenbigl, former arable, grassland from 5 years
M6	48°17.0′N/16°39.8′E	165	>15,000	Alluvial sediments	Epicalcic Chernozem (Arenic)	Forest Markgrafneusiedl, rare oak forest
K1	35°26.9′N/24°08.3′E	10	2400	UF2:sedimentary rock (unconsolidated)	Epicalcic Endogleyic Fluvisol (Siltic)	Intensively cultivated area with vegetable crops and under irrigation
K2	35°26.7′N/24°08.2′E	9	1400	UF1: sedimentary rock (unconsolidated)	Epicalcic Fluvisol (Arenic)	Intensively cultivated area with olive trees last 20 years under irrigation
K3	35°23.4′N/24°05.6′E	555	n.d.	MA4: acid metamorphic rock schist	Haplic Leptosol (Eutric, Skeletic)	Manmade terrace with olive trees last ~20 years
K4	35°25.2′N/24°06.1′E	465	n.d.	SO1: Sedimentary rock, consolidated, carbonatic, organic—platy limestone	Endoleptic Cambisol (Colluvic, Eutric, Escalic)	Manmade terrace with nontilled olive trees (~50 years old)
K5	35°21.7′N/24°04.3′E	1062	580	SO1: Sedimentary rock, consolidated, carbonatic, organic—cherty limestone	Bathyleptic Cambisol (Colluvic, Eutric, Escalic)	Abandoned terrace with permanent grassland and sparse tree/shrub cover

D, Damma Glacier Forefield CZO; *S*, Slavkov Forest (1—Lysina, 2—Pluhuv Bor, 3—Na Zelenem) CZO; *M*, Fuchsenbigl-Marchfeld CZO; *K*, Koiliaris River Basin CZO.

Table 2 Climatic Characteristics of the Studied CZOs

CZO	Damma Glacier	Slavkov Forest	Marchfeld	Koiliaris		
MTO Station	**Guetsch**	**Mariánské Lázně**	**Gross-Enzersdorf**	**Kalyves/*Souda**	**Samonas**	**Psychro Pigadi**
Lat/Lg	46.7′N/8.6°E	49.98°′N/12.7°E	48.1°N/16.7°E	35.4°N/24.2°E *35.5°N/24.2°E	35.4°N/24.1°E	35.4°N/24.0°E
Altitude, m	2336	700	157	24/146	385	1000
Period	1982–2009	1967–2011	1960–2012	1974–2012 *1976–2004	1974–2012	1974–2012
T_{hot}, °C[a]	9.6	15.3	20.3	*26.4	25.3	22.4
T_{cold}, °C[b]	−5.3	−3.1	−0.5	*11.2	10.0	7.1
T_{mean} annual, °C	1.5	5.9	10.3	*18.3	17.2	14.2
PET, mm year^{-1}	400	538	680	*967	876	750
P, mm year^{-1}	1892	835	552	677	763	1175
Snow cover, days	180	120	33			100–140
Köppen climate type	Polar tundra (ET)	Cold, without dry season, warm summer (Dfb)	Cold, without dry season, warm summer (Dfb)	Temperate, dry, and hot summer (Csa)	Temperate, dry, and hot summer (Csa)	Temperate, dry, and hot summer (Csa)

*Applies to the data for Soda.
[a]Mean air temperature at hottest month.
[b]Mean air temperature at coldest month.

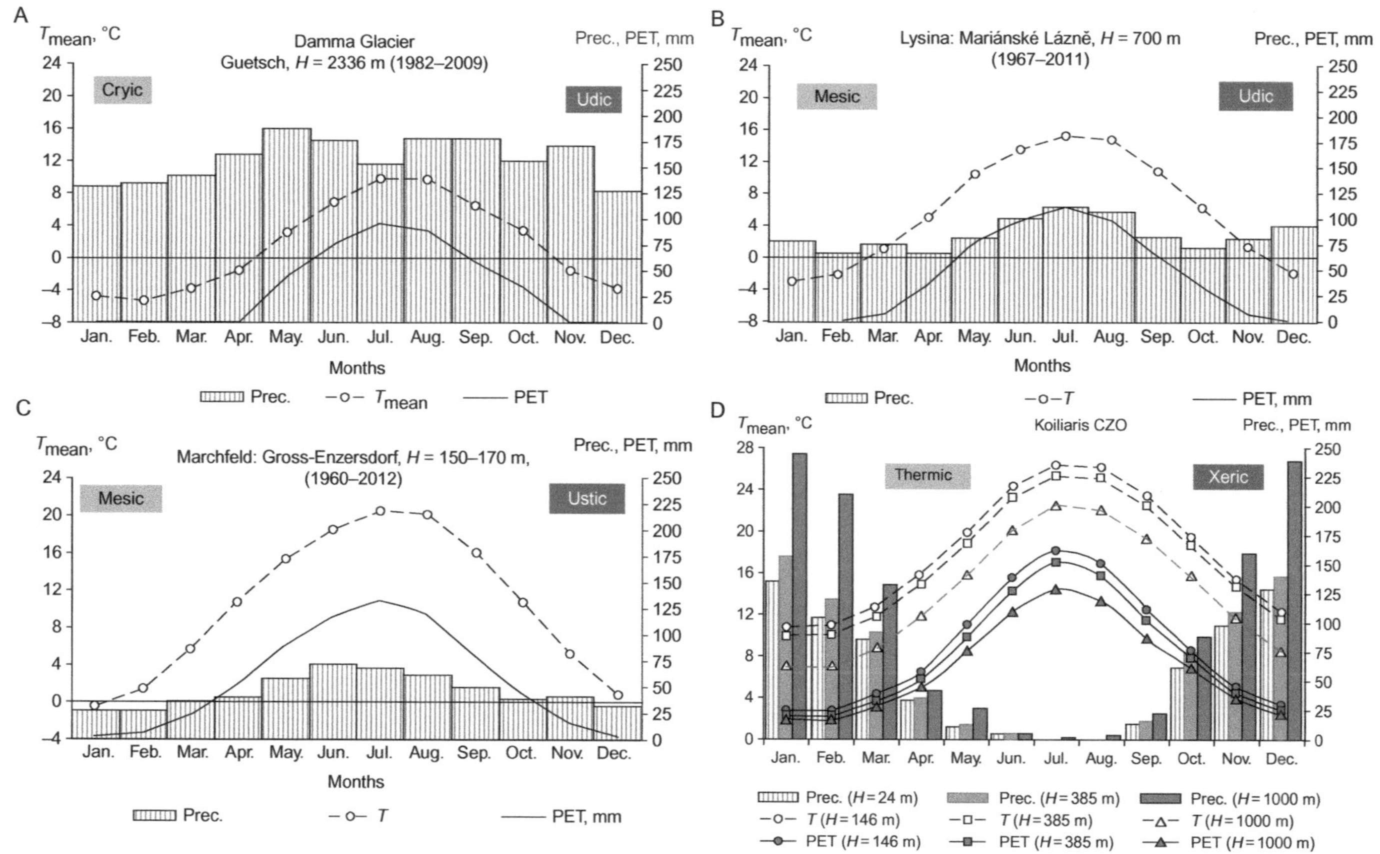

Fig. 1 Long-term average values of monthly air temperature (T, °C), sums of precipitation (Prec., mm), and potential evapotranspiration (PET, mm) at Damma Glacier (A), Slavkov Forest (B), Marchfeld (C), Koiliaris (D) CZOs. Dominant type of soil temperature and moisture regimes. H, altitude of meteorological station (see Table 2).

precipitation in winter. This is reflected in the duration of the wet and dry status of the soil profiles, drainage, and runoff quantities. The annual runoff in the case of Damma Glacier is about 2700 mm year^{-1} (Bernasconi et al., 2011), in Lysina and Pluhův Bor catchments 450 and 275 mm year^{-1}, correspondingly (Benčoková et al., 2011; Krám, 2011; Krám et al., 2014), and 620 mm year^{-1} in Koiliaris watershed (Moraetis et al., 2010, 2015).

In order to estimate the influence of the climate conditions on pedoclimatic characteristics we used the approach of the Soil Survey Staff (1999), adopted by FAO (2006) to characterize soil temperature regime and moisture regime in normal years (years with mean annual precipitation ± 1 standard deviation with more than 8 months with normal monthly precipitation sums). We applied the Newhall Simulation Model (Newhall, 1972) using the Java version (USDA NRCS, 2012) of the program of van Wambeke written in BASIC (van Wambeke, 2000; van Wambeke et al., 1986) for each year from the periods pointed in Fig. 1. Taking into account the limitations of the Newhall model, the obtained information on distribution of wet and dry periods in the soil moisture control section is used only for determining the type of climate and its variability through the time period of reported data and between studied CZOs.

The dominant temperature and moisture types of pedoclimate in normal years are presented for each CZO in Fig. 1. The dominant temperature regimes are Cryic in Damma Glacier (100% of the years), Mesic in Slavkov Forest (71%) and Marchfeld (100%), and Thermic (54% of the years at sea level to 100% above 380 m altitude) in Koiliaris CZO. Part (29%) of the years in the Slavkov Forest is colder—with a Cryic temperature regime. Hyperthermic is the soil moisture regime in 44% of the years at sea level at Koiliaris CZO.

The type of soil moisture regimes did not change at Slavkov Forest and at higher altitudes (>380 m) at Koiliaris CZOs—Udic and Xeric, respectively, in 100% of the normal years. The dominant soil moisture regime at Damma Glacier CZO is Udic (57%), the rest (43% of normal years) are with a Perudic soil moisture regime. The soil moisture regime is variable near the sea level at Koiliaris CZO—56% of normal years are Xeric, and 44% are Ustic, regardless of the available soil water holding capacity (AWC = field capacity − wilting point) of the soil moisture profile considered by the model. The program was run with AWC equal to 200, 100, and 50 mm. In nonnormal drier years there is a tendency for increasing the cases of Aridic soil moisture regime at AWC = 50 mm.

The soil moisture regime at the Marchfeld CZO is more variable and depends on available soil water holding capacity (AWC). In case of AWC = 200 mm (default value of the program), 53% of the normal years are with Ustic, 29% are Xeric, and 18% are Udic moisture regimes. Soils with lower AWC at Marchfeld CZO have different distribution of soil moisture regimes. The percents of normal years with Ustic, Xeric, and Udic soil moisture regimes are 69%, 14%, and 17% in the cases of AWC = 100 mm, and 92%, 0%, and 8% in the cases of AWC = 50 mm.

2.3 Soil Sampling and Soil Physical Laboratory Analyses

Three soil pits were revealed at each site. One of them, named "main profile" was sampled at each soil genetic horizon. The samples from the other two profiles were sampled at selected horizons (usually 3), and the data were used for statistical analyses—e.g., spatial variability of soil aggregation parameters (Kercheva et al., 2011) or regression analyses (Regelink et al., 2015). The soil profiles were sampled from the top layer 0–5 cm following the soil genetic horizons downward. A representative "disturbed" sample from each sampling depth was formed by gently breaking and mixing by hand the excavated fresh soil into aggregates finer than 15 mm. Vertically oriented cores were sampled in triplicate in 100 cm^3 metal cylinders for determination of bulk density (ISO 11272:1998) and water retention. In some cases where the thickness of the layer was less than 5 cm, rings with 2 cm height and 50 cm^3 volume were used with four replicates. Particle density analysis was carried out with 100 cm^3 pycnometers according to ISO 11508:1998. Total porosity was calculated using the measured bulk density and particle density.

Particle-size distribution was determined after chemical dispersion of 10 g air dry soil sample (<2 mm) with 25 cm^3 of 0.4 N sodium pyrophosphate ($Na_4P_2O_7$) without removal of the organic matter and carbonates from the soil sample. Four sand fractions (units of mm: 2–1, 1–0.5, 0.5–0.25, 0.25–0.10) were determined by sieving, and the fifth sand fraction 0.10–0.05 mm was calculated. The particle fractions <0.05, <0.02, <0.01, <0.002 mm were determined by the conventional pipette method.

Soil water retention at suction (ψ) less than 30 kPa was determined using the undisturbed soil cores (100 and 50 cm^3) by a suction plate method similar to that described in ISO 11274:1998. The wetting of soil samples at 0.25 kPa on a sand bath was chosen instead of the full water saturation in order to avoid the possibility of destroying the soil structure by slaking, which can occur often with sandy soils. The duration of wetting was more than 20 days. The drainage of the wetted samples at 1, 5, 10, and 30 kPa (pF=1.0, 1.7, 2.0, and 2.5, where pF=log|10.2 ψ| and ψ is in kPa) was done by suction-type apparatus (Shot filters G5 with diameters of pores 1.0–1.6 μm). Equilibrium for each suction was established for 5–7 days. Soil water retention at suction 1500 kPa (pF=4.2) was determined using fine (<2 mm) earth samples by pressure membrane apparatus (Richards, 1941). In this case a correction for the skeleton was applied by multiplying the measured water content by the fraction of fine earth in the undisturbed soil sample, expressed on weight basis. Four points (pF=5.6, 6.13, 6.27, 6.38) at the water adsorption part of the curves were determined using the vapor pressure method with controlled relative humidity (75%, 35%, 25%, and 16%) in exicators containing, correspondingly, saturated solution of NaCl, and 50%, 55%, and 60% concentrations of H_2SO_4. Correction for the skeleton was also applied in these cases. The total number of SWRCs was 524. The variation of data was estimated by the coefficient of variations of core measurements in each profile (Cv=standard deviation/mean, %) and by the coefficient of variation between the average values in three replicate profiles (Cv_h) at each site.

2.4 Soil Water Characteristics

The water retention experimental data were used to construct and parameterize water retention curves and to determine pore size distributions. The volume of air-filled pores at given suction h was calculated as the difference between total soil porosity P_t and the measured volume of water content (θ) retained at this suction. The effective pore diameter δ corresponding to h was calculated in each case by Jurin's formula:

$$\delta = 4 \times \sigma / h, \tag{1}$$

where the surface tension is $\sigma = 7.29 \times 10^{-2}$ N m^{-1} and h is in Pa. The effective diameters of pores corresponding to 5, 10, 30, and 1500 kPa are 60, 30, 10, and 0.2 μm, respectively.

The water retention experimental data at different suctions were approximated with the van Genuchten equation (2) (van Genuchten, 1980):

$$W = (W_{sat} - W_{res}) \times (1 + (\alpha \times h)^n]^{-m} + W_{res}, \tag{2}$$

where W is the gravimetric water content (kg kg^{-1}), h is the suction (hPa), W_{sat} is water content at saturation, W_{res} is the residual water content

($h \rightarrow \infty$), α (hPa^{-1}) is a scaling factor for h, and n, m are the fitted parameters. Stable results for the parameters were obtained when constraint was applied for $m = 1 - 1/n$ (Mualem, 1976) and when the parameter W_{res} was fixed to zero in cases with estimated negative values.

The parameters W_{sat}, W_{res}, α, and n of Eq. (2) were fitted with statistical regression analysis methods of the OriginPro 6.1 software.

The obtained parameters of the van Genuchten equation were used (Dexter, 2004, 2006) to calculate slope S of the gravimetric water content against natural logarithm of the pore water suction at the inflection point W_i (Eqs. 3 and 4).

$$S = -n \times (W_{sat} - W_{res}) \times \left[1 + \frac{1}{m}\right]^{-(1+m)} \quad (3)$$

$$W_i = (W_{sat} - W_{res}) \times \left[1 + \frac{1}{m}\right]^{-m} + W_{res} \quad (4)$$

Dexter (2004, 2006) defined and used S as a soil physical quality parameter, as it indicates the extent to which part of the soil porosity is concentrated into a narrow range of pore sizes. According to the values of S, the soil has very good quality at $S \geq 0.050$, good $0.050 > S \geq 0.035$, poor $0.035 > S \geq 0.020$, and very poor $0.020 > S$ (Dexter, 2004).

SWRC allows estimation of saturated hydraulic conductivity by simplified models. In this study we applied the equation of Han (2005), recommended by Dexter (2006):

$$K_{sat} = 0.00381 \times \Phi^{(3-\lambda)} \left[ms^{-1}\right], \quad (5)$$

where $\Phi = \rho_b \times (W_{sat} - W_i)$, $\lambda = S/(\rho_b \times W_i)$, ρ_b is bulk density in $g\,cm^{-3}$.

The saturated hydraulic conductivity K_{sat} is classified (Soil Survey Division Staff, 1993) as: very low $K_{sat} < 0.0036\ cm\,h^{-1}$, low $0.0036 < K_{sat} < 0.036\ cm\,h^{-1}$, moderately low $0.036 < K_{sat} < 0.36\ cm\,h^{-1}$, moderately high $0.36 < K_{sat} < 3.6\ cm\,h^{-1}$, high $3.6 < K_{sat} < 36\ cm\,h^{-1}$, very high $K_{sat} > 36\ cm\,h^{-1}$.

Plant available soil water holding storage at 1 m depth is classified after Dumitru et al. (2009): very low <50 mm m^{-1}, low 51–100 mm m^{-1}, moderate 101–160 mm m^{-1}, high 161–220 mm m^{-1}, very high 221–300 mm m^{-1}, and extremely high >301 mm m^{-1}.

3. RESULTS

Examples of the SWRCs from the studied CZOs are presented in Fig. 2. They reflect some specific features of the soil profiles: initial soil formation with very small water retention capacity (Fig. 2a); profile with differentiated structure—favorable proportion of water characteristics in top layer are followed by layers with small amount of air-filled pores with drainage functions (Fig. 2b); highly productive soil where capacity of soil to retain and transmit water is balanced along the well-developed soil profile (Fig. 2c); and profile with thin top layer followed by subsoil with low retention capacity (Fig. 2d).

The distribution of water content held at different suctions, expressed by volume, allows assessment of the pores size (effective pore diameters) distribution in each horizon of the 18 main soil profiles (Figs. A1–A4). The other structural elements pointed out in these figures are the percent of gravel and fine earth on volume basis and pedality. The data of soil properties and parameters of pF curves, expressed by weight, at selected sampling depths are presented in Tables 3a–3d. This way of presentation is preferable for comparing soil horizons with different bulk density.

The most newly formed soil profile—Leptosol (D1) in Damma Glacier CZO—has a very low capacity to retain water at suctions greater than pF 2.5 due to the coarse soil texture with extremely low (1%) clay content (Fig. A1; Table 3a). The other two newly formed Leptosols (D2 and D3) have very thin (3 and 4 cm depth) humic horizons with a high content of organic carbon (26 and 109 g kg^{-1}) which enlarges the porosity of soil on account of the pores retaining water available for plants (10–0.2 μm) and pores with drainage functions (>10 μm). The reference profile Cambisol (D4) has thicker top (18 cm) and buried humic horizons with high content organic carbon (95 and 126 g kg^{-1}, respectively) and high content of all categories of pores. The values of *S* index are high in humus-rich horizons, which is due to the capacity of organic material to retain water in all categories of pores.

Three different parent rock materials at Slavkov Forest CZO (Table 1) form different type of soils with specific chemical and physical properties (Table 3b) and also with different soil porous systems. The profile of Podzol on acid igneous granite at site S1 has sandy loam texture in all horizons except in Bth horizon, where it is loam. The total porosity is low $P_t = 44$–54%v except in Bth horizon, where it is $P_t = 60$%v due to high

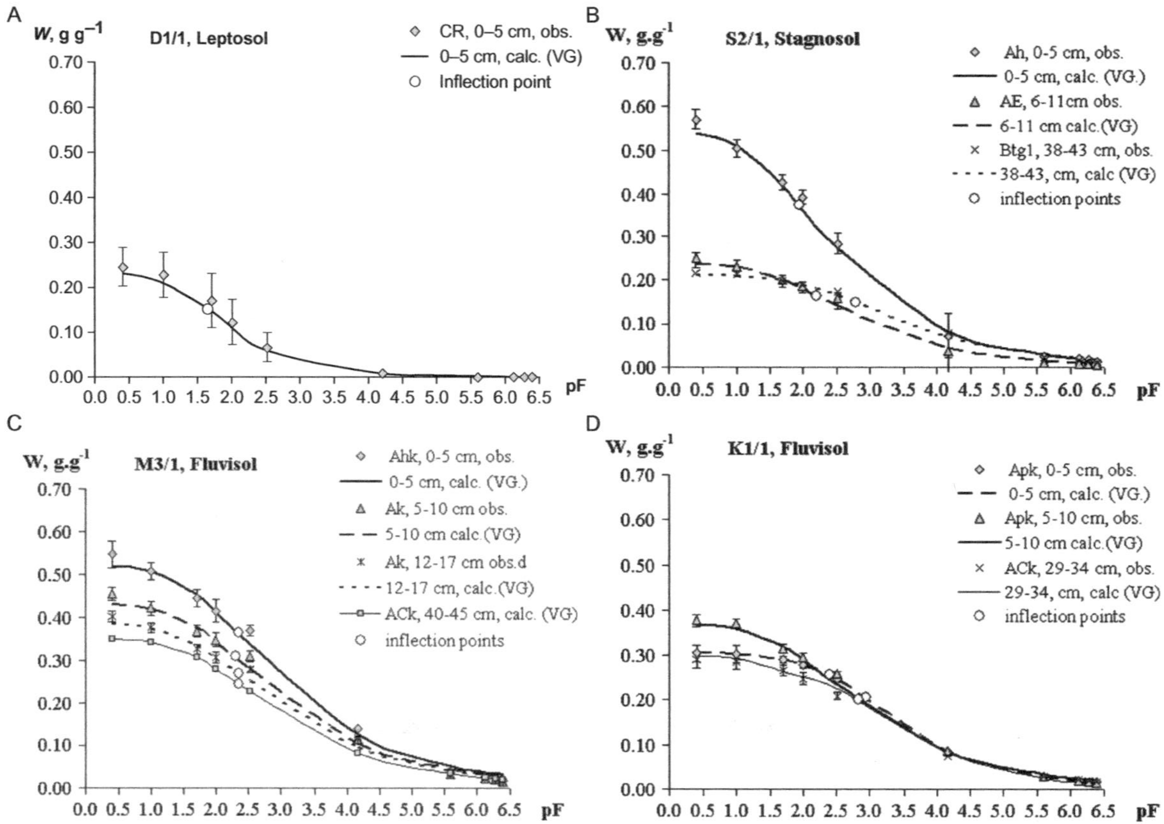

Fig. 2 Examples of soil water retention curves from the Damma Glacier (A), Slavkov Forest (B), Marchfeld (C), Koiliaris (D) CZOs.

Table 3a Soil Characteristics and Hydraulic Properties of Selected Top- and Sublayers of the Main Soil Profiles Studied at Damma Glacier CZO

ID Site	Soil Group	Hor.	Depth, cm	Gravel, %	Texture Class	Sand, %	Silt, %	Clay, %	OC, $g\ kg^{-1}$	BD, $g\ cm^{-3}$	P_t, $cm^3\ g^{-1}$	FC, $cm^3\ g^{-1}$	WP, $cm^3\ g^{-1}$	PAWC, $cm^3\ g^{-1}$	$AP_{pF1.7}$, $cm^3\ g^{-1}$	S (–)	W_i, $g\ g^{-1}$
D1	Leptosol	CR	0–5	16	LS	79	20	1	2.2	1.46	0.31	0.07	0.01	0.06	0.14	0.07	0.15
D2	Leptosol	AC	0–3	0	SL	53	46	2	25.9	0.82	0.83	0.28	0.08	0.20	0.33	0.16	0.56
		CR	3–12	44	SL	68	31	1	5.8	1.54	0.27	0.10	0.02	0.17	0.13	0.04	0.15
D3	Leptosol	Ah	0–4	0	SL	59	37	4	108.8	0.32	2.96	0.98	0.24	0.74	1.24	0.53	1.97
		AC	4–7	24	LS	84	15	1	8.3	1.39	0.34	0.07	0.03	0.04	0.17	0.07	0.17
		C	7–15	20	LS	79	21	1	4.9	1.67	0.22	0.04	0.01	0.03	0.11	0.05	0.10
D4	Cambisol	Ah	0–5	4	SL	53	39	8	94.9	0.36	2.43	0.93	0.34	0.59	1.13	0.33	1.51
		AC	5–11	31	LS	74	23	3	30.0	1.03	0.63	0.21	0.08	0.13	0.25	0.10	0.38
		2A	11–18	9	L	44	42	14	125.7	0.63	1.17	0.46	0.25	0.21	0.53	0.11	0.61

$AP_{pF1.7}$, air-filled pores at tension pF=1.7; *BD*, bulk density; *FC*, field capacity (pores filled with water at pF=2.5); P_t, total porosity; *PAWC*, plant available water holding capacity; *OC*, organic carbon; *S*, slope of water retention curve at inflection point; W_i, water content at inflection point; *WP*, wilting point (pores filled with water at pF=4.18).

Table 3b Soil Characteristics and Hydraulic Properties of Selected Top- and Sublayers of the Main Soil Profiles Studied at Slavkov Forest CZO

ID Site	Soil Group	Hor.	Depth, cm	Gravel, %	Texture Class	Sand, %	Silt, %	Clay, %	OC, g kg^{-1}	BD, g cm^{-3}	P_t, cm^3 g^{-1}	FC, cm^3 g^{-1}	WP, cm^3 g^{-1}	PAWC, cm^3 g^{-1}	$AP_{pF1.7}$, cm^3 g^{-1}	S (–)	W_i, g g^{-1}
S1	Podzol	AE	0–5	41	SL	72	25	3	25.6	1.34	0.36	0.15	0.03	0.12	0.17	0.04	0.14
		E	8–13	44	SL	70	27	4	8.4	1.48	0.30	0.09	0.01	0.08	0.17	0.03	0.09
		BE	20–25	42	SL	55	28	17	28.6	1.44	0.31	0.13	0.05	0.08	0.15	0.03	0.12
		Bth	35–40	37	L	47	30	23	43.8	1.03	0.60	0.30	0.08	0.22	0.22	0.07	0.30
		Bts	50–54	40	SCL	54	26	20	36.0	1.20	0.46	0.25	0.07	0.18	0.16	0.06	0.24
		B	60–65	44	SL	59	25	16	14.3	1.35	0.37	0.14	0.04	0.10	0.19	0.04	0.18
S2	Stagnosol	Ah	0–5	4	SiL	33	54	13	43.8	0.82	0.85	0.30	0.07	0.22	0.39	0.10	0.37
		AE	5–12	25	SiL	38	52	10	19.1	1.49	0.31	0.16	0.04	0.12	0.11	0.04	0.16
		BE	12–32	23	L	43	49	8	7.1	1.76	0.21	0.14	0.04	0.10	0.04	0.04	0.14
		Btg1	38–43	22	L	45	38	17	4.7	1.76	0.22	0.17	0.08	0.11	0.02	0.03	0.15
		Btg2	57–62	29	SL	53	29	19	2.1	1.81	0.21	0.18	0.09	0.09	0.02	0.03	0.16
S3	Cambisol	Ah	0–7	6	L	44	41	15	254.7	0.40	2.75	0.91	0.54	0.37	1.19	0.42	1.53
		A	18–23	28	L	42	46	13	31.2	1.01	0.62	0.25	0.12	0.17	0.24	0.09	0.39
		B	40–45	35	L	45	39	17	13.8	1.25	0.45	0.15	0.08	0.09	0.24	0.06	0.27

$AP_{pF1.7}$, air-filled pores at tension pF = 1.7; *BD*, bulk density; *FC*, field capacity (pores filled with water at pF = 2.5); P_t, total porosity; *PAWC*, plant available water holding capacity; *OC*, organic carbon; *S*, slope of water retention curve at inflection point; W_i, water content at inflection point; *WP*, wilting point (pores filled with water at pF = 4.18).

Table 3c Soil Characteristics and Hydraulic Properties of Selected Top- and Sublayers of the Main Soil Profiles Studied at Marchfeld CZO

ID Site	Soil Group	Hor.	Depth, cm	Gravel, %	Texture Class	Sand, %	Silt, %	Clay, %	OC, g kg^{-1}	BD, g cm^{-3}	P_t, cm^3 g^{-1}	FC, cm^3 g^{-1}	WP, cm^3 g^{-1}	PAWC, cm^3 g^{-1}	$AP_{pF1.7}$, cm^3 g^{-1}	S (–)	W_i, g g^{-1}
M1	Fluvisol	ACkg	0–5	0	SiL	31	61	8	24.8	0.94	0.70	0.45	0.08	0.37	0.17	0.12	0.37
		Ckg	15–20	0	SiL	24	60	6	17.6	1.09	0.56	0.26	0.04	0.22	0.09	0.15	0.32
M2	Fluvisol	Ahk	0–5	0	SiL	27	65	8	36.7	0.93	0.71	0.43	0.12	0.31	0.20	0.10	0.37
		Ak	5–10	0	SiL	29	63	9	28.4	1.10	0.53	0.37	0.10	0.27	0.10	0.09	0.32
		ACk	35–40	0	L	45	46	9	15.4	1.23	0.45	0.21	0.06	0.15	0.16	0.06	0.24
		Ckg	60–65	0	SL	65	30	5	11.0	1.22	0.45	0.13	0.03	0.10	0.18	0.09	0.22
M3	Fluvisol	Ahk	0–5	0	L	43	47	10	43.5	0.95	0.69	0.37	0.14	0.23	0.24	0.08	0.36
		Ak	12–17	0	SiL	30	56	14	27.8	1.07	0.56	0.28	0.12	0.16	0.24	0.06	0.27
		ACk	40–45	0	SiL	14	68	18	18.1	1.19	0.47	0.26	0.09	0.17	0.17	0.05	0.22
		Ck1	65–70	0	SiL	32	54	15	10.7	1.36	0.37	0.18	0.06	0.12	0.12	0.05	0.21
M4	Chernozems	Apk	5–10	0	L	43	38	19	16.1	1.50	0.30	0.18	0.08	0.10	0.08	0.04	0.19
		ACk1	30–35	0	L	45	37	19	20.7	1.52	0.29	0.17	0.08	0.09	0.08	0.04	0.19
		Ck	70–75	0	SL	72	21	7	10.5	1.38	0.36	0.07	0.03	0.04	0.11	0.12	0.20
M5		Ak	2–7	0	SL	58	26	16	19.1	1.66	0.23	0.15	0.07	0.08	0.02	0.03	0.16
		A′k	25–30	0	SL	56	27	18	14.8	1.58	0.26	0.15	0.08	0.07	0.04	0.04	0.18
		ACk	45–50	0	L	39	36	24	6.9	1.40	0.35	0.19	0.10	0.09	0.10	0.04	0.21
		Ck	70–75	0	SL	61	28	11	5.6	1.40	0.34	0.09	0.04	0.05	0.10	0.08	0.19

M6	Ah	0–5	0	SL	57	32	12	23.4	1.40	0.34	0.25	0.08	0.17	0.05	0.05	0.20
	A′	5–10	0	SL	56	30	14	14.0	1.46	0.31	0.20	0.06	0.14	0.06	0.05	0.19
	BCk1	58–63	4	SL	62	24	14	8.7	1.33	0.38	0.11	0.05	0.06	0.19	0.05	0.20
	Ck1	85–90	23	SL	70	20	10	5.6	1.37	0.37	0.09	0.04	0.05	0.19	0.05	0.18

$AP_{pF1.7}$, air-filled pores at tension pF = 1.7; *BD*, bulk density; *FC*, field capacity (pores filled with water at pF = 2.5); *PAWC*, plant available water holding capacity; P_t, total porosity; *OC*, organic carbon; *S*, slope of water retention curve at inflection point; W_i, water content at inflection point; *WP*, wilting point (pores filled with water at pF = 4.18).

Table 3d Soil Characteristics and Hydraulic Properties of Selected Top- and Sublayers of the Main Soil Profiles Studied at Koiliaris CZO

ID Site	Soil Group	Hor.	Depth, cm	Gravel, %	Texture Class	Sand, %	Silt, %	Clay, %	OC, g kg^{-1}	BD, g cm^{-3}	P_t, cm^3 g^{-1}	FC, cm^3 g^{-1}	WP, cm^3 g^{-1}	PAWC, cm^3 g^{-1}	$AP_{pF1.7}$, cm^3 g^{-1}	S (–)	W_i, g g^{-1}
K1	Fluvisol	Apk	0–5	1	SiL	23	62	15	15.2	1.47	0.31	0.26	0.09	0.17	0.01	0.05	0.21
			5–10	1	SiL	22	62	15	14.4	1.34	0.38	0.26	0.09	0.17	0.06	0.06	0.26
		ACk	29–34	3	SiL	21	60	19	7.8	1.48	0.31	0.22	0.08	0.14	0.05	0.05	0.20
		Ck1	45–50	5	SiL	20	60	20	4.6	1.60	0.26	0.21	0.07	0.14	0.02	0.04	0.17
K2	Fluvisol	ACpk	0–5	0	SL	64	30	6	16.5	1.24	0.44	0.17	0.04	0.13	0.10	0.09	0.28
		Ck1	5–10	0	SL	66	29	4	14.3	1.24	0.43	0.16	0.04	0.12	0.08	0.11	0.28
		Ck2	55–60	0	SL	64	30	6	4.4	1.38	0.36	0.10	0.03	0.07	0.10	0.10	0.22
K3	Leptosol	ACp	0–5	23	SL	53	40	7	15.8	1.34	0.37	0.20	0.06	0.14	0.11	0.06	0.23
			5–10	24	SL	56	36	8	19.4	1.27	0.42	0.19	0.05	0.14	0.17	0.05	0.23
K4	Cambisol	Ah	0–5	23	CL	20	52	28	55.7	1.10	0.52	0.24	0.12	0.12	0.22	0.06	0.24
		AC	5–10	24	SiCL	16	50	34	37.9	1.20	0.44	0.18	0.10	0.08	0.23	0.05	0.23
K5	Cambisol	Ah	0–5	0	SiCL	10	60	30	50.1	1.06	0.56	0.36	0.18	0.18	0.14	0.07	0.36
		AC1	7–12	0	SiCL	12	57	31	21.6	1.21	0.46	0.27	0.15	0.12	0.14	0.05	0.28
		AC2	40–45	0	SiCL	9	58	33	20.8	1.07	0.57	0.29	0.17	0.12	0.24	0.06	0.34

$AP_{pF1.7}$, air-filled pores at tension pF = 1.7; *BD*, bulk density; *FC*, field capacity (pores filled with water at pF = 2.5); *OC*, organic carbon; *PAWC*, plant available water holding capacity; P_t, total porosity; *S*, slope of water retention curve at inflection point; W_i, water content at inflection point; *WP*, wilting point (pores filled with water at pF = 4.18).

organic C = 44 g kg^{-1} (Fig. A2). More than half of the porosity in A horizons is occupied by macropores (>60 μm), while the volume of pores with effective diameters between 10 and 60 μm is small (4–6%). The content of finer pores increases in the illuvial Bt horizons. The textural differentiation in Stagnosol, which is formed on ultrabasic metamorphic serpentinite at site S2, is much better pronounced than in the case of Podzol. Total porosity diminishes from 68%v in top Ah horizon to 20%v in Btg horizon on account of the decreased volume of the large pores. These horizons are characterized with high values of bulk density—1.8 g cm^{-3}—and particle density—2.9 g cm^{-3}. The volume of pores filled with air (AP) at suction pF 1.7 is critically low in the BE horizon (7%v) and especially in the Btg horizon (3%). The olive gray color (Munsell color moist 5Y4/2) confirms the waterlogging occurrence at this site. The Cambisol (S3) formed on basic metamorphic amphibolite has the largest total soil porosity (80–56%v), with high drainage aeration pores—33–23%v (Fig. A2).

The different age for formation of studied Fluvisols (M1–M3 sites) and Chernozems (M4–M6 sites) at the Marchfeld CZO as well as the agricultural land use (M4 and M5) impacts the soil structure at this typically highly productive region (Table 3c; Fig. A3). The surface horizons of Fluvisols have about 15–20% greater total porosity than the humus (A) horizons of Chernozems. The plant available water is more than 1/3 of total porosity in all sites. The finer texture of Chernozem soils at the Fuchsenbigl experimental station (M5) leads to an increase of water retained at suctions greater than pF 4.18 (unavailable water for plants) and diminishes the drainage porosity of the upper horizons. The high porosity of parent material for the Chernozems (C horizons) diminishes the risk of prolonged surface waterlogging.

The studied Fluvisols at Koiliaris CZO (K1 and K2 sites) exhibit different soil water characteristics (Fig. A4; Table 3d). This can be explained with the different periods for formation of the soil profiles and respectively different soil textures—silt loam and sandy loam (Moraetis et al., 2015). The Fluvisol (K1 site) with more advanced development is formed by sedimentation of finer particles in more arid conditions, while the Fluvisol at K2 site is relatively younger and formed under more intense hydrodynamic conditions (Moraetis et al., 2015). The total porosity of Fluvisol at site K1 is less than 50%v, most of which (>50% of P_t) are pores retaining water available for plants (more than 50% of total porosity), on account of the relatively low content (<10%) of pores with drainage functions. The other Fluvisol exhibits high capacity to transmit water—more than 60% of pores are with drainage functions and very low capacity to retain water—the plant available water is less than 15%v. The soils at sites K3 and K4 are shallow Leptosol and Cambisol with organic matter content 16 and 56 g kg^{-1}, respectively. The maximum content of water by weight, which can fill all pores, is 0.37 and 0.52 $cm^3\ g^{-1}$, correspondingly (Table 3d). This is due to the higher content of large (>10 μm) and small (<0.2 μm) pores of Cambisol (K4 site) due to the finer soil texture (SiCL) and higher organic content. The Cambisol at (K5 site) located in the uplands with the highest altitude is similar to K4 regarding soil texture, organic content, and total porosity. The water retention properties at pF 2.5 and 4.18 are higher at site K5 than at K4 and are distributed throughout the soil profile.

An integrated parameter used for comparison of SWRCs is the *S* index (slope of the curves at the inflection point W_i) (Dexter, 2004, 2006). The obtained parameters of the van Genuchten equation (Eq. 2) for each curve are not presented in this paper as they are used only for calculation of *S* (Eq. 3) and W_i (Eq. 4). As it is shown in Fig. 2a–d, the data are well fitted by the van Genuchten equation (2). The *S* values are presented in Tables 3a–3d and Figs. 3 and 4. It is established that there is good agreement between the average of the *S* values of the individual curves for each soil sample and the *S* values obtained on averaged water retention data per horizon (as presented in Tables 3a–3d).The estimation of *S* for each individual curve allows assessing the variation of this parameter due to micro (within horizon) variability (Fig. 3). The standard deviation in values for *S* is very high in surface horizons of profiles (D2, D3, D4) in Damma Glacier CZO and in Ah horizons of S2 and S3 profiles in Slavkov Forest CZO. The values for *S* are much less variable in the deeper horizons. The microvariation reflected in the coefficient of variation (Cv) of the soil quality index *S* in each horizon of the soil profiles is on average Cv=32%, 15%, 7%, and 6% for Damma Glacier, Slavkov Forest, Marchfeld, and Koiliaris CZOs, and it is similar to the variability between the three replicate profiles at each site, which is Cv_h=29%, 19%, 13%, and 7%, correspondingly for these four CZOs.

The variability of *S* parameter values is related to variability of water retention data, total porosity, bulk density, soil texture fractions, and organic matter content. The soil bulk density measurements varied on average between Cv=3% for Marchfeld and Koiliaris sites to Cv=6% and 9% for Slavkov Forest and Damma Glacier sites. The variation between the three replicate profiles of each site is Cv_h=19%, 10%, 7%, and 4% on

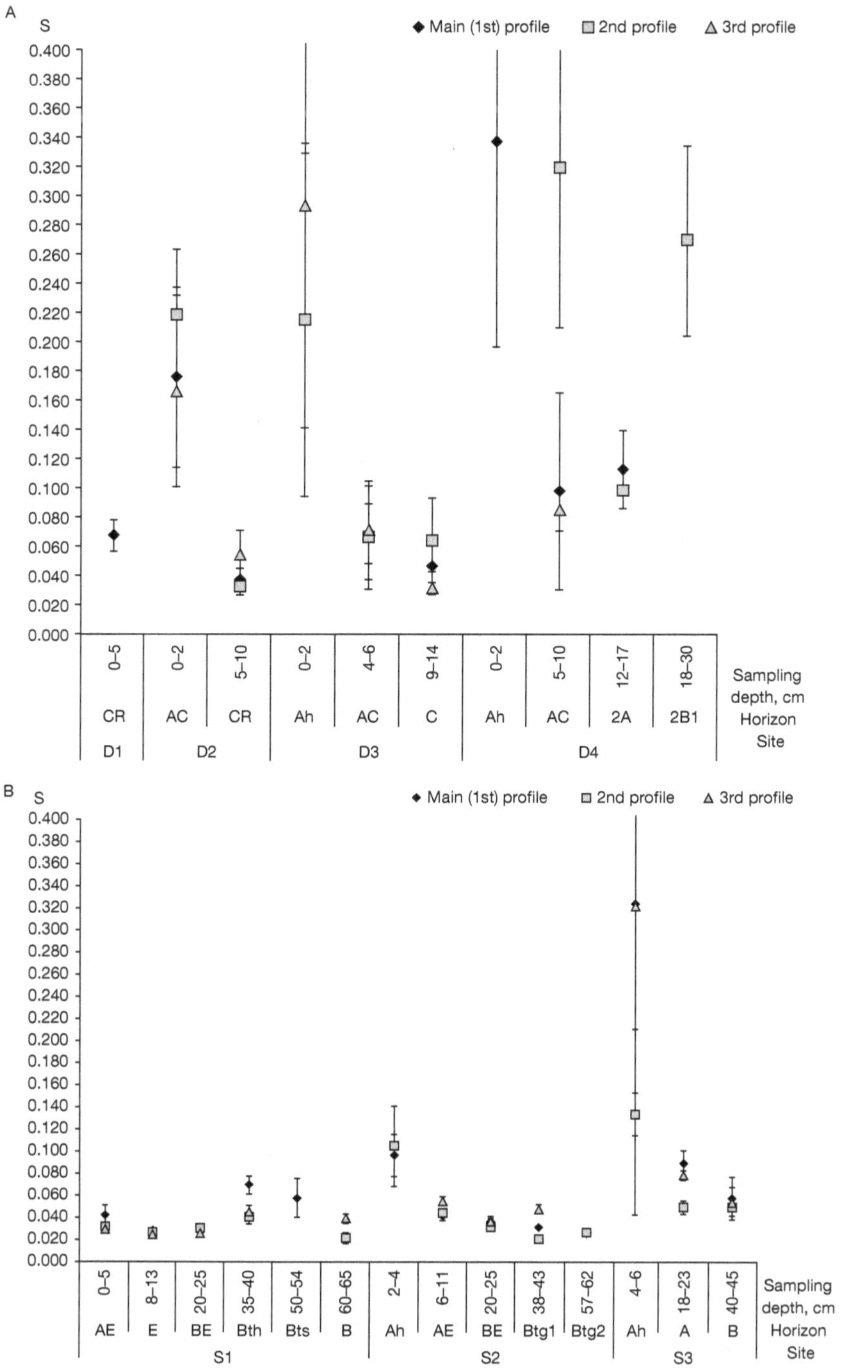

Fig. 3 Soil quality index *S* (slope at the inflection point of soil water retention curves) along the soil profiles at Damma Glacier (A), Slavkov Forest (B), Marchfeld (C), Koiliaris (D) CZOs.

(Continued)

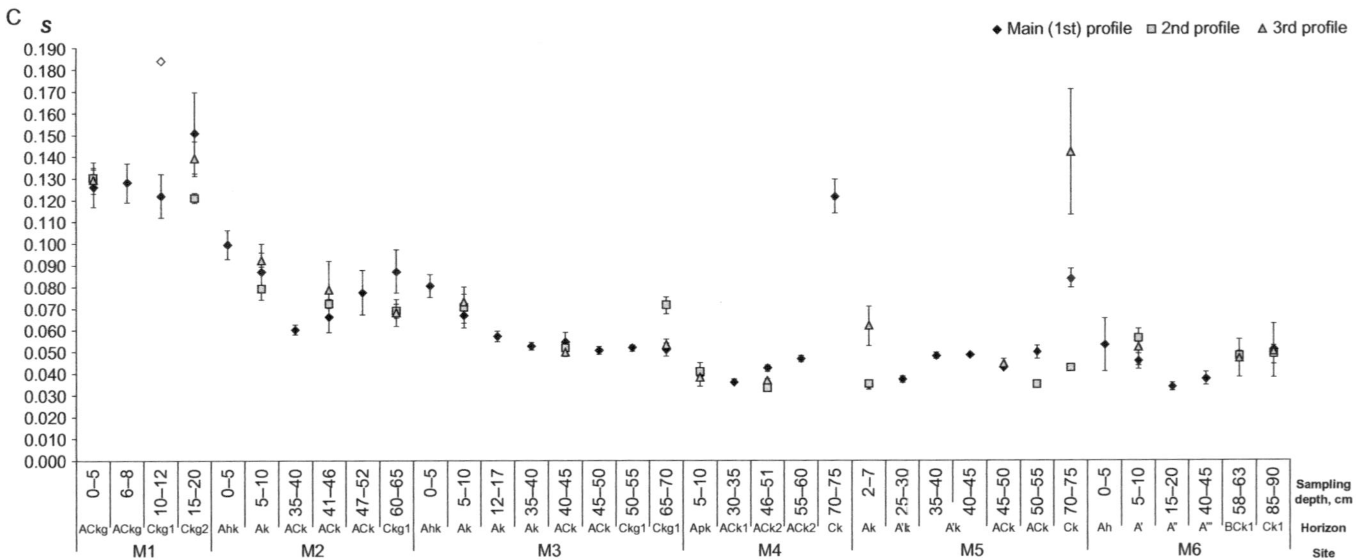

C s
Main (1st) profile
2nd profile
3rd profile
0.190
0.180
0.170
0.160
0.150
0.140
0.130
0.120
0.110
0.100
0.090
0.080
0.070
0.060
0.050
0.040
0.030
0.020
0.010
0.000
Sampling depth, cm
Horizon
Site
0–5 6–8 10–12 15–20 0–5 5–10 35–40 41–46 47–52 60–65 0–5 5–10 12–17 35–40 40–45 45–50 50–55 65–70 5–10 30–35 46–51 55–60 70–75 2–7 25–30 35–40 40–45 45–50 50–55 70–75 0–5 5–10 15–20 40–45 58–63 85–90
ACkg ACkg Ckg1 Ckg2 Ahk Ak ACk ACk ACk Ckg1 Ahk Ak Ak Ak ACk ACk Ckg1 Ckg1 Apk ACk1 ACk2 ACk2 Ck Ak A'k A'k ACk ACk Ck Ah A' A'' A''' BCk1 Ck1
M1
M2
M3
M4
M5
M6

Fig. 3—Cont'd

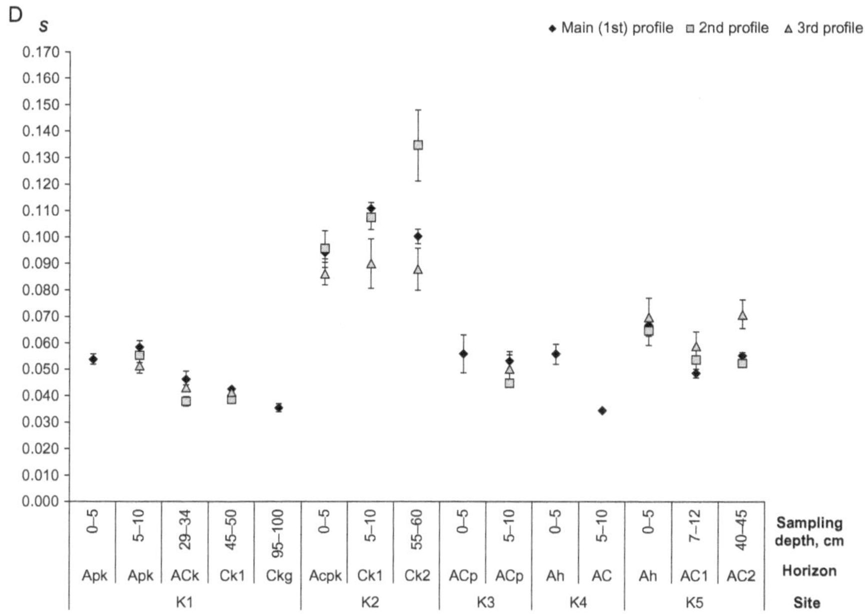

Fig. 3—Cont'd

average, respectively, for Damma Glacier, Slavkov Forest, Marchfeld, and Koiliaris CZOs. Variations in soil structure, presence of gravel, and organic substance lead to even more variation in measurements of the water content retained at pF 2.5 (field capacity), with Cv=28%, 11%, 5%, 3% on average for combined data at a site, and between the three replicate profiles for each site Cv_h=36%, 17%, 17%, 7% on average, respectively, for Damma Glacier, Slavkov Forest, Marchfeld, and Koiliaris CZO. The standard deviation (std) of laboratory analyses of water content at pF 4.18 (wilting point) of fine earth was std=0.2% (Cv=1.5%) on average, but the variation between the three replicate profiles is estimated to be Cv_h=41%, 29%, 15%, 15%, which is close to the variation of clay content Cv_h=40%, 21%, 17%, 18% correspondingly for Damma Glacier, Slavkov Forest, Marchfeld, and Koiliaris CZO. The coefficient of variation of plant available water content (by weight) is Cv=41%, 18%, 9%, 4% in average and Cv_h=44%, 26%, 19%, 10% on average between the three profiles at each particular site, respectively, for Damma Glacier, Slavkov Forest, Marchfeld, and Koiliaris CZO. This ordering of variability across the four CZO sites most likely reflects in broad terms the difference in topographical features and parent materials on soil profile and soil structure development

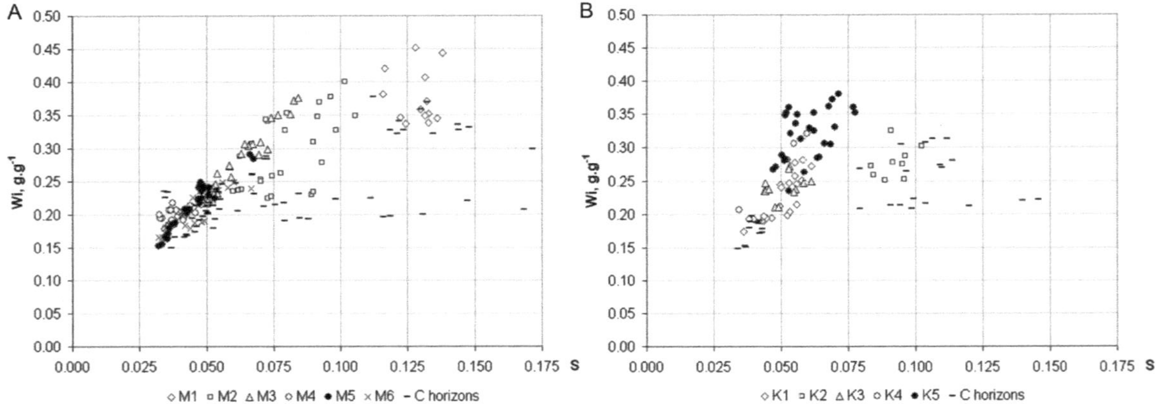

Fig. 4 Slope (S) and water content (W_i) at inflection points of soil water retention curves of A horizons at each site at Marchfeld (A) and Koiliaris (B) CZOs. Data for C horizons are presented for all sites with one symbol (—).

(Wilding et al., 1994). The reduced variability can be explained also with the domination of one of soil forming factor for a long-term period (Burrough, 1993). Plausibly the dominant factor responsible for the low variation at the Koiliaris CZO can be the long-term (millennia) history of traditional agricultural use, low humification of plant residues in the relatively drier Mediterranean climate, and high mineralization rates of organic matter in the relatively warmer Mediterranean climate.

The values of the *S* parameter increase with water content at the inflection point (W_i) as is shown in the case of Marchfeld and Koiliaris CZO (Fig. 4), but the relationship is different for C horizons. The high values of *S* for these coarse texture C horizons are due to the narrower distribution of pore sizes and usually do not correlate with improvement of other soil physical properties (Dexter et al., 2008). It appears also that for well-structured horizons soil water content at the inflection point is very close to field capacity (pF = 2.52).

The values of *S* parameters are related to water transmission properties of the soil connected with the presence of structural pores. This is confirmed by the estimated (Eq. 5) saturated hydraulic conductivity (Fig. 5a–d), which predicts high values of K_{sat} for surface horizons of younger soils, e.g., Leptosols, Cambisols, and Fluvisols, where *S* values are also high ($S > 0.06$). In the case of Leptosols the deepwater percolation is restricted by the rock parent material. The older Fluvisol at site M3 at Marchfeld CZO becomes more compacted (moderately high K_{sat}) at depths below 30–40 cm, while the Chernozems (M4–M6) have moderately high K_{sat} in the surface horizons and high K_{sat} in C horizons. The K_{sat} decreases somewhat with depth in the upper 25 cm (till BE horizon) of the Podzol (S1) at Slavkov Forest CZO but remains moderately high. Below the BE horizon the value for K_{sat} increases. The horizons below the high conductive surface (Ah) horizon of the Stagnosol (S2) form a thicker, compacted ($K_{sat} < 1.5$ cm h^{-1}) layer. The upper (AO) horizons with thicknesses of 12 and 7 cm at sites S1, S2 Slavkov Forest contribute to the accumulation of water and restrict drying of the soil profile by evaporation. The predicted high K_{sat} (9–13 cm h^{-1}) and moderately high K_{sat} (1–4 cm h^{-1}) of Fluvisols at K2 and K1 sites stem from the presence of pores with larger diameters at the younger and with coarser texture site K2.

The average plant available water capacity (% by volume) throughout the soil profile is very high (24%v) for the Fluvisol at site M1, high (19%)

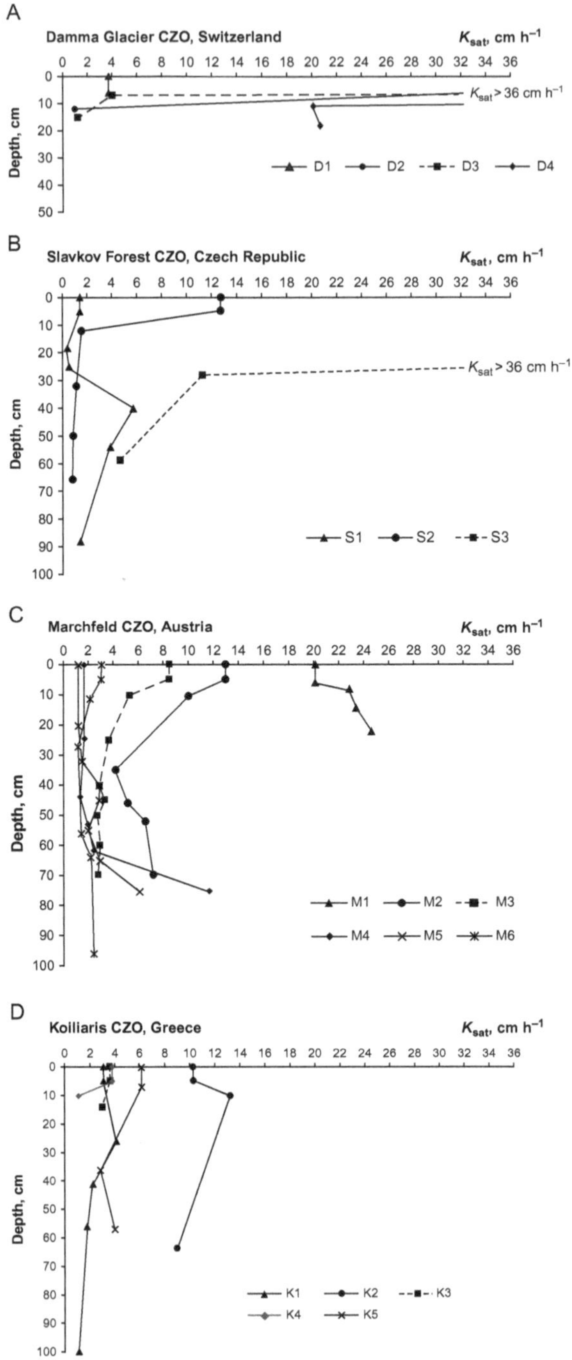

Fig. 5 Values of saturated hydraulic conductivity (K_{sat}, cm h^{-1}) estimated by Eq. (5) at Damma Glacier (A), Slavkov Forest (B), Marchfeld (C), Koiliaris (D) CZOs.

for the Fluvisols at the M2 and M3 sites, and moderate (10–13%) for the Chernozems (M4–M6 sites) at the Marchfeld CZO. Profiles with high plant available water capacity (PAWC) at Koiliaris CZO are the Fluvisols at site K1 (20%v) and the Leptosols at site K3 (18%v), with moderate PAWC exhibited for the Cambisols at sites K4 (11%v) and K5 (15%), and with low PAWC for the Fluvisol at site K2 (9%v). The Podzol (S1 site), Stagnosol (S2 site), and Cambisol (S3 site) at Slavkov Forest CZO exhibit moderate (15.9%v), high (17%v), and low (10.4%v) PAWC on average for the soil profiles. The young Leptosols at Damma Glacier CZO show low PAWC at sites D1 (9.5%v) and D3 (8.5%), and moderate (14%v) PAWC at site D2. The PAWC for the Cambisol at site D4 site is also moderate.

The vertical profile distributions of water, stored at field capacity and at wilting point, are shown in Fig. 6a–d for each site at the studied four CZOs. These characteristics are important for closing the soil water balance. The differences between the soils are better understood when comparing the maximum quantities of available water which can be stored at rooting depth (Table 4). As it is shown in Table 4 the possibility of each soil profile to retain water is very different due to the differences in thickness of soil profiles and differences in the soil structure characteristics described above. The profiles with maximum plant available water stored at rooting depth are Cambisols at site D4 of Damma Glacier CZO (40 mm), the Stagnosol at site S2 of Slavkov Forest (114 mm), Fluvisols at site M2 and M3 of Marchfeld CZO (136 and 130 mm), the Fluvisol at site K1 of Koiliaris CZO (92 mm), and Cambisols at site K5 (126 mm). Shallow soils (D1–D3, K2–K4) retain less than 25 mm available water for plants.

4. DISCUSSION

The soil hydraulic functions studied at 18 sites from 4 CZOs were compared on the basis of soil hydraulic properties and climate conditions. According to the genetic climatic classification (Alisov et al., 1952) Damma Glacier, Slavkov Forest, and Marchfeld CZOs are located in the belt of moderate continental type of climate at different altitudes. The hydrothermal conditions at these CZOs create three combinations of the dominant

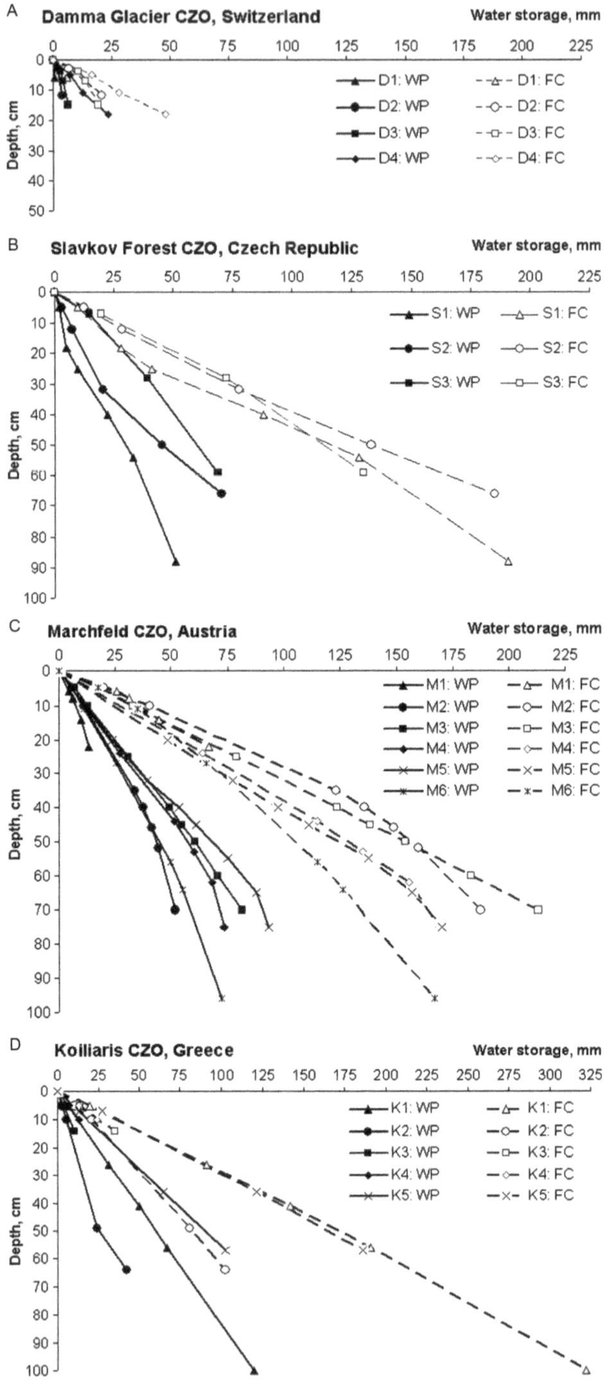

Fig. 6 Water storage in main soil profiles at wilting point (pF = 4.2) and at field capacity (pF = 2.5) at Damma Glacier (A), Slavkov Forest (B), Marchfeld (C), Koiliaris (D) CZOs.

Table 4 Plant Available Water Holding Capacity (PAWC, mm) in Rooting Depth of Main Profiles at Each CZO

ID Site	Soil Group	Horizons	Depth, m	PAWC, mm
D1	Leptosol	CR	0–0.06	5
D2	Leptosol	AC, CR	0–0.12	17
D3	Leptosol	Ah, AC, C	0–0.15	13
D4	Cambisol	Ah, AC, 2A	0–0.18	40
S1	Podzol	AE, E, BE, Bth, Bts	0–0.54	95
S2	Stagnosol	Ah, AE, BE, Btg	0–0.66	114
S3	Cambisol	Ah, A, B, BC	0–0.50	83
M1	Fluvisol	ACkg, Ckg	0–0.22	53
M2	Fluvisol	Ahk, Ak, ACkg, Ckg	0–0.70	136
M3	Fluvisol	Ahk, Ak, ACkg, Ckg	0–0.70	132
M4	Chernozem	Apk, ACk	0–0.44	63
M5	Chernozem	A, Ak	0–0.45	50
M6	Chernozem	A	0–0.56	65
K1	Fluvisol	Apk, ACk	0–0.41	92
K2	Fluvisol	Acpk	0–0.05	8
K3	Leptosol	ACp	0–0.14	25
K4	Cambisol	Ah, AC	0–0.10	16
K5	Cambisol	Ah, AC	0–0.90	126

soil temperature and moisture regimes: Cryic–Udic, Mesic–Udic, and Mesic–Ustic, respectively. The water balance at Damma Glacier and Slavkov Forest CZO is positive throughout the year, while in Marchfeld CZO it is negative in summer time. The steep slope of the terrain and the positive water balance create high runoff during snow free seasons at Damma Glacier CZO and Slavkov Forest CZO. The Marchfeld CZO is characterized by a more variable soil moisture regime through the normal years. This variability of climate conditions increases the risk for drought impacts especially for soils with lower available water holding capacity. The sites at

Koiliaris watershed are also situated at different altitudes but all are in the conditions of subtropical, Mediterranean, type of climate. The dominant soil temperature regime is Thermic and the dominant soil moisture regime is Xeric. More variable is the pedoclimate at sea level sites. The Xeric soil moisture regime in Koiliaris CZO creates conditions for runoff and drainage in wintertime at sites with shallow soil profiles. As the selected sites are natural and manmade terraces where the soil water erosion is restricted, the hydraulic properties of the soil profiles reflect the influence of parent material, vegetation/land use, and other pedogenetic factors (Moraetis et al., 2015).

The characteristics of soil hydraulic properties related to the different age of soil development (chronosequences at Damma Glacier and Marchfeld CZOs), vegetation cover/land use (Marchfeld and Koiliaris CZOs), and parent materials (Slavkov Forest and Koiliaris CZOs) were presented and discussed above. The texture classes of studied soil horizons differ significantly—loamy sand, sandy loam, sandy clay loam, loam, silt loam, and silt clay loam. Some of them contain gravel. The organic matter content, as the other factor which influences the hydraulic properties, is well represented from very low to very high values. The surveyed soil profiles in each CZO were aligned to more than one soil group according to WRB. Occurrence of Leptosols, Cambisols, and Fluvisols studied at least in two CZOs from different climatic regions provides the evidence to compare the similarities and differences of soil hydraulic properties in each of these groups.

4.1 Leptosols: Damma Glacier CZO (D1, D2, D3) and Koiliaris CZO (K3)

The studied Leptosols represent different stages of soil formation. The initial soil formation in Damma Glacier CZO is localized in a small area between the exposed rocks and cliffs as a fine layer (profiles D1, D2, and D3). This stage of soil evolution is manifested with an almost complete lack of clay fraction in the soil texture, domination of sand, and gravel. This explains the very low soil water uptake and retention capacity which in the case of the youngest profile is manifested via the CR horizon. The increasing age of soil evolution along the chronosequence (from 7–13 years till about 110–130 years) extending away from the face of the glacier is manifested by increasing depths of the soil profiles, formation of a thin surface (AC) horizon in the topsoil, capable to retain more water, and a slight increase of the fine texture particle fractions. All studied properties have high spatial variation. The studied Leptosol (profile K3) in Koiliaris CZO was developed during a longer period (about 550 years) under different environmental conditions with human interference, as it is on a manmade cultivated terrace. The spatial variability in soil physical properties of the K3 site in Koiliaris CZO is lower than that of the Leptosols in Damma Glacier CZO. This reduced variability is explained by the domination of long-term degradation processes related to the history of intensive agriculture use and high mineralization rates at K3 site. The profiles differ regarding the amount and size distribution of pores. The soil pore space is very high (0.8–2.9 $cm^3\ g^{-1}$) in the thin surface layers of D2 and D3, and drops to about 0.30 $cm^3\ g^{-1}$ in the C horizons. In the top AC horizon of K3 the total porosity is about 0.40 $cm^3\ g^{-1}$. The proportions of pores with drainage functions ($\delta > 10$ μm), pores holding plant available water (0.2 μm $< \delta <$ 10 μm), and pores holding water unavailable to plants ($\delta < 0.2$ μm), compared to the total porosity of the top Ac horizons, are 7:2:1 in Damma Glacier and 4:4:2 in Koiliaris. This reflects the very high values of S (>0.05) for the coarse texture horizons of Damma Glacier due to the narrower distribution of pore sizes, while S in top horizons is 0.05 in K3 at Koiliaris CZO. This is also reflected in the higher capacity of K3 soils to store water (25 mm) at the rooting depth, while the Leptosols at Damma Glacier store less than 17 mm plant available water.

4.2 Cambisols: Damma Glacier CZO (D4), Slavkov Forest CZO (S3), and Koiliaris CZO (K4, K5)

The comparison of Cambisols studied in three of the CZOs shows that the texture is coarser in the D4 site at Damma Glacier and finer at the K4 and K5 sites at Koiliaris CZO. The organic matter content of surface Ah horizons content is highest in S3 (Slavkov Forest CZO), followed by D4 (Damma Glacier CZO), and lowest in K4 and K5 (Koiliaris CZO). Total porosity is more than 0.50 $cm^3\ g^{-1}$. The proportions of pores with drainage functions ($\delta > 10$ μm), pores holding plant available water (0.2 μm $< \delta <$ 10 μm), and pores holding unavailable to plant water ($\delta < 0.2$ μm) to total porosity are almost the same in all the studied Cambisols, around 6:2:2, with the exception of K5, where they are 4:3:3. The top horizons of these soils are also characterized by very high values of S due to their transmission properties. The difference in rooting depth is of primary importance for water retention capacity of the studied profiles. The plant available water holding capacity of the Cambisol (S3) at Slavkov Forest CZO is 83 mm, more than twice that of the reference profile D4 at Damma Glacier. The profile with the greatest thickness and plant water holding capacity (126 mm) is profile K5—the manmade terrace at noncultivated condition. It is interesting to speculate if the agronomically beneficial rooting depth and water holding capacity at K5 are evidence of beneficial human intervention in soil management through terrace construction.

4.3 Fluvisols: Marchfeld CZO (M1, M2, M3) and Koiliaris CZO (K1, K2)

The Fluvisols at Marchfeld CZO exhibit the most favorable soil hydraulic characteristics as result of environmental conditions that maintain a high intensity of both humification and aerobic status for biological activity. The total porosity of the top horizons is about 0.7 $cm^3\,g^{-1}$ and decreases with depth to about 0.40 $cm^3\,g^{-1}$. The proportions of pores with drainage functions ($\delta > 10$ μm), pores holding plant available water (0.2 μm $< \delta <$ 10 μm), and pores holding water unavailable to plants ($\delta < 0.2$ μm) to total porosity show well-balanced hydraulic functions of structured soils 4:5:1 (M1), 4:4:2 (M2), and 5:3:2 (M3). The total porosities of Fluvisols at Koiliaris CZO are about 0.3–0.4 $cm^3\,g^{-1}$. The size distribution of soil pores is different for profiles K1 and K2. The Fluvisol at K1 has greater water retention capacity (the respective proportions are 1:6:3 and $S \approx 0.05$ in top horizon, and 3:5:2, $S = 0.04$ in depth), while the coarser texture of profile K2 determines better water transmission properties (proportions 6:3:1 and $S > 0.05$). The climate and vegetation/land use interactions can be explored with profiles M3 and K1 which have the same texture class—SiL, but the content of organic matter at K1 is more than two times lower than at M3. This correlates with a significantly lower quantity of macropores in K1 compared to M3. The cause of low organic matter content may be high mineralization rates during the summer as well as low humification of plant residues under intensive agricultural use of the lowlands, which has continued over centuries at Koiliaris CZO. On the other hand the forest trees at site M3 create root biopores which also contribute to formation of larger pores.

The integrated *S* parameter is suitable for comparison of land use impact on soil quality for the top horizons with similar soil texture, such as the Chernozems in this study. The soil quality of A horizons under forest is quantified as very good, while the soil quality under agricultural land is of lower class, i.e., good.

5. CONCLUSION

This study presents a summary of the soil water characteristics of 18 locations within in the Damma Glacier, Slavkov Forest, Marchfeld, and Koiliaris CZOs, surveyed in the framework of the SoilTrEC project. The data and parameter values of the SWRCs were used for assessing and comparing the water storage capacity, pore size distribution, and water permeability characteristics of soil profiles formed under different environmental conditions and evolution stages. The van Genuchten equation fits well all sets of the soil water retention data. The integrated index *S*—the slope at the inflection point of the water retention curve—is used for comparison of the soil physical quality of the top horizons and assessment of their spatial variation. Soil water functions at different sites were analyzed regarding soil hydraulic properties and climate characteristics. Soil moisture regime was assessed on long-term basis by the Newhall model. The obtained information can be used for parameterizing soil water flow and transport models, for impact assessment of natural and anthropogenic factors on soil water functions under different land use and pedoclimatic scenarios—particularly of relevance to impact assessment of soil quality under environmental change, and for extending this applications through comparison of these results with those of analogous environmental data in other locations.

ACKNOWLEDGMENT

This work was supported by the 7th Framework Programme of the European Commission as a Large Integrating Project (SoilTrEC, www.soiltrec.eu, Grant Agreement No. 244118).

APPENDIX

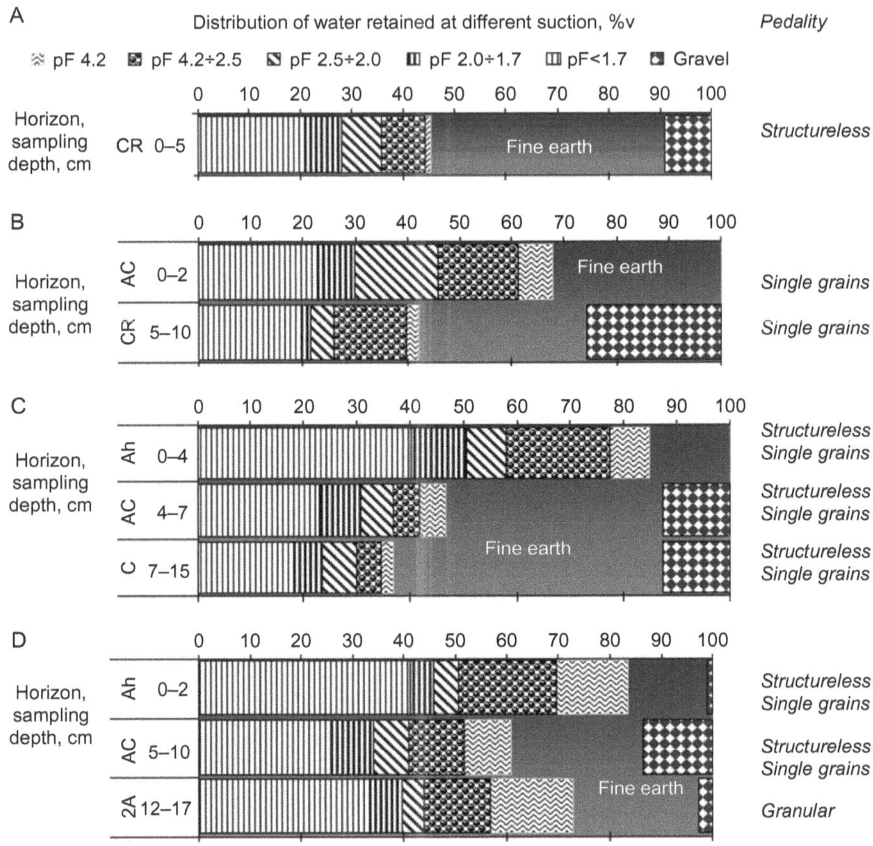

Fig. A1 Distribution (%) of water hold at different suctions and pedality of main profiles' horizons. Damma Glacier Forefield CZO, Switzerland. Pedoclimate Cryic, Udic. (A) *D1*, Damma Glacier CZO. $H=2059$ m. Nudlithic Leptosol; (B) *D2*, Damma Glacier CZO. $H=2015$ m. Lithic Leptosol (Dystric); (C) *D3*, Damma Glacier CZO. $H=1964$ m. Haplic Leptosol (Dystric); (D) *D4*, Damma Glacier CZO. $H=2019$ m. Haplic Cambisol (Ruptic, Dystric, Skeletic).

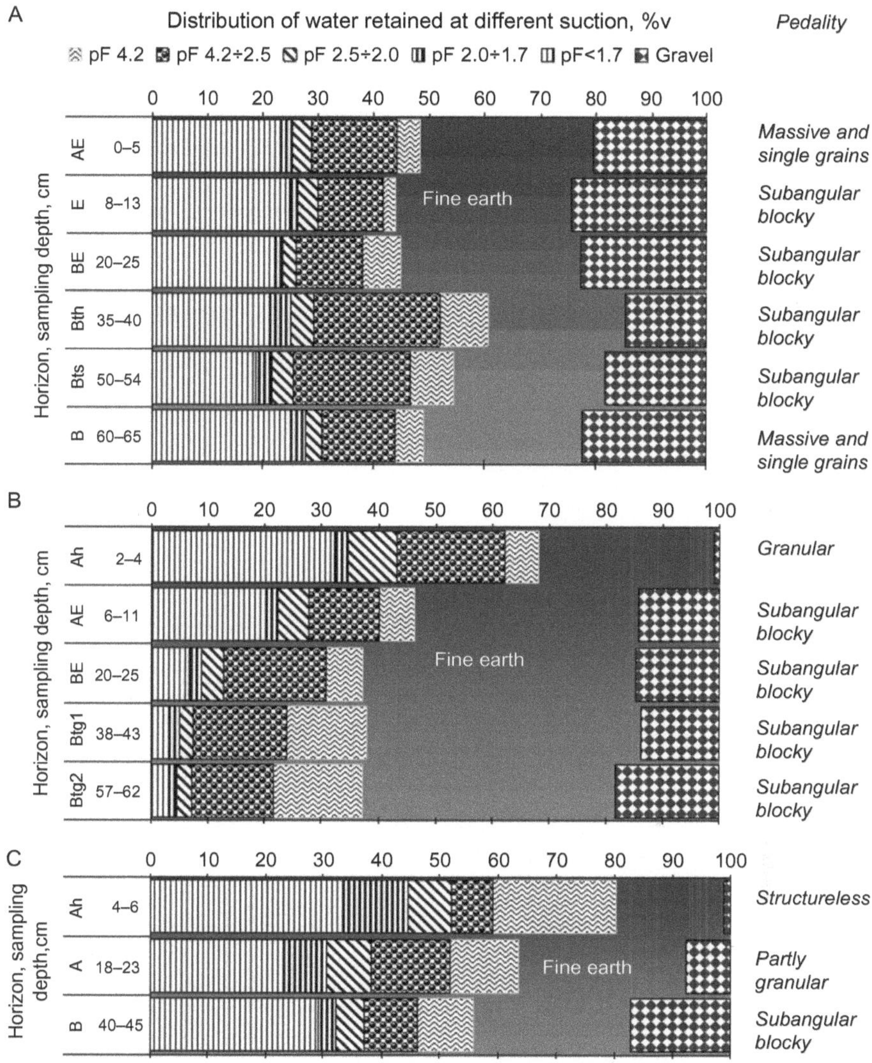

Fig. A2 Distribution (%) of water hold at different suctions and pedality of main profiles' horizons. Slavkov Forest CZO, Czech Republic. Pedoclimate Mesic, Udic. (A) *S1*, Slavkov Forest CZO. Lysina catchment. $H = 891$ m. Albic Podzol formed on granite; (B) *S2*, Slavkov Forest CZO. Pluhův Bor catchment. $H = 772$ m. Luvic Endogleyic Stagnosol formed on serpentinite; (C) *S3*, Slavkov Forest CZO. Na Zeleném catchment. $H = 759$ m. Leptic Cambisol formed on amphibolite.

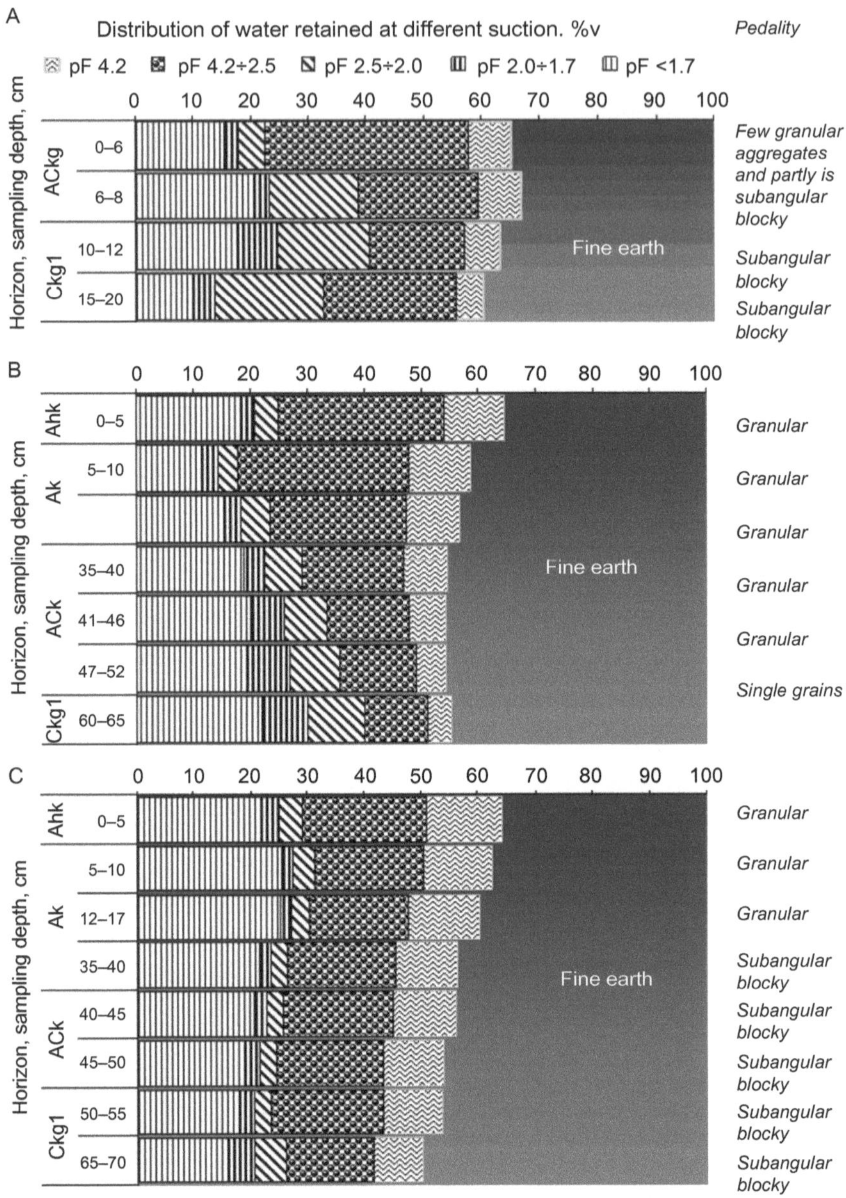

Fig. A3 Distribution (%) of water held at different suctions and pedality of main profiles' horizons. Fuchsenbigl-Marchfeld CZO, Austria. Pedoclimate Mesic, Ustic. (A) *M1*, Marchfeld CZO. $H = 157$ m. Epigleyic Fluvisol under rare willow trees; (B) *M2*, Marchfeld CZO. $H = 156$ m. Haplic Endogleyic Fluvisol under mixed deciduous forest; (C) *M3*, Marchfeld CZO. $H = 157$ m. Mollic Endogleyic Fluvisol under mixed deciduous forest; (D) *M4*, Marchfeld CZO. $H = 155$ m. Haplic Chernozem under old cropland; (E) *M5*, Marchfeld CZO. $H = 148$ m. Epicalcic Chernozem under cropland Fuchsenbigl, former arable, grassland from 5 years; (F) *M6*, Marchfeld CZO. $H = 165$ m. Epicalcic Chernozem under rare oak forest.

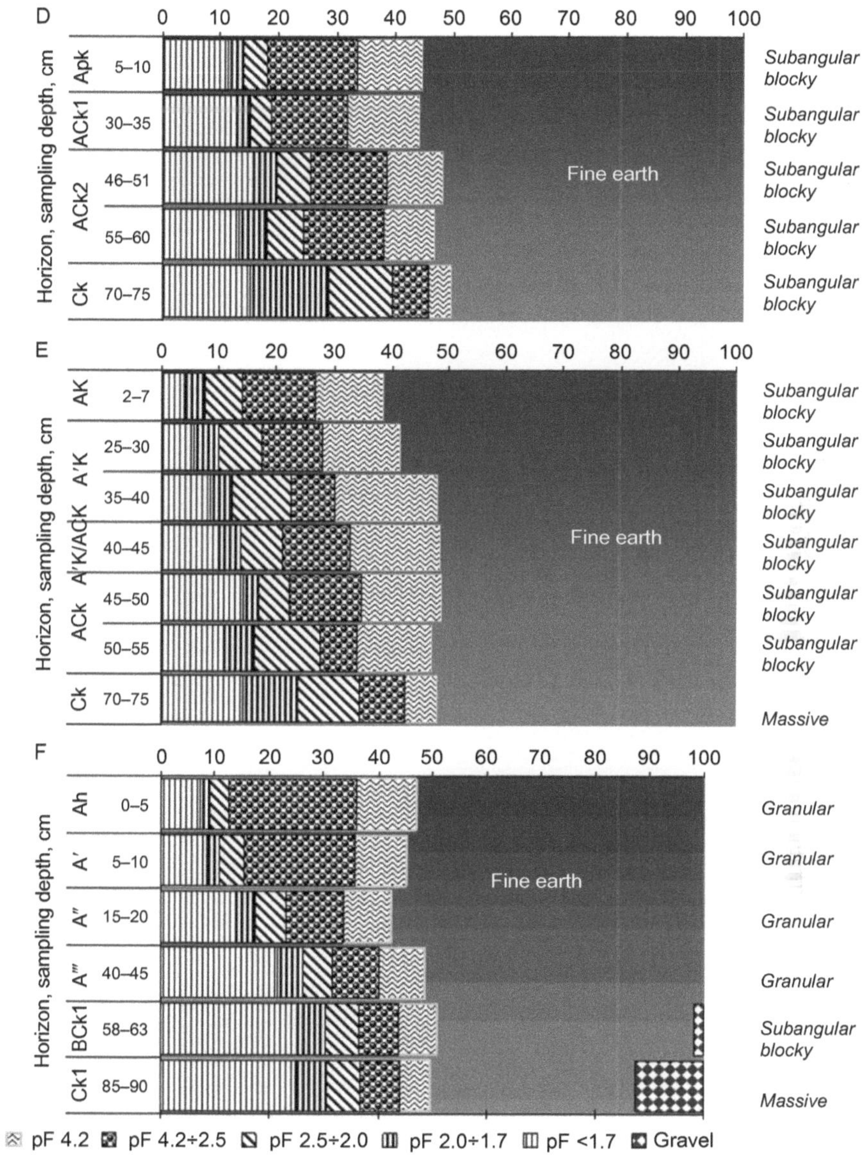
D
0 10 20 30 40 50 60 70 80 90 100
Horizon, sampling depth, cm
Apk 5–10
ACk1 30–35
ACk2 46–51
55–60
Ck 70–75
Fine earth
Subangular blocky
Subangular blocky
Subangular blocky
Subangular blocky
Subangular blocky
E
0 10 20 30 40 50 60 70 80 90 100
Horizon, sampling depth, cm
AK 2–7
A'K 25–30
35–40
A'K/ACK 40–45
ACk 45–50
50–55
Ck 70–75
Fine earth
Subangular blocky
Subangular blocky
Subangular blocky
Subangular blocky
Subangular blocky
Subangular blocky
Massive
F
0 10 20 30 40 50 60 70 80 90 100
Horizon, sampling depth, cm
Ah 0–5
A' 5–10
A'' 15–20
A''' 40–45
BCk1 58–63
Ck1 85–90
Fine earth
Granular
Granular
Granular
Granular
Subangular blocky
Massive
pF 4.2
pF 4.2÷2.5
pF 2.5÷2.0
pF 2.0÷1.7
pF <1.7
Gravel

Fig. A3—Cont'd

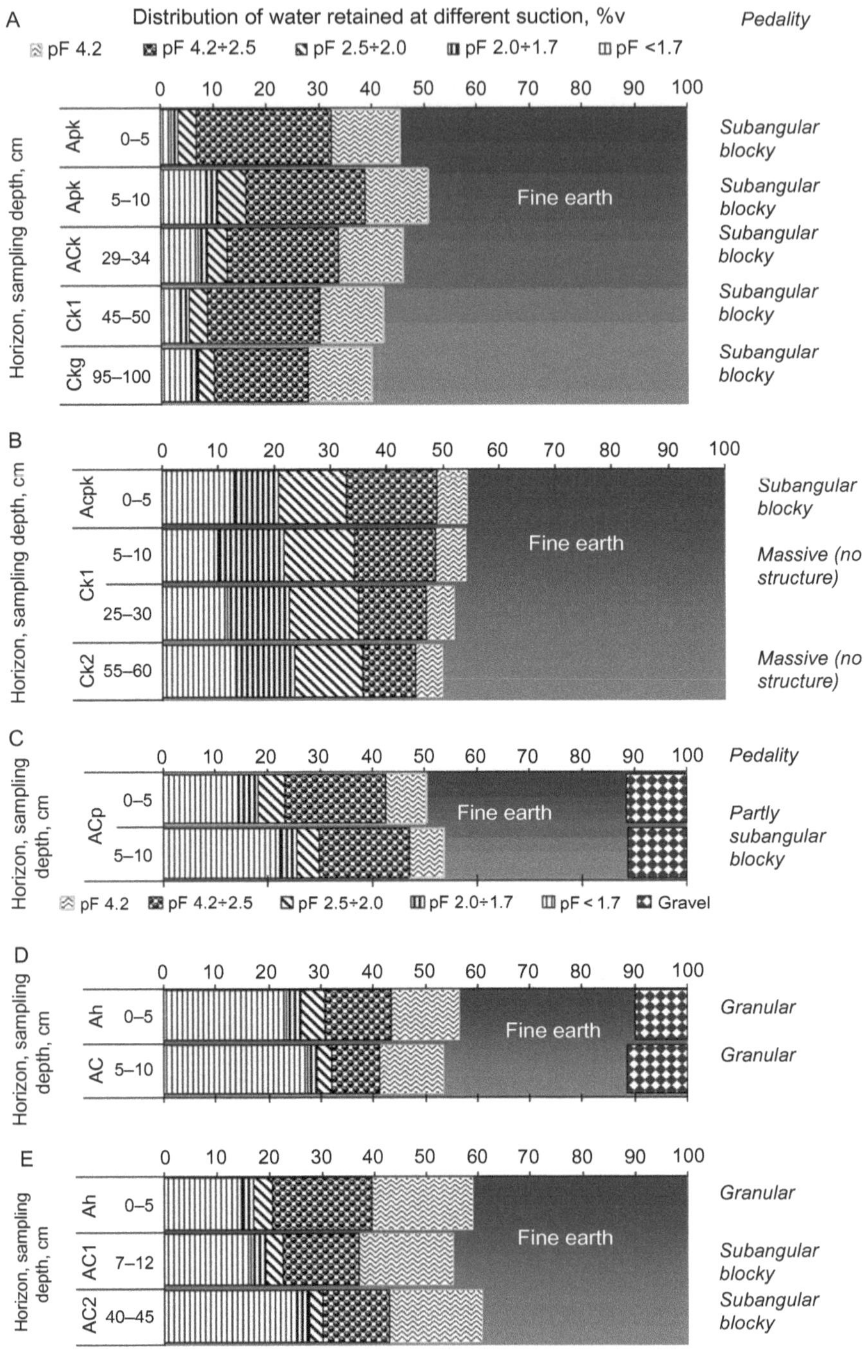

Fig. A4 See figure legend on opposite page.

Fig. A4 Distribution (%) of water hold at different suctions and pedality of main profiles' horizons. Koiliaris CZO, Crete, Greece. Pedoclimate Thermic, Xeric. (A) *K1*, Koiliaris CZO. $H=10$ m. Epicalcic Endogleyic Fluvisol under intensively cultivated irrigated area with vegetable crops. Parent material—sedimentary rock; (B) *K2*, Koiliaris CZO. $H=9$ m. Epicalcic Fluvisol under intensively cultivated irrigated area with olive trees last 20 years. Parent material—sedimentary rock; (C) *K3*, Koiliaris CZO. $H=555$ m. Haplic Leptosol. Manmade terrace with olive trees last ~20 years. Parent material—acid metamorphic rock (schist); (D) *K4*, Koiliaris CZO. $H=465$ m. Endoleptic Cambisol. Manmade terrace with nontilled olive trees (~50 years old). Parent material—platy limestone; (E) *K5*, Koiliaris CZO. $H=1062$ m. Bathyleptic Cambisol. Abandoned terrace with permanent grassland and sparse tree/shrub cover. Parent material—cherty limestone.

REFERENCES

Alisov, B.P., Drozdov, O.A., Rubinstein, E.S., 1952. Climatology Course. Part 1 and 2. Gidrometeoizdat, Leningrad. 482 pp. (in Russian).

Banwart, S., Menon, M., Bernasconi, S.M., Bloem, J., Blum, W.E.H., de Souza, D.M., Davidsdotir, B., Duffy, C., Lair, G.J., Kram, P., Lamacova, A., Lundin, L., Nikolaidis, N.P., Novak, M., Panagos, P., Ragnarsdottir, K.V., Reynolds, B., Robinson, D., Rousseva, S., de Ruiter, P., van Gaans, P., Weng, L., White, T., Zhang, B., 2012. Soil processes and functions across an International Network of Critical Zone Observatories: introduction to experimental methods and initial results. C. R. Geosci. 344, 758–772.

Banwart, S.A., Chorover, J., Gaillardet, J., Sparks, D., White, T., Anderson, S., Aufdenkampe, A., Bernasconi, S., Brantley, S.L., Chadwick, O., Dietrich, W.E., Duffy, C., Goldhaber, M., Lehnert, K., Nikolaidis, N.P., Ragnarsdottir, K.V., 2013. Sustaining Earth's Critical Zone. Basic Science and Interdisciplinary Solutions for Global Challenges. The University of Sheffield, United Kingdom.

Benčoková, A., Krám, P., Hruška, J., 2011. Future climate and flow patterns changes in Czech headwater catchments. Climate Res. 49, 1–15.

Bernasconi, S.M., Bauder, A., Bourdon, B., Brunner, I., Bunemann, E., Christl, I., Derungs, N., Edwards, P., Farinotti, D., Frey, B., Frossard, E., Furrer, F., Gierga, M., Göransson, H., Guelland, K., Hagedorn, F., Hajdas, I., Hindshaw, R.S., Ivy-Ochs, S., Jansa, J., Jonas, T., Kiczka, M., Kretzschmar, R., Lemarchand, E., Luster, J., Magnusson, J., Mitchell, E.A.D., Olde Venterink, H., Plötze, M., Reynolds, B.C., Smittenberg, R.H., Stähli, M., Tamburini, F., Tipper, E.T., Wacker, L., Welc, M., Wiederhold, J.G., Zeyer, J., Zimmermann, S., Zumsteg, A., 2011. Chemical and biological gradients along the Damma Glacier soil chronosequence (Switzerland). Vadose Zone J. 10, 867–883.

Beven, K., Germann, P., 1982. Macropores and water flow in soils. Water Resour. Res. 18, 1311–1325.

Blum, W.E.H., 2006. Functions of soil for society and the environment. Rev. Environ. Sci. Biotechnol. 4, 75–79.

Bonfante, A., Basile, A., Manna, P., Terribile, F., 2010. Soil moisture regime classification of soil taxonomy through physically based soil water balance modelling. In: 19th World Congress of Soil Science, Soil Solutions for a Changing World, 1–6 August 2010, Brisbane, Australia. Published on DVD.

Brewer, R., 1964. Fabric and Mineral Analysis of Soils. John Wiley & Sons, New York.

Burrough, P.A., 1993. Soil variability: a late 20th century view. Soil. Fertil. 56, 529–562.

Corey, A.T., 1977. Mechanics of Heterogeneous Fluids in Porous Media. Water Resources Publications, Fort Collins, CO.
Dexter, A.R., 1988. Advances in characterization of soil structure. Soil Tillage Res. 11, 199–238.
Dexter, A.R., 2004. Soil physical quality. Part I. Theory, effects of soil texture, density, and organic matter, and effects on root growth. Geoderma 120, 201–214.
Dexter, A.R., 2006. Application of S-theory in tillage research. In: 17th Triennial Conference. Proceedings of ISTRO 17. Christian-Albrechts-University, Kiel, Germany, pp. 429–442. CD-ROM. (ISBN 3-9811134-0-3).
Dexter, A.R., Czyz, E.A., Richard, G., Reszkowska, A., 2008. A user-friendly water retention function that takes account of the textural and structural pore spaces in soil. Geoderma 143, 243–253.
Dimitrov, E., Kercheva, M., Stoinov, St, 2014. Variability of moisture regime of clayey soils in normal years for Sofia region. In: Eurasian Congress of Soil Science, Antalya, October 2014. Antalya.
Dumitru, E., Calciu, I., Carabulea, V., Canarache, A., 2009. Metode de analiza in laboratorul de fizica a solului. Editura Sitech, Craiova. p. 341.
FAO, 2006. Guidelines for Soil Description, fourth ed. Food and Agriculture Organization of the United Nations, Rome. 97 pp.
Fiebig, M., Preusser, F., Steffen, D., Thamo-Bozso, E., Grabner, M., Lair, G.J., Gerzabek, M.H., 2009. Luminescence dating of historical fluvial deposits from the Danube and Ebro. Geoarchaeol. Int. J. 24 (2), 224–241.
Greenland, D.J., 1981. Soil management and soil degradation. J. Soil Sci. 32 (3), 301–322.
Han, H., 2005. Predicting saturated hydraulic conductivity from particle and pore information. M.S. thesis, Rutgers University, New Brunswick, NJ.
Helliwell, R.C., Wright, R.F., Jackson-Blake, L.A., Ferrier, R.C., Aherne, J., Cosby, B.J., Evans, C.D., Forsius, M., Hruska, J., Jenkins, A., Kram, P., Kopáček, J., Majer, V., Moldan, F., Posch, M., Potts, J.M., Rogora, M., Schopp, W., 2014. Assessing recovery from acidification of European surface waters in the year 2010: evaluation of projections made with the MAGIC model in 1995. Environ. Sci. Technol. 48, 13280–13288.
Hillel, D., 1980. Fundamentals of Soil Physics. Academic Press, New York. 413 pp.
Huber, S., Prokop, G., Arrouays, D., Banko, G., Bispo, A., Jones, R.J.A., Kibblewhite, M.G., Lexer, W., Moller, A., Rickson, R.J., Shishkov, T., Stephens, M., Toth, G., Van den Akker, J.J.H., Varallyay, G., Verheijen, F.G.A., Jones, A.R. (Eds.), 2008. Environmental Assessment of Soil for Monitoring: Volume I Indicators and Criteria. EUR 23490 EN1. In: Office of the Official Publications of the European Communities, Luxembourg. 339 pp.
ISO 11272:1998 named: Soil quality—determination of dry bulk density.
ISO 11274:1998 named: Soil quality—determination of the water retention characteristics—laboratory methods.
ISO 11508:1998 named: Soil quality—determination of particle density.
IUSS Working Group WRB, 2006. World reference base for soil resources 2006. A framework for international classification, correlation and communication: Worlds soil resources reports no. 103. FAO, Rome.
Jarvis, N.J., Moeys, J., Koestel, J., Hollis, J.M., 2012. Preferential flow in a pedological perspective. In: Lin, H.S. (Ed.), Hydropedology. Synergistic Integration of Soil Science and Hydrology. Elsevier Academic Press, San Diego, CA, pp. 75–120. Chapter 3.
Kercheva, M., Rousseva, S., Dimitrov, E., Nenov, M., Shishkov, T., 2011. Soil aggregation estimates in CZO-Fuchsenbigl. Appl. Geochem. 26, S57–S59.
Köppen, W., 1936. Das geographische System der Klimate. In: Köppen, W., Geiger, G. (Eds.), Handbuch der Klimatologie. Gebrüder Borntraeger, Berlin, pp. 1–44.

Kosugi, K., 1994. Three-parameter lognormal distribution model for soil water retention. Water Resour. Res. 30, 891–901.
Krám, P., 2011. Hydrological balance of the long-term monitored Lysina catchment (in Czech). In: Šír, M., Tesař, M. (Eds.), Conference Proceedings Hydrology of a Small Catchment 2011. Institute for Hydrodynamics of the Czech Academy of Sciences, Prague, pp. 259–265.
Krám, P., Hruška, J., Shanley, J.B., 2012. Streamwater chemistry in three contrasting monolithologic Czech catchments. Appl. Geochem. 27, 1854–1863.
Krám, P., Čuřík, J., Veselovský, F., Myška, O., Lamačová, A., Hruška, J., Štědrá, V., 2014. Hydrology and hydrochemistry of the long-term monitored ultrabasic Pluhův Bor catchment (in Czech). In: Brych, K., Tesař, M. (Eds.), Conference Proceedings Hydrology of a Small Catchment 2014. Institute for Hydrodynamics of the Czech Academy of Sciences, Prague, pp. 238–245.
Kutílek, M., 2004. Soil Water in the System of Hydropedology. In: EUROSOIL 2004. http://www.bodenkunde2.uni-freiburg.de/eurosoil/abstracts/id1116_Kutilek_full.pdf.
Kutílek, M., Jendele, L., 2008. The structural porosity in soil hydraulic functions—a review. Soil Water Res. 3 (special issue 1), S7–S20.
Kutílek, M., Nielsen, D., 1994. Soil Hydrology. GeoEcology Textbook. Catena Verlag, Cremlingen-Destedt, Germany. 370 pp.
Lair, G.J., Zehetner, F., Hrachowitz, M., Franz, N., Maringer, F.-J., Gerzabek, M.H., 2009. Dating of soil layers in a young floodplain using iron oxide crystallinity. Quat. Geochronol. 4 (3), 260–266.
Lin, H.S., 2012. Chapter 1: Hydropedology: addressing fundamentals and building bridges to understand complex pedologic and hydrologic interactions. In: Lin, H.S. (Ed.), Hydropedology. Synergistic Integration of Soil Science and Hydrology. Elsevier, Amsterdam, pp. 3–39.
Lin, H., Bouma, J., Wilding, L., Richardson, J., Kutílek, M., Nielsen, D.R., 2005. Advances in hydropedology. Adv. Agron. 85, 1–89.
Menon, M., Rousseva, S., Nikolaidis, N.P., van Gaans, P., Panagos, P., de Souza, D.M., Ragnarsdottir, K.V., Lair, G.J., Weng, L., Bloem, J., Kram, P., Novak, M., Davidsdottir, B., Gisladottir, G., Robinson, D.A., Reynolds, B., White, T., Lundin, L., Zhang, B., Duffy, C., Bernasconi, S.M., de Ruiter, P., Blum, W.E.H., Banwart, S.A., 2014. SoilTrEC: a global initiative on critical zone research and integration. Environ. Sci. Pollut. Res. 21, 3191–3195.
Moraetis, D., Efstathiou, D., Stamati, F., Tzoraki, O., Nikolaidis, N.P., Schnoor, J.L., Vozinakis, K., 2010. High frequency monitoring for the identification of hydrological and biochemical processes in a Mediterranean river basin. J. Hydrol. 389, 127–136.
Moraetis, D., Paranychianakis, N., Nikolaidis, N., Banwart, S., Rousseva, S., Kercheva, M., Nenov, M., Shishkov, T., de Ruiter, P., Bloem, J., Blum, W.E.H., Lair, G., van Gaans, P., Verheul, M., 2015. Sediment provenance, soil development, and carbon content in fluvial and manmade terraces at Koiliaris River Critical Zone Observatory. J. Soil. Sediment. 15, 347–364.
Mualem, Y., 1976. A new model for predicting the hydraulic conductivity of unsaturated porous media. Water Resour. Res. 12 (3), 513–522.
Newhall, F., 1972. Calculation of Soil Moisture Regimes From the Climatic Record, Revision 4. USDA Soil Conservation Service, Washington, DC.
Pachepsky, Ya, Rawls, W.J. (Eds.), 2004. Development of Pedotransfer Functions in Soil Hydrology. In: Elsevier, Amsterdam, p. 514.
Peel, M.C., Finlayson, B.L., McMahon, T.A., 2007. Updated world map of the Köppen-Geiger climate classification. Hydrol. Earth Syst. Sci. 11, 1633–1644.
Regelink, I.C., Stoof, C.R., Rousseva, S., Wenga, L., Lair, G.J., Kram, P., Nikolaidis, N.P., Kercheva, M., Banwarth, S., Comans, R.N.J., 2015. Mechanistic linkages between soil aggregates, soil porosity and soil chemical properties. Geoderma 247–248, 24–37.

Richards, L.A., 1941. A pressure-membrane extraction apparatus for soil solution. Soil Sci. J. 51 (5), 377–386.
Rousseva, S., Kercheva, M., Blum, W., Lair, G., Shishkov, T., Regelink, I., van Gaans, P., Ilieva, R., Bloem, J., 2010. Dataset on soil physical, geochemical and biological properties, characterizing the CZOs. Deliverable D 1.1. Project 244118 SoilTrEC—Soil Transformations in European Catchments. European Commission, Brussels.
Saxton, K.E., Rawls, W.J., 2006. Soil water characteristic estimates by texture and organic matter for hydrologic solutions. Soil Sci. Soc. Am. J. 70, 1569–1578.
Schjønning, P., van den Akker, J.J.H., Keller, T., Greve, M.H., Lamandé, M., Simojoki, A., Stettler, M., Arvidsson, J., Breuning-Madsen, H., 2015. Chapter five—driver-pressure-state-impact-response (DPSIR) analysis and risk assessment for soil compaction—a European perspective. Adv. Agron. 133, 183–237.
Soil Survey Division Staff, 1993. Soil Survey Manual. Soil Conservation Service. U.S. Department of Agriculture. Agriculture Handbook 18.
Soil Survey Staff, 1999. Soil taxonomy, second ed. USDA, NRCS, Agricultural Handbook No. 436, Washington, DC.
Šútor, J., Kutílek, M., Majerčák, J., 1999. Evaluation of water storage in the aeration zone of soil. J. Hydrol. Hydromech. 47 (4), 288–295.
Thornthwaite, C.W., 1948. An approach towards a rational classification of climate. Geogr. Rev. 38, 55.
Tóth, B., Weynants, M., Nemes, A., Makó, A., Bilas, G., Tóth, G., 2015. New generation of hydraulic pedotransfer functions for Europe. Eur. J. Soil Sci. 66, 226–238.
USDA NRCS, 2012. Java Newhall simulation model. Release 1.6.0. Available at: http://www.nrcs.usda.gov/wps/portal/nrcs/detail/soils/survey/class/?cid=nrcs142p2_053559. (accessed 10.09.13).
van Genuchten, M.T., 1980. A closed-form equation for predicting the hydraulic conductivity of unsaturated soils. Soil Sci. Soc. Am. J. 44, 892–898.
van Wambeke, A.R., 2000. The Newhall Simulation Model for Estimating Soil Moisture and Temperature Regimes. Department of Crop and Soil Sciences. Cornell University, Ithaca, NY.
van Wambeke, A., Hastings, P., Tolomeo, M., 1986. Newhall Simulation Model: A BASIC Program for the IBM PC. Department of Agronomy, Cornell University. Diskette and Booklet, Ithaca, NY.
Wilding, L.P., Bouma, J., Goss, D., 1994. Impact of spatial variability on modelling. In: Bryant, R., Hoosbeek, M.R. (Eds.), Quantitative Modeling of Soil Forming Processes. Soil Science Society of America, Madison, WI, pp. 61–75. Soil Sci. Soc. Am. Special Publ. #39.

CHAPTER THREE

Reduced Subsurface Lateral Flow in Agroforestry System Is Balanced by Increased Water Retention Capacity: Rainfall Simulation and Model Validation

Y. Wang*, B. Zhang†,1, S.A. Banwart‡

*Key Laboratory of Agro-ecological Processes in Subtropical Regions, Changsha Research Station for Agricultural and Environmental Monitoring, Institute of Subtropical Agriculture, Chinese Academy of Sciences, Hunan, China

†National Engineering Laboratory for Improving Fertility of Arable Soils, Institute of Agricultural Resources and Regional Planning, Chinese Academy of Agricultural Sciences, Beijing, China

‡Integrating Chair in Soil/Agriculture/Water Research, School of Earth and Environment, University of Leeds, Leeds, United Kingdom

[1]Corresponding author: e-mail address: zhangbin01@caas.cn

Contents

Abstract

Soil hydrology controls the terrestrial water cycle and the transport of substances to influence the environmental quality of Earth's critical zone (CZ). Soil and water management in agroforestry systems (AF) is able to reduce soil nitrogen losses and to alleviate secondary salinity in some regions of the world by reducing subsurface lateral flow. Compared to monocropping (MC) system, the reduction of subsurface lateral flow in AF has been attributed not only to the enhanced evapotranspiration and canopy

Advances in Agronomy, Volume 142
ISSN 0065-2113
http://dx.doi.org/10.1016/bs.agron.2016.10.005

interception but also to changes in soil structure and related hydraulic properties. However, for AF, it remains unclear how changes in soil structure and hydraulic properties occur and can act to reduce the subsurface later flow. Rainfall simulation experiments were conducted in the field and soil matric potential was measured to determine the effect of AF and MC on the dynamics of rainfall infiltration, subsurface lateral flow, and soil water storage in the soil profile. The calculated isolines of soil matric potential showed that a physical domain of water saturation occurred in the subsoils during the rainfall and diminished after the rainfall. The water saturation domain was larger during the rainfall and drained more slowly after the rainfall in AF than in MC. These results illustrated that AF increased vertical preferential flow and retarded the subsurface lateral flow, resulting in increased water retention capacity in the soil profile, compared to MC. The changed water mass and flow distribution was attributed to the deep roots, which increase macropores oriented in the vertical direction and modify micro- and mesopores in the lateral direction, resulting in changes in anisotropy of soil hydraulic properties along transects of slope. These proposed mechanisms were successfully verified by mathematical modeling. Numerical experiments using the Hydrus-2D mathematical modeling code at the virtual condition of the same antecedent soil moisture condition along the slope at different rainfall events ruled out the effect of antecedent soil moisture or evapotranspiration on generation of subsurface flow. These findings suggest that land use has strong effects on water distribution not only above the ground but also in the subsurface. The changes in soil structure and hydraulic properties need to be considered in understanding landscape hydrology related to agricultural practices and their impacts on Earth's CZ.

1. INTRODUCTION

Soil is the central interface at the heart of Earth's critical zone (CZ), representing a natural geomembrane across which water is actively exchanged with the atmosphere, biosphere, hydrosphere, and lithosphere (Brantley et al., 2007; Lin, 2010). Subsurface lateral flow is generally induced when sufficient infiltrating water is impeded over a less permeable layer or bedrock within soil profiles (Wang et al., 2011). The resulting changes in the water flow field significantly contribute to the transport and transformation of energy and material within the CZ (Lin, 2010; Schaik, 2010; Wang et al., 2011). As subsurface lateral flow occurs within soil, the governing processes are ruled by soil properties (Lin, 2010; Schaik, 2010). Land use, evolving with different natural and anthropogenic processes, has strong influences on soil structure and hydraulic properties (Bottinelli et al., 2016; Schaik, 2010; Zhang et al., 2015a,b). Once a small-scale change in soil structure and hydraulic properties occurs, a relatively large-scale variation in subsurface lateral flow can be induced (Schaik, 2010; Zhang et al., 2015a,b). Although numerous studies have intensively investigated subsurface lateral flow under various land uses (Gerrits, 2010; Wang et al., 2011), few have explored the relationship between land use and subsurface lateral flow from the view of a controlling role for soil structure and hydraulic properties.

Soil structure is the physical architecture that arises from the physical arrangement of the pore spaces among closely packed soil particles, including minerals, decaying biomass, and living organisms. There are many possible natural and anthropogenic factors that influence soil structure and hydraulic properties under various land uses. Vegetation species significantly influence the development of soil structure through the inputs of belowground biomass and exudates (Nair et al., 2010), which influences the microbiome and aggregation of soil particles and alters the pore space architecture. Through this, vegetation species influence the soil hydraulic conductivity and the soil water retention capacity (Nath, 2015; Seobi et al., 2005; Wang et al., 2013). Plant root penetration and tillage may create soil macropores with low water entry pressures, high connectivity, and large cross section which accelerates soil water flow along the orientation of the macropore (Bogner et al., 2010; Schaik, 2010; Zhang et al., 2015a,b). Plant root shape may result in significant anisotropy in soil hydraulic properties, reflecting an increase in the spatial variability in vertical preferential flow and subsurface lateral flow (Jing et al., 2008; Schaik, 2010). The impacts of land use change on soil structure and hydraulic properties can occur rapidly, within a single season, in natural and agricultural ecosystems (Schaik, 2010; Seobi et al., 2005). These changes can have immediate influences on subsurface lateral flow process (Bogner et al., 2010; Schaik, 2010), although quantitative effects on the flow field are not understood.

Agroforestry system (AF) can reduce subsurface lateral flow in the downward direction along slopes (Dunin, 2002; Wang et al., 2011). The mechanisms behind this effect have been intensively explored with respect to atmospheric and biospheric processes, i.e., the characteristics of precipitation, evapotranspiration, and canopy interception (Dunin, 2002; Ticehurst et al., 2001; Wang et al., 2011), but rarely with consideration of soil hydrological properties and processes. Mapa (1995) has observed the greatest soil water retention at any given pore entry pressure in reforested soil profiles, in comparison with the observations on cultivated and grassland soils. Other studies also suggest that tree roots are able to create well-connected macropores or channels to accelerate soil water flow (Bogner et al., 2010). Recently, some studies suggest that tree-based ecosystems increase soil carbon sequestration in deeper soil due to greater root density and larger carbon inputs compared to tree-free ecosystems (Nair et al., 2010), which thereby leads to changes in soil porosity and associated soil hydraulic properties in tree-based ecosystems (Nath, 2015). Zhang et al. (2015a,b) also reported that a high tree root density had a pivotal influence on preferential flow in forest soils. These findings imply that tree-based agroforestry may strongly influence subsurface lateral flow by altering soil structure and hydraulic properties.

Numerous studies have attempted to explore the mechanisms of subsurface lateral flow using virtual modeling, or numerical experiments with a model that is driven by collective field intelligence (Gerrits, 2010; Keim et al., 2006; Wang et al., 2011). For example, Keim et al. (2006) applied the Hydrus-2D modeling code to assess the impacts of tree canopy on subsurface lateral flow by using measured rainfall and synthetic and smoothed throughfall. Gerrits (2010) separated the impacts of interception between tree canopy and forest floor on subsurface lateral flow by simulating the conditions of no interception, canopy interception only, and both canopy and forest floor interception. However, virtual modeling often lacks field validation, and hence its application does not fully develop insight into the fundamental physical mechanisms of subsurface flow. Experimental rainfall simulation is commonly used for furthering understanding on subsurface lateral flow process, because it allows predefining some key parameters like rainfall duration and quantity (Wildhaber et al., 2012). Examples of this approach include recent work by Zhao et al. (2014) and Fu et al. (2015). In the study reported here, a combination of field rainfall simulation experimentation and numerical modeling is applied together to determine the mechanism of reduced subsurface lateral flow in an AF in compared to a monocropping (MC) system. The AF consisting of citrus (*Citrus reticulata*) trees with intercropped peanut (*Arachis hypogaea*) was compared with an adjacent MC system of peanut. The objectives of this study were (i) to characterize the mechanistic connections between vertical preferential flow, subsurface lateral flow, and plant root systems and (2) to explore using numerical modeling the mechanisms through which AF can reduce subsurface lateral flow. The numerical modeling with the Hydrus-2D code was developed using field-collected soil hydraulic properties, slope structure, and rainfall intensities and virtually designed antecedent soil matric potentials. The hypothesis of this study is that changes in soil structure and hydraulic properties between AF and MC lead to variation in subsurface lateral flow in the downward direction along slopes.

2. MATERIALS AND METHODS

2.1 Study Site and Land Uses

The field experiment was carried out in a small agricultural catchment (Tang et al., 2011; Zepp et al., 2005), about 5 km from the Ecological Experimental Station of Red Soil, Chinese Academy of Sciences (EESRS, CAS) in Yingtan, Jiangxi province, China (28°15′N, 116° 55′E). The elevation at the catchment is 41–55 m above mean sea level. The study area has a subtropical warm humid climate, with an annual mean air temperature of 17.8°C, a minimum of 5.3°C in January, a maximum of 29.6°C in July, and a frost-free period of 262 days. The annual rainfall average is 1706 mm during from 1954 to 1999, with approximately 50% concentrated during the period from April to July. The hillslope gradient was 6% and the slope length was 130–150 m. The soil depth along the slope ranged from 0.4 to 2.0 m overlying an impermeable parent material horizon. The soils developed from Quaternary clay are classified as a loam clay Ultisol according to the USDA soil taxonomy (Soil Survey Staff, 2010).

The two agroecosystems were established after a clearance of tea plants (*Camellia sinensis* L.) in 1982. The two systems were established adjacently along the downward trajectory of the hillslope. The citrus trees were spaced at 4×4 m^2 along the slope and peanut crops were sown in a 1.5 m wide alley between the tree rows. For this study, the trees were approximately 3.0 m high, with a canopy width of 3–4 m. The root depth was about 1.1 m (Wang et al., 2011). The peanut crops were sown horizontally along the slope at a spacing of 0.2×0.3 m^2 in rows in both systems. Peanut crops grew from the middle of April to the middle of August and were managed in the same way in both the agroecosystems.

Soil pits were dug in both systems and characterized following soil survey protocols (FAO/SRIC/ISSS, 1998). The tree root density in a 0.5×0.5 m^2 area was counted and soil samples were taken for physical analysis at each soil horizon. Soil particle size distribution was measured with the pipette method (Lu, 1999). Five bulk soil cores were taken at each soil horizon, and then air-dried and weighed in the laboratory to determine bulk soil density. In AF, four horizontal frames of 1.0×0.5 m^2 in each soil horizon were exposed vertically, sequentially during pit digging, and the roots in each horizontal frame were counted. The selected soil properties and root distribution are presented in Table 1.

Table 1 Selected Soil Physical Properties at Different Depths in the Monocropping (MC) and Agroforestry (AF) Systems

Land Use	Soil Horizon[a]	Depth (m)	Bulk Density (Mg m^{-3})	Particle Size Distribution (%) 0.05–2.0 mm	0.002–0.05 mm	<0.002 mm	Comments[b]
Monocropping (MC)	A_p	0–0.2	1.36	42.49	24.65	32.86	Prismatic/blocky; rich in biopores formed from decayed root
	AB	0.2–0.5	1.43	38.19	25.96	35.85	Prismatic/blocky; some biopores formed from decayed root
	B_t	0.5–0.9	1.51	36.45	25.75	37.80	Granular/crumbs
	BC_v	0.9–1.5	1.60	34.04	26.79	39.17	Coherent and single grain
Agroforestry (AF)	A_p	0–0.2	1.35	41.81	25.80	32.39	Prismatic/blocky; tree root density of 1522 m^{-2}
	AB	0.2–0.5	1.47	39.56	26.18	34.26	Prismatic/blocky, tree root density of 725 m^{-2}
	B_t	0.5–0.9	1.51	35.69	26.71	37.60	Granular/crumbs; tree root density of 307 m^{-2}
	BC_v	0.9–1.5	1.60	32.07	26.93	41.00	Coherent and single grain; tree root density of 175 m^{-2}

[a]The small letters for soil horizons are: t, accumulation of silicate clay and v, plinthite.
[b]Root density, number of roots per square meter.

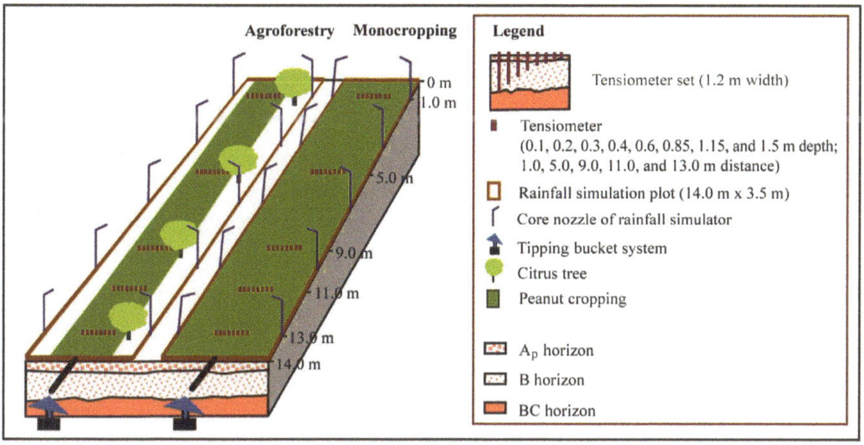

Fig. 1 The equipment layout from rainfall simulation experiment in the erosion plots under the monocropping and agroforestry systems.

2.2 Rainfall Simulation

The rainfall simulation was conducted by using rainfall sprinklers in two erosion plots located at the same slope position under both land uses (Fig. 1). The erosion plots enclosed with metal plates were 3.5 m wide and 14.0 m long along the slope. The metal plates were inserted into 0.2 m soil depth and projected 0.2 m above the ground. The sprinklers made in Germany, consisted of 10 core spraying nozzles that were positioned at 1.0, 4.0, 7.0, 10.0, and 13.0 m downslope from the upper boundary of each plot and mounted 3.0 m above the ground. The rainfall simulations were performed during periods without wind; water was supplied to control rainfall intensity by changing the number of nozzles. At the lower end of each erosion plot, one tipping bucket system was installed to collect overland flow during rainfall simulation and the number of bucket tips were recorded using an event data logger (Onset Computer Corporation, USA), then calibrated for flow rate using the total overland flow volume that was collected. Rainfall simulation experiments were performed over 2 h from 17:00 to 19:00 on six different days during the period 1st–18th April, 2008 (Eq. 1). Before rainfall simulation, the soil was tilled with moldboard plow and peanuts were not sown. Each simulated rainfall event was applied for 2 h to both plots at the same time. During rainfall simulation, rainfall volume was measured by using eight cups randomly distributed in the peanut cropping area and another eight cups placed at different positions underneath the tree canopy. Water from the cups was weighed to determine rainfall in both plots and canopy throughfall in the AF. The six rainfall simulations were initiated when the soil matric potential was approximately −80 hPa at the 0.1 m depth and −0 hPa at the 1.50 m depth.

Two weeks before the rainfall simulation, five sets of tensiometers were installed at the 1.0, 5.0, 9.0, 11.0, and 13.0 m distance downslope from the upper boundary of each plot to determine the spatial and temporal variations of soil matric potential. The sets of tensiometers were 0.1, 0.2, 0.3, 0.4, 0.6, 0.85, 1.15, and 1.5 m deep and installed at a width of 1.2 m (Fig. 1). Therefore, a grid of 40 tensiometers was installed in each plot. Readings of the soil matric potential were recorded at 10-min intervals using a data logger (DL2e, Delta T Inc., UK) connected with pressure transducers installed in the tensiometers. The soil matric potential was negative in the unsaturated zone, positive in the saturated zone, and zero at the interface. Soil water storage was demonstrated by the difference and the average of soil matric potential before and after each simulated rainfall in the soil profiles for both agroecosystems. For the rainfall events of 95.6 mm that was equal in both plots (see in Table 2), catenary variations of soil moisture were presented for 10 h continuously from the beginning of each 2-h simulated rainfall. These variations were represented by isolines of soil matric potential based on conventional Kriging interpolation on a rectangular grid using horizontal and vertical georeference points.

Table 2 Estimated Saturated Soil Hydraulic Conductivities (K_s) and van Genuchten Parameters ($\theta(\psi_m)$) at Different Soil Depths in the Monocropping (MC) and Agroforestry Systems (AF) Through Hydrus-2D Inversion Modeling for the Rainfall Events of 95.6 mm

			van Genuchten Parameters of $\theta(\psi_m)^a$			
Land Use	**Depth (m)**	K_s^a (m h^{-1})	θ_r (m^3 m^{-3})	θ_s^a (m^3 m^{-3})	a (m^{-1})	n
Monocropping (MC)	0–0.2	10	0.33	0.50	14.5	2.68
	0.2–0.9	80	0.30	0.44	80	1.33
	0.9–1.5	18	0.41	0.43	34	2.68
Agroforestry (AF)	0–0.2	10	0.30	0.50	19.5	2.68
	0.9–0.5	80	0.25	0.45	68	1.38
	0.9–1.5	13	0.31	0.44	34	2.66

[a]K_s, saturated hydraulic conductivity; $\theta(\psi_m)$, water retention curve; θ_s, saturated soil water content were measured by Jing et al. (2008).

2.3 Estimation of Subsurface Lateral Flow by Water Balance

Assuming that the horizon of parent material below the soil profiles was impermeable, such that the infiltrating water would produce subsurface lateral flow along the slope, the subsurface lateral flow (Q_{inter}, mm) was quantified by the water mass balance (Eq. 1).

$$Q_{inter} = P - P_{int} - R - ET - \Delta S \quad (1)$$

where P is total precipitation (mm), calibrated using plastic cups in the field; P_{int} is canopy interception (mm) measured in AF; R is overland flow (mm), measured using the tipping bucket system; ET is evapotranspiration (mm), which was assumed zero as the rainfall simulation was conducted in the evening under humid air and low temperature; and ΔS is change in soil water storage (mm), calculated based on measured soil moisture in the soil profile and the soil water retention curve reported for the same slope position by Jing et al. (2008).

2.4 Inverse Modeling of Subsurface Lateral Flow

Saturated soil hydraulic conductivities (K_s) and van Genuchten parameters ($\theta(\psi_m)$) in each agroecosystem were estimated through inverse modeling capabilities of the Hydrus-2D code (Simunek et al., 1996), based on the soil matric potentials measured during the rainfall events. To reduce the uncertainty in the estimated parameter values, the saturated soil water contents (θ_s) were fixed to the observed field values (Jing et al., 2008). The dimensions of the calculation domain were set to the size of the rectangular plots, being 14.0 m long and 2.0 m deep (Fig. 2). The slope gradient was 6%. The upper

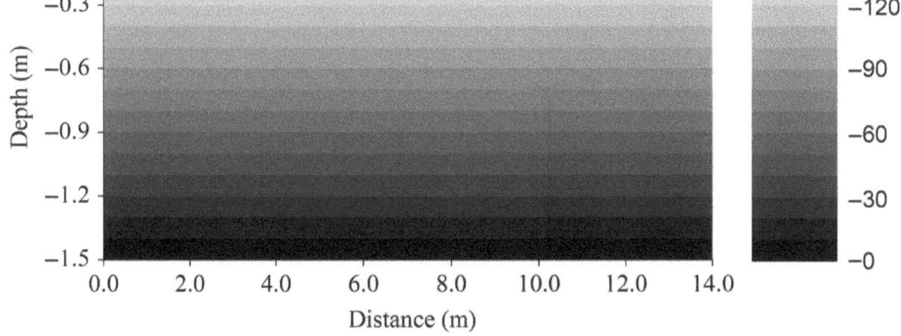

Fig. 2 The slope domain for Hydrus-2D modeling, with the 6% slope gradient georeferenced into rectangular. The isolines of soil matric potential in the soil profile illustrate the equilibrium along the slope, set up for the virtual modeling.

boundary was the atmospheric boundary. The bottom boundary was defined as zero flux due to the impermeable horizon of the parent material. The upper vertical boundary was defined as a no-flux boundary and the lower vertical boundary was defined as a free drainage boundary. The horizons of the soil profiles, i.e., layers of depths 0–0.2, 0.2–0.9, and 0.9–1.5 m, were defined as homogenous in terms of soil physical and hydraulic properties. Soil matric potentials, measured at the 0.1, 0.4, and 1.15 m depth at the 13.0 m distance from the upslope vertical boundary, were chosen as the soil horizons to perform the inverse modeling. To avoid possible numerical instability in the case of saturation water excess as overland flow, actual infiltration water (I, mm) (Eq. 2) was used as the input variable instead of rainfall during modeling. The actual infiltration was calculated with Eq. (2):

$$I = P - P_{int} - R \tag{2}$$

where P is total precipitation (mm), calibrated using plastic cups in the field; P_{int} is canopy interception (mm) measured in the AF plot; and R is overland flow (mm), measured using the tipping bucket system.

The goodness of fit was evaluated by comparing the simulated and measured soil matric potentials for the simulated rainfall events. The estimated values for saturated soil hydraulic conductivities and van Genuchten parameters were further validated by comparing the modeled and the calculated subsurface lateral flow. These comparisons for goodness of fit were carried out for all rainfall events. The coefficient of determination (R^2) and relative error (RE, %) were calculated to quantify the goodness of fit and the validity of the estimated parameter as follows (Eqs. 3 and 4):

$$R^2 = 1 - \frac{\sum_i (M_i - \overline{M_i})^2}{\sum_i (M_i - S_i)^2} \tag{3}$$

$$RE = \frac{M_i - S_i}{M_i} \times 100\% \tag{4}$$

where M_i is measured soil matric potential (hPa) or calculated subsurface lateral flow (mm); S_i is modeled soil matric potential or subsurface lateral flow simulated by Hudrus-2D; and $\overline{M_i}$ is the mean value of M_i.

The antecedent soil matric potential showed strong spatial heterogeneity (Fig. 3). This heterogeneity could create uncertainty in determining the effects of soil structure and hydraulic properties on subsurface lateral flow process along hillslopes. Therefore, a numerical experiment with Hydrus-2D was

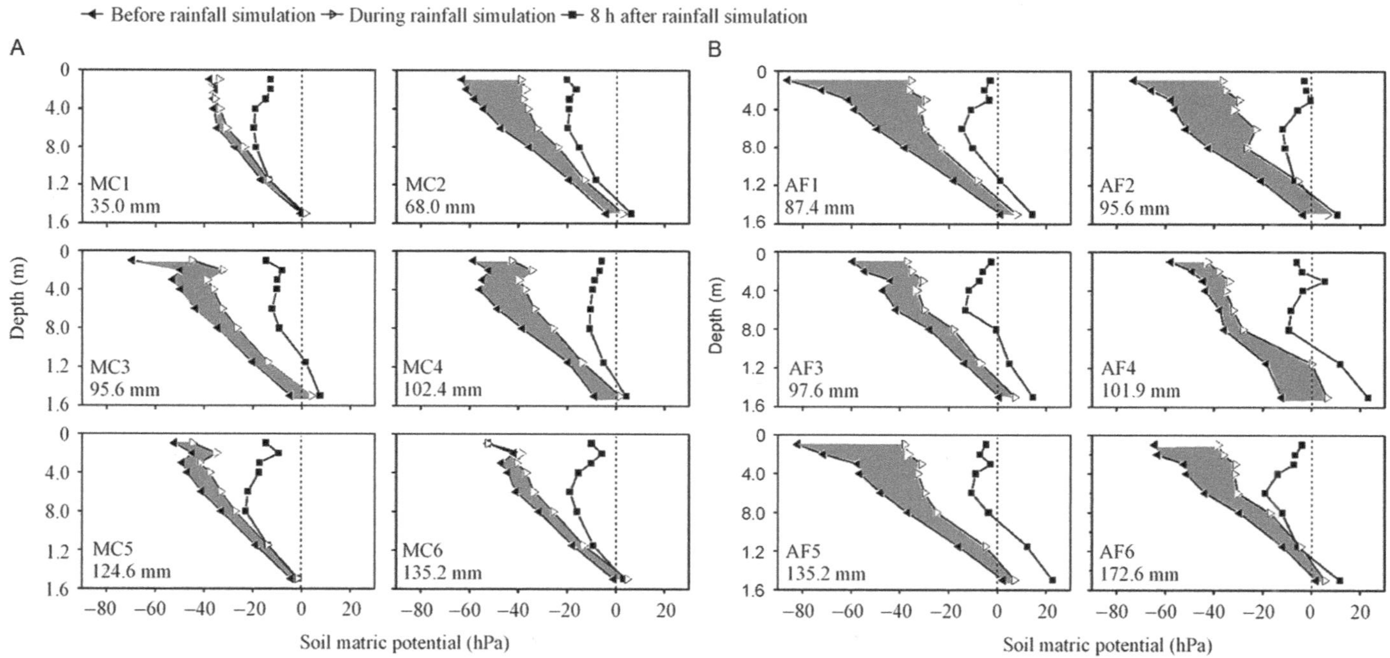

Fig. 3 Average soil matric potential before, during, and after rainfall simulation in the (A) monocropping (MC) system and (B) agroforestry (AF) system. The *shadow* indicates variation of soil matric potential between before and after rainfall simulation. The numbers following MC and AF were the sequence numbers of simulation rainfall events listed in Table 2.

performed under the virtual condition of the same antecedent soil matric potential along the slope for all simulated rainfall events to determine the effect of land use on subsurface lateral flow, the changes in soil water storage, and temporary water table elevation. The virtual soil matric potential profile was set up as equal distribution from −150 to 0 hPa at intervals of 10 cm in the 1.5-m depth soil profile (Fig. 2). When the two land uses have similar responses to different rainfall intensities and the effects of antecedent soil moisture can be ruled out, the results are designed to shed light on mechanisms of reduced subsurface lateral flow in AF systems, in comparison to MC through modification of soil structure and hydraulic properties.

3. RESULTS

The profiles of soil matric potential are plotted for all six simulated rainfall events (Fig. 3). The increase in soil matric potential before and 8 h after rainfall simulation represents the variation in soil water storage. After the rainfall simulation, the increased magnitude of the water storage in the soil profiles was larger for the agroforestry (AF) treatment than for the MC treatment in all six rainfall events. The soil matric potential in the lower soil layer 8 h after rainfall was near zero in MC and more positive, or demonstrating evidence of a higher temporary water table, in AF. Moreover, the soil matric potential during rainfall showed a different distribution in the upper soil profiles between the two agroecosystems, and it was greater at the 0.3–0.4 m depth than adjacent soil depths in AF, deeper than at the 0.2 m depth in MC.

The catenary variation of soil matric potential over time for the rainfall events of 95.6 mm demonstrates the effects of agroecosystems on subsurface water distribution (Fig. 4). The isolines of soil matric potential show that antecedent soil moisture before the rainfall (0 h) was similarly evenly distributed along the slope in both agroecosystems. There was a slightly drier condition at the downslope location in AF, compared with this location in MC. The isoline of zero soil matric potential shows that water saturation domains occurred not only in the surface soil but also in the deepest soil layers studied. At 0.5 h after the beginning of the rainfall event, the domains with water saturation were disconnected between the surface soil and the deep soil and between the upslope and the downslope locations. At 1.0 and 1.5 h after the beginning of rainfall, the water saturation domains expanded in the deep soil and became connected between the upslope

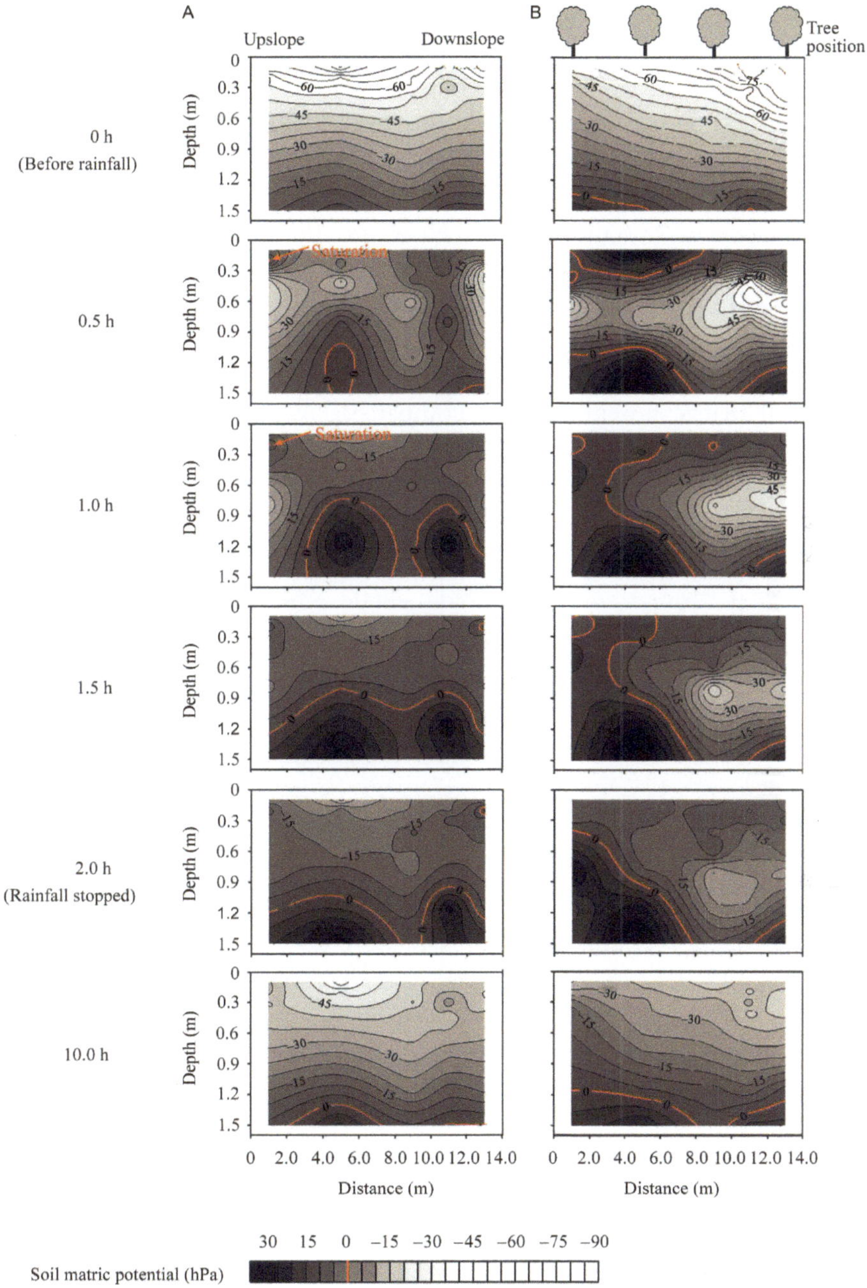

Fig. 4 Ordinary Kriging interpolation of soil matric potential (hPa) measured at different depths along slope for the simulated rainfall events of 95.6 mm in the monocropping system (A) and agroforestry system (B). The *red lines* show zero soil matric potential.

and downslope locations for the MC treatment. For the AF treatment, the water saturation domain expanded at the upslope position and became connected between the surface soil and the deep soil. For both agroecosystems, the water saturation domain became smaller following the end of the rainfall (after 2.0 h) compared to that at 1.5 h.

The modeled and measured soil matric potentials agreed well for the rainfall events of 95.6 mm (Fig. 5). However, the modeled values showed somewhat of a delay compared to the measured responses at the beginning of the rainfall, particularly at the 0.4 and 1.15 m depths in AF and at the 1.15 m depths in MC. The mean hourly determination coefficients (0.78–0.79) and RE values (14.6–15.7%) between the modeled and measured soil matric potentials were comparable in the two agroecosystems

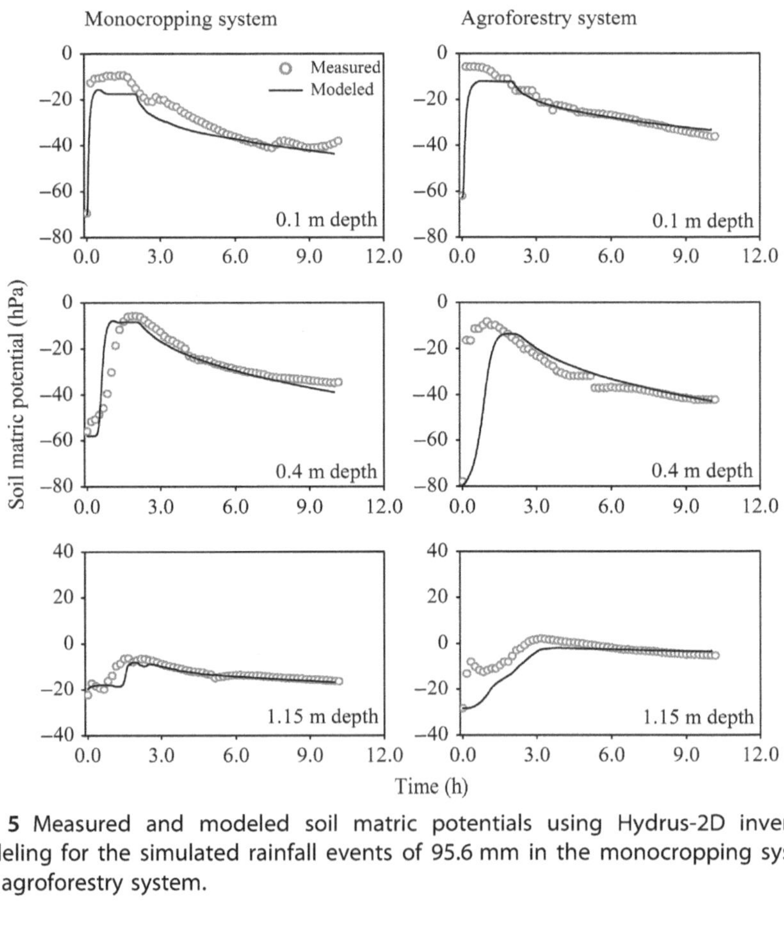

Fig. 5 Measured and modeled soil matric potentials using Hydrus-2D inversion modeling for the simulated rainfall events of 95.6 mm in the monocropping system and agroforestry system.

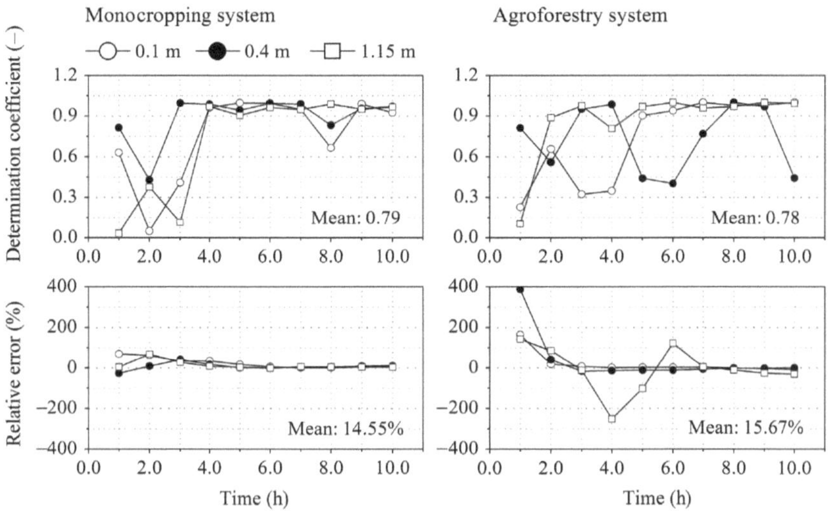

Fig. 6 Determination coefficient (R^2) and RE of soil matric potential at different soil depths between the values measured and modeled using Hydrus-2D for the simulated rainfall events of 95.6 mm in the monocropping system (MC) and agroforestry system (AF). The R^2 and RE was hourly calculated using Eqs. (3) and (4).

(Fig. 6). The estimated values for the parameters of the van Genuchten water retention curve were similar for both agroecosystems (Table 2). There was a smaller value for α at the 0–0.2 m depth (14.5 vs 19.5 m^{-1}) and at the 0.2–0.9 m depth (80 vs 68 m^{-1}), respectively, in MC compared to AF. By using estimated soil hydraulic parameters, soil water matric potentials were simulated for all rainfall events and the water mass balance was calculated (Table 3). The contributions of the water balance demonstrated that the overland flow accounted for approximately 50% the total rainfall in both agroecosystems. The infiltration water was about 16% less in AF than in MC due to approximately 12% greater canopy interception in AF. On average, the soil water storage and subsurface lateral flow accounted, respectively, for 38.7% and 61.3% of the infiltration water in AF and 11.8% and 88.2% the infiltration water in MC. The measured and simulated subsurface lateral flow was well correlated for all the rainfall events, showing high goodness of fit as indicated by the high values of determination coefficient (R^2) and low RE (Figs. 6 and 7).

Under the virtual condition of even distribution of antecedent soil matric potential, the numerical modeling results show that the two systems exhibited similar responses under different rainfall intensities in terms of predicted subsurface lateral flow and changes in soil water storage and

Table 3 Components of Water Budget for All Simulated Rainfall Events in the Monocropping (MC) and Agroforestry (AF) Systems

			Interception[b] P_{int}		Overland Flow[b] R		Infiltration I		Soil Water Storage[b] ΔS		Subsurface Lateral Flow, Q_{inter}	
Land Use	**Event[a]**	**Total Rainfall[b] P (mm)**	**mm**	**(E_{int}/P) %**	**mm**	**(R/P) %**	**mm**	**(I/P) %**	**mm**	**($\Delta S/I$) %**	**mm**	**(Q_{inter}/I) %**
Monocropping (MC)	MC1	35.0	–	–	12.3	35.1	22.7	64.9	0.9	4.0	21.8	96.0
	MC2	68.0	–	–	29.7	43.7	38.3	56.3	10.5	27.4	27.8	72.6
	MC3	95.6	–	–	52.6	55.0	43.0	45.0	4.4	10.2	38.6	89.8
	MC4	102.4	–	–	52.6	51.4	49.8	48.6	8.4	16.9	41.4	83.1
	MC5	124.6	–	–	61.3	49.2	63.3	50.8	5.5	8.7	57.8	91.3
	MC6	135.2	–	–	73.7	54.5	61.5	45.5	2.2	3.6	59.3	96.4
Agroforestry (AF)	AF1	95.6	12.5	13.1	50.9	53.2	32.2	33.7	15.8	49.1	16.4	50.9
	AF2	97.6	11.7	12.0	46.1	47.2	39.8	40.8	9.8	24.6	30.0	75.4
	AF3	87.4	14.4	16.5	51.9	59.4	21.1	24.1	16.9	80.1	4.2	19.9
	AF4	101.9	15.9	15.5	54.1	48.9	31.9	31.3	11.2	35.1	20.7	64.9
	AF5	135.2	14.0	10.4	66.2	54.7	55.0	40.7	15.3	27.8	39.7	72.2
	AF6	172.6	12.1	7.0	82.8	48.0	77.7	45.0	12.1	15.6	65.6	84.4

[a]The number following MC and AF indicates the sequence number of simulation rainfall events.
[b]Overland flow, determined by tipping bucket system; rainfall and interception, determined by plastic monitoring cups; evapotranspiration, defined as zero here associated with the humid air moisture and low temperature during rainfall simulation period; and soil water storage, calculated based on the measured soil matric potentials during the simulation rainfall and the van Genuchten parameters determined by Jing et al. (2008).

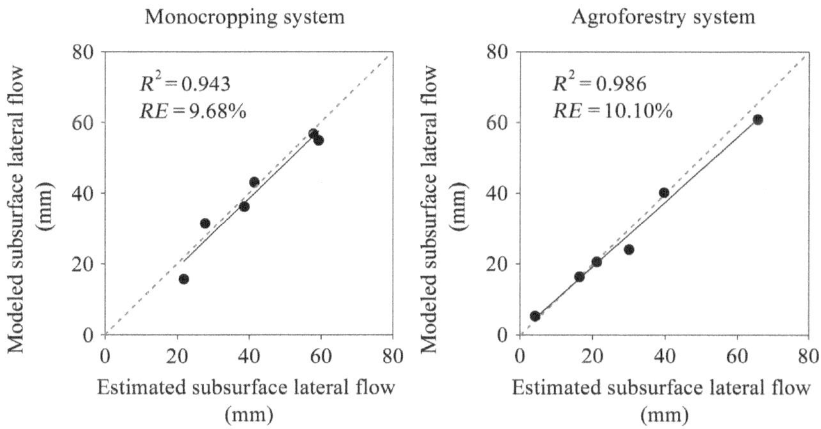

Fig. 7 The modeled subsurface lateral flow using Hydrus-2D and the calculated using water balance equation (Eq. 1) in the monocropping system and agroforestry system. R^2 indicates the determination coefficient, and RE indicates the relative error (Eqs. 3 and 4).

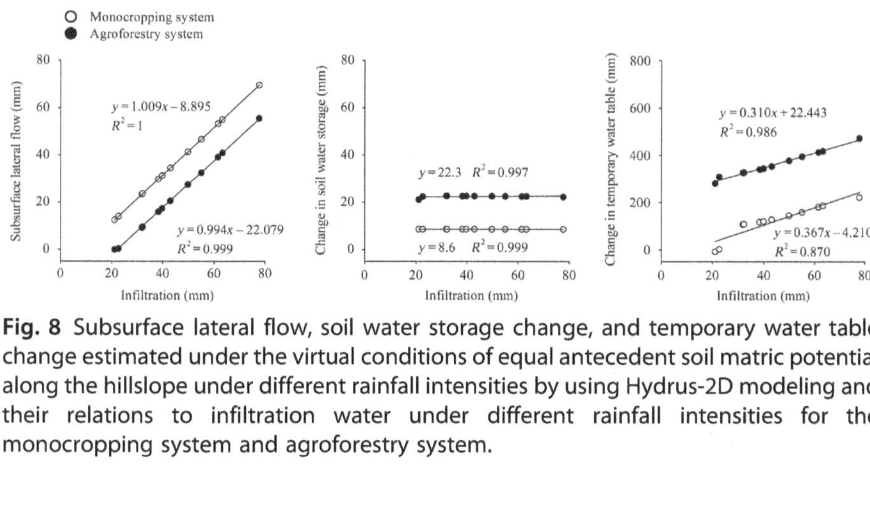

Fig. 8 Subsurface lateral flow, soil water storage change, and temporary water table change estimated under the virtual conditions of equal antecedent soil matric potential along the hillslope under different rainfall intensities by using Hydrus-2D modeling and their relations to infiltration water under different rainfall intensities for the monocropping system and agroforestry system.

temporary water table (Fig. 8). The predicted subsurface lateral flow and the change in temporary water table increased linearly with the increasing amount of infiltration water (rainfall minus runoff), while the predicted changes in soil water storage were constant under amount of different infiltration water. Comparing results between the two systems at the same infiltration, the AF treatment generated less subsurface lateral flow with greater soil water storage and a higher temporary water table elevation, compared with MC.

4. DISCUSSION

4.1 Generation of Subsurface Lateral Flow

The water budget estimated using Hydrus-2D code shows that the subsurface later flow was less in AF than in MC at all the comparable rainfall intensities (Table 2). This result is consistent with our previous study based on field monitoring data obtained over 3 years (Wang et al., 2011). Both studies demonstrate a larger domain and a longer resident time of water saturation in the deeper soil layers in AF than in MC during rainstorms. However, the previous study suggests that the reduced lateral flow was largely related to the increased evapotranspiration and tree canopy interception in AF (Wang et al., 2011). Here, we show that the reduced subsurface lateral flow was related to the increased tree canopy interception and soil water storage (by 12.4% and 6.4%, respectively, for the comparable 95.6 mm rainfall events) in AF compared to MC (Table 2). The increased canopy interception will lower infiltration water, while the increased soil water storage will increase the residence time of soil water in the soil profile, both of which will then reduce the subsurface lateral flow during rainstorms. The discrepancy on evapotranspiration and soil water storage is discussed later.

The evapotranspiration is higher in AF than in MC which will influence antecedent soil moisture before rainstorms and then may reduce subsurface lateral flow. The rainfall simulations have been carried out under conditions with negligible evapotranspiration and therefore are vital to test the hypothesis that subsurface lateral flow may change through modification of belowground processes by roots. With the simulation rainfall experiments, the effect of altered soil properties in AF on the reduction in subsurface lateral flow compared to MC was confirmed, though the effect was smaller than that of the tree canopy interception.

The rainfall stimulation experiments show negligible differences in the average overland flow values between AF and MC (Table 2), which is inconsistent with many other studies, which have reported that AF reduces the overland flow compared to MC (Udawatta et al., 2002; Wang et al., 2011). The different results of this study can be attributed to the high intensities of the simulated rainfalls, which may have caused saturation excess runoff as reported by Verbist et al. (2007). In addition, the plow pan found in MC may reduce vertical preferential flow and enhance the subsurface lateral flow above the plow pan compared with AF (Wang and Zhang, 2016).

When the plow layer is saturated in MC, overland flow may occur (Verbist et al., 2007). This will result in larger overland flow in MC than in AF even though infiltration can be increased after soil tillage which is included in the experimental conditions of this study. Therefore, the smaller overland flow in AF does not appear to be the reason for the reduced subsurface lateral flow compared to MC.

The spatial and temporal variations of the soil moisture during the rainfall events of 95.6 mm in this study show a large water saturation domain in deeper soil layers in both agroecosystems and disconnected water saturation domains between the top and deep soil layers in AF soon after the beginning of the rainfall (0.5 h) (Fig. 4), indicating vertical preferential flow in both agroecosystems, but likely with different pathways. The dye tracing experiments conducted before and after the rainfall in the same fields confirmed that the preferential flow through the plow pan below the plowed layer occurred mainly through well-connected cracks in MC and through relative disconnected large biopores developed by senile or decaying roots (Wang and Zhang, 2016). Field excavation from the studied fields showed that majority of peanut roots were distributed within the top 0.20 m soil, while majority of citrus roots were concentrated within the top 0.90 m soil (Table 1). Large of macropores from decaying roots are often reported in other ecosystems with vegetation dominated by trees (Schaik, 2010; Weiler and McDonnell, 2007; Zhang et al., 2015a,b). The existence of macropores is consistent with the disconnected water saturation domains in the soil profile of the AF system in this study. With the prolongation of the rainfall (0.5–2.0 h after the beginning of the rainfall), the water saturation domains became larger and connected along the slope and between the top and deep soil layers in MC and AF, respectively (Fig. 4). These results indicate that more soil water was retained in the soil profile in AF than in MC, resulting in the reduced subsurface lateral flow in AF. After the end of rainfall, the water saturation domain decreased in both systems, but was still larger in deep soil layers along the slope in AF than in MC, indicating slower drainage in AF. Therefore, the simulation rainfall experiments explicitly illustrate that the reduced subsurface lateral flow was attributed to the increased canopy interception and the increased water retention capacity in the soil profile due to different pathways of preferential vertical flow in AF compared to MC. The findings of this study and a dye tracing study (Wang and Zhang, 2016) provide useful insights to explore the mechanisms of subsurface lateral flow in Earth's CZ, which will benefit land use planners and land managers for supporting and sustaining terrestrial ecosystems (Banwart et al., 2012).

4.2 The Role of Soil Structure on Subsurface Lateral Flow Generation

The selected soil properties do not show significant differences between the two agroecosystems (Table 1), considering the short antecedent period for the differing agricultural practices and the small dimensions of the experimental plots. Therefore, the reduced subsurface lateral flow in AF could not be attributed to soil texture, but to altered soil structure and the resulting changes in hydraulic properties. Forests can retain more infiltrated water within soil profiles during rainfall (Bruijnzeel, 2004; Myers, 1983), as a part of the reservoir function of trees described in the literatures (Bruijnzeel, 2004; Myers, 1983). Mapa (1995) found higher soil water retention at any given water potential at all of the studied depths in reforested soils than in cultivated and grassland soils. For our research fields, the lab-determined soil water retention was higher for the 015–0.20 m soil depth in AF compared to MC and not different for the deeper soil layers between the two agroecosystems (Jing et al., 2008; Wang et al., 2013) (Fig. 9). However, at 0.2–0.9 m depth, the estimated van Genuchten parameter values using inverse modeling describe a water retention curve with lower water retention at lower soil matric potentials in AF than in MC. This is despite greater water retention in AF at near-saturation soil matric potentials (Fig. 9). These results indicate that the changed soil water retention curve cannot explain the increased water retention capacity and the reduced subsurface lateral flow in AF. The saturated hydraulic conductivity was greater at the 0.2–0.9 m depths compared to the top and deep soil layers in both agroecosystems (Table 2), suggesting presence of macropores at the 0.2–0.9 m depths. In AF, the subsurface later flow may be retarded due to the deeper root system, which may have not only intensified the structural connectivity through macropores in vertical direction, but also reduced the structural connectivity in the lateral direction. As if the subsurface lateral flow was retarded in AF, quick preferential flow would be quicker due to more macropores compared to MC. This would then generate a temporary water table over the impermeable layer generates (Fig. 4). Therefore, the increased water retention capacity in AF can be explained by the raised water table due to the presence of macropores and the reduced subsurface lateral flow.

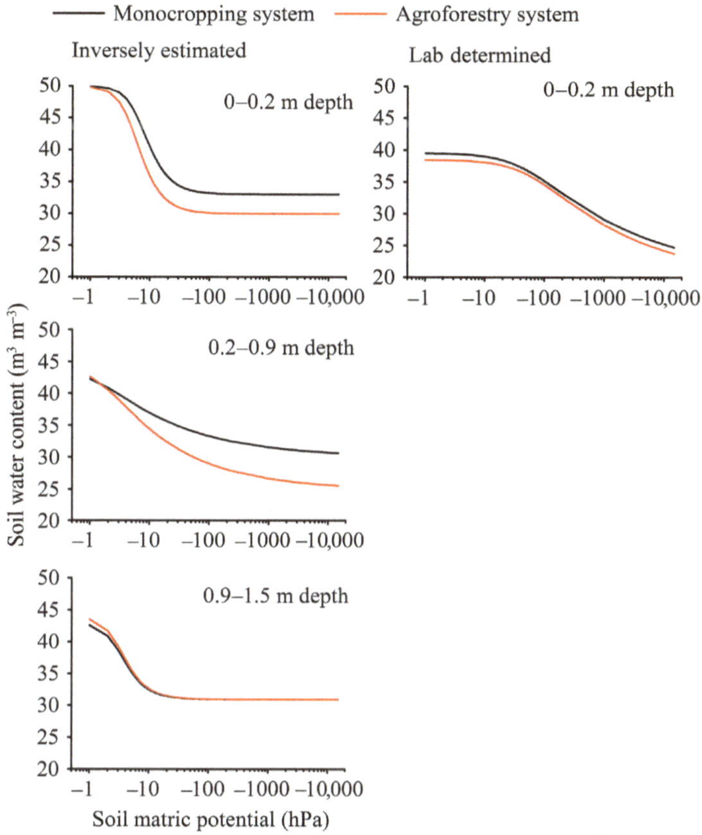

Fig. 9 Soil water retention curve at different soil depths in the monocropping system and agroforestry system, inversely estimated based on the van Genuchten parameters listed in Table 2 (*left*) and measured through pressure–membrane meter method using 100 cm^3 soil cores (Wang et al., 2013) (*right*).

There are few studies on the effect of land use on soil hydraulic properties in the lateral direction. The effects may be related to the development of soil micro- and mesopores due to the presence of fine roots or increased inputs of soil carbon as below-ground biomass and exudates, which promote the formation of large soil aggregates (Bogner et al., 2010; Mapa, 1995; Nair et al., 2010). Many studies have shown anisotropy of soil structure and hydraulic properties at landscape scale (Bottinelli et al., 2016; Jing et al., 2008; Soracco et al., 2010). The effects of macropores on the reduced subsurface lateral flow is indirectly reflected by the time delay of modeled peak flow rate compared to the observed values in different soil layers for both agroecosystems, but which is more profound in AF than in MC (Fig. 5).

The weaker goodness of fit in AF is likely due to the altered macropore system (Fig. 6), which is not represented by the altered soil water retention curve in the Hydrus-2D model.

To isolate the effects of soil structure change on the generation of subsurface lateral flow from the effect of antecedent soil moisture, a numerical experiment with Hydrus-2D was performed under the virtual condition of the same antecedent soil matric potential along the hillslope. The two agroecosystems showed a similar increasing trend with increasing effective rainfall intensity (the amount of infiltration water) in terms of the temporal water table and the subsurface lateral flow and no relation in terms of the soil water storage (Fig. 8). These results suggest that the effect of antecedent soil moisture or its related evapotranspiration can be ruled out as the reason for the reduced subsurface lateral flow in AF compared to MC. That is, the reduced subsurface lateral flow in AF is attributed to the increased tree canopy interception (or reduced infiltration) and the altered soil structure, which controls the anisotropy of soil hydraulic properties in vertical and lateral directions. Therefore, further studies are needed to elucidate the mechanisms how the tree root system shapes the anisotropy of soil hydraulic properties and consequently controls hillslope subsurface hydrological processes in Earth's CZ.

5. CONCLUSIONS

The present study coupled pedological and biological processes to explore the mechanisms through which agroforestry practices reduce subsurface lateral flow compared to monocropping system. The rainfall simulation experiments show lower infiltration water due to tree canopy interception, greater water retention capacity, and lower subsurface lateral flow in the agroforestry than in the monocropping system. The water partitioning and water movement is attributed to the deep roots of the agroforestry, which increase macropores in the vertical direction and alter micro- and mesopores in the lateral direction, resulting in more vertical preferential flow and less lateral flow within the soil profile and then enlarging soil water retention capacity. The numerical modeling at the virtual condition of the same antecedent soil moisture along the hillslope for all different rainfall intensities rules out the effect of antecedent soil moisture on the generation of subsurface lateral flow. Thus, although land use impacts evapotranspiration and then influences soil moisture conditions before rainfall, evapotranspiration may not be related to the effect of land use on generation of subsurface lateral flow.

This study suggests that the land use has strong effects on soil water mass and flow distribution due to the interaction between soils and roots. Therefore, understanding coupled pedological and biological processes in plant–soil–water systems are essential to understand the complexity of not only land surface but also subsoil ecological processes in Earth's CZ. Further studies need to elucidate the mechanisms by which these physical and biochemical processes connect to subsurface lateral flow processes, and also to quantify the magnitude of their relationship in Earth's CZ.

ACKNOWLEDGMENTS

This work was funded by EC-FP7 (SoilTrEC, Grant No. 244118), the National Key Technology Research and Development Program of the Ministry of Science and Technology of China (No. 2014BAD14B01), the Commonwealth Research Project of the Chinese Ministry of Water Resources (No. 201501055), and the National Science Foundation of China (NSFC) (No. 41201220). Mr. Q.T. Zhang, Z.B. Zhang, and L. Lin are acknowledged for their contribution to the field work.

REFERENCES

Banwart, S., Menon, M., Bernasconi, S.M., Bloemc, J., Blumd, W.E.H., de Souza, D.M., Davidsdotir, B., Duffy, C., Lair, G.J., Kram, P., Lamacova, A., Lundin, L., Nikolaidis, N.P., Novak, M., Panagos, P., Ragnarsdottir, K.V., Reynolds, B., Robinson, D., Rousseva, S., de Ruiter, P., van Gaansm, P., Weng, L.P., White, T., Zhang, B., 2012. Soil processes and functions across an international network of critical zone observatories: introduction to experimental methods and initial results. C. R. Geosci. 344, 758–772.

Bogner, C., Gaul, D., Kolb, A., Schmiedinger, I., Huwe, B., 2010. Investigating flow mechanisms in a forest soil by mixed effects modeling. Eur. J. Soil. Sci. 61, 1079–1090.

Bottinelli, N., Zhou, H., Boivin, P., Zhanga, Z.B., Jouquet, P., Hartmannc, C., Peng, X.H., 2016. Macropores generated during shrinkage in two paddy soils using X-ray micro-computed tomography. Geoderma 265, 78–86.

Brantley, S.L., Goldhaber, M.B., Ragnarsdottir, K.V., 2007. Crossing disciplines and scales to understand the critical zone. Elements 3, 307–314.

Bruijnzeel, L.A., 2004. Hydrological functions of tropical forests: not seeing the soil for the trees? Agri. Ecos. Environ. 104, 185–228.

Dunin, F.X., 2002. Integrating agroforestry and perennial pastures to mitigate water logging and secondary salinity. Agri. Water Manag. 53, 259–270.

FAO/SRIC/ISSS, 1998. World reference base for soil resources: World Soil Resources Report Nr. 84. FAO, Rome.

Fu, Z.Y., Chen, H.S., Zhang, W., Xu, Q.X., Wang, S., Wang, K.L., 2015. Subsurface flow in a soil-mantled subtropical dolomite karst slope: a field rainfall simulation study. Geomorphology 250, 1–14.

Gerrits, A.M.J., 2010. The Role of Interception in the Hydrological Cycle (PhD thesis). Faculty of Civil Engineering and Geosciences. Delft University of Technology, Netherlands.

Jing, Y.S., Zhang, B., Thimm, A., Zepp, H., 2008. Anisotropy of soil hydraulic properties along arable slopes. Pedosphere 18, 353–362.

Keim, R.F., Tromp-van Meerveld, H.J., McDonnell, J.J., 2006. A virtual experiment on the effects of evaporation and intensity smoothing by canopy interception on subsurface stormflow generation. J. Hydrol. 327, 352–364.
Lin, H., 2010. Earth's critical zone and hydropedology: concepts, characteristics, and advances. Hydrol. Earth Syst. Sci. 14, 25–45.
Lu, R.K., 1999. Analysis Methods for Soil and Agricultural Chemistry. Chinese Agricultural Science and Technology Press, Beijing, China (in Chinese).
Mapa, R.B., 1995. Effect of reforestation using Tectona grandis on infiltration and soil water retention. For. Ecol. Manage. 77, 119–125.
Myers, N., 1983. Tropical moist forests: over-exploited or under-utilized? For. Ecol. Manage. 6, 59–79.
Nair, P.K.R., Nair, V.D., Kumar, B.M., Showalter, J.M., 2010. Carbon sequestration in agroforestry systems. Adv. Agron. 108, 237–307.
Nath, T.N., 2015. Soil texture and total organic matter content and its influence on soil hydraulic conductivity of some selected tea cultivated soils in Sivasagar district of Assam, India. J. Chem. Cheml. Sci. 5, 241–252.
Schaik, L.V., 2010. The role of macropore flow from plot to catchment scale: a study in a semi-arid area. PhD thesis. Faculty of Geosciences, Utrecht University, Netherlands.
Seobi, T., Anderson, S.H., Udawatta, R.P., Gantzer, C.J., 2005. Influences of grass and agroforestry buffer strips on soil hydraulic properties. Soil Sci. Soc. Am. J. 69, 893–901.
Simunek, J., Othmer, H., van Genuchten, M.T., 1996. HYDRUS-2D, Simulation Water Flow and Solute Transport in Two Dimensional Variably Saturated Media, Version 2.0. IGWMC-TPS-53, International Groundwater Modeling Center, Colorado School of Mines, Goden, CO.
Soil Survey Staff, 2010. Keys to Soil Taxonomy, 11th ed. USDA-Natural Resources Conservation Service, Washington, DC.
Soracco, C.G., Lozano, L.A., Sarli, G.O., Gelati, P.R., Filgueira, R.R., 2010. Anisotropy of saturated hydraulic conductivity in a soil under conservation and no-till treatments. Soil Till. Res. 109, 18–22.
Tang, J.L., Zhang, B., Gao, C., Zepp, H., 2011. Subsurface lateral flow from slope and its contribution to nitrate loading in the streams during typical storm events in an agricultural catchment. Hydrol. Earth Syst. Sci. 15, 3153–3170.
Ticehurst, J.L., Cresswell, H.P., McKenzie, N.J., Jakeman, A.J., 2001. Using simulation to understand the occurrence of subsurface lateral flow in south-east Australia. In: International Congress on Modelling and Simulation (MODSIM 2001), vol. 2. pp. 561–566.
Udawatta, R.P., Krstansky, J.J., Henderson, G.S., Garrett, H.E., 2002. Agroforestry practices, runoff, and nutrient loss: a paired watershed comparison. J. Environ. Qual. 31, 1214–1225.
Verbist, K., Cornelis, W.M., Schiettecatte, W., Oltenfreiter, G., Van Meirvenne, M., Gabriels, D., 2007. The influence of a compacted plow sole on saturation excess runoff. Soil Till. Res. 96, 292–302.
Wang, Y., Zhang, B., 2016. Interception of subsurface lateral flow through enhanced vertical preferential flow in an agroforestry system observed using dye tracing and rainfall simulation experiments. Adv. Agron. 142, 99–118.
Wang, Y., Zhang, B., Lin, L., Zepp, H., 2011. Agroforestry system reduces subsurface lateral flow and nitrate loss in Jiangxi Province, China. Agri. Ecos. Environ. 140, 441–453.
Wang, L.L., Zhang, Z.B., Peng, X.H., 2013. Soil hydraulic properties under different land uses in hilly red soil region of southern China. Soils 45, 875–882 (in Chinese).
Weiler, M., McDonnell, J.J., 2007. Conceptualizing lateral preferential flow and flow networks and simulating the effects on gauged and ungauged hillslopes. Water Resour. Res. 43. http://dx.doi.org/10.1029/2006WR004867.

Wildhaber, Y.S., Bänninger, D., Burri, K., Alewell, C., 2012. Evaluation and application of a portable rainfall simulator on subalpine grassland. Catena 91, 56–62.
Zepp, H., Tang, J.L., Zhang, B., 2005. A methodological framework for a multi-scale study on hydrological process and soil erosion in subtropical southeast China. Pedosphere 15, 695–709.
Zhang, Y.H., Niu, J.Z., Zhu, W.L., Du, X.Q., Li, J., 2015a. Effects of plant roots on soil preferential pathways and soil matrix in forest ecosystems. J. For. Res. 26, 397–404.
Zhang, Z.B., Lin, L., Wang, Y., Peng, X.H., 2015b. Temporal change in soil macropores measured using tension infiltrometer under different land uses and slope positions in subtropical China. J. Soils Sediments 16, 854–863.
Zhao, N.N., Yu, F.L., Li, C.Z., Wang, H., Liu, J., Mu, W.B., 2014. Investigation of Rainfall-Runoff processes and soil moisture dynamics in grassland plots under simulated rainfall conditions. Water 6, 2671–2689.

CHAPTER FOUR

Interception of Subsurface Lateral Flow Through Enhanced Vertical Preferential Flow in an Agroforestry System Observed Using Dye-Tracing and Rainfall Simulation Experiments

Y. Wang*, B. Zhang[†,1]

*Key Laboratory of Agro-ecological Processes in Subtropical Regions, Changsha Research Station for Agricultural and Environmental Monitoring, Institute of Subtropical Agriculture, Chinese Academy of Sciences, Hunan, China

[†]National Engineering Laboratory for Improving Fertility of Arable Soils, Institute of Agricultural Resources and Regional Planning, Chinese Academy of Agricultural Sciences, Beijing, China

[1]Corresponding author: e-mail address: zhangbin01@caas.cn

Contents

Abstract

Surface soil hydrology is a major control on the terrestrial water cycle in Earth's Critical Zone (CZ). Partitioning of vertical preferential flow and subsurface lateral flow is commonly attributed to the heterogeneity of slope, soil profile horizon, and soil structure, but the influences of land-use types are largely unknown. Agroforestry systems (AF) can

Advances in Agronomy, Volume 142
ISSN 0065-2113
http://dx.doi.org/10.1016/bs.agron.2016.10.014

intercept subsurface lateral flow for reducing nitrogen losses to drainage water and for alleviating secondary salinity in central China and southeast Australia. These effects have been attributed to enhanced evapotranspiration and canopy interception in the agroforestry systems compared to monocropping systems (MC). Here, we show the differences in lateral and vertical soil hydrological pathways between AF and MC with dye-tracing experiments before and during a simulated rainfall event. Before the rainfall, the vertical and horizontal dye-staining patterns demonstrated that preferential flow occurred through isolated macropores with fine tree roots in AF and through connected cracks in MC. The dye coverage area and depth indicated greater vertical preferential flow in AF than in MC. During the 2-h rainfall event, the dye-staining area at different depths indicated that the preferential flow contributed to greater near-surface lateral flow in MC than in AF. The changes in the hydrological pathways were attributed to deep roots and no physical barrier from plough pan in AF and the presence of the plough pan in MC. These results suggest that land use has strong water partitioning effects not only above ground but also in the subsurface, and that understanding the landscape hydrology in the Earth's Critical Zone required quantification of the considering coupled pedological and biological processes.

1. INTRODUCTION

Earth's Critical Zone (CZ), extending from the top of the vegetative canopy down to the deepest fresh groundwater circulation, involves the flows of energy, water, carbon, and nutrients, thereby sustaining the natural resources that upon which mankind depends (Brantley et al., 2007; Lin, 2010). Vertical preferential flow and subsurface lateral flow are not only common surface soil hydrological processes (Wang et al., 2011; Zhang et al., 2011) but are also major physical controls on major environmental and sustainability challenges (i.e., drought, flood, soil erosion, water pollution). These challenges facing mankind today involve important linkages with vertical preferential flow and subsurface lateral flow in Earth's Critical Zone (Lin, 2010; Zhang et al., 2011). These water flow pathways transport water and pollutants vertically to deep soil as recharge to groundwater, and laterally along slopes into low-lying water bodies (Schaik, 2010; Wang et al., 2011; Zhang et al., 2011). Vertical preferential flow and subsurface lateral flow usually interact with each other due to soil heterogeneity in soil physical properties and process rates (Alletto and Coquet, 2009; Schaik, 2010; Zhang et al., 2015a). Therefore, the hydrological interactions may drastically change under various land uses that are a major source of soil heterogeneity in Earth's Critical Zone. However, the effects of land use on the hydrological interactions are still poorly understood, presenting a major barrier to solutions to these environmental and sustainability challenges.

Vertical preferential flow generally occurs when infiltrating water flows through soil macropores and bypasses most of soil matrix (Schaik, 2010). Subsurface lateral flow is often gravity-induced along the slope when sufficient water is impeded over a less permeable soil layer or bedrock within a soil profile (Shaw et al., 2001). Land use, subjected to natural and man-induced processes, is a source of soil heterogeneity (Alletto and Coquet, 2009; Schaik, 2010; Zhang et al., 2015b) that may influence the spatial variability in soil water flow (Flury et al., 1994; Forrer et al., 2000). Compaction of subsoil by using heavy machinery (Horn and Smucker, 2005) and the presence of plough pan beneath surface-cultivated soil (Heppell et al., 2000) can create a less permeable layer to reinforce infiltration resistance, which then induces subsurface lateral flow (Shaw et al., 2001). Macropores created by tillage (Zhang et al., 2015b), deeper penetration of plant roots (Schaik, 2010; Zhang et al., 2015a), and earthworm's activities (Trojan and Linden, 1992) can trigger vertical preferential flow in soils. Part of the flow variability can become negligible through flow mixing in the different directions (i.e., vertical and lateral). However, if soil heterogeneity is large and the flow mixing is limited, the land use will result in significant variability in vertical preferential flow and subsurface lateral flow (Allaire et al., 2009). Until now, as a source of soil heterogeneity, the connection between land-use practices and vertical preferential flow and subsurface lateral flow processes is unclear.

Knowledge of land use and subsurface lateral flow has been widely applied at the practical level for addressing the environmental and sustainability issues in Earth's Critical Zone. For instance, agroforestry systems were proposed in southeast Australia to alleviate secondary salinity and in central China to reduce nitrogen losses from fields in drainage water by intercepting subsurface lateral flow and nitrate transport (Dunin, 2002; Wang et al., 2011). However, the function of agroforestry systems to intercept subsurface lateral flow has been largely attributed to the enhanced evapotranspiration and canopy interception (Dunin, 2002; Wang et al., 2011). Preferential flow in agroforestry system may occur due to increased saturated conductivity (Seobi et al., 2005) and more biopores due to decayed roots (Schaik, 2010). Wang et al. (2011) found that more soil water can be retained in the soil profile, and reduced subsurface lateral flow can be generated, during rainstorm in an agroforestry system compared with a monocropping system. These studies implied that agroforestry systems may modify soil hydrological pathways to intercept subsurface lateral flow along hillslopes.

Dye-tracing techniques allow soil hydrological pathways to be both visualized and quantified at different physical scales of observation (Edwin et al., 2009; Forrer et al., 2000; Zehe and Flühler, 2001). Also, the interactions between morphological observations and soil hydrological process can be characterized when dye-tracing is used in conjunction with image analysis (Morris and Mooney, 2004; Sander and Horst, 2007). Until recently, most dye-tracing experiments were carried out under ponding condition to determine vertical hydrological pathway, while fewer experiments have been performed under rainfall conditions to study lateral flow pathways. Therefore, with dye-tracing experiments before and during a simulated rainfall, the objectives of this study were to characterize the vertical and lateral soil hydrological pathways along the slope and to determine the effects of an agroforestry system (AF) on the interaction between lateral and vertical soil water flows compared with a monocropping system (MC).

2. MATERIALS AND METHODS

2.1 Experimental Site

The study was carried out in a small agricultural catchment (Zepp et al., 2005), about 5 km from the Ecological Experiment Station of Red Soil, Chinese Academy of Sciences (EESRS, CAS) in Yingtan, China. The research area had a typical subtropical climate with a mean annual temperature of 17.7°C, a maximum daily temperature of around 40°C in summer, and an annual rainfall of 1786 mm. The elevation at the catchment ranges from 41 to 55 m above the sea level and the soil depth ranges from 0.4 to 2.0 m. The upland soils were developed from Quaternary clay and are classified as loam clay Ultisols according to the Soil Taxonomy (Soil Survey Staff, 2010).

Because this study aimed to differentiate the agricultural land-use effects on lateral and vertical soil hydrological pathways, two cropping systems were compared on the same slope. One was agroforestry system (AF) with citrus trees (*Citrus reticulata*) intercropped with peanut crops (*Arachis hypogaea* L.) in a downslope alley and another was monocropping of peanut system (MC) located in parallel to AF. The trees and crops were planted in rows down the slope, which was about 6% in gradient and 130–150 m long. The 20-year-old citrus trees were planted in rows of 4 m × 4 m along the slope and intercropped for 10 years. The citrus trees were about 3 m high, with the roots penetrating to 1.1 m soil depth and the leaf area index ranging from 4.4 to 5.5 during the growth period. Peanut crops were sown at a spacing of 0.20 m × 0.30 m in the middle of April and harvested in the middle of August in both systems. The peanut crops in AF were planted in 1.5-m-wide strips between the citrus trees, and all the field management practices were identical for the peanut crops.

One soil profile under each cropping system was excavated and characterized following soil survey protocols in the field (FAO/ISRIC/ISSS, 1998). Soil samples were taken by soil horizon for physical and chemical analysis. Soil particle size distribution was measured with the pipette method (Lu, 1999). Five soil cores taken from each soil horizon were used to measure soil bulk density. Soil organic carbon was measured after wet digestion with $K_2Cr_2O_7$–H_2SO_4 (Kalembasa and Jenkinson, 1973). The selected soil properties are presented in Table 1.

2.2 Dye-Tracing Experiment Without Rainfall Simulation

The dye-tracing experiment without rainfall simulation was conducted at the same slope position under both land uses following the method of Sander and Horst (2007). Briefly, two 1.0 m × 1.0 m steel frames as replicates were inserted to the soil depth of 0.10 m in the two systems. The frames were placed 0.5 m away from two tree rows in the untilled alley in AF and in the tilled area in MC. The soil around the frame walls was compacted to prevent bypass flow, and a plastic sheet was used to cover the frames before adding Brilliant blue (CAS number: 6104-59-2) solution (5 g L^{-1}). Fifty-liter dye solution was carefully poured over the plastic sheet, forming a 50 mm ponding water head. The plastic sheet was rapidly removed from one side of the frame to start infiltration, and then the frame was covered with plastic sheet to prevent evaporation. Twenty-four hours after infiltration, a 1.0 m × 1.0 m access pit was excavated along one side of the frame after removing the frame. The dyed area within the frame was separated into two parts from the middle line, for analysis of vertical and horizontal profiles. For vertical profile analysis, photos were taken within a rectangular frame (1.0 m × 1.0 m) as a spatial reference. The vertical profile was photographed and moved at an internal distance of 0.1 m from the

Table 1 Selected Chemical and Physical Properties of Soil Profiles in the Agroforestry System (AF) Consisting of Peanut and Citrus and Monocropping System (MC)

Land Use	Horizon[a]	Depth (m)	Bulk Density ($Mg\ m^{-3}$)	Soil Organic Carbon ($g\ kg^{-1}$)	Particle Size Distribution (%)			Comments
					2–0.05 mm	0.05–0.002 mm	<0.002 mm	
Agroforestry	A_p	0–0.2	1.36	10.85	42.49	24.65	32.86	Prismatic/blocky; surface compacted, rich in fine root, biopores formed from decayed root
	B_1	0.2–0.5	1.43	2.705	38.19	25.96	35.85	Prismatic/blocky; rich in root; some biopores formed from decayed root
	B_2	0.5–0.9	1.51	2.59	36.45	25.75	37.80	Granular/crumbs; root distributed
	B_{3t}	0.9–1.5	1.60	—	34.04	26.79	39.17	Coherent and single grain; rare root
Monocropping	A_p	0–0.2	1.35	11.22	41.81	25.80	32.39	Prismatic/blocky, surface crust; loose; rich in fine root, biopores formed from decayed root
	B_1	0.2–0.5	1.47	2.9	39.56	26.18	34.26	Prismatic/blocky; some biopores formed from decayed root
	B_2	0.5–0.9	1.51	2.44	35.69	26.71	37.60	Granular/crumbs
	B_{3t}	0.9–1.5	1.60	—	32.07	26.93	41.00	Coherent and single grain

[a]t, accumulation of silicate clay.

edge to the middle of the metal frame, and five vertical profiles were photographed in each plot. For the horizontal profile analysis, horizontal photos were taken within a rectangular frame (0.5 m × 1.0 m) as a spatial reference, and the surface soil was excavated at an interval of 0.05 m downward. Before photographing, the soil surfaces were brushed to remove the crumbs carefully using a knife. The photos were taken using a digital camera (Olympus E510, 3264 by 2448 pixels) under a black sheet to avoid strong contrasts by unequal sunlight exposure of the photographs. Altogether, 54 horizontal and 20 vertical profiles were photographed in both systems.

2.3 Dye-Tracing Experiment With Rainfall Simulation

Dye-tracing experiment with rainfall simulation was conducted in confined erosion plots (3.5 m × 14.0 m) at the downslope position in both systems during the period from 15th to 30th, April, when is the start month of rainy season in study region. The plots were plowed to 0.15 m to prepare peanut sowing. A set of tensiometers were installed in the middle of each plot at the depths of 0.10, 0.20, 0.30, 0.40, 0.60, 0.85, 1.15, and 1.50 m to monitor soil matric potential during rainfall simulation. A steel ring (0.3 m diameter × 0.4 m in height) was inserted to the soil depth of 0.10 m at the same slope position in the erosion plots, where the soil conditions (approximately −80 hPa at the 0.1 m depth and −20 hPa at the 1.50 m) were similar to those as described for the dye experiment under ponding infiltration. The ring was used to conduct dye-tracing experiment during rainfall simulation. Three point fifty three liter Brilliant blue solution (5 g L^{-1}) was poured into the ring carefully before the rainfall simulation, resulting in a 50 mm ponding water head. The rainfall simulation lasted for 2 h. During the rainfall simulation, the steel ring was covered with a plastic sheet to prevent rainfall falling into the ring. The rainfall intensity was designed differently for the two systems to ensure the same infiltration, considering the difference in through-fall in AF. Ten plastic cups of 0.15 m in diameter and 0.2 m in height were randomly placed in open area in both erosion plots to calibrate rainfall intensity. And another ten cups were placed under the tree canopy in AF to weigh through-fall. Overland flow was measured by a tipping bucket system that was installed at the lower end of each plot. The number of bucket tips was recorded using an event data logger (Onset Computer Corporation, USA) to calculate the overland flow volume. The recorded rainfall in 2 h was 97.6 mm in AF and 68.0 mm in MC. The overland flow was 46.7 mm in AF and 29.7 mm in MC; the through-flow was 11.7 mm in AF. Thus, the amount of infiltration into soil during the rainfall simulation was 39.2 mm in AF and 38.3 mm in MC.

When the soil matric potential decreased back to the initial values before rainfall simulation (about 23 h after the cease of rainfall simulation) in both systems, the steel ring was removed to take horizontal photos within a 1.5 × 1.5-m frame that had the ring as the center for 0.1 and 0.2 m depths following the similar procedures as described above. Briefly, the soil surface was brushed and crumbs were removed after the soils were excavated. The distance of dye movement upslope and downslope from the center for the ring was measured using a ruler. The horizontal soil surfaces were photographed, together with a wooden spatial reference frame (1.5 m × 1.5 m) and four Kodak color and gray scales onto the each side of the frame, following Forrer et al. (2000) and Kasteel et al. (2007). A black cover sheet was used to avoid strong contrasts during photographing.

During the excavation, 10 bulk soil cores (100 cm^3) were collected from the top soil apart from the ring in each plot. In the laboratory, the soil cores were saturated for 6 days in solution with different dye concentration: 0.1, 0.5, 1, 2, 4, 6, 20, 40, 80, and 150 g L^{-1}. Thereafter, the soil cores were freely drained at 25°C for 24 h, brushed with a knife on the surface, and used to take photographs using the same camera as for the field observations, in order to establish quantitative calibration between dye concentration and soil color intensity captured in the digital images (Alaoui and Goetz, 2008).

2.4 Image Analysis

The images obtained from the dye-tracing experiment without rainfall simulation were orthogonal with WGEO software (WASY GmbH, 2006) by transforming the four corners of photographed reference frames from internal raster to real coordinates. The undyed parts of the orthogonal images were removed with the Photoshop software (Photoshop 7.0). The images were transformed into binary format to calculate dye coverage in ImageJ software (National Institute of Mental Health, USA). To quantify the variability of soil water flow, the concept of actual penetration depth (H_a) was employed (Wang et al., 2006). An orthogonal three-dimensional coordinate system (X, Y, Z) was set into the soil columns for the analysis of horizontal dye-staining pattern (0.5 m × 1.0 m), with an X axis along the length direction (1.0 m), a Y axis along the width direction (0.5 m), a Z axis along the depth, and an origin (0,0,0) located at a corner of the soil surface. Each horizontal profile was split into 100 segments along the X axis direction and 50 segments along the Y axis direction; hence, the entire horizontal profile was separated into 5000 nodes, and each node covered an area of 0.01 m × 0.01 m. The actual penetration depth was calculated from Eq. (1):

$$H_a(x, y) = \sum_{i=1}^{k} \alpha(x, y, z)\Delta h_i \tag{1}$$

where, $H_a(x, y)$ is the actual penetration depth of the plot; $\alpha(x, y, z)$ is the coordinate of α, when the soil was dyed at the coordinate of (x, y, z), then $\alpha(x, y, z) = 1$, if not, then $\alpha(x, y, z) = 0$; Δh_i is the distance of the two adjacent nodes in vertical direction (Z axis direction).

The images obtained from the dye-tracing experiment under rainfall simulation condition were analyzed with a rigorous and repeatable method on the basis of calibrated dye concentrations. This method was employed in many other studies (Forrer et al., 2000; Kasteel et al., 2007). The first step was to orthogonalize the images with WGEO software. The second step was to adjust the color of images to warrant the comparability among images by applying a background subtraction technique through which the color and illumination were corrected by using Kodak gray scale applied to the each color channel (R, G, and B values). The third step was to determine the relationship between the dye concentrations and their associated mean values of every color channel ($\bar{R}$, $\bar{G}$, and $\bar{B}$ values) on the images by fitting a second-order polynomial using the standardized mean variables $\bar{R}$, $\bar{G}$, and $\bar{B}$ as independent variables, based on the correlation between dye concentrations and the value of color channels obtained in the laboratory. The fourth step was applied with the determined second-order polynomial function and the R, G, and B values fitted for each pixel in the processed horizontal images, for estimating the dye concentrations. To separate dyed and undyed soil in the images, the threshold concentrations were set between the highest measured concentration in the nonstained samples and the lowest measured concentration in the stained samples (Kasteel et al., 2007).

During the rainfall simulation, subsurface lateral flow can cause dye movement out of the surface area of the ring containing dye solution.

Assuming that preferential flow was the only flow through macropores in soil and that no chemical and physical reactions occurred between the soil matrix and dye solution, the area with dye concentration of >5 g L^{-1} during the rainfall simulation indicated that preferential flow occurred in the area. Therefore, the preferential flow contribution to subsurface lateral flow (C_p) was calculated based on Eqs. (2) and (3):

$$C_P = \frac{D_{md} - D_{ad}}{D_{md}} \times 100\% \tag{2}$$

$$D_{ad} = \sqrt{\frac{A_{stain}}{\pi}} \tag{3}$$

where D_{md} is the maximum distance of dye movement from the central point of the steel ring; d_{ring} is the radius of the steel ring (0.15 m); and D_{ad} is the actual distance of dye movement. It is assumed that when no subsurface lateral flow occurs, the wetness front of dye solution is uniform in the soil and the dye-staining area should be a circulate section in the horizontal images. Then, D_{ad} is the radius of the circle, which is calculated based on circle area (A_{stain}).

3. RESULTS

3.1 Vertical and Horizontal Dye-Staining Patterns Without Rainfall

Geometrically corrected photos of the vertical soil profiles are shown in Fig. 1. The fingering pattern of dye-staining fronts showed a high spatial density of vertical preferential flow in both systems. Comparing the two systems, the dye coverage was significantly larger in AF than in MC ($p < 0.05$) (Table 2). With the increasing soil depths, the dye coverage decreased evenly in AF, while it showed an abruptly decrease at the 0.15–0.17 m depth in MC, suggesting the presence of plough plan in MC. The dye-staining paths were more spatially related in MC than in AF, in which most of the dye staining was concentrated in the surface-ploughed layer and the preferential flow pattern started below the plough pan in MC. The photos taken horizontally with the metal frame demonstrated the dye-staining pattern at the 0.10 m soil depth (Fig. 2). The dye-stained flow paths were well connected along crack networks in the both systems, but the stained area along the cracks was wider and more extensive in MC than in AF. In addition, the

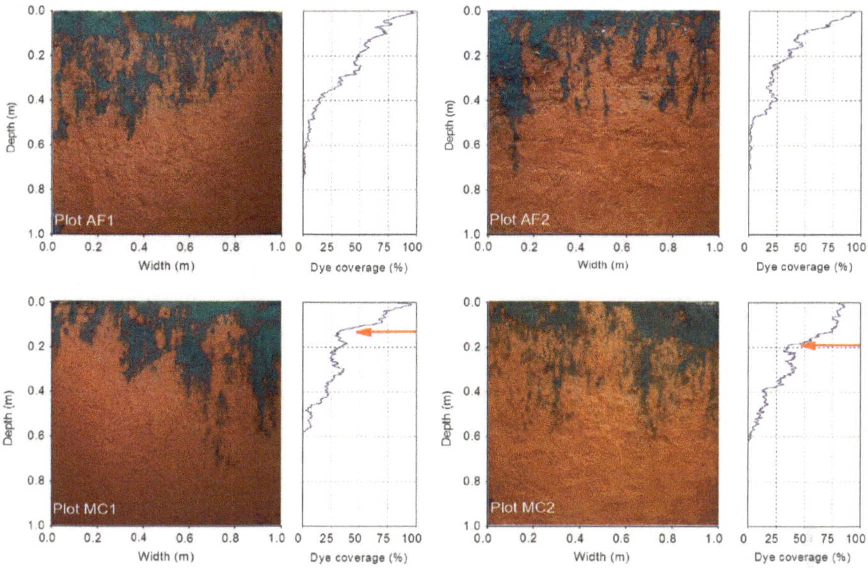

Fig. 1 Typical photos showing vertical dying patterns in the soil profiles under agroforestry system (Plot AF1 and AF2) and monocropping system (Plot MC1 and MC2) under the ponding infiltration without rainfall simulation. Dye coverage by soil depth was presented on the right hand side of each of the photos.

Table 2 The Dye Coverage and Plough Pan Depths Without Rainfall Simulation in the Vertical Images of Agroforestry (AF) and Monocropping System (MC)

Plot	Image Number (n)	Dye Coverage (%)		Plough Pan Depth (m)	
		Mean	SD	Mean	SD
Agroforestry AF1	5	15.1	1.6	—	—
Agroforestry AF2	5	15.9	1.6	—	—
Monocropping MC1	5	12.2	2.0	0.14	0.04
Monocropping MC2	5	13.0	1.5	0.17	0.04

dye-stained areas were 86.5–91.2% associated with sites having roots and were mainly distributed along the cracks in MC.

The actual penetration depth (H_a) and its frequency distribution demonstrated the variability of vertical preferential flow (Fig. 3). H_a in a single node (0.01 m × 0.01 m) was as deep as 0.55–0.65 m in the AF plots, 37–43% of

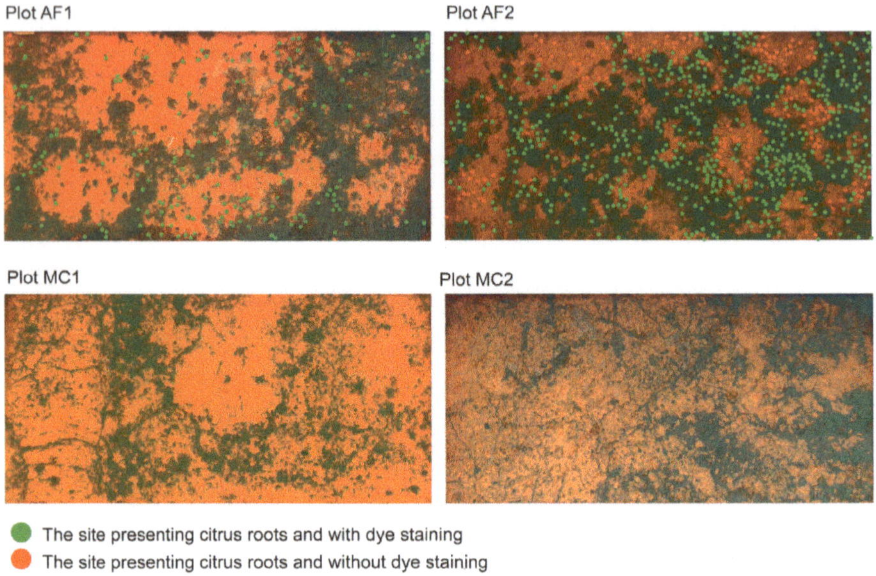

Fig. 2 Horizontal photographs showing dye-staining patterns at 0.10 m depth: within the frame (1.0 m × 1.0 m) in the agroforestry system (Plot AF1 and Plot AF2) and the monocropping system (Plot MC1 and Plot MC2) under the ponding infiltration without rainfall simulation. The *dots* in the photos of AF1 and AF2 indicated the sites presenting citrus roots, with the *green* showing stained dye and the *red* showing no stained dye.

which was distributed within the 0–0.10 m soil depth, while it was as shallow as 0.25–0.45 m in MC, 65–67% of which was concentrated within the 0–0.10 m soil depth. The mean H_a indicates the average percolation depth of vertical preferential flow in the metal frame. The mean H_a values were 0.13–0.17 m in AF, being deeper than 0.08 m in MC.

3.2 Horizontal Dye-Staining Patterns After a Simulated Rainfall

The horizontal photos taken after the rainfall simulation demonstrated that the dye-staining area extended out of the ring area upslope and downslope obviously at the 0.10 and 0.20 m soil depths in MC and only at 0.10 m soil depth in AF (Fig. 4). Differently, the dye staining diffused in a small domain around the steel ring at the 0.10 m soil depth in AF, but spread over the whole horizontal image at the 0.10 m and 0.20 m soil depth in MC. According to the dye coverage area change, the maximum distance of dye movement and the preferential flow contribution to

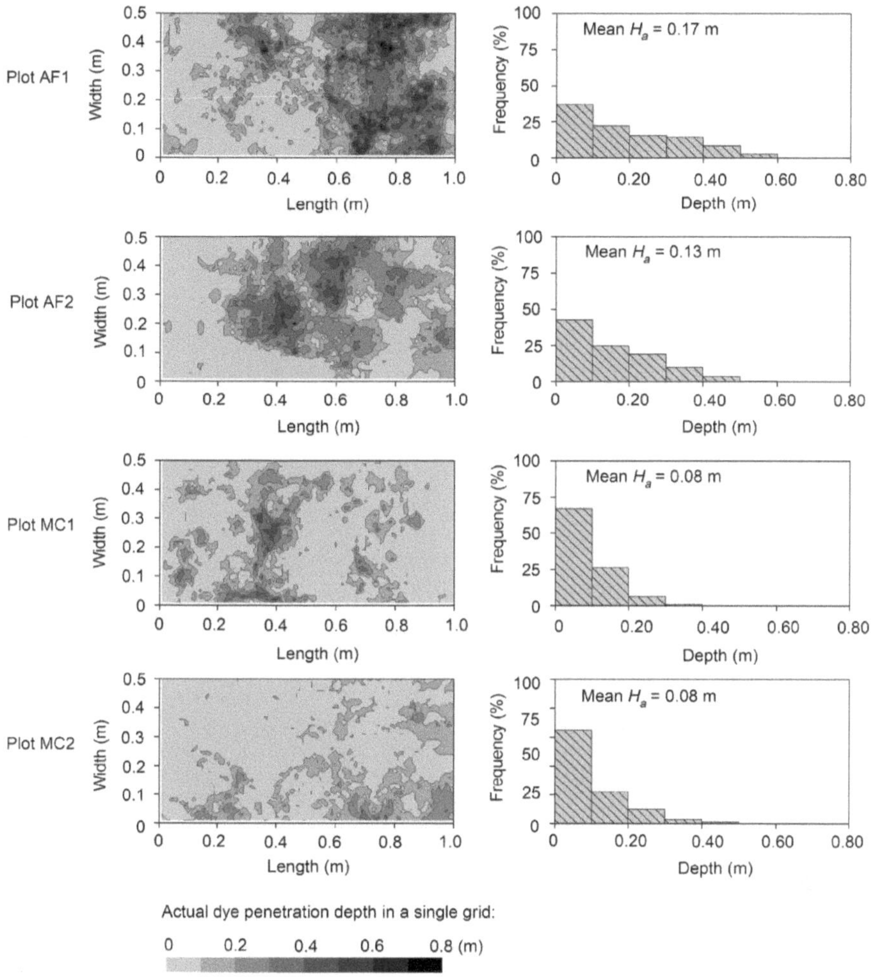

Fig. 3 Ordinary Kriging interpolation of the actual dye penetration depth in a single node (0.01 m × 0.01 m) (*left*) and its frequency distribution (*right*) based on horizontal photos from the dye-tracing experiment under ponding infiltration without rainfall simulation in the agroforestry system (Plot AF1 and AF2) and monocropping system (Plot MC1 and MC2). The *gray* intensity indicated the actual dye penetration depth in a single node (from 0 to 0.8 m), and mean H_a indicated the actual dye penetration depth at the whole plot scale (0.5 m width, 1.0 m length).

subsurface lateral flow were estimated (Table 3). The maximum movement distance was further downslope than upslope in the both systems and further in MC than in AF. The preferential flow contribution to subsurface lateral flow (C_P) was 21.3–54.6% in AF and 58.0–59.8% in MC.

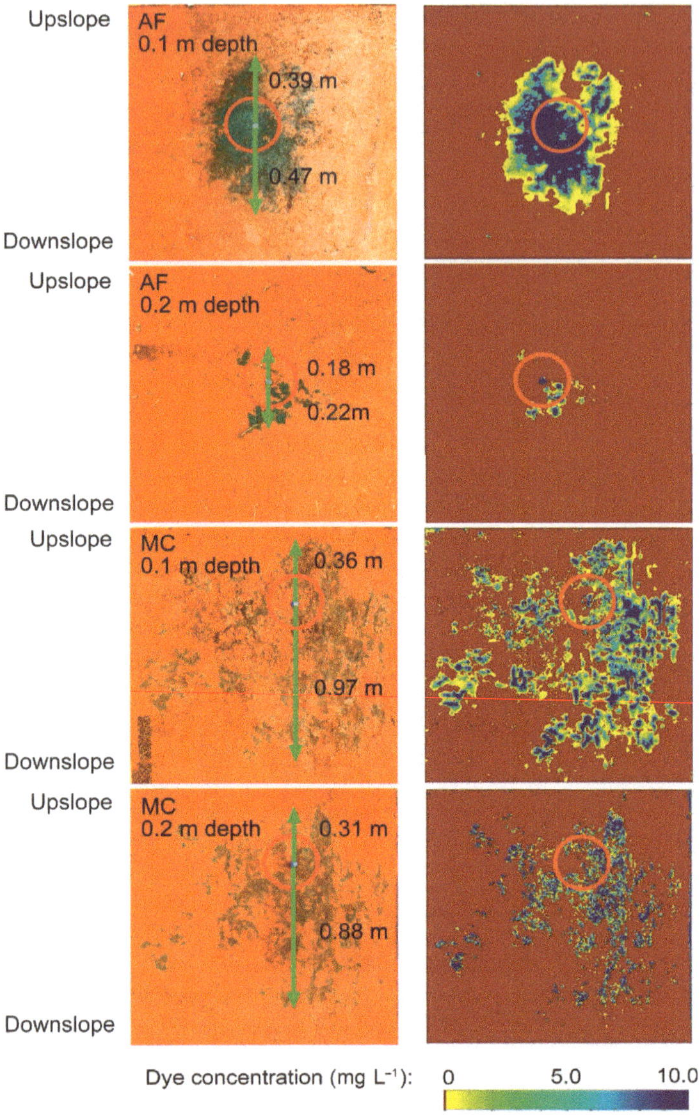

Fig. 4 Horizontal photographs and digitized images showing dye-strained areas surrounding the ring (*red circle*) at 0.1 and 0.2 m depth after rainfall simulation in the agroforestry (AF) and monocropping system (MC). *Different colors* in the images indicated a range of dye concentrations from 0 to 10.0 mg L^{-1}. The *green arrows* indicated the dye movement distance along upslope and downslope direction.

Table 3 Maximum Distance of Dye Movement (D_m) Upward and Downward, Dye Coverage (D_c), Dye Coverage of Preferential Flow ($D_{c\text{-}pf}$), Actual Distance of Dye Movement (D_{ad}), Preferential Flow Contribution to Subsurface Lateral Flow (C_p) in the Agroforestry (AF) and Monocropping System (MC) Applying the Dye-Tracing Experiment With Rainfall Simulation

Land Use	Depth (m)	D_m (m) Upward	Downward	D_c (%)	$D_{c\text{-}pf}$[a] (%)	D_{ad}[b] (m)	C_p[c] (%)
Agroforestry	0.1	0.39	0.47	19.3	7.8	0.37	21.3
	0.2	0.18	0.22	1.5	0.3	0.10	54.6
Monocropping	0.1	0.36	0.97	21.6	8.5	0.39	59.8
	0.2	0.31	0.88	19.4	8.6	0.37	58.0

[a]Dye coverage of preferential flow estimated with dye concentration of >5 g L^{-1}.
[b]Actual distance of dye movement calculated using Eq. (3).
[c]Preferential flow contribution to subsurface lateral flow calculated using Eq. (2).

4. DISCUSSION

4.1 Agroforestry System Influences on Vertical Preferential Flow Pathways

The dye staining illustrates the hydrological pathways for soil water flow in the upper part of soil profiles in Earth's Critical Zone. The dye-staining patterns under no rainfall condition show that vertical preferential flow occurred in both cropping systems, but the vertical preferential flow pattern differed (Figs. 1 and 2). The vertical preferential flow likely developed more in the surface soil (above 0.15–0.17 m) but less in the subsoil (below 0.15–0.17 m) in MC than AF, which corresponded well to the plough pan in MC. This result is consistent with the results of Andreini and Steenhuis (1990), who observed the same phenomenon between no-tilled and tilled soils, which differed in the connectivity of macropores in the surface and subsoils. The horizontal images suggested that the hydrological pathways of vertical preferential flow in the plough layer were more related to identified roots, particularly senile or decaying roots in AF, and were more related to well-connected cracks in MC (Fig. 2). Cracks form due to wetting and drying cycles (Hendrickx and Flury, 2001; Zhang et al., 2014). The stained areas along the cracks were well associated with the presence of tree roots in AF, suggesting that the citrus roots developed vertical preferential flow in soils. Therefore, the interaction between soil structure and tree roots, particularly in deep soil, is the key triggering mechanism for vertical preferential flow in AF (Seobi et al., 2005; Weiler and

McDonnell, 2007). There were few roots that were associated with the stained cracks in the plough layer in MC (Fig. 2), demonstrating that tillage destroyed the root-generated macropores and made the soil more homogenous (Andreini and Steenhuis, 1990; Zhang et al., 2015b).

The actual dye penetration depth was deeper in AF than in MC (Fig. 3), indicating that the preferential flow pathways were better connected from the top to deeper soil in AF than in MC. Soil texture has been considered to be a major reason for spatial heterogeneity; however, it showed negligible differences between the contrasting land uses (Table 1). These results suggested that the distinguished dye penetration depths in AF and MC were not induced by different soil texture, but from different soil structure. Literature reported that plough pan in shallow-rooted agroecosystems may prevent growth of crop roots, resulting in a decrease in connectivity of macropores between the surface and subsurface soils if any (Bertolino et al., 2010; Verbist et al., 2007). Field excavation carried out in the field showed that 88% of peanut roots were distributed within the top 0.20 m soil layer, while 90% of the citrus roots concentrated within the top 0.85 m (Wang et al., 2011). These results suggested that tree roots may have penetrated through the plough pan in the peanut growth alley. Therefore, soil infiltration and vertical preferential flow increase when tree roots increase (Seobi et al., 2005; Weiler and McDonnell, 2007) or when the size and connectivity of macropores increase (Schaik, 2010; Trojan and Linden, 1992). This is also the reason why the actual dye penetration depth in a single node (0.01 m × 0.01 m) was mostly concentrated within the 0–0.10 m soil depth in MC (Fig. 3).

4.2 Agroforestry System Influences on Subsurface Lateral Flow

The interaction between vertical preferential flow and subsurface lateral flow determines the spatial variability of water flow and material transport in the upper part of soil profiles in Earth's Critical Zone. The dye-staining domain and the diffusion distance downslope were smaller in AF than in MC after the rainfall (Fig. 4), demonstrating less subsurface lateral flow in the plough layer in AF than in MC. This is confirmed by the smaller contribution of preferential flow to the subsurface lateral flow in AF than in MC (Table 3). The difference can be also attributed to the difference in the presence of plough pan between AF and MC. The absence of plough pan allowed free percolation of water toward deeper soils without flow resistance (Shaw et al., 2001). The plough pan, however, reduced the hydrological conductivity of the soils in the plough pan and reduced the hydrological connectivity between the surface and deep soils (Heppell et al., 2000; Horn and Smucker, 2005) and along the top and bottom positions along the slope transect. The well-connected cracks and identified decaying roots in the plough layer of MC (Fig. 2) likely facilitated the subsurface lateral flow through preferential flow along lateral roots (Schaik, 2010; Verbist et al., 2007). Therefore, the heavy rainfall may contribute more to subsurface lateral flow and less to vertical preferential flow in MC compared to AF. This is consistent with our previous study based on long-term monitoring in the same field site, which demonstrated that the subsurface lateral flow monitored during rainstorms was less, but the saturated water table was higher and more stable in AF than in MC (Wang et al., 2011). Here, it is confirmed that the subsurface lateral flow generated above the plough pan may be an important flow pathway in MC. When the plough layer is saturated in MC, overland flow may occur (Verbist et al., 2007). This agrees with higher overland flow in MC than in AF, also observed from erosion plots located in the same field (Tang et al., 2011).

Soil is the central interface of Earth's Critical Zone, representing a geomembrane across which water is actively exchanged with the atmosphere, biosphere, hydrosphere, and lithosphere (Lin, 2010). Once a small change occurred in soil surface, surface soil hydrological conditions and processes can dramatically vary in the upper soil profiles in Earth's Critical Zone. The traditional hydrology theory usually considers the pedosphere as a static porous medium, but neglects the dynamic influences of the biosphere on it (Lin, 2010). This is the reason why the function of AF systems to reduce subsurface lateral flow was largely attributed to increased evapotranspiration and enhanced canopy interception in previous studies (Dunin, 2002; Wang et al., 2011). This study suggested that the reduced subsurface lateral flow along the slope in AF can also be attributed to enhanced vertical connectivity and reduced lateral connectivity of hydrological pathways compared with MC, where tillage caused formation of a homogenous plough layer and compacted plough pan. Such small-scale soil heterogeneity will result in significant variability in soil flow pattern at a large scale (Allaire et al., 2009). However, this study provides qualitative results on the relationship of land-use practices to the interaction between vertical preferential flow and subsurface lateral flow as controlling hydrological processes of Earth's Critical Zone. Further studies are needed to quantify the relationship and controlling mechanisms as it is crucial for addressing major environmental and sustainability issues such as providing sustainable agricultural practices.

5. CONCLUSIONS

This study explored the land-use effects on vertical preferential flow and subsurface lateral flow as controlling hydrological processes of Earth's Critical Zone. The dye patterns before and after rainfall demonstrated that vertical preferential flow and subsurface lateral flow were generated in different pathways in the agroforestry and monocropping systems. The vertical preferential flow started from the surface soil to a deeper soil layer through pathways associated with tree root growth in the agroforestry system and below plough pan to a shallower layer through cracks in the monocropping system. The results suggest greater vertical preferential flow in the agroforestry system than in the monocropping system. During a rainfall event, the vertical preferential flow was redirected into near-surface lateral flow along the slope. This effect was greater through the plough layer in the monocropping system than in the agroforestry system. These differences in the hydrological pathways were due to different isotropy and heterogeneity of porous soil structure. These results suggest that land use has strong impacts on the distribution of water mass and flow paths not only overland but also in the subsurface, and that understanding the landscape hydrology in Earth's Critical Zone needs to simultaneously consider the coupled pedological and biological processes.

ACKNOWLEDGMENTS

This work was funded by the EC FP7 (SoilTrEC, Grant No. 244118), the Commonwealth Research Project of the Chinese Ministry of Water Resources (No. 201501055), and the Technology Research and Development Program of the Ministry of Science and Technology of China (No. 2014BAD14B01). Mr. Q.T. Zhang, Z.B. Zhang, and L. Lin are acknowledged for their contribution to the field work.

REFERENCES

Alaoui, A., Goetz, B., 2008. Dye tracer and infiltration experiments to investigate macropore flow. Geoderma 144, 279–286.

Allaire, S.E., Roulier, S., Cessna, A.J., 2009. Quantifying preferential flow in soils: a review of different techniques. J. Hydrol. 378, 179–204.

Alletto, L., Coquet, Y., 2009. Temporal and spatial variability of soil bulk density and near-saturated hydraulic conductivity under two contrasted tillage management systems. Geoderma 152, 85–94.

Andreini, M.S., Steenhuis, T.S., 1990. Preferential paths of flow under conventional and conservation tillage. Geoderma 46, 85–102.

Bertolino, A.V.F.A., Fernandes, N.F., Miranda, J.P.L., Souza, A.P., Lopes, M.R.S., Palmieri, F., 2010. Effects of plough pan development on surface hydrology and on soil physical properties in Southeastern Brazilian plateau. J. Hydrol. 393, 94–104.

Brantley, S.L., Goldhaber, M.B., Ragnarsdottir, K.V., 2007. Crossing disciplines and scales to understand the critical zone. Elements 3, 307–314.

Dunin, F.X., 2002. Integrating agroforestry and perennial pastures to mitigate water logging and secondary salinity. Agric. Water Manag. 53, 259–270.

Edwin, E.C., David, L.R., Passmore, J., 2009. Influence of macroporosity on preferential solute and colloid transport in unsaturated field soils. J. Contam. Hydrol. 107, 45–57.

FAO/ISRIC/ISSS, 1998. World Reference Base for Soil Resources. FAO/ISRIC/ISSS, Rome.

Flury, M., Fleuhler, H., Jury, W.A., Leuenberger, J., 1994. Susceptibility of soils to preferential flow of water: a field study. Water Resour. Res. 30, 1945–1954.

Forrer, I., Papritz, A., Kasteel, R., Flühler, H., Luca, D., 2000. Quantifying dye tracers in soil profiles by image processing. Eur. J. Soil Sci. 51, 313–322.

Hendrickx, J.M.H., Flury, M., 2001. Uniform and Preferential Flow Mechanisms in the Vadose Zone. National Academic Press, Washington, DC.

Heppell, C.M., Burt, T.P., Williams, R.J., 2000. Variations in the hydrology of an underdrained clay hillslope. J. Hydrol. 227, 236–256.

Horn, R., Smucker, A., 2005. Structure formation and its consequences for gas and water transport in unsaturated arable and forest soils. Soil Tillage Res. 82, 5–14.

Kalembasa, S.J., Jenkinson, D.S., 1973. A comparative study of titrimetric and gravimetric methods for the determination of organic carbon in soil. J. Sci. Food Agr. 24, 1085–1090.

Kasteel, R., Garnier, P., Vachier, P., Coquet, Y., 2007. Dye tracer infiltration in the plow layer after straw incorporation. Geoderma 137, 360–369.

Lin, H., 2010. Earth's critical zone and hydropedology: concepts, characteristics, and advances. Hydrol. Earth Syst. Sci. 14, 25–45.

Lu, R.K., 1999. Analysis Methods for Soil and Agricultural Chemistry. Chinese Agricultural Science and Technology Press, Beijing, China (in Chinese).

Morris, C., Mooney, S.J., 2004. A high-resolution system for the quantification of preferential flow in undisturbed soil using observations of tracers. Geoderma 118, 133–143.

Sander, T., Horst, H.G., 2007. Preferential flow patterns in paddy fields using a dye tracer. Vadose Zone J. 6, 105–115.

Schaik, L.V., 2010. The role of macropore flow from plot to catchment scale: a study in a semi-arid area. PhD thesis, Faculty of Geosciences, Utrecht University, The Netherlands.

Seobi, T., Anderson, S.H., Udawatta, R.P., Gantzer, C.J., 2005. Influences of grass and agroforestry buffer strips on soil hydraulic properties. Soil Sci. Soc. Am. J. 69, 893–901.

Shaw, J.N., Bosch, D.D., West, L.T., Truman, C.C., Radcliffe, D.E., 2001. Lateral flow in loamy to sandy Kandiudults of the Upper Coastal Plain of Georgia (USA). Geoderma 99, 1–25.

Soil Survey Staff, 2010. Keys to Soil Taxonomy, 11th ed. USDA-Natural Resources Conservation Service, Washington, DC.

Tang, J.L., Zhang, B., Gao, C., Zepp, H., 2011. Subsurface lateral flow from slope and its contribution to nitrate loading in the streams during typical storm events in an agricultural catchment. Hydrol. Earth Syst. Sci. 15, 3153–3170.

Trojan, M., Linden, D., 1992. Micro relief and rainfall effects on water and solute movement in earthworm burrows. Soil Sci. Soc. Am. J. 56, 727–733.

Verbist, K., Cornelis, W.M., Schiettecatte, W., Oltenfreiter, G., Van Meirvenne, M., Gabriels, D., 2007. The influence of a compacted plow sole on saturation excess runoff. Soil Tillage Res. 96, 292–302.

Wang, K., Zhang, R., Yasuda, H., 2006. Characterizing heterogeneity of soil water flow by dye infiltration experiments. J. Hydrol. 328, 559–571.
Wang, Y., Zhang, B., Lin, L., Zepp, H., 2011. Agroforestry system reduces subsurface lateral flow and nitrate loss in Jiangxi Province, China. Agric. Ecosyst. Environ. 140, 441–453.
Weiler, M., McDonnell, J.J., 2007. Conceptualizing lateral preferential flow and flow networks and simulating the effects on gauged and ungauged hillslopes. Water Resour. Res. 43. http://dx.doi.org/10.1029/2006WR004867.
Zehe, E., Flühler, H., 2001. Slope scale variation of flow patterns in soil profiles. J. Hydrol. 247, 116–132.
Zepp, H., Tang, J.L., Zhang, B., 2005. A methodological framework for a multi-scale study on hydrological process and soil erosion in subtropical southeast China. Pedosphere 15, 695–709.
Zhang, B., Tang, J.L., Gao, C., Zepp, H., 2011. Subsurface lateral flow from hillslope and its contribution to nitrate loading in streams through an agricultural catchment during subtropical rainstorm events. Hydrol. Earth Syst. Sci. 15, 3153–3170.
Zhang, Z.B., Zhou, H., Zhao, Q.G., Lin, H., Peng, X.H., 2014. Characteristics of cracks in two paddy soils and their impacts on preferential flow. Geoderma 228–229, 114–121.
Zhang, Y.H., Niu, J.Z., Zhu, W.L., Du, X.Q., Li, J., 2015a. Effects of plant roots on soil preferential pathways and soil matrix in forest ecosystems. J. For. Res. 26, 397–404.
Zhang, Z.B., Lin, L., Wang, Y., Peng, X.H., 2015b. Temporal change in soil macropores measured using tension infiltrometer under different land uses and slope positions in subtropical China. J. Soils Sediments 16, 854–863.

CHAPTER FIVE

Effects of Dry and Wet Sieving of Soil on Identification and Interpretation of Microbial Community Composition

A. Blaud*[,1,4], M. Menon*[,2], B. van der Zaan†, G.J. Lair‡, S.A. Banwart*[,3]

*Kroto Research Institute, The University of Sheffield, Sheffield, United Kingdom
†Deltares, Subsurface and Groundwater Systems, Princetonlaan, Utrecht, The Netherlands
‡Institute of Soil Research, University of Natural Resources and Life Sciences (BOKU), Vienna, Austria
[4]Corresponding author: e-mail address: aimeric.blaud@gmail.com

Contents

Abstract

Soil aggregates are microhabitats for microorganisms, and directly influence microorganisms that live within and are influenced by microorganisms in return. Two methods are used to isolate soil aggregates by their size: dry sieving (sieving air-dried soil) and

---

[1] Current address: Agroecology Department, Rothamsted Research, Harpenden, Hertfordshire AL5 2JQ, UK.
[2] Current address: Department of Geography, The University of Sheffield, Sheffield S10 2TN, UK.
[3] Current address: School of Earth and Environment, University of Leeds, Leeds LS2 9JT, UK.

Advances in Agronomy, Volume 142
ISSN 0065-2113
http://dx.doi.org/10.1016/bs.agron.2016.10.006

wet sieving (sieving soil in water). Wet-sieving methods are generally considered to represent separation of aggregate classes that are stable to physical disaggregation in water, a condition considered favorable for protecting soil structure over time. However, little is known about the effect of sieving methods on microbial abundance, diversity, and functions, hindering the understanding of the relationship between soil structure and soil aggregates as habitat and soil microorganisms. In this study, the effect of dry and wet sieving on bacterial diversity, and abundance of microorganisms involved in N fixation (*nifH* gene), nitrification (*amoA* bacteria and archaea), and denitrification (*narG*, *nirS* and *nosZ* genes), was determined for four sizes of soil aggregates from a cropland and grassland. Quantitative-PCR (Q-PCR) showed little differences in relative gene abundance between size fractions of soil aggregates, but wet-sieving method significantly increased gene abundance for *amoA* bacteria, *nirS* and *nosZ* genes. When the N functional genes were expressed as percentage of the bacterial 16S rRNA genes, the wet sieving resulted in significantly higher genes percentage for all the genes (except for *narG* gene), and significant differences between soil aggregate size fractions at the grassland site. The different sieving methods resulted in different bacterial community compositions, but only the wet-sieving method was able to reveal significant differences in bacterial community composition between soil fractions in grassland. The results demonstrate significantly different quantitative and qualitative interpretation of soil microbial community depending on whether aggregate samples were obtained from wet or dry sieving, highlighting the importance in the choice of the sieving method.

1. INTRODUCTION

Soil is an extremely complex and heterogeneous environment, due to the complexity of its structure (i.e., 3D architecture of pores and particles), the large vertical spatial heterogeneity across the different horizons of soil profiles, and a huge and largely unknown microbial genetic diversity. Soil aggregates, composed of soil mineral fragments, decaying biomass, gases, water and solutes, and living organisms bound together as porous particles, represent the complexity of the soil structure and also the microhabitats for the microorganisms. Both the soil structure and soil microorganisms are central soil features that determine many key functions such as soil water retention and transmission, C, N, P, K sequestration, and nutrient transformations that ultimately sustain soil fertility. Different sizes of soil aggregates were shown to harbor different bacterial community structure (Blaud et al., 2012; Fall et al., 2004; Helgason et al., 2010; Kandeler et al., 2000; Sessitsch et al., 2001; Vaisanen et al., 2005), different bacterial diversity (Davinic et al., 2012; Kravchenko et al., 2014; Sessitsch et al., 2001), bacterial abundance and biomass (Helgason et al., 2010; Mendes et al., 1999; Sainju, 2006;

Schutter and Dick, 2002), and microbial activity (Bach and Hofmockel, 2014; Lensi et al., 1995; Sey et al., 2008). These differences are linked to the specific environmental conditions which exert biological selection pressures and are highly variable within aggregates.

To study soil aggregates, sieving methods are used to isolate different size classes of soil aggregates. The separation of soil aggregates is mainly done by dry- or wet-sieving method. The wet-sieving method, first described by Yoder (1936), is the most commonly used method to study microbial communities in soil aggregates and involves immersing soil for several minutes in water to breakdown aggregates. This occurs by increasing the surrounding static water pressure on the air trapped inside immersed particle pores, followed by vertical strokes in water to create shear forces to separate the soil particles that are initially placed on the top of a nest of subsequently immersed sieves. Dry sieving involves shaking usually air-dried soil, on top of a nest of sieves. Thus, the energy applied to the soil differs greatly between dry and wet sieving which affects directly the amount of stable soil aggregates that are obtained. Furthermore, wet sieving affects the aqueous colloidal forces at particle surfaces that can enhance or diminish the cohesive forces between aggregated particles. Thus, these two methods are expected to have direct effect on microbial communities due to the different sizes of soil aggregates which are isolated, i.e., the "washing" effect during wet sieving coupled with potential cross contamination between soil fractions, the effect of drying soil before dry sieving, and the different mechanical and physical–chemical forces applied on soil aggregates.

Only few studies have investigated the impact of dry and wet sieving to separate soil aggregates. Most of these studies focused on the effects of sieving methods on the physicochemical characteristics of soil aggregates. Dry sieving maintains large soil aggregates sizes (>2 mm) but is usually limited to the size fractions >250 μm. In contrast, wet sieving can separate soil aggregates from various size classes and in particular smaller sizes (<250 μm). The proportion of soil aggregates with size <2 mm mainly increase with wet sieving while soil aggregates >2 mm decrease due to the breakdown of the macroaggregates into smaller aggregates, and inversely for dry sieving (Bach and Hofmockel, 2014; Beauchamp and Seech, 1990; Sainju, 2006). Wet sieving leads to a loss of total C or total N, especially for soil fractions <250 μm, although no change or sometimes an increase in C content (for either >250 and <250 μm soil fractions) was found for wet sieving in comparison to dry sieving (Sainju, 2006). Seech and Beauchamp (1988) concluded that wet-sieving methods result in underestimating C and N pools.

The impact of aggregate fractionation procedures on microbial communities is not well studied. Sainju (2006) showed that the wet-sieving method decreases the nitrogen microbial biomass in comparison to dry sieving. In contrast, the carbon microbial biomass can decrease or increase depending on the soil type (Sainju, 2006). However the C or N microbial biomass is a gross indicator of microbial biomass, and no study has investigated the effect of sieving methods on microbial abundance, community structure, or diversity using DNA-based approaches (e.g., quantitative-PCR (Q-PCR), next-generation sequencing). A recent study comparing the effect of dry and wet sieving on microbial enzymatic activity showed that wet sieving overestimated the potential microbial enzymatic activity in comparison to dry sieving (Bach and Hofmockel, 2014). However, only the enzymatic activity differed between sizes of soil aggregates with wet sieving and not with dry sieving. This study also showed that drying the samples at 4°C to reach 10–20% of soil gravimetric water content did not affect the enzymatic activities before dry sieving.

The effect of sieving methods on microbial communities and resulting microbial characterization data and their interpretation remain largely unknown. This gap in understanding may represent a major factor influencing the results of any study investigating microbial communities in soil aggregates and is limiting the understanding of the relationship between soil structure, soil functions, and microbial diversity. Thus, the objective of this study was to determine the effect of dry and wet sieving on microbial community abundance and diversity within different size classes of soil aggregates. Four sizes of soil aggregates from a cropland and grassland were obtained by dry and wet sieving. Then, the abundance of bacteria, fungi, and microbial communities involved in N fixation, nitrification, and denitrification, and bacterial diversity were determined by Q-PCR and amplicon sequencing, respectively, for each size class of soil aggregates and for the bulk soil.

2. MATERIALS AND METHODS

2.1 Study Sites and Soil Sampling

The study sites, a cropland and grassland are located east of Vienna, Austria, in the National Park "Donau-Auen" on a floodplain of the Danube River. The cropland site was a grassland since 1781 and was converted to intensive cropland in the first half of the 20th century. The grassland site was converted from forest to grassland (presently *Onobrychido*

Table 1 Soil Characteristics of Bulk Soil Samples on a Dry Mass Basis

	Cropland	**Grassland**
Location	48°09′N, 16°41′E	48°11′N, 16°44′E
Water content (%)	22.0 ± 2.9	15.8 ± 2.0
Soil pH (H_2O)	7.7 ± 0.14	7.4 ± 0.09
Organic C (%)	2.4 ± 0.36	5.0 ± 0.60
Total N (%)	0.13 ± 0.01	0.33 ± 0.04
C_{org}/N	18.1 ± 1.83	15.0 ± 0.52
N-NH_4^+ (mg kg^{-1})	1.59 ± 0.29	4.77 ± 0.98
N-NO_3^- (mg kg^{-1})	20.3 ± 3.07	1.5 ± 0.66
P-PO_4^{3-} (g kg^{-1})	0.35 ± 0.10	0.59 ± 0.04
$CaCO_3$ (%)	19.0 ± 1.90	21.1 ± 1.41
Sand, 63–2000 μm (%)	32.7	8.2
Silt, 2–63 μm (%)	43.8	63.0
Clay, <2 μm (%)	23.5	28.8

Mean value ± one standard deviation ($n = 3$) are shown.

viciifoliae-Brometum) between 1809 and 1859 and is currently cut twice a year. The topsoil (0–10 cm) age is approximately 250–350 years since deposition of fluvial sediments as parent material forming a terrace above the down cutting river channel (Lair et al., 2009). The soils are classified as Mollic Fluvisols (IUSS Working Group WRB, 2006). The soil characteristics for cropland and grassland are shown in Table 1. Both sites were sampled on the 27th September 2013. Three distinct soil samples (500 g) were sampled at each site (grassland and cropland sites) from 5 to 10 cm depth and store at 4°C until soil fractionation.

2.2 Soil Fractionation

The soil samples were sieved at 2 mm before dry or wet sieving to homogenize the samples and to remove large roots and stones. Dry and wet sieving was performed on all the replicate samples for each site. Twenty grams of soil were used for each soil fractionation by size, recovered from the sieves of specific screen sizes. Henceforth, the term "soil fraction" is preferred to "soil aggregates" because this study did not separate soil aggregates from single mineral particles.

2.2.1 Dry Sieving

Prior to dry sieving, the 2 mm sieved soils were air dried at 4°C for 7 days until they reached a gravimetric water content of ~80 g kg^{-1} (Sainju et al., 2003). The air drying was required to obtain the soil fraction <53 μm from grassland soil and any soil fractions <250 μm from cropland soil. The dry-sieving protocol consisted of shaking by hand the soil samples placed on top of a nest of sieves (1000, 250, and 53 μm; 10 cm Ø) for 3 min at ~200 rotation min^{-1} (Sainju, 2006; Sainju et al., 2003). Soil retained on the 1000, 250, and 53 μm sieves were considered as 1000–2000, 250–1000, and 53–250 μm soil fractions, respectively. The soil collected in the cup under the 53 μm sieve was the <53 μm soil fraction. Soil aliquots were taken directly from each sieve for DNA extraction and stored at −20°C, and the rest of the soil fractions were dried at 55°C and used to measure soil fractions' mass distributions.

2.2.2 Wet Sieving

The wet-sieving fractionation method was adapted from Yoder (1936) and Blaud et al. (2012). Fresh soil samples were place on the top of a nest of sieves (1000, 250, and 53 μm; 10 cm Ø) and immersed in ~1.3 l ultrapure sterile water (4°C) tank for 5 min. Then, the sieves were raised and lowered during 10 min (stroke length ~30 mm, frequency 30 cycles min^{-1}). Soil retained on the 1000, 250, and 53 μm sieves were considered as 1000–2000, 250–1000, and 53–250 μm soil fractions, respectively. The water and soil left in the tank were centrifuged at 4500 × *g* for 10 min. The centrifugation was repeated to reduce the volume of water as much as possible and collect the soil particles, which represented the soil fraction <53 μm. Two soil aliquots were taken directly in each sieve: one for DNA extraction placed at −20°C and one for soil water content measurement dried at 55°C. The rest of the soil fractions for each sieve were washed in tubes and dried at 55°C and used to measure soil fractions' mass distributions.

The pore liquid collected after each round of centrifugation was filtered at 0.22 μm (47 mm Ø GTTP filter, Wathman) in order to collect and quantify the microorganisms washed from soil fractions during the sieving method. For each sample, five filters were required to filter the entire volume of water (due to the clogging of the filter), except for two replicates of cropland that required six and seven filters. The filters were kept at −20°C before DNA extraction.

2.3 DNA Extraction

DNA was extracted from 0.25 g of fresh soil for each soil fraction and bulk soil (i.e., 2 mm sieved soil) using the PowerSoil® DNA Isolation Kit

(Mo-Bio Laboratories, Carlsbad, CA, USA) according to manufacturer's instruction, except for the final step where the nucleic acids were eluted in 100 μL of sterile nuclease-free water.

DNA was extracted from the water used for wet sieving (after centrifugation to obtain <53 μm soil fraction) to determine the relative abundance of microorganisms lost during wet sieving. The same amount of water without soil was also filtered and used as control to ensure that the result obtained came from the wet sieving and not from contamination of the water or filter. The water for each sample was filtered and DNA was extracted from the filter using the PowerWater® DNA Isolation Kit (Mo-Bio Laboratories, Carlsbad, CA, USA) according to manufacturer's instruction, except for the final step where the nucleic acids were eluted in 100 μL of sterile nuclease-free water. DNA was extracted for each filter (i.e., 33 filters in total) and the DNA extracts were pooled for each sample.

2.4 Quantitative-PCR

Variation in microbial gene abundance was determined by Q-PCR targeting specific genes or genetic regions. Bacterial community was targeted via the 16S rRNA gene while the fungal community abundance was investigated by targeting the ITS region. The different communities involved in most steps of the N-cycle were investigated: the nitrogen fixing microorganisms were quantified based on the *nifH* gene; nitrification was investigated by targeting the ammonia oxidizing bacteria (AOB) and archaea (AOA) via the *amoA* gene; and denitrifiers were targeted via the *narG* gene coding for the nitrate reductase, the *nirS* gene coding for the nitrite reductase, and the *nosZ* gene coding for the nitrous oxide reductase. The details of the primers used to amplify the different amplicons are given in Table S1 (http://dx.doi.org/10.1016/bs.agron.2016.10.006).

Q-PCR standards for each molecular target were obtained using a 10-fold serial dilution of plasmids carrying a single cloned target gene, constructed by cloning PCR product of environmental samples (pCR2.1 TOPO vector), isolating cloned inserts (Qiagen Plasmid Mini Kit), and checking for the presence of gene of interest by sequence analysis. Standard curves and the no template control were amplified in triplicate in the same plate as the environmental samples. Q-PCR amplifications were performed in 25 μL volumes containing 12.5 μL of iQ™ SYBR® Green Supermix (Bio-Rad, Hemel Hempstead, UK), 8.5 μL of nuclease-free water (Ambion, Warrington, UK), 1.25 μL of each primer (10 μM), and 1 μL of template DNA using a CFX96™ Real-Time System (Bio-Rad, Hemel

Hempstead, UK). Standard amplification was used for all Q-PCR assays except AOA, starting with an initial denaturation at 95°C for 3 min, followed by 40 cycles of 30 s at 95°C, 0.5–1 min of annealing (annealing temperature and time for each primers pairs are given in Table S1), and 30 s at 72°C (Tsiknia et al., 2013). The fluorescence was measured at the end of each synthesis step (i.e., at 81°C for AOA and at 72°C for all other genes).

Threshold cycle (Ct) values and amplicon numbers were determined automatically using the Bio-Rad CFX Manager™ software. The efficiency of the Q-PCR assays varied between 70% and 98%. The r^2 were >0.99 for all the genes, except for *nifH* gene (0.984). The presence of Q-PCR inhibitors was tested for bacterial 16S rRNA gene, by running a Q-PCR with DNA extracts 10 times diluted or mixed with a known amount of the standard. No inhibition was detected.

The specificity of the Q-PCR was assessed via a melting curve analysis (increase of temperature from annealing temperature to 95°C by 0.5°C per step of 0.05 s) at the end of each Q-PCR amplification (Ririe et al., 1997). The melting curves for the bacterial 16S rRNA, *nifH*, *amoA*, *narG*, *nirS*, and *nosZ* genes Q-PCR assays showed specificity for the amplified targeted genes (i.e., single peak). As expected, the melting curve of the Q-PCR for fungal ITS showed the amplification of products of different lengths, due to the variability in length of ITS regions between different fungal taxa (Manter and Vivanco, 2007).

2.5 Amplicon Sequencing

The bacterial diversity of the different soil fractions obtained by dry and wet sieving, bulk soil and microbial suspension from water of the wet sieving, for the cropland and grassland was determined using the Ion Torrent® platform. The bacterial 16S rRNA gene V4 variable region was amplified using the primers 515F (5′-GTGCCAGCMGCCGCGGTAA-3′) and 806R (5′-GGACTACHVG GGTWTCTAAT-3′) (Caporaso et al., 2011) in a single-step 30 cycle PCR using the HotStarTaq Plus Master Mix Kit (Qiagen, USA) and the following conditions: 94°C for 3 min, followed by 28 cycles (5 cycles used on PCR products) of 94°C for 30 s, 53°C for 40 s, and 72°C for 1 min, followed by a final elongation step at 72°C for 5 min. Amplicon sequencing was performed at MR DNA (www.mrdnalab.com, Shallowater, TX, USA) on an Ion Torrent PGM following the manufacturer's guidelines.

The PGM data were analyzed following the pipeline developed by Pylro et al. (2014) that uses UPARSE (Edgar, 2013) and QIIME (Caporaso et al., 2010). Briefly, strip barcode, quality filtering, dereplication, abundance sort, and discard singletons were done using USEARCH 1.8. Chimera filtering was done using the *rdp_gold.fa* dataset. Then, taxonomy was assigned to operational taxonomic unit (OTU) using *uclust* method on QIIME 1.8 and Greengenes database (13_8) as a reference. The number of bacterial sequences per sample was on average 9183 ± 1443. Few archaeal sequences were found with on average 174 ± 99 per sample.

2.6 Statistical Analysis

To determine differences in the means of aggregate distribution, microbial gene abundance, or bacterial phylum relative abundance, ANOVA tests were performed with sites, sieving methods, and soil fractions as factors. The normality of the model residuals and the homoscedasticity of the variances were checked before statistical analysis. Log transformations of the Q-PCR data were applied to meet these criteria, except for *narG* gene abundance. When significant differences were found by ANOVA, the post hoc test of Newman–Keuls was performed to reveal the significant differences between class pairs. To test the differences between sites of the loss of genes in the wet-sieving water, the Student test was used.

The bacterial community composition was visualized by principal coordinate analysis (PCoA) based on the relative abundance of the OTU and generated using Bray–Curtis distance. ANOSIM (Analysis of SIMilarity; 10,000 maximum permutations) was used to investigate potential differences between bacterial community composition due to the sieving method, site fraction, or soil fraction (Clarke and Green, 1988). Two-way ANOSIM was used to compare one factor against the other factors and one-way ANOSIM to investigate the influence of an individual factor. ANOSIM analysis yields an R value, whereby ANOSIM values close to $R=1$ indicate a high separation between groups (e.g., between soil fractions), while ANOSIM values close to $R=0$ indicate a low group separation.

ANOVA and PCoA were performed using R v3.2.1 (R Development Core Team, 2015) and the package Phyloseq for PCoA (McMurdie and Holmes, 2013), while the ANOSIM tests were performed using PRIMER software (v6, PRIMER-E Ltd, Plymouth, UK).

3. RESULTS

3.1 Aggregates Distribution

The soil fractionation procedure resulted in mean mass recovery ~100% of the original unfractionated soil. The aggregates mass distribution showed similar pattern within both sites. The soil fractions >250 μm represented 35–50% of the aggregate distribution, while the soil fractions <250 μm were significantly lower and represented 2–20% (Fig. 1; Table S2 (http://dx.doi.org/10.1016/bs.agron.2016.10.006)). In contrast, the aggregate distribution from the cropland obtained by wet sieving showed the opposite distribution compared to any other distribution, with an increase in the mass of soil fractions with decreasing size of soil factions. The soil fraction 1000–2000 μm in cropland was significantly lower than <53 μm fraction (~16% and ~35% of the aggregate distribution, respectively). The mass distribution of each soil fraction for cropland was significantly different between sieving methods, except for the 250–1000 soil fraction. The aggregate distribution for cropland obtained by wet sieving showed large standard

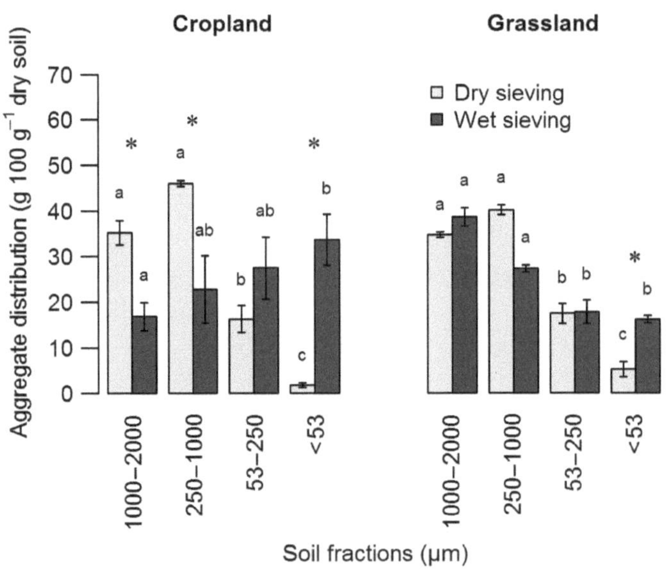

Fig. 1 Weight distribution of soil fractions (g 100 g^{-1} dry soil) obtained by dry- or wet-sieving method of soils from cropland and grassland. Means values ± standard error ($n = 3$) are shown. * indicates significant ($P < 0.05$) difference between dry and wet sieving for a specific soil fraction and site. *Different letters* indicate significant ($P < 0.05$) difference between soil fractions for a specific sieving method and site.

errors in comparison to any other aggregate distribution. The sieving methods had also some effect for the grassland, with an increase by ~10% of the 250–1000 μm fraction with dry sieving, and a significant increase by ~10% of the <53 μm fraction with wet sieving.

3.2 Microbial Gene Abundance

Microbial gene abundance showed significant differences ($P < 0.01$) between sites for all the genes except for *narG* gene (Fig. 2, Table S3 (http://dx.doi.org/10.1016/bs.agron.2016.10.006)). The genes abundance were higher in the grassland site for bacterial 16S rRNA gene, fungal ITS amplicon, *nifH*, *nirS*, and *nosZ* genes. In contrast, *amoA* bacteria (AOB) gene showed higher abundance in cropland, while *amoA* archaea (AOA) showed slightly higher abundance in grassland. Only the bacterial 16S rRNA gene showed significant differences ($P = 0.027$; Table S3) between soil fractions, and the post hoc test revealed significant differences in grassland and dry sieving between 1000–2000 μm and the fractions 250–1000 and 53–250 μm (Fig. 2). A significant effect ($P < 0.001$) of the sieving methods was found for the relative abundance of AOB, *nirS* and *nosZ*, with higher relative genes abundance found in fractions obtained by wet sieving in grassland (Fig. 2; Table S3). However, the post hoc test did not reveal significant pair-wise differences.

The proportion of microbial gene, expressed as percentage of bacterial 16S rRNA gene copies, was significantly ($P < 0.001$) different between sites, with higher *nifH* gene proportion found in grassland than cropland, while higher proportions for AOB and *narG* genes were found in cropland (Fig. 3; Table S4 (http://dx.doi.org/10.1016/bs.agron.2016.10.006)). Significant differences between soil fractions and sieving methods were found for all the genes except for *narG* gene. The post hoc test revealed a similar trend between soil fractions for grassland obtained by dry sieving, with the 1000–2000 μm fraction showing significantly ($P < 0.05$) higher proportion of microbial genes in comparison to most soil fractions and bulk soil (Fig. 3). The soil fractions from grassland obtained by wet sieving showed higher proportions of AOB, *nirS* and *nosZ* genes than bulk soil, but no significant differences between soil fractions were found. The effect of sieving methods, showed higher genes proportions with wet sieving by ~0.5%, except for the 1000–2000 μm fraction for grassland that showed higher proportion of *nifH*, AOB, *nirS* and *nosZ* gene with dry sieving by 0.5–2%. The post hoc test revealed significant ($P < 0.05$) differences in gene proportions between

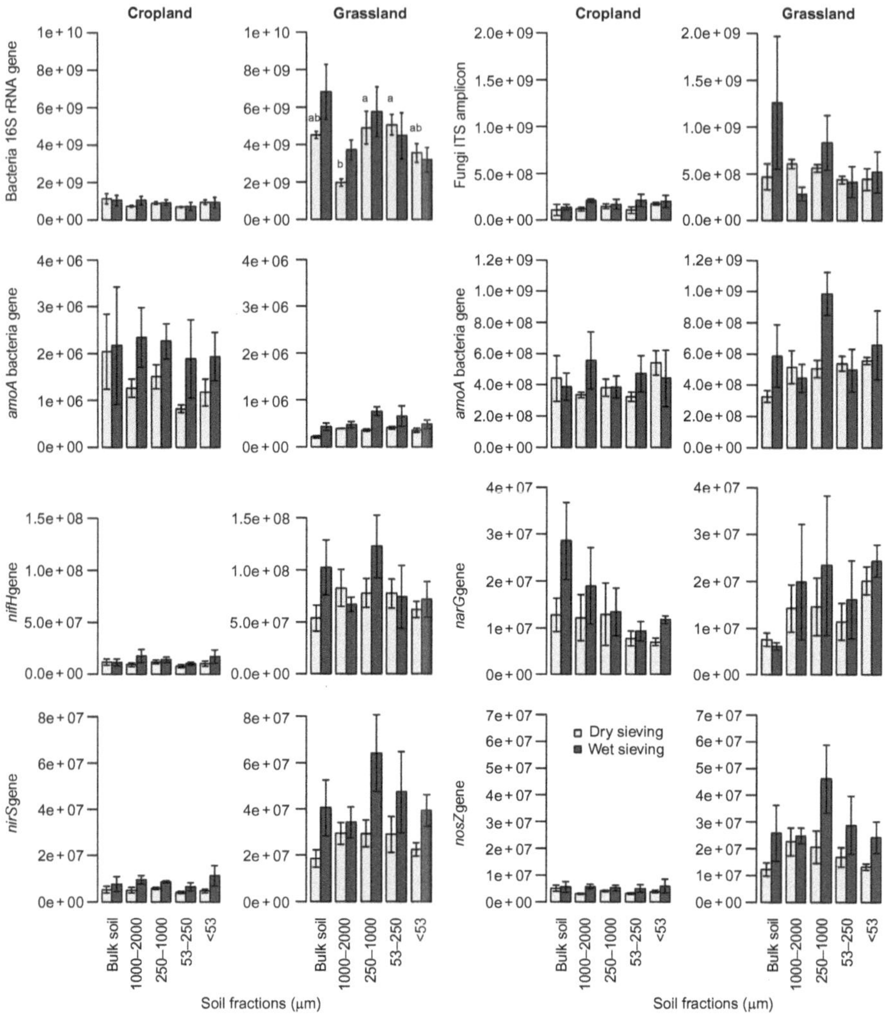

Fig. 2 Variation in gene abundance of bacteria (16S rRNA gene), fungi (ITS amplicon), N fixating (*nifH* gene), ammonia oxidizing bacteria and archaea (*amoA* gene), nitrate reductase (*narG* gene), nitrite reductase (*nirS* gene), and nitrous oxide reductase (*nosZ* gene) between four soil fractions obtained by dry- or wet-sieving method from cropland and grassland. All abundances are expressed on the basis of 1 g of dry mass of soil fraction or bulk soil. Means values ± standard error ($n = 3$) are shown. * indicates significant ($P < 0.05$) different between dry and wet sieving for a specific soil fraction and site. *Different letters* indicate significant ($P < 0.05$) difference between soil fractions for a specific sieving method and site.

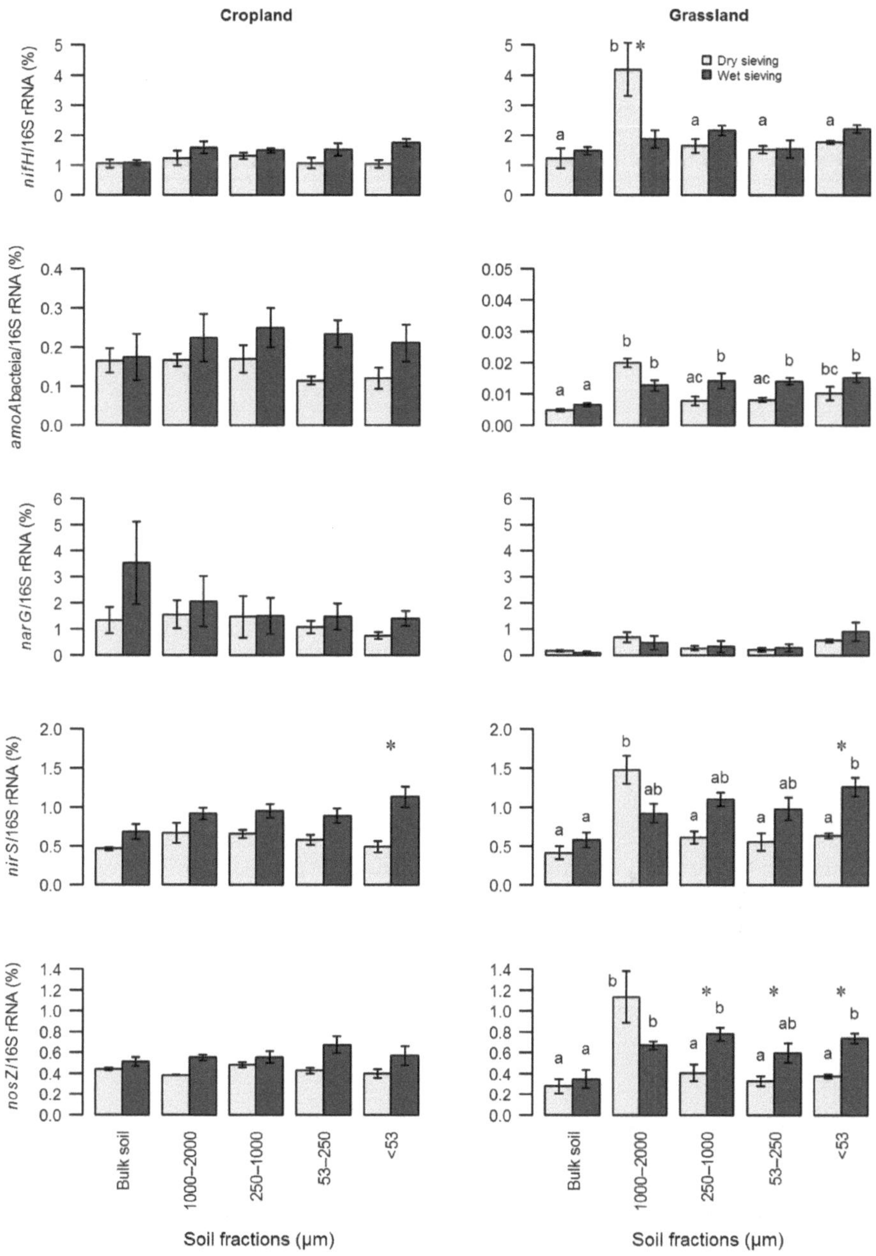

Fig. 3 Variation in N functional gene/bacterial 16S rRNA (%), of the N fixating (*nifH* gene), ammonia oxidizing bacteria (*amoA* gene), nitrate reductase (*narG* gene), nitrite reductase (*nirS* gene), and nitrous oxide reductase (*nosZ* gene) between four soil fractions obtained by dry- or wet-sieving method from cropland and grassland. Means values ± standard error ($n = 3$) are shown. * indicates significant ($P < 0.05$) different between dry and wet sieving for a specific soil fraction and site. *Different letters* indicate significant ($P < 0.05$) difference between soil fractions for a specific sieving method and site.

Table 2 Proportion of Genes (%) Lost in the Water During Soil Fractionation Using Wet Sieving

Gene	Cropland	Grassland
Bacterial 16s rRNA	1.55 ± 0.43	0.75 ± 0.30
Fungal ITS	0.48 ± 0.11	0.71 ± 0.52
nifH	2.31 ± 0.84	1.90 ± 0.85
amoA bacteria	0.33 ± 0.12	2.14 ± 0.63
amoA archaea	0.83 ± 0.09	1.83 ± 0.60
narG	1.16 ± 0.41 A	6.97 ± 0.80 B
nirS	0.85 ± 0.31	0.57 ± 0.17
nosZ	0.45 ± 0.14	0.60 ± 0.18

The loss of gene number in the water is express as a percentage of the number of the same gene present in 1 g of bulk soil. Mean value ± one standard error ($n = 3$) are shown. *Different letters* indicate significant ($P < 0.01$) differences between cropland and grassland for a specific gene.

sieving methods for *nifH*, *nirS*, and *nosZ* genes for grassland, and *nirS* gene for cropland (Fig. 3).

The microbial genes abundance lost in the water during wet sieving were expressed as percentage of the same gene present in 1 g of bulk soil. The proportion of microbial genes found in the sieving water varied between 0.3% and 2.3% (Table 2). Only *narG* gene showed ~7% of gene copies lost in sieving water for grassland and was also the only gene with a significant ($P = 0.0075$) difference between sites. The microbial gene abundance in the sieving water was consistently higher in grassland than cropland and significant ($P < 0.05$) for bacteria, fungi, *nifH*, *narG*, and *nosZ*, and marginally significant for AOA and AOB ($P = 0.06$ and 0.053, respectively; Fig. S1 (http://dx.doi.org/10.1016/bs.agron.2016.10.006)).

3.3 Bacterial Diversity

The PCoA showed that the bacterial community composition from the water of wet sieving differed greatly in comparison to any other samples (Fig. 4). The PCoA and ANOSIM also showed that the bacterial composition differed significantly ($R = 0.45$, $P = 0.0001$) between dry and wet sieving although some samples were mixed within each group. Then, significant differences between cropland and grassland were found, showing similar ANOSIM values compared to those reflecting the effect of the sieving methods ($R = 0.45$, $P = 0.0007$). The ANOSIM also revealed

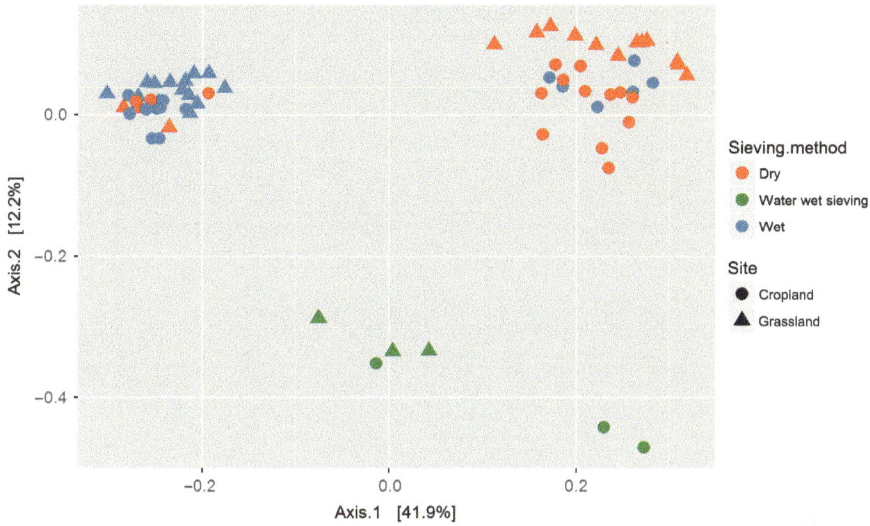

Fig. 4 PCoA of bacterial community of four soil fractions obtained by dry- or wet-sieving method and bulk soil from cropland and grassland. The PCoA was based on relative abundance of OTU and generated using Bray–Curtis distance. The six samples (in *green*) isolated from the rest of the samples correspond to water from the wet sieving.

significant differences between soil fractions, bulk soil, and water fractions but with a lower R value than those obtained for sieving methods and sites ($R=0.32$, $P=0.0001$).

The PCoA and ANOSIM were also performed on soil fractions and bulk soil for each site to reveal how the sieving methods affected the bacterial community composition between soil fractions at each site, which was not visible on the global analysis (Fig. 5). Significant differences between sieving methods and soil fractions were found for grassland (sieving: $R=0.82$, $P=0.0001$; fractions: $R=0.56$, $P=0.0001$) but not for cropland ($P>0.2$). The ANOSIM analysis performed on grassland for each sieving method revealed significant differences between soil fractions or bulk soil with both sieving methods (dry sieving: $R=0.57$, $P=0.0001$; fractions: $R=0.58$, $P=0.0001$). The bulk soil showed clear differences with the soil fractions especially for dry sieving. Interestingly, bulk soil from dry sieving grouped closely to bulk soil from wet sieving and soil fractions. However, the PCoA revealed differences between soil fractions with the wet-sieving method, and high variation between replicates with dry sieving (Fig. 5). This was confirmed when the ANOSIM was performed without the bulk soil, showing only significant and relatively strong differences between soil

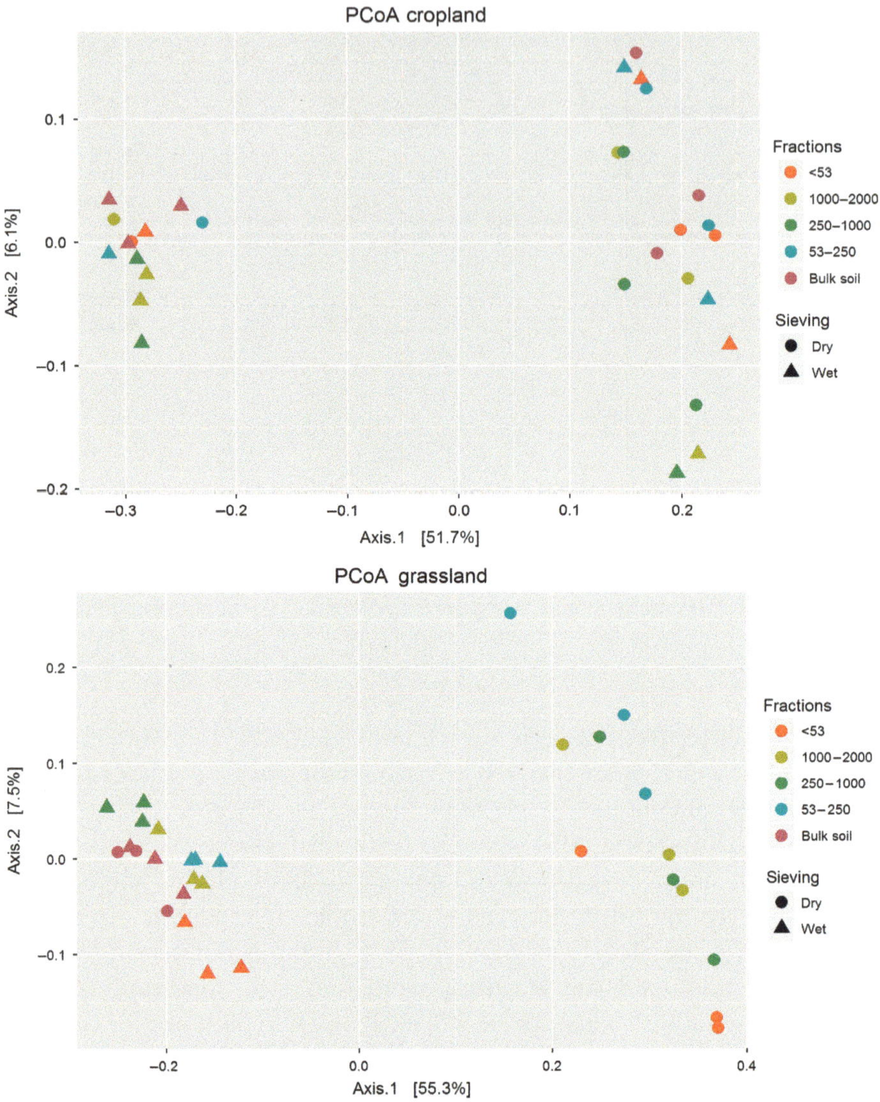

Fig. 5 PCoA of bacterial community of four soil fractions obtained by dry- or wet-sieving method and bulk soil from cropland (*top*) and grassland (*bottom*). The PCoA were based on relative abundance of OTU and generated using Bray–Curtis distance.

fractions when obtained by wet sieving ($R=0.44$, $P=0.0001$) and no difference with dry sieving ($R=0.1$, $P=0.108$).

The relative abundances of most of the dominant phyla were strongly affected by the sieving methods with a decrease with wet sieving for most of them except for *Actinobacteria*, *Cyanobacteria*, and *Verrucomicrobia* that

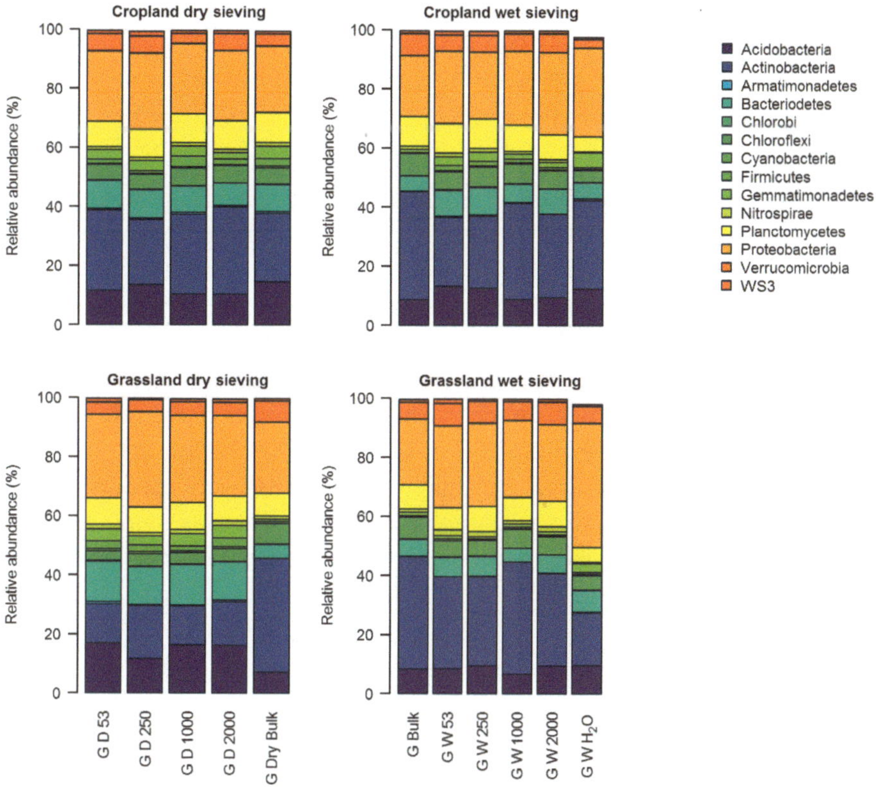

Fig. 6 Relative abundance (%) of bacterial phyla of four soil fractions obtained by dry- or wet-sieving method, bulk soil and water from wet sieving from cropland and grassland. Means values ($n=3$) are shown. Only the dominant phyla ($\sim>0.2\%$) are shown.

increased with wet sieving (Fig. 6; Table S5 (http://dx.doi.org/10.1016/bs.agron.2016.10.006)). The different sizes of soil fractions also affected the relative abundance of most phyla. The differences between sieving methods and soil fractions size were more visible and statistically significant for the grassland than cropland. The differences between cropland and grassland were related to only few of the dominant phyla, with *Chloroflexi*, and *Planctomycetes* that were higher in cropland, while *Nitrospirae*, and *Proteobacteria* were higher in grassland (Fig. 6; Table S5). The water from wet sieving in grassland showed a significant decrease in *Actinobacteria* and *Planctomycetes* in comparison to the soil fractions, while *Proteobacteria* increased.

PCoA were also performed on the archaeal community composition, showing strong differences between the water from wet sieving and the rest of the samples although water samples from grassland grouped with the soil fractions (Figs. S2 and S3 (http://dx.doi.org/10.1016/bs.agron.2016.10.006)).

Then strong differences in archaeal community composition were also found between sieving methods but not between soil fractions.

4. DISCUSSION

The study of the distribution of microbial communities diversity, abundance, and activities between different sizes of soil aggregates size classes started more than 2 decades ago (Chotte et al., 1993; Gupta and Germida, 1988; Jocteur Monrozier et al., 1991; Kanazawa and Filip, 1986; Lensi et al., 1995). The study of microbial distribution in soil aggregates starts from the premise that the vast variation in the size of aggregates, as well as their physicochemical properties, provides a huge diversity of habitats for micro-organisms influencing carbon and nutrients dynamics within the soil. Subsequently, it implies that each soil aggregate size class could harbors specific microbial communities and activities. However, little is known about the effects of size fractionation methods such as sieving on the isolation and interpretation of microbial community data from soil aggregates. The current study clearly shows that dry- or wet-sieving method affects the acquisition and interpretation of microbial data from different soil aggregates. Furthermore, the effects of sieving methods vary with the site/soil studied, and also which component of the microbial community was studied (i.e., diversity vs abundance).

Differences in bacterial community composition between sizes of soil aggregates were only revealed in grassland and only when using wet sieving. Dry-sieving method resulted in high variation between replicates, hindering potential differentiation between sizes. The higher disruption energy introduced by water movement and washing effect provided during wet sieving in comparison to dry sieving are likely to be the main factors explaining such differences in the results obtained by both sieving methods (Cambardella and Elliott, 1993; Chotte et al., 1993). This result implies that the different spatial domains of microbial diversity within soil are distinguished by patterns in the adhesive forces within soil that bind organisms, minerals, and fluids together. This suggests that some factors that are important in the spatial variation in particle binding to form aggregates may be also important as selective pressures to establish differences in microbial diversity. Similar results were found with the potential enzyme activity, with only the wet-sieving method that revealed significant differences between soil aggregate sizes in comparison to two dry-sieving methods (i.e., soil either air dry or dry to 10–15% of soil gravimetric water content) (Bach and Hofmockel, 2014). These results highlight the fact that wet sieving might be a better method over dry sieving to isolate the different microbial communities within each size fraction and is relevant for different microbial characteristics: diversity and activity. The bacterial gene abundances showed overall less clear variation between soil aggregates sizes regardless of the sieving methods, although wet sieving resulted showed more variation in genes abundance between sizes than dry sieving.

The washing of soil aggregates during wet sieving did not result in significant cross contamination between aggregate sizes, at least for grassland where significant differences were found. In contrast, dry sieving and its rubbing effect on the outer part of aggregates may result in stronger cross contamination due to the absence of water carrying the soil particles into the soil fraction $<53\ \mu m$, which represent a patchwork of the different soil fractions, and its mass is directly affected by the disruption strength energy (Chotte et al., 1993). This was supported by the high variation between soil fractions replicates for cropland. Relatively low percentages of bacterial genes, often below 1%, were lost in the wet-sieving water, although this percentage was likely to be underestimated. Interestingly, high *narG* gene percentage and relative abundance of *Proteobacteria* in water from grassland may indicate that this gene and phylum might be located in the outer part of the aggregates or interaggregates space, where the washing effect was high. In contrast, *Actinobacteria* and *Planctomycetes* decreased in relative abundance, indicating a location within aggregates or high adhesion to soil particles. Thus, water from wet sieving might give some indication on the location of some bacterial community.

In cropland no differences in bacterial diversity between aggregate size fractions were found regardless of the sieving method, highlighting that differences between soil aggregates sizes are not always expected but clearly depend on the soil type and land use. Previous studies also showed no difference between microbial community in different size fractions from cropland, likely due to the high turnover of soil aggregates because of anthropogenic activities (e.g., soil tillage, plant harvest) leading to high physical instability of the microenvironment hindering the differentiation of the microhabitats and microbial communities (Blaud et al., 2014). Thus, the absence of difference in microbial diversity between aggregate sizes at a site could potentially be used as an indicator of the instability of the systems and soil health.

Wet sieving extracted higher gene abundance than dry sieving. Wetting dry soil was shown to increase the amount of DNA extracted from soil (Clark and Hirsch, 2008), and a physical effect rather than biological might explain the difference for wet sieving within the ~30 min that the fractionation last. The same trend was found for potential enzyme activity, with fourfold greater activity found with wet sieving in comparison to dry sieving (Bach and Hofmockel, 2014). This can either reflect an overestimation of the measured variable due to wetting effect (i.e., biological), or accessing a hidden microbial community protected within the pores of the aggregates. In contrast, dry sieving could lead to underestimating the microbial gene abundance. Bach and Hofmockel (2014) suggested that wetting the soil leads to overestimation of potential enzyme activity due to the contact between microorganisms and soluble C compounds and the potential short-term microbial metabolic changes. However, there is also a large number of slow growing microorganisms in soil. Most studies showing a rapid response of the microbial community to changes in moisture (<30 min), were done only on a few microbial strains in optimal laboratory conditions far from in situ conditions (Halverson et al., 2000; Lamarre et al., 2008). Nevertheless, the biological effect of wetting on microorganisms cannot be discarded.

Chotte et al. (2002) suggested that studying the microbial community within soil aggregates gives access to changes in microbial community that would not be visible in the bulk soil, and a greater diversity of *Azospirillum*. Most studies assessing microbial community composition within soil aggregates found significant differences with the bulk soil (Blaud et al., 2012; Chotte et al., 2002; Davinic et al., 2012; Ranjard et al., 2000). Despite the bacterial diversity found not to be higher in each soil aggregates size in comparison to the bulk soil, in the study using next-generation sequencing (Davinic et al., 2012) and in the current study, taken all together, the different soil aggregates harbor a greater bacterial diversity than the bulk soil. It remains unclear if pooling the same number of DNA extractions from bulk soil as the number of those from soil fractions plus replicates (e.g., 12 DNA extract in the current study) would lead to an increase in bacterial diversity harvested in the bulk soil. This issue could be partly a methodological constraint, as DNA extraction usually uses an extremely small amount of soil; 0.25 g is commonly used, which reduces the representation of the different soil aggregate sizes within the extraction. The recent study from Penton et al. (2016) showed that higher bacterial diversity was found when 10 g of soil was used, which could be related to higher representation of the different soil aggregates sizes and in general the heterogeneous structure of the soil. Similar issue could also be relevant when studying microbial activity that uses often only 1 g of soil (Bach and Hofmockel, 2014).

Overall, it should not be expected that the bulk soil provides a summary of the different soil fractions when working on small amounts of soil.

Furthermore, isolated soil fractions are likely to behave in a different way to those in situ because of exposure for example to oxygen and high concentration of soil fractions in comparison to their dispersion within a soil horizon. These characteristics could be major limitations when trying to link microbial diversity, abundance, and activity between bulk soil and soil fractions, or to model these variables taking into consideration the soil structure.

5. CONCLUSIONS

Sieving methods clearly affect the resulting observed bacterial diversity and abundance found in soil aggregates, and there is a need to carefully choose the methods used prior to their study. Wet sieving was potentially the most adapted method to study microbial community diversity and abundance in soil aggregates in comparison to dry sieving, although it is time consuming and difficult to perform. Furthermore, studies are needed to assess if wet sieving is the relevant method across a larger number of land use and soil types, and also to assess if is relevant for the measure of other microbial variables (e.g., RNA). Aggregates isolated with sieving methods are the products of sieving and it might be difficult to relate the microbial results to in situ reality. However, aggregates are real units of greater cohesion in the soil formed by biogeochemical processes. Overall, this study raises the question on how to consider soil structure in the study of soil microbial communities, in order to address important question such as the biological mechanisms controlling soil fertility or the stabilization of organic matter in soil.

ACKNOWLEDGMENT

This work is supported by the European Commission 7th Framework Program as a Large Integrating Project, SoilTrEC (www.soiltrec.eu), Grant Agreement No. 244118. We thank Winfried E.H. Blum and Jasmin Schiefer from University of Natural Resources and Life Sciences Vienna (BOKU) for their help with soil sampling in the CZO Marchfeld/Fuchsenbigl.

REFERENCES

Bach, E.M., Hofmockel, K.S., 2014. Soil aggregate isolation method affects measures of intra-aggregate extracellular enzyme activity. Soil Biol. Biochem. 69, 54–62. http://dx.doi.org/10.1016/j.soilbio.2013.10.033.

Beauchamp, E.G., Seech, A.G., 1990. Denitrification with different sizes of soil aggregates obtained from dry-sieving and from sieving with water. Biol. Fertil. Soils 10, 188–193. http://dx.doi.org/10.1007/BF00336134.

Blaud, A., Lerch, T.Z., Chevallier, T., Nunan, N., Chenu, C., Brauman, A., 2012. Dynamics of bacterial communities in relation to soil aggregate formation during the decomposition of 13C-labelled rice straw. Appl. Soil Ecol. 53, 1–9. http://dx.doi.org/10.1016/j.apsoil.2011.11.005.

Blaud, A., Chevallier, T., Virto, I., Pablo, A.-L., Chenu, C., Brauman, A., 2014. Bacterial community structure in soil microaggregates and on particulate organic matter fractions located outside or inside soil macroaggregates. Pedobiologia 57, 191–194. http://dx.doi.org/10.1016/j.pedobi.2014.03.005.

Cambardella, C.A., Elliott, E.T., 1993. Methods for physical separation and characterization of soil organic matter fractions. Geoderma 56, 449–457. http://dx.doi.org/10.1016/0016-7061(93)90126-6.

Caporaso, J.G., Kuczynski, J., Stombaugh, J., Bittinger, K., Bushman, F.D., Costello, E.K., Fierer, N., Peña, A.G., Goodrich, J.K., Gordon, J.I., Huttley, G.A., Kelley, S.T., Knights, D., Koenig, J.E., Ley, R.E., Lozupone, C.A., McDonald, D., Muegge, B.D., Pirrung, M., Reeder, J., Sevinsky, J.R., Turnbaugh, P.J., Walters, W.A., Widmann, J., Yatsunenko, T., Zaneveld, J., Knight, R., 2010. QIIME allows analysis of high-throughput community sequencing data. Nat. Methods 7, 335–336. http://dx.doi.org/10.1038/nmeth.f.303.

Caporaso, J.G., Lauber, C.L., Walters, W.A., Berg-Lyons, D., Lozupone, C.A., Turnbaugh, P.J., Fierer, N., Knight, R., 2011. Global patterns of 16S rRNA diversity at a depth of millions of sequences per sample. Proc. Natl. Acad. Sci. 108, 4516–4522. http://dx.doi.org/10.1073/pnas.1000080107.

Chotte, J.-L., Jocteur Monrozier, L., Villemin, G., Albrecht, A., Mulongoy, K., Merckx, R., 1993. Soil microhabitats and the importance of the fractionation method. Soil Organic Matter Dynamics and Sustainability of Tropical Agriculture. IITA, Ibadan, pp. 39–45.

Chotte, J.L., Schwartzmann, A., Bally, R., Jocteur Monrozier, L., 2002. Changes in bacterial communities and Azospirillum diversity in soil fractions of a tropical soil under 3 or 19 years of natural fallow. Soil Biol. Biochem. 34, 1083–1092. http://dx.doi.org/10.1016/S0038-0717(02)00041-X.

Clark, I.M., Hirsch, P.R., 2008. Survival of bacterial DNA and culturable bacteria in archived soils from the Rothamsted Broadbalk experiment. Soil Biol. Biochem. 40, 1090–1102. http://dx.doi.org/10.1016/j.soilbio.2007.11.021.

Clarke, K.R., Green, R.H., 1988. Statistical design and analysis for a "biological effects" study. Mar. Ecol. Prog. Ser. 46, 213–226.

Davinic, M., Fultz, L.M., Acosta-Martinez, V., Calderón, F.J., Cox, S.B., Dowd, S.E., Allen, V.G., Zak, J.C., Moore-Kucera, J., 2012. Pyrosequencing and mid-infrared spectroscopy reveal distinct aggregate stratification of soil bacterial communities and organic matter composition. Soil Biol. Biochem. 46, 63–72. http://dx.doi.org/10.1016/j.soilbio.2011.11.012.

Edgar, R.C., 2013. UPARSE: highly accurate OTU sequences from microbial amplicon reads. Nat. Methods 10, 996–998. http://dx.doi.org/10.1038/nmeth.2604.

Fall, S., Nazaret, S., Chotte, J.L., Brauman, A., 2004. Bacterial density and community structure associated with aggregate size fractions of soil-feeding termite mounds. Microb. Ecol. 48, 191–199. http://dx.doi.org/10.1007/s00248-003-1047-2.

Gupta, V.V.S.R., Germida, J.J., 1988. Distribution of microbial biomass and its activity in different soil aggregate size classes as affected by cultivation. Soil Biol. Biochem. 20, 777–786. http://dx.doi.org/10.1016/0038-0717(88)90082-X.

Halverson, L.J., Jones, T.M., Firestone, M.K., 2000. Release of intracellular solutes by four soil bacteria exposed to dilution stress. Soil Sci. Soc. Am. J. 64, 1630. http://dx.doi.org/10.2136/sssaj2000.6451630x.

Helgason, B.L., Walley, F.L., Germida, J.J., 2010. No-till soil management increases microbial biomass and alters community profiles in soil aggregates. Appl. Soil Ecol. 46, 390–397. http://dx.doi.org/10.1016/j.apsoil.2010.10.002.

IUSS Working Group WRB, 2006. World Reference Base for Soil Resources 2006. World Soil Resources Rep., 103rd ed. FAO, Rome.

Jocteur Monrozier, L., Ladd, J.N., Fitzpatrick, R.W., Foster, R.C., Rapauch, M., 1991. Components and microbial biomass content of size fractions in soils of contrasting aggregation. Geoderma 50, 37–62. http://dx.doi.org/10.1016/0016-7061(91)90025-O.

Kanazawa, S., Filip, Z., 1986. Distribution of microorganisms, total biomass, and enzyme activities in different particles of brown soil. Microb. Ecol. 12, 205–215. http://dx.doi.org/10.1007/BF02011205.

Kandeler, E., Tscherko, D., Bruce, K.D., Stemmer, M., Hobbs, P.J., Bardgett, R.D., Amelung, W., 2000. Structure and function of the soil microbial community in microhabitats of a heavy metal polluted soil. Biol. Fertil. Soils 32, 390–400. http://dx.doi.org/10.1007/s003740000268.

Kravchenko, A.N., Negassa, W.C., Guber, A.K., Hildebrandt, B., Marsh, T.L., Rivers, M.L., 2014. Intra-aggregate pore structure influences phylogenetic composition of bacterial community in macroaggregates. Soil Sci. Soc. Am. J. 78, 1924. http://dx.doi.org/10.2136/sssaj2014.07.0308.

Lair, G.J., Zehetner, F., Hrachowitz, M., Franz, N., Maringer, F.-J., Gerzabek, M.H., 2009. Dating of soil layers in a young floodplain using iron oxide crystallinity. Quat. Geochronol. 4, 260–266. http://dx.doi.org/10.1016/j.quageo.2008.11.003.

Lamarre, C., Sokol, S., Debeaupuis, J.-P., Henry, C., Lacroix, C., Glaser, P., Coppée, J.-Y., François, J.-M., Latgé, J.-P., 2008. Transcriptomic analysis of the exit from dormancy of Aspergillus fumigatus conidia. BMC Genomics 9, 417. http://dx.doi.org/10.1186/1471-2164-9-417.

Lensi, R., Clays-Josser, A., Jocteur Monrozier, L., 1995. Denitrifiers and denitrifying activity in size fractions of a mollisol under permanent pasture and continuous cultivation. Soil Biol. Biochem. 27, 61–69. http://dx.doi.org/10.1016/0038-0717(94)00132-K.

Manter, D.K., Vivanco, J.M., 2007. Use of the ITS primers, ITS1F and ITS4, to characterize fungal abundance and diversity in mixed-template samples by qPCR and length heterogeneity analysis. J. Microbiol. Methods 71, 7–14. http://dx.doi.org/10.1016/j.mimet.2007.06.016.

McMurdie, P.J., Holmes, S., 2013. Phyloseq: an R package for reproducible interactive analysis and graphics of microbiome census data. PLoS One 8, e61217. http://dx.doi.org/10.1371/journal.pone.0061217.

Mendes, I.C., Bandick, A.K., Dick, R.P., Bottomley, P.J., 1999. Microbial biomass and activities in soil aggregates affected by winter cover crops. Soil Sci. Soc. Am. J. 63, 873. http://dx.doi.org/10.2136/sssaj1999.634873x.

Penton, C.R., Vadakattu, V.S.R.G., Yu, J., Tiedje, J.M., 2016. Size matters: assessing optimum soil sample size for fungal and bacterial community structure analyses using high throughput sequencing of rRNA gene amplicons. Front. Microbiol. 7, 824. http://dx.doi.org/10.3389/fmicb.2016.00824.

Pylro, V.S., Roesch, L.F.W., Morais, D.K., Clark, I.M., Hirsch, P.R., Tótola, M.R., 2014. 3Data analysis for 16S microbial profiling from different benchtop sequencing platforms. J. Microbiol. Methods 107, 30–37. http://dx.doi.org/10.1016/j.mimet.2014.08.018.

R Development Core Team, 2015. R: A Language and Environment for Statistical Computing.

Ranjard, L., Nazaret, S., Gourbiere, F., Thioulouse, J., Linet, P., Richaume, A., 2000. A soil microscale study to reveal the heterogeneity of Hg(II) impact on indigenous bacteria by quantification of adapted phenotypes and analysis of community DNA fingerprints. FEMS Microbiol. Ecol. 31, 107–115. http://dx.doi.org/10.1111/j.1574-6941.2000.tb00676.x.

Ririe, K.M., Rasmussen, R.P., Wittwer, C.T., 1997. Product differentiation by analysis of DNA melting curves during the polymerase chain reaction. Anal. Biochem. 245, 154–160. http://dx.doi.org/10.1006/abio.1996.9916.

Sainju, U.M., 2006. Carbon and nitrogen pools in aoil aggregates seperated by dry and wet sieving methods. Soil Sci. 171, 937–949. http://dx.doi.org/10.1097/01.ss0000228062.30958.5a.

Sainju, U.M., Terrill, T.H., Gelaye, S., Singh, B.P., 2003. Soil aggregation and carbon and nitrogen pools under Rhizoma peanut and perennial weeds. Soil Sci. Soc. Am. J. 67, 146–155.

Schutter, M.E., Dick, R.P., 2002. Microbial community profiles and activities among aggregates of winter fallow and cover-cropped soil. Soil Sci. Soc. Am. J. 66, 142–153.

Seech, A.G., Beauchamp, E.G., 1988. Denitrification in soil aggregates of different sizes. Soil Sci. Soc. Am. J. 52, 1616–1621.

Sessitsch, A., Weilharter, A., Gerzabek, M.H., Kirchmann, H., Kandeler, E., 2001. Microbial population structures in soil particle size fractions of a long-term fertilizer field experiment. Appl. Environ. Microbiol. 67, 4215–4224. http://dx.doi.org/10.1128/AEM.67.9.4215-4224.2001.

Sey, B.K., Manceur, A.M., Whalen, J.K., Gregorich, E.G., Rochette, P., 2008. Small-scale heterogeneity in carbon dioxide, nitrous oxide and methane production from aggregates of a cultivated sandy-loam soil. Soil Biol. Biochem. 40, 2468–2473. http://dx.doi.org/10.1016/j.soilbio.2008.05.012.

Tsiknia, M., Tzanakakis, V.A., Paranychianakis, N.V., 2013. Insights on the role of vegetation on nitrogen cycling in effluent irrigated lands. Appl. Soil Ecol. 64, 104–111. http://dx.doi.org/10.1016/j.apsoil.2012.10.010.

Vaisanen, R.K., Roberts, M.S., Garl, J.L., Frey, S.D., Dawson, L.A., 2005. Physiological and molecular characterisation of microbial communities associated with different water-stable aggregate size classes. Soil Biol. Biochem. 37, 2007–2016. http://dx.doi.org/10.1016/j.soilbio.2005.02.037.

Yoder, R.E., 1936. A direct method of aggregate analysis of soils and a study of the physical nature of erosion losses. J. Am. Soc. Agron. 28, 337–351.

CHAPTER SIX

Soil Mineralogy Changes With Different Agricultural Practices During 8-Year Soil Development From the Parent Material of a Mollisol

Y.-L. Liu[*,†,1], S.-H. Yao[‡,1], X.-Z. Han[†,2], B. Zhang[†,‡,2], S.A. Banwart[§]

[*]National Engineering Laboratory for Efficient Utilization of Soil and Fertilizer, College of Resources and Environment, Shandong Agricultural University, Taian, PR China
[†]National Observation Station of Hailun Agro-ecology System, Key Laboratory of Mollisols Agroecology, Northeast Institute of Geography and Agroecology, Chinese Academy of Sciences, Harbin, PR China
[‡]National Engineering Laboratory for Improving Fertility of Arable Soils, Institute of Agricultural Resources and Regional Planning, Chinese Academy of Agricultural Sciences, Beijing, China
[§]Integrating Chair in Soil/Agriculture/Water Research, School of Earth and Environment, University of Leeds, Leeds, United Kingdom
[2]Corresponding authors: e-mail address: xzhan@iga.ac.cn; zhangbin01@caas.cn

Contents

[1] These authors contributed equally to this work.

Advances in Agronomy, Volume 142
ISSN 0065-2113
http://dx.doi.org/10.1016/bs.agron.2016.10.015

Abstract

Clay minerals are involved in the formation of soil structure and soil cation exchange capacity through interaction with soil organic matter. Through these interactions, clay minerals contribute to soil functions and the delivery of soil ecosystem services. Although clay minerals are considered stable over long periods of time, recent studies have demonstrated that the crystallographic structures and types of clay minerals can undergo spontaneous modification and transformation with changes in environmental conditions. However, little is known whether and how soil organic matter affects the evolution of clay minerals during these transitions. We examined changes in soil mineralogy during the 8 years of soil development from the parent material (PM) of a Mollisol under different agricultural practices in Northeastern China. Mineralogical changes were determined by X-ray diffraction and data analysis through spectra decomposition. The mineral composition varied with initial particle size classes and showed rapid changes, including transformation between illite and vermiculite or illite and smectite, depending on K^+ availability and transformations between smectite and vermiculite under acidic conditions. The mineral composition changed with soil depth due to transport of vermiculite and well-crystallized illite as fine particles. Soil organic matter was more strongly bound to vermiculite and smectite than to kaolinite and illite and therefore influenced the transport of these mineral particles. However, the soil organic matter–mineral association did not affect transformation between illite and vermiculite or smectite. Therefore, the effects of particle size, organic matter, and soil depth on soil mineral composition were exhibited to different extents and interacted differently under different agricultural practices. These results of the 8-year field experiment suggest that soil mineral composition of the more intensely weathered PM of the studied Mollisols can change after agricultural restoration more quickly than previously known. The rate and extent of these transformations must be considered in agricultural practices that are used to manage soil functions and associated ecosystem services of soil systems.

1. INTRODUCTION

Soils are at the center of Earth's Critical Zone, which is defined as the surface layer from treetop to bedrock and which plays a central role in providing life-sustaining ecosystem functions and services (Menon et al., 2014). Soil formation is classically considered as a slow process that occurs under natural conditions. Soil formation is generally considered to take millennia to develop soil functions to a state where they provide essential ecosystem services for human benefits (Jorgensen and Fath, 2008). Once soils are degraded, it is difficult to restore soil functions to their original productive state (Chazdon, 2008).

Human pressures from burgeoning population, land use change, and increasing intensity of human intervention all raise significant concerns about whether soil formation and the delivery of soil ecosystem services can continue indefinitely (Haygartha and Ritz, 2009; Wilkinson and Humphreys, 2005). Therefore, understanding the life cycle of soil functions starting from initial soil formation to soil degradation and potential loss through physical erosion is crucial to develop strategies for restoration, sustainable use practices, and protection of soil functions. Despite the urgency of these challenges, there have been remarkably few studies that examine soil formation and soil restoration compared to those that examine soil degradation. Moreover, compared with successful restoration of above-ground ecosystems, our understanding of soil formation and development during vegetation restoration is still very poor (Griffiths et al., 2008; Harris, 2009). Some studies have demonstrated that soil properties and fertility can change rapidly under agricultural practices in contrast to natural condition (Knops and Tilman, 2000; Li et al., 2014, 2016). It is still unclear whether and how agricultural practices as restoration techniques per se can accelerate soil formation and the associated development of soil functions.

Clay minerals with the characteristics of small physical particle sizes and resulting high-specific mineral surface area are key components in soil ecosystems. They are involved in the formation of soil structure and soil cation exchange capacity (CEC) through interaction with soil organic matter. Therefore, soil minerals are fundamental in determining soil functions and ecosystem services (Cébron et al., 2015; Owliaie et al., 2006; Velde and Meunier, 2008). All clay minerals contribute to soil processes, but each type of clay mineral does not contribute to the same extent (e.g., Barré et al., 2007a; Denef and Six, 2005; Wattel-Koekkoek et al., 2001). Formation of clay minerals with different particle sizes results from the interaction between bedrock, soil biota, and rain water, the intensity of which can be modified by climate, topography, time, and vegetation (Birkeland, 1999; Cornu et al., 2008; Mavris et al., 2011). It has long been considered that clay minerals in soils are stable over century to millennium time scales. However, recent advances in X-ray diffraction (XRD) in combination with the studies of soil chronosequences (Ruffell and Wiltshire, 2004; Viennet et al., 2015), controlled experiments (Ramirez-Perez et al., 2013; Sayen and Guillon, 2014), and field experiments (Beck et al., 2014; Manhaes et al., 2002) have demonstrated that the crystallographic structures and the types of clay minerals can undergo spontaneous modification and transformation with changes in environmental conditions. Therefore, clay mineral particle sizes may change not only due to physical weathering but also due to mineral chemical transformations.

Isomorphic alteration of the interlayer cations, dissolution and precipitation, structural transformation, and transport of clay mineral particles can modify clay mineralogy within soil profiles (Barré and Velde, 2010; Cornu et al., 2012). Some 2:1 clay minerals such as smectite or vermiculite can absorb anhydrous potassium ion (K^+) in a K^+-rich environment, forming illite-like layers that collapse to 1 nm interlayer spacing, and release K^+ in a K^+-poor environment through exchange with hydrated cations during plant uptake, resulting in reopening the interlayers of 2:1 smectite or vermiculite. These phenomena of adsorption and release of K^+ can occur reversibly without any modification of the clay sheet structure as indicated by the effects of K^+ and K^+-free fertilization during crop growth (Barré et al., 2007a,b, 2008; Liu et al., 2008; Paola et al., 2016; Velde and Barré, 2010). In other cases, chemical weathering of chlorite or illite can produce smectite due to substitution of interlayer calcium (Ca^{2+}) or magnesium (Mg^{2+}) by aluminum (Al^{3+}) released in acidic conditions (pH 4–5). The soil acidification will occur due to the production of organic acids in the rhizosphere (Drever and Stillings, 1997), the release of hydrogen ions to maintain soil–water charge balance by vegetation during uptake of base cations for biomass production (Caner et al., 2010), the decomposition of plant litter (Zanellia et al., 2007), the presence of mycorrhizal fungi (Calvaruso et al., 2009; Caner et al., 2010; Mavris et al., 2011; Paola et al., 2016), or the application of ammonium sulfate (Tye et al., 2009). Soil clay minerals may change with soil depth due to selective transport of minerals in colloidal size fractions, such as Al- and Fe oxides, and clay particles, such as mixed-layer illite and smectite (Cornu et al., 2012; Frison et al., 2009). This selective transport may leave a greater mass fraction of larger particle-sized minerals such as those of illite and chlorite in the surface soil.

Agricultural practices can substantially influence clay mineralogy. These influences are through fertilization effects on K^+ availability, ion-exchange, and acidification, soil tillage that may physically disrupt aggregates, potentially increasing the mobility of clay minerals. Agricultural practices can also substantially influence soil organic matter. It has long been recognized that organic–mineral association is affected by clay mineral surfaces and can induce changes in mineral properties. However, information on the effects of organic–mineral association on the composition of clay minerals is very limited. There is greater complexation capacity for soil organic matter with smectite and vermiculite that have smaller size and larger specific surface area than other minerals (Balabane and Plante, 2004; Wagai and Mayer, 2007). When soil organic carbon (SOC) concentration is low, organic association with smectite and vermiculite surface reduces the available specific surface area of the minerals due to preferential association at micropores (Kaiser and Gugenberger, 2003) and more favorable surface metal-binding sites. Meanwhile, organic coatings on mineral surfaces and refilling at the microsites on the surfaces or in interlayers may affect mineral transformation, e.g., through K^+ adsorption/release (Claret et al., 2004; Dümig et al., 2012; Wagai et al., 2009; Zhang et al., 2013) and ion substitutions (Szymański et al., 2014; Velde et al., 2003). Furthermore, organic–mineral association and mineral particle aggregation may retard mineral transport and therefore can influence mineral composition at different soil depths. The effects of agricultural practices on soil mineralogy can be particularly significant at the early stages of pedogenesis.

In previous studies, we have demonstrated that different agricultural practices changed the soil microbial community composition (Li et al., 2014, 2015) and soil food web (Li et al., 2016) during the 8-year development from the parent material (PM) of a Mollisol (MO). Using this field experiment, here we aimed to (1) determine the effects of clay particle size, organic carbon, and soil depth on soil clay mineral composition; (2) identify the shifts in the soil mineral composition under different agricultural practices compared to the PM and an arable MO; and (3) elucidate the main soil processes that control the shifts in soil mineral composition.

2. MATERIAL AND METHODS

2.1 Study Site and Experimental Design

The study site is located at the State Key Experimental Station of Agroecology, Chinese Academy of Sciences, Hailun, Heilongjiang Province (47°26′N, 126°38′E), and the field experiment started from June, 2004. The region has a typical temperate continental monsoon climate, with the mean annual temperature of 2.2°C, the highest monthly temperature (35°C) in July, and the lowest in January (−38°C). The mean annual precipitation is 550 mm, with 65% occurring from June to August. The soil is classified as Pachic Haploborolls according to the USDA Soil Taxonomy (Survey Staff, 2010). The PM is clay loess-like material (Xiong and Li, 1987), containing texture class mass distributions of 42% clay, 35.6% silt, and 22.4% sand.

The field experiment was established to determine how soil ecosystems and soil functions develop under different agricultural practices. The basic information of the field experiment was given in Li et al. (2014, 2015).

Briefly, PM was taken from the C-horizon (>2 m) from an adjacent field and was used to replace the surface soil down to 0.8-m depth. The refilled surface layer was separated from the soil layer below using nylon nets with <75-μm mesh. The soil bulk density in the refilled plots was approximately 1.3 mg m^{-3}. The field experiment was laid out by the randomized blocks design of six treatments with four replicates. The field treatments were (1) Natural fallow (NatF) dominated with *Poa annua* L., *Equisetum arvense* L., and *Spodiopogon sibiricus* Trin; (2) leguminous alfalfa (*Medicago sativa* L.) (Alfa); (3) an annual rotation system between soybean (*Glycine max* (Merrill.) L.) and maize (*Zea mays* L.) without chemical fertilization and without the above-ground biomass incorporation (F0C0); (4) the same treatment but with chemical fertilization and without biomass incorporation (F1C0); (5) the same treatment but with chemical fertilization and the above-ground biomass partially incorporated (F1C1); and (6) totally incorporated (F1C2) into soil. The amounts of mineral fertilizers and organic carbon input are presented in Table 1. Soybean and maize were planted in alternative years. The first crop was soybean in 2004, resulting in 2012 with maize as the crop before the sampling. The crops were generally sown by row in May and harvested in October. During the time of crop harvest, the above-ground biomass was cut and weighed for each plot to determine the dry mass. The chopped above-ground biomass (0.2–0.5 m in length) in NatF and Alfa was mulched onto the ground surface. Tillage was carried out to 0.2-m depth using a spade to incorporate the chopped crop residue into soil and ridges (0.75 m wide and 0.25 m high) after the organic amendment.

2.2 Soil Sampling and Analysis

Soil samples were collected by soil depth from three replicate plots before sowing in May, 2012. Another block remained intact for additional observations. Four soil cores (0.01 m^3) were taken to measure soil bulk density. Thereafter, all the soil in each plot was excavated by soil depth and mixed thoroughly for homogenous composite samples. After the sampling, the soils were refilled to their original depths. The sampling depths were different in the no-tilled perennials plots and for the tilled soils under cropping because more roots were found below the surface soil (0–0.2 mm). The sampling depths were 0–0.2, 0.2–0.3, and 0.3–0.4 m. Archived PM taken from >2 m depth and an arable MO taken from the surface soil of an adjacent experimental field were used as controls to characterize the reference soil state. The mature MO was managed under the same cropping system as

Table 1 Land Use, Crop Residue, Inputs of Organic C, and Mineral Fertilizers in Different Field Treatments

Field Treatments	Land Use[a]	Crop Residue[b]	Tillage[c]	Organic C Input[d] (Mg ha^{-1})	Mineral Fertilization (kg ha^{-1})[e]		
					N	P	K
NatF	Natural fallow	Shoot and roots	No	0.95 d	0		0
Alfa	Alfalfa	Shoot and roots	No	1.17 c	0		0
F0C0	Cropping	Roots only	Yes	0.86 d	0		0
F1C0	Cropping	Roots only	Yes	1.31 c	54[f]/192[g]	56	50
F1C1	Cropping	Roots, baked soybean, and maize straw	Yes	2.84 b	54[f]/192[g]	56	50
F1C2	Cropping	All crop shoot, seeds, and roots	Yes	3.70 a	54[f]/192[g]	56	50

[a]Crop rotation is in a sequence of soybean and maize in different years since 2004. The crop densities were 70,000 and 270,000 ha^{-1} for maize and soybean, respectively.
[b]The amendment rates in F1C1 were 2250 kg ha^{-1}of baked soybean powder and 4500 kg ha^{-1}of maize straw mixed homogeneously.
[c]Tillage was performed manually down to 200 mm with ridges made in October after harvest.
[d]The amount of organic C input were estimated based on the sources of organic incorporation. The average C concentrations were 428, 443, 442, and 431 g kg^{-1} in above-ground biomass and root of the grasses, alfalfa, maize, and soybean, respectively. The root: shoot ratios were 2.0 and 0.4 in the grasses and alfalfa, and the root to yield ratio was 0.19 and 0.13 in maize and soybean, respectively.
[e]Diammoniumphosphate and potassium sulfate were applied as source of N, P, and K, respectively.
[f]Is for soybean.
[g]Is for maize.
Cited from Li, N., Pan, F.J., Han, X.Z., Zhang, B., 2016. Development of soil food web of microbes and nematodes under different agricultural practices during the early stage of pedogenesis of a Mollisol. Soil Biol. Biochem. 98, 208–216.

did in the field treatments, but receiving only mineral N and P fertilizers since 1993. All above-ground biomass was removed from the field of the reference MO and soil tillage was also incorporated to 0.2-m depth using a spade.

All the composite soil samples were divided into three parts. One part was ground to pass a 2-mm sieve for measuring soil texture and soil pH. Another part was ground to pass a 0.25-mm sieve for measuring CEC and exchangeable K^+ (EK). The third part was ground to pass a 0.15-mm sieve for measuring SOC and the elements of Si, Al, and K. Soil texture was measured using a laser particle size analyzer (Malvern Mastersizer 2000, Worcester, UK). Soil pH was measured in a 1:2.5 mass ratio of soil to water suspension with a combination reference glass electrode. CEC was determined using the buffered ammonium acetate method (Rhoades, 1982). Exchangeable potassium was determined by extraction with 0.1 M $BaCl_2$ for 2 h. SOC was measured by the $K_2Cr_2O_7$ oxidation method (Nelson and Sommers, 1996). Oxalate-, Na-pyrophosphate-, and dithionite-extractable fractions were measured for the elements Si and Al in <0.25 mm soil samples according to McKeague et al. (1971), and the concentration of Si and Al was determined using ICP-MS (VG PQ Excell; Thermo Elemental, Winsford, UK) (Zhang et al., 2016).

2.3 Soil Clay Mineralogy Analysis and Quantification

After removing carbonates by hydrochloric acid, soil samples were dispersed in deionized water at pH 7.3 adjusted by 0.5 M NaOH and agitated by ultrasonic energy. The <2- and 2- to 5-μm clay size fractions were extracted by sedimentation according to Stokes' Law. The clay fractions were repeatedly treated with 30% H_2O_2 until completion of the reaction to remove soil organic matter by oxidation. After the H_2O_2 oxidization treatment, the soil samples were rinsed with distilled water to remove remaining H_2O_2 and dried at 60°C. After the H_2O_2 treatments, the content of SOC decreased from 4.8–29.4 to 1.7–3.2 g kg^{-1}, accounting for 65–90% of its initial values (data not shown). The H_2O_2 treatment has been applied elsewhere, showing no mineral structural changes, but changes in the expansibility of illite–smectite minerals (Frost et al., 1998; Velde, 2001; Velde and Peck, 2002) and surface characteristics (Wissing et al., 2014). In this study, the H_2O_2 treatment was done in order to subsequently determine the effect of organic matter on clay mineralogy compared with those without the H_2O_2 treatment, and how mineral and organic carbon are associated and then influence the abundance, transform, and transportability of minerals.

Soil clay mineralogy was determined by XRD (Dümig et al., 2012; Moore and Reynolds, 1997) with quantitative analysis of mineralogical composition determined by numerical decomposition of the XRD spectra (Lanson, 1997). First, the soil samples were treated by dithionite–citrate–bicarbonate to remove extractable $Fe_2O_{3(s)}$ and treated with 1 mol L^{-1} KCl and 0.5 mol L^{-1} $MgCl_2$ solutions to achieve K^+ and Mg^{2+} cation exchange saturation (Huang et al., 2011). After the addition of the solutions, the soils were thoroughly shaken, centrifuged for 5 min, and rinsed with deionized water, and these procedures were repeated five times. The samples with Mg^{2+} saturation treatment were further saturated with glycerol solution. After the treatments, the <2- and 2- to 5-μm particle size classes were mounted as oriented subsamples on glass slides. The glass slides with K^+ saturation treatment were step scanned from 2 to 30 degree 2θ, with stepintervals of 0.02 degree 2θ at 3 s using a D/max-III C (Cu–Kα) diffractometer. After the initial XRD analysis at 25°C, the oriented glass slides were further scanned with the diffractometer after sample treatment with sequential heating for 2 h at 300°C and 550°C using the samples with K^+ saturation treatment. Thereafter, the XRD spectra were decomposed after subtracting the background and then smoothing the patterns on a seven-point interval. Convergence of the fitting routine to 98% agreement was used as the criterion for identifying the peak position and intensity. The relative surface areas under the peak for each identified clay mineral were calculated by multiplying the peak intensity and full-width at half-maximum intensity and expressing the value as a percentage of the total sum of all calculated areas.

2.4 Statistical Analysis

Two-way analysis of variance (ANOVA) was performed to determine the effect of field treatment and soil layer and their interaction on SOC content, EK, Al, Si, CEC, soil pH, and mechanical composition. Multiple comparisons were used to determine the significant differences among the field treatments by least significant difference at $P < 0.05$, unless otherwise noted. Principal component analysis was performed to interpret the contributions of field treatment and soil layer properties on mineral composition, with ANOVA applied to principal component scores to assess the significance of effects. Statistics were conducted with the SPSS Statistics software for Windows.

3. RESULTS

3.1 Soil Physiochemical Properties

The amount of organic carbon input ranged from 861 to 3696 kg ha^{-1} year^{-1} among the field treatments (Table 1). After 8 years, the selected soil physiochemical properties at different soil depths varied among the field treatments and in comparison to the reference soils (Table 2). At the 0–0.2 m depth, the soils from field treatments contained more SOC and exhibited soil texture with a greater sand size fraction. The bulk soils also exhibited less Al and clay contents, and had lower CEC and soil pH than for the PM. Meanwhile, the field treatments contained less SOC and Al, more silt, and sand size fractions and had higher soil pH than the arable MO. In addition, the field treatments had lower CEC and greater EK than MO. There were no differences in Si content among the surface soils except for the treatment F0C0, which contained less Si than other soils, including the reference soils PM and MO. Among the field treatments, most of the soil properties varied down to 0.4-m depth, and the variations were larger at the 0–0.2 m depth than at the deeper depths. Among the field treatments, SOC content at the 0–0.2 m depth was highest in F1C1 and F1C2, followed by NatF and Alfa, and lowest in F0C0 and F1C0; the Al and Si contents were lower in F0C0 than in other soils; the EK content and CEC were greater in F1C1 compared to F1C2 and to the other field treatments; soil pH was higher in the treatments without mineral fertilizers (NatF, Alfa, and F0C0) than in those with mineral fertilizes (F1C0, F1C1, F1C2); the clay and silt contents tended to be lower in the arable soils than in the soils under perennial crops. At the 0.2–0.3 m depth, the EK content was larger in F1C1 than in F1C2, and the CEC was larger in the soils under perennial crops than in the arable soils. Soil pH was higher at the 0.2–0.4 m depths in the soils under perennial crops (NatF and Alfa) compared to the arable soils. With increasing soil depth, the contents of SOC and EK decreased in all field treatments. In addition, the following soil properties all increased with depth in the following field treatments: CEC, pH, and the contents of Al, Si, clay, and silt in F0C0' soil pH, and the contents of Si and silt in F1C0; soil pH, CEC, and the contents of Al and silt in F1C1; and CEC and soil pH in F1C2. The EK content below the 0.2-m depth in Alfa was lower than in the other field treatments.

Table 2 Selected Physico-chemical Properties of Soils Sampled at Different Soil Depths Under Different Field Treatments (Abbreviation Seen in Table 1) and the Reference Soils (parent material, *PM;* matured Mollisol, *MO*)

Soil Properties	Soil Depths	PM	MO	NatF	Alfa	F0C0	F1C0	F1C1	F1C2
SOC (g kg^{-1})	0–0.2	5.1 f	29.4 a	7.6 Ade	9.0 Acd	7.0 Ae	7.9 Ade	10.8 Abc	12.3 Ab
	0.2–0.3			4.8 Bd	5.0 Bcd	5.1 Bc	5.6 Bb	6.3 Ba	6.2 Ba
	0.3–0.4			4.8 Bb	5.0 Bab	4.8 Bb	4.8 Bb	5.2 Ba	4.9 Bb
Al (g kg^{-1})	0–0.2	77.28 a	76.9 a	68.9 Aab	69.7 Aab	59.1 Bc	72.9 Aab	64.8 Bbc	72.7 Aab
	0.2–0.3			69.0 Ab	71.4 Aab	68.7 Ab	73.8 Aab	70.5 ABab	76.7 Aa
	0.3–0.4			72.9 Ab	70.5 Ab	69.9 Ab	77.5 Aa	71.7 Ab	76.1 Aa
Si (g kg^{-1})	0–0.2	171.5 ab	195.3 a	195.4 Aa	143.9 Aab	116.3 Bb	175.0 Ba	137.1 Aab	180.8 Aa
	0.2–0.3			134.7 Bb	159.6 Aab	164.4 Aab	191.1 ABa	154.1 Aab	205.0 Aa
	0.3–0.4			161.3 ABb	159.2 Aab	163.8 Aab	218.5 Aa	153.3 Aab	203.9 Aa
EK (mg kg^{-1})	0–0.2	114.6 d	136 cd	188.3 Ab	176.5 Abc	173.7 Abc	166.2 Abc	270.2 Aa	268.4 Aa
	0.2–0.3			113.5 Bbc	100.8 Bc	127.3 Bb	130.5 ABab	153.2 Ba	126.8 Bb
	0.3–0.4			112.1 Bb	105.0 Bb	114.7 Bb	117.1 Bb	131.3 Ba	136.0 Ba
CEC (cmol (+) kg^{-1})	0–0.2	23.3 ab	25.3 a	22.6 Ab	22.1 Abc	21.6 Bbc	22.2 Abc	20.0 Cc	20.1 Cc
	0.2–0.3			22.1 Aa	21.8 Aa	21.5 Bab	20.9 Ab	20.7 Bb	21.2 Bab
	0.3–0.4			22.5 Aab	22.8 Aa	22.3 Aab	21.9 Aab	21.4 Ab	22.6 Aa

Continued

Table 2 Selected Physico-chemical Properties of Soils Sampled at Different Soil Depths Under Different Field Treatments (Abbreviation Seen in Table 1) and the Reference Soils (parent material, *PM;* matured Mollisol, *MO*)—cont'd

Soil Properties	Soil Depths	PM	MO	NatF	Alfa	F0C0	F1C0	F1C1	F1C2
pH (water)	0–0.2	6.9 a	5.4 f	6.7 Aab	6.4 Ac	6.5 Bbc	6.0 Cd	5.7 Bde	5.6 Bef
	0.2–0.3			6.8 Aa	6.5 Ab	6.8 Aa	6.3 Bb	6.1 ABb	6.2 Ab
	0.3–0.4			6.8 Aa	6.4 Ab	6.7 Aab	6.5 Aab	6.5 Aab	6.4 Ab
Clay (<2 μm) ($g\ kg^{-1}$)	0–0.2	421 a	386 b	399 Aab	395 Ab	396 Bb	403 Aab	383 Ab	386 Ab
	0.2–0.3			401 Aa	394 Aab	403 Ba	386 Ab	391 Aab	385 Ab
	0.3–0.4			401 Aa	400 Aa	416 Aa	406 Aa	399 Aa	401 Aa
Silt (2–20 μm) ($g\ kg^{-1}$)	0–0.2	356 a	299 c	358 ABa	353 Bab	350 Bab	351 Bab	344 Bb	341 Ab
	0.2–0.3			369 Aa	367 Aab	368 Aab	355 Bbc	352 Bc	346 Ac
	0.3–0.4			342 Bb	368 Aa	368 Aa	372 Aa	368 Aa	361 Aa
Sand ((20–2000 μm) ($g\ kg^{-1}$)	0–0.2	224 c	315 a	243 Abc	252 Abc	255 Abc	246 Abc	274 Ab	273 Ab
	0.2–0.3			230 Ac	238 Bbc	230 Bc	259 Aab	257 ABab	269 Aa
	0.3–0.4			257 Aa	232 Bab	216 Ba	222 Aab	233 Bab	238 Aab

Same capital letters indicate no significant difference among soil depths; same lowercase letters indicate no significant difference among field treatments at $P < 0.05$. *CEC*, cation exchange capacity; *EK*, exchange capacity; *SOC*, soil organic carbon.

3.2 Identification of Clay Minerals

In all soil samples with Mg^{2+}–glycerol saturation, relatively sharp and clear peaks were identified at 1.4, 1.0, 0.7, 0.5, 0.33, and 0.35 nm for the 2- to 5-μm particle size class and 1.8, 1.4, 1.0, 0.7, 0.5, 0.33, and 0.35 nm for the <2-μm particle size class. This pattern was observed both with (+) H_2O_2 and without (−) H_2O_2 treatment as exemplified using the soil of NatF (Fig. 1A and B). Smectite and vermiculite were identified at 1.8 and 1.4 nm peak, respectively, in the Mg^{2+}–glycerol saturation sample, both of which collapsed to the 1.0 nm peak following K^+ saturation treatment. The peaks at 1.0, 0.5, and 0.33 nm remained unchanged when the sample was subjected to heating treatment at various temperatures (25°C, 300°C, and 550°C) following K^+ saturation treatment, indicating the presence of illite. The sharp peak of quartz was at 0.33 nm in the 2- to 5-μm particle size class. The peaks at 0.7 and 0.35 nm were identified as kaolinite because the peak appeared in the K^+ saturation and 300°C heating treatments. Asymmetrical peaks, with a tailing in the low-angle region for each mineral indicated the possible presence of interstratification or poor crystallization (Lanson, 1997). To further elucidate the clay mineral composition, the Mg^{2+}–glycerol saturation sample of clay particles was analyzed by numerical decomposition of the recorded spectra, indicating the presence of vermiculite, poorly and well-crystallized illite, and poorly crystallized kaolinite in both the 2- to 5- and <2-μm particle size class. Results for these sample treatments also showed the presence of mixed-layer minerals and the absence of smectite in the 2- to 5-μm particle size class and vice versa in the <2-μm particle size class (Fig. 1C–F). After the oxidization with H_2O_2 treatment, the types of clay mineral composition did not change, but the intensities of the peaks were reduced at 0.71 nm for kaolinite and at 1.0 nm for illite, and were increased at 1.42 nm for vermiculite (Fig. 1A and B). In addition, the peak was increased at 1.8 nm for smectite in the Mg^{2+}–glycerol saturation treatment.

3.3 Clay Mineral Types and Content

The clay mineral types and their contents were affected by field treatment, soil depth, H_2O_2 treatment, and clay particle size (Figs. 2 and 3). In the 2- to 5-μm particle size class (Fig. 2), the dominant clay mineral was vermiculite (48–68%), followed by total illite (19–50%) and then mixed-layer mineral (0–20%) contents in the soil samples except for PM. PM contained only 16% vermiculite. In the <2-μm particle size class, the dominant clay mineral

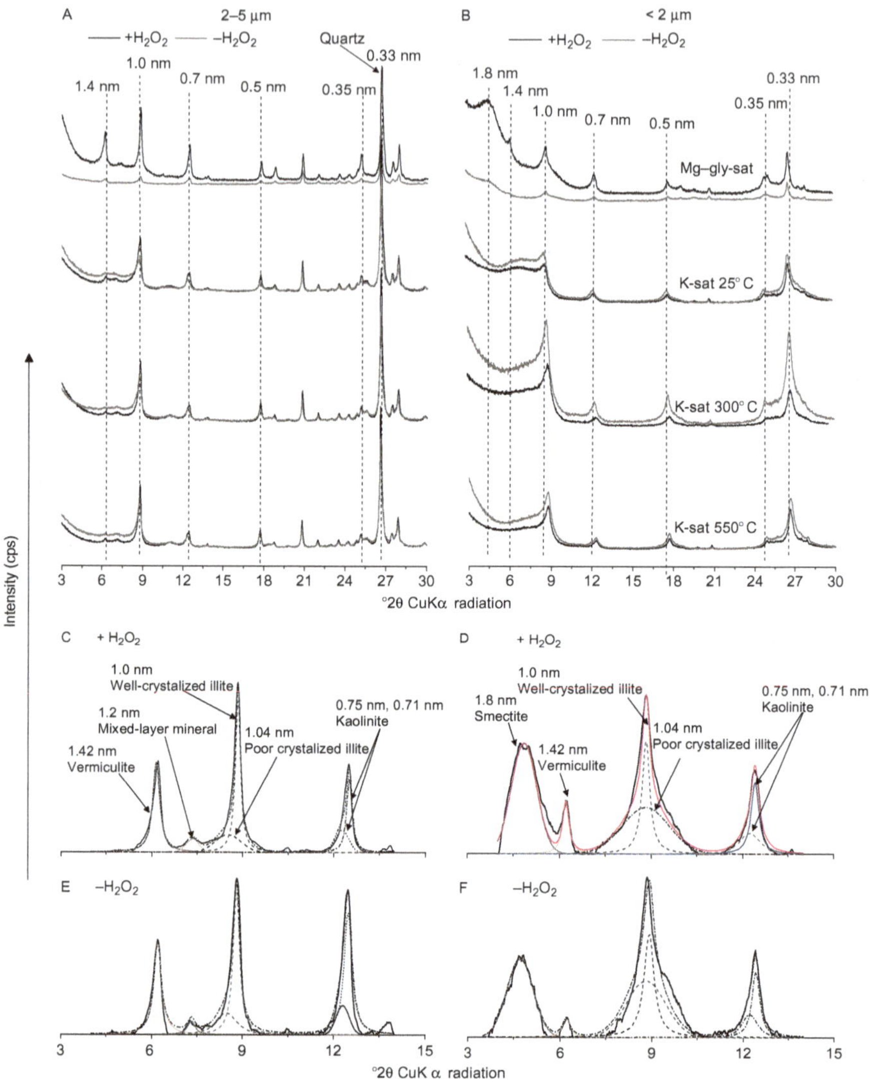

Fig. 1 Original X-ray diffraction patterns obtained using an oriented specimen slide under Mg–glycerol saturation (Mg-gly-sat), K saturation (K-sat) followed sequential heating at 25°C, 300°C, and 550°C in an exemplified surface soil (NatF) with (+) and without (−) H_2O_2 treatment in the 2- to 5-μm particle size class (A), the <2-μm particle size class (B), the corresponding decomposed X-ray patterns of Mg–glycerol saturation with (+) H_2O_2 treatment in the 2- to 5-μm particles (C), the <2-μm particles (D), without (−) H_2O_2 treatment in the 2- to 5-μm particle size class(E), and the <2-μm particle size class (F). The *black lines* in (C–F) represent the measured diffractogram, and the *dotted lines* represent a population of each identified clay minerals.

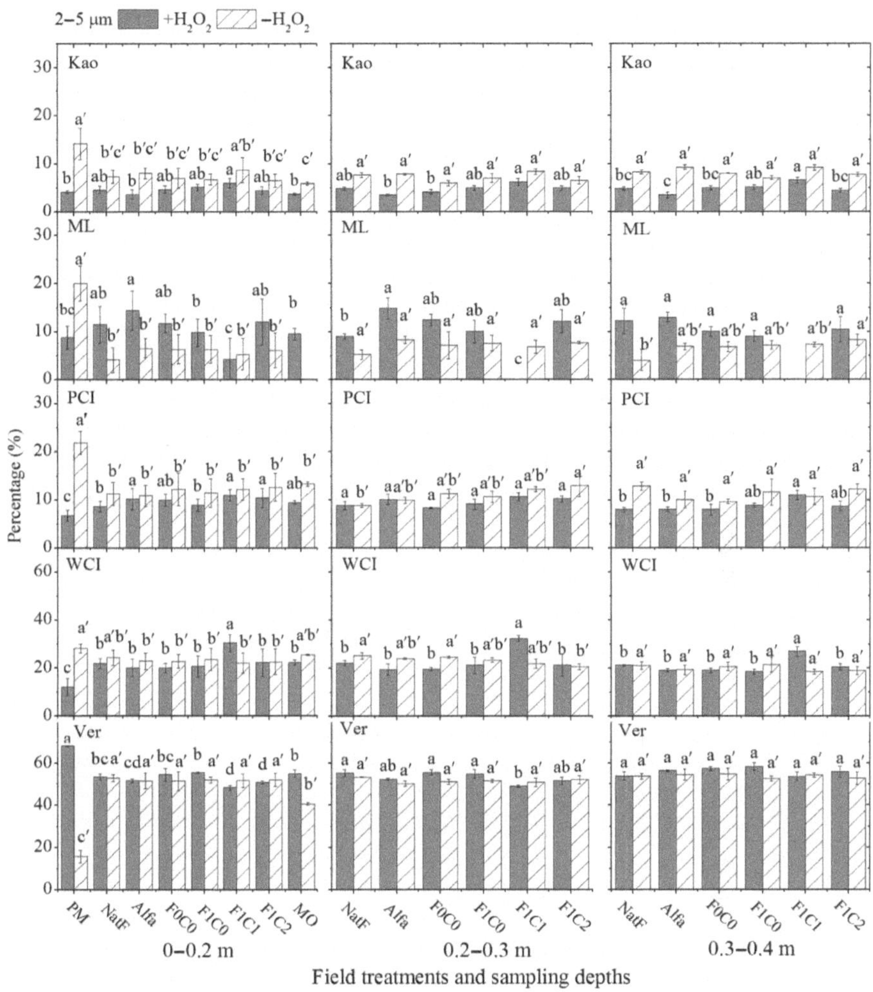

Fig. 2 Peak area percentage of the clay minerals in the 2- to 5-μm particle size class without (−) and with (+) H_2O_2 treatment in the field treatments sampled from different soil depths. *Kao*, kaolinite; *ML*, mixed-layer mineral, *PCI and WCI*, poorly and well-crystallized illite; *Ver*, vermiculite. The minerals are arranged in an ascending order by the percentage as indicated by the difference in the scale of the *Y*-axis. The *bars* over the columns represent standard error of means of three field replicates. The same *small letters* with and without apostrophes indicate no significant differences among the field treatments at the same sampling soil depth for the with (+) and without (−) H_2O_2 treatment, respectively, at $P < 0.05$.

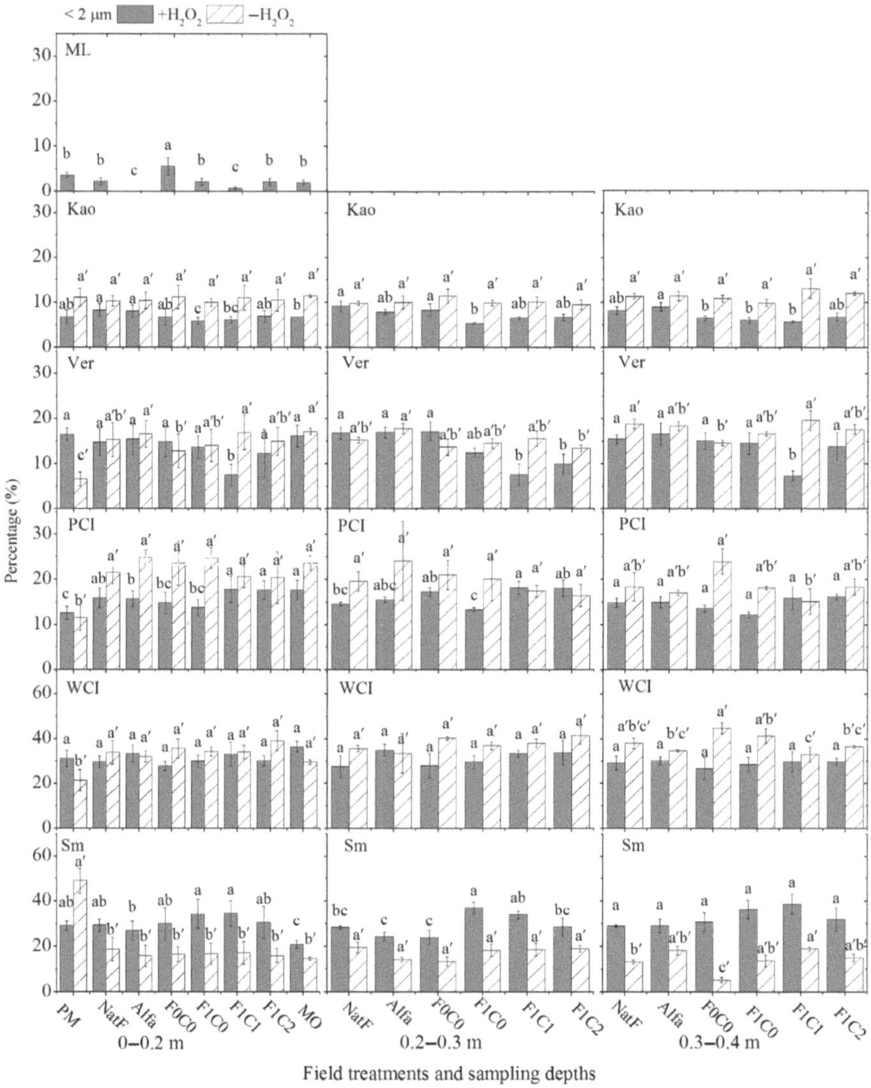

Fig. 3 Peak area percentage of clay minerals in the <2-μm particle size class without (−) and with (+) H_2O_2 treatment in the field treatments sampled from different soil depths. *ML*, mixed-layer mineral; *PCI and WCI*, poorly and well-crystallized illite; *Sm*, smectite; *Ver*, vermiculite. The minerals are arranged in an ascending order by the percentage as indicated by the difference in the scale of the *Y*-axis. The *bars* over the columns represent standard error of means of three field replicates. The same *small letters* with and without apostrophes indicate no significant differences among the field treatments at the same sampling soil depth for with (+) and without (−) H_2O_2.

was illite (33–70%), followed by smectite (20–49%) and vermiculite (7–20%) (Fig. 3). Mixed-layer mineral phases were detected only for some surface soils after oxidation with the H_2O_2 treatment.

For the 2- to 5-μm particle size class without H_2O_2 treatment, the field treatments at the 0–0.2 m depth demonstrated greater amounts of vermiculite and less of other clay minerals compared to the PM (Fig. 2). Compared to MO, the field treatments at the 0–0.2 m depth also had greater amounts of vermiculite and mixed-layer minerals. There were no consistent changes in the content of other minerals. There were no significant differences in any mineral content among the field treatments at the 0–0.2 m depth. Compared to NatF, the F1C2 treatment contained more poorly crystallized illite at the 0.2–0.3 m depth and more mixed-layer mineral at the 0.3–0.4 m depth. Moreover, the amount of well-crystalized illite increased in the subsoils (0.2–0.4 m) more than in the topsoil (0–0.2 m) in all field treatment except for F1C1 and F1C2 at the 0.3–0.4 m depth. For F1C1 and F1C2, the amount of vermiculite increased more at the 0.3–0.4 m depth than at the 0–0.3 m depth.

For the <2-μm particle size class without H_2O_2 treatment, the field treatments at the 0–0.2 m depth contained greater amounts of illite and vermiculite, contained less smectite, and were similar in kaolinite content compared to the PM. Only F0C0 showed less vermiculite content compared to MO (Fig. 3). Among the field treatments, F0C0 contained the least vermiculite at the 0–0.2 m depth, with the differences significant from Alfa and F1C1, and contained least vermiculite and smectite and more well- and poorly crystallized illite at the 0.3–0.4 m depth, with the significant differences from F1C1. In addition, F1C2 contained the least vermiculite at the 0.2–0.3 m depth, with the differences significant from Alfa. Moreover, the amount of smectite was greater in the fertilized soils and in Alfa compared to NatF. Also, the amount of well-crystallized illite was greater in F1C0 than in F1C1 and likewise was greater in F0C0 than in F1C2 and Alfa. For F1C1 and F1C2, the amount of vermiculite increased more at the 0.3–0.4 m depth than at the 0–0.3 m depth.

The oxidization with the H_2O_2 treatment altered the soil minerals as well. The alterations varied between the two particle size classes and between the PM and the soils (Figs. 2 and 3). The exception was for kaolinite, which showed few differences among the treatments, and there were few differences with soil depths. The oxidization decreased the kaolinite contents in both soil particle sizes for all soil samples. After the oxidization of the PM samples, the amount of both poorly and well-crystallized illite decreased in the 2- to 5-μm particle size class (Fig. 2) but increased in the <2-μm particle size class (Fig. 3). In addition, the oxidization of the PM samples decreased the mixed-layer mineral contents in the 2- to 5-μm particle size class and the smectite contents in the <2-μm particle size class. On the other hand, the oxidation treatment increased the vermiculite contents in both the 2- to 5- and <2-μm particle size classes. The H_2O_2 treatment decreased the contents of both the poorly and well-crystallized illite in both particle size classes of the arable soils. Meanwhile, the H_2O_2 treatment increased the contents of vermiculite and mixed-layer minerals in the 2- to 5-μm particle size class, and the contents of smectite in the <2-μm particle size class for all of the arable soils with a few exceptions. Specifically, there were no significant changes to the contents of well-crystallized illite in both the particle sizes in MO and to the content of the mixed-layer mineral in the 2- to 5-μm particle size class in F1C1. The vermiculite content of the <2-μm particle size class, after the oxidation, remained constant except for F1C1 and F1C2. The magnitudes of the changes were much larger in the 2- to 5-μm particle size class than in the <2-μm particle size class for illite in the PM and were otherwise very small among the soils for the well-crystallized illite in both of the particle sizes.

For the <2-μm particle size class, the oxidization created great differences between the soils including MO, particularly at the 0–0.2 m depth (Fig. 3). For these differences, Alfa was lowest in kaolinite content, F1C1 was the lowest in mixed-layer mineral and vermiculite contents, while F1C0 was the lowest in vermiculate content. The mixed-layer mineral was not detectable in F1C1 at the 0.2–0.4 m depth. The poorly and well-crystallized illite showed the greatest content in F1C1 at all the soil depths.

For the 2–5-μm particle size class (Fig. 2), the oxidization caused greater differences among the soils including the deeper soils. The kaolinite was greater in NatF and Alfa than in other soils for all soil depths. At all three soil depths, the kaolinite and poorly crystallized illite were the least abundant in F1C0; the vermiculite was the least abundant in F1C1, and the smectite was more abundant in F1C0 and in F1C1 compared to Alfa.

3.4 Composition of Clay Minerals

The principal component analysis was applied for the detected clay mineral composition using all data set together (Fig. 4). The loadings of different clay minerals were also specified for the PC1 and PC2 components. The first two components accounted for 38.9% and 32.5% of total variance. There was a

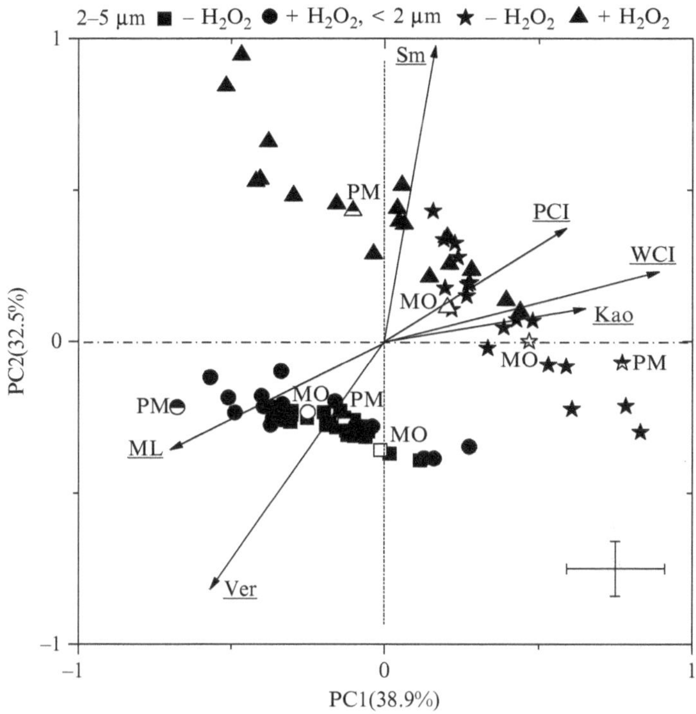

Fig. 4 First two principal components of principal component analysis using all clay mineral data in two particle sizes and with (+) and without (−) H_2O_2 treatment for all the soil samples, showing the largest separations between the 2- to 5- and <2-μm particle size class. Loadings of the identified clay minerals are presented. The reference soils, parent material (PM, *semisolid legend*) and matured arable Mollisol (MO, *hollow legend*), were identified in different symbols. *Kao*, kaolinite; *ML*, mixed-layer mineral; *PCI*, poorly crystallized illite; *Sm*, smectite; *Ver*, vermiculite; *WCI*, well-crystallized illite. *Bars* represent a least significant difference ($P < 0.05$, $n = 3$).

striking separation between the 2- to 5- and <2-μm particle size classes along the diagonal in the first and third quadrants and visible separation between the soils with and without H_2O_2 treatments with <2-μm particle size class through the second and fourth quadrants. The separations between the soils with the H_2O_2 treatment were mainly via PC2. The separations among the soils with 2- to 5-μm particle size class were much smaller than those in the <2-μm particle size class. The separations across the first and third quadrants were driven by the illite and smectite associated with the <2-μm particle size class and by the vermiculite and mixed-layer mineral associated with the 2- to 5-μm particle size class. The separations across the second and fourth quadrants were driven by the kaolinite, well-crystallized illite associated with the <2-μm particle size class with H_2O_2 treatment, and by the smectite associated with the <2-μm particle size class without H_2O_2 treatment. For the soils in the 2- to 5-μm particle size class, the composition of clay minerals was similar to that of PM, while for those in the <2-μm particle size class, the composition of clay minerals was similar to that of MO (Fig. 4).

The principal component analysis was applied for the detected clay mineral composition using all data separated into data sets by particle size classes and H_2O_2 treatments (Fig. 5). The clay composition was significantly differentiated by soil depth across the second and fourth quadrants for both the clay particle size classes without H_2O_2 treatment of all the field treatments (Fig. 5A and B), except for the <2-μm particle size class of Alfa (Fig. 5B). After the H_2O_2 treatment, there is no significant separation by soil depth soils for both the clay particle size classes of all the field treatments (Fig. 5A and B), except for the 2- to 5-μm particle size class of F1C1 and F1C2 (Fig. 5C). The differentiations by soil depth for the 2- to 5-μm particle size class were driven by the associations of well-crystallized illite associated with the 0–0.2 m depth, and of mixed-layer mineral and vermiculite contents associated with the 0.3–0.4 m depth (Fig. 5A). The differentiations by soil depth for the <2-μm particle size class were driven by the associations of smectite and poorly crystallized illite associated with the 0–0.2 m depth, and more well-crystallized illite associated with the 0.3–0.4 m depth (Fig. 5B). There were differentiations among the field treatments, which were less significant for the <2-μm particle size class than for the 2- to 5-μm particle size class.

4. DISCUSSION

Classically, soil mineralogy has been quantitatively studied and observed to change within soil profiles (Margaret et al., 1997; Zhang et al., 2016) along soil sequences for long-term pedogenesis (e. g., Egli et al., 2001; Fritsch et al., 2011; Righi et al., 1999) and after decades-long periods of soil management practices and land use changes (Cornu et al., 2008, 2012; Li et al., 2011; Liu et al., 2008). The soil processes modifying soil minerals are poorly addressed because of lack of land use history and statistical analysis to identify dominant environmental and land use factors that relate to the observed change. Here, the results of statistical analysis using PCA demonstrated that the soil mineralogy changed significantly with clay particle size more than with organic association or with soil depth (Fig. 4). These results were obtained on soil samples characterized after an 8-year period of restoration applied to the well-weathered PM of a MO under

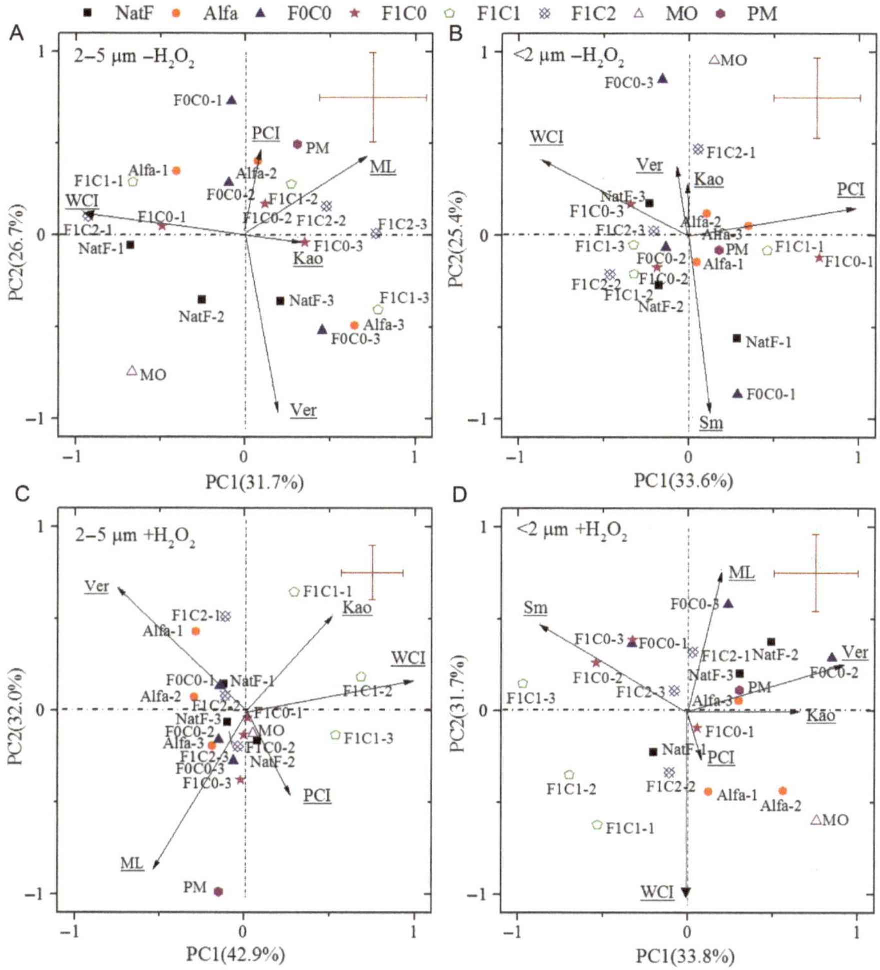

Fig. 5 First two principal components of principal component analysis using the clay mineral data for all the soil samples in the 2- to 5-μm (A) and <2-μm (B) clay particle size classes without (−) H_2O_2 treatment and in the 2- to 5-μm (C) and <2-μm (D) clay particle size classes with (+) H_2O_2 treatment. Loadings of the identified clay minerals are presented in each diagraph. *Kao*, kaolinite; *ML*, mixed-layer mineral; *PCI* and *WCI*, poorly and well-crystallized illite; *Sm*, smectite; *Ver*, vermiculate. *Bars* represent a least significant difference ($P < 0.05$, $n = 3$).

different land uses and agricultural practices. The evolution in soil mineralogy from the PM and the quantitative variations in mineralogy under different land use and agricultural practices provided opportunities to identify soil processes modifying soil mineralogy—as discussed later. Such rapid (within a decade) changes in soil mineralogy under agricultural practices may have considerable impacts on restoration practices. Such practices are highly relevant because of the critical role of soil clay minerals to successfully improve degraded soil functions and ecosystem services such as SOC sequestration, formation of soil structure, water retention, and nutrient supply for plant production. This new knowledge is essential for the successful development of sustainable agriculture practices, as well as for mitigating the impact of climate change on soils.

4.1 The Relation of Clay Mineralogy to Clay Particle Size

The clay mineral composition was significantly different between the <2- and 2- to 5-μm particle size class irrespective of field treatments, soil depths, and H_2O_2 treatments (Fig. 4). The larger differences in mineral content among the field treatments, soil depths, and between H_2O_2 treatments in the <2-μm particle size class suggested that the fine clay minerals are more reactive and are affected significantly by agricultural practices. The differences between the clay particle sizes were driven by smectite associated with the <2-μm particle size class and mixed-layer mineral and vermiculate associated with the 2- to 5-μm particle size class (Fig. 4). The mineral types in each particle size classes were the same in the field treatments and the reference soils of the PM and the mature MO (Figs. 2 and 3), indicating the effect of inherent variation between soil particle sizes in the PM. The inherent variation can be attributed to particle sizes of different mineral types. For example, the clay mineral size is <0.5 μm for smectite and 0.5–4 μm for kaolinite (Moore and Reynolds, 1997), while the primary mineral mainly exists in 0.45–2 μm size particles (Grim, 1953; Jepson, 1984; Muir et al., 1957; Tsao et al., 2013). Therefore, the presence of smectite only in the <2-μm particle size class, mixed-layer mineral only in the 2- to 5-μm particle size class and vermiculated mainly in the 2- to 5-μm particle size class demonstrated that there is a general distinction of mineralogy by particle size in the order of size mixed-layer mineral > vermiculate > kaolinite and illite > smectite.

The reversible changes between mixed-layer minerals/illite and vermiculite/smectite (Figs. 2 and 3) and their similar correlation between the particle size classes (Fig. 6) indicated that the changes in clay mineral composition with particle size can be attributed to chemical transformation processes These correlations demonstrated that illite and vermiculate in larger size fraction could transform into the smaller size fraction through K^+ exchange between the illite and smectite or vermiculite, or between smectite and vermiculite. Many recent studies reported spontaneous changes

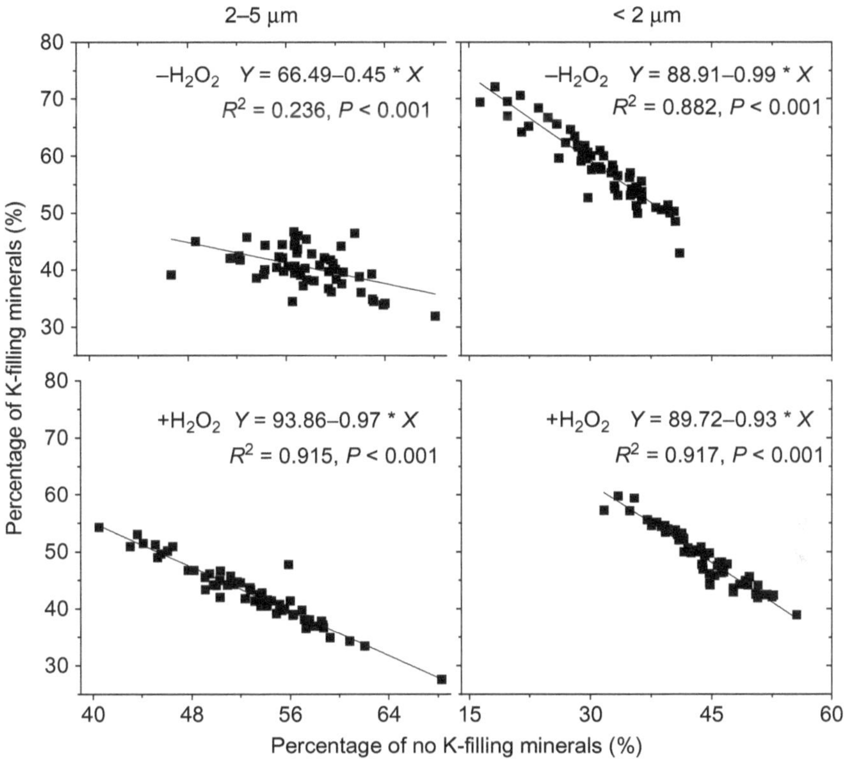

Fig. 6 Correlation between the percentage of K-filling minerals (illite plus mixed-layer mineral) and no K-filling minerals (vermiculite or vermiculite plus smectite) for the 2- to 5- and <2-μm particle size classes with $+H_2O_2$ and $-H_2O_2$ treatments in all the field treatment and the reference soils (parent material and the matured Mollisol).

between illite and smectite or vermiculite through K^+ adsorption in a K-rich environment to form illite-like layers, and K^+ release in a K^+-poor environment through plant uptake to expand the layers of 2:1 smectite or vermiculite by Mg^{2+} and Ca^{2+} substitution (Barré et al., 2007a,b, 2008; Paola et al., 2016). The EK content at the 0–0.2 m depth also increased greatly in the soils without K fertilization (Alfa, NatF, and F0C0) compared to PM (Table 2). This K^+ accumulation in the surface soil was attributed to K^+ taken up from the subsoils, released on the surface soil after the decomposition of plant litter and incorporated into illite (Li et al., 2011), which is referred to as K^+ uplift. The structural interlayer of illite and smectite can also be destroyed or the interlayer cation, such as K^+, Ca^{2+}, and Mg^{2+} can be substituted by Al^{3+} released in acidic conditions (pH 4–5) (Calvaruso et al., 2009; Cornu et al., 2008; Dixon and Weed, 1989;

Wilson, 1999). It was notable that the MO had a lower smectite content than the other treatments, irrespective of H_2O_2 treatments (Figs. 2 and 3). This result confirmed that smectite was weathered to vermiculite and illite due to acidification. Smectite is often charged with Ca^{2+} in the interlayer of its structure, which can be easily exchanged and transported due to soil acidification. Smectite has been reported to transform into vermiculite with the dissolution of Al-hydroxide in the smectite interlayer due to low soil pH under ammonium sulfate fertilization for grasses (Tye et al., 2009). Weathering of smectite can be inhibited due to the presence of soil organic matter, which can also occupy the swelling layers (Szymański et al., 2014) or adsorb as a coating on mineral surfaces as discussed later. The formation of vermiculite under acidic conditions was corroborated by the higher content of vermiculate and illite in the acidic soils (MO, F1C1, and F1C2) than in F0C0 in the <2-μm particle size class at the 0–0.2 m depth. Zhang et al. (2016) suggested that differences in smectite and vermiculite are more strongly reflected by particle size than in illite. The same study attributed these effects to higher specific surface area and more active substitution in the interlayers of smectite and vermiculite than in illite. However, the results of this study are the reverse. This indicates that the transformation of illite and vermiculate from large particles to small particles occurred in all of the young soils that were developed from the well-weathered PM, which had substantial amounts of illite in the larger size class of particles.

4.2 Effects of Agricultural Practices on Clay Mineralogy

Soil mineralogy in different clay particle size classes (Figs. 2 and 3) changed under different field treatments and the changes were related to the changes in soil properties such as the amounts of Al, Si, EK, and soil pH (Table 2). For the 2- to 5-μm particle size class, the soil mineral content was significantly different only between F1C2 and NatF at the 0.2–0.4 m depths (Fig. 2). The larger content of poorly crystallized illite and mixed-lay minerals in F1C2 than in NatF corresponded to the lower EK content in NatF than in F1C2 (Table 2). The different EK content indicated more K^+ uptake from the subsoils for NatF and more K^+ leached to the subsoils from excessive K fertilization for F1C2. As discussed earlier, illite in the mixed-layer mineral can transform into vermiculite or smectite due to K^+ uptake and, vice versa, vermiculite or smectite can transform into illite and mixed-layer mineral due to K^+ release from the vermiculite or smectite. This reversible transformation is able to explain the greater content in poorly crystalized illite and mixed-layer mineral in F1C2 than in NatF. The soil pH was lower, but for the 0.2–0.4 m depth, the Al and Si content was greater in F1C2 than in NatF (Table 2). The differences with depth and pH indicated that soil acidification was stronger in F1C2 than in NatF, presumably due to NO_3 transformations (Colin-Belgrand et al., 2003; Zeller et al., 2007). The soil acidification could have destroyed mineral structures by dissolution and then enhanced the transformation into poorly crystallized illite and mixed-layer minerals for the conditions of F1C2. The difference in the K balance between root uptake and K leaching in the subsoils would increase with time between the fertilized arable soils (F1C0, F1C1, and F1C2) and the perennial soils (Alfa and NatF). Therefore, the differences in the amounts of illite and mixed-layer minerals would become larger and more significant in the 2- to 5-μm particle size class in the subsoils between the fertilized arable soils and the perennial soils.

For the <2-μm particle size class, the clay mineral contents varied mainly among F0C0, F1C0, F1C1, and Alfa at all the three soil depths (Fig. 3). The vermiculite content was lower in F0C0 and F1C2 than in other soils at the 0–0.2 m depth and at the 0.2–0.3 m depth, respectively, and the differences from Alfa were significant (Fig. 3). F0C0 received no K fertilization and no straw incorporation, which was expected to favor the formation of vermiculite with a reduction in the amount of illite due to more K^+ uptake and less K^+ uplift than the deeper rooted Alfa. Therefore, the lowest vermiculite content in F0C0 at the 0–0.2 m depth suggested other pathways of vermiculite formation rather than from illite through K^+ substitution. Soil pH at the 0–0.2 m depth decreased in all the field treatments compared with PM, indicating that vermiculite could have formed under acidic condition (Bain and Griffen, 2002; Bain et al., 1990). The lower soil pH corresponded to the greater contents of Al and Si in Alfa and F1C1 than F0C0 at the 0–0.2 m depth (Table 2), indicating that a decrease in soil pH can destroy the Al-O/Si-O structure by dissolution (Dümig et al., 2012; Zanelli et al., 2007), leading to an increase in Al^{3+} mobilization. Substitution of interlayer water with hydroxyl Al^{3+} can transform smectite into vermiculate (Mouhamad et al., 2016; Wilson, 1999). Therefore, lower soil pH in Alfa and F1C1 at the 0–0.2 m depth would have caused formation of vermiculite content, resulting in the lowest vermiculate content in F0C0, which had the highest soil pH. The decrease in soil pH can be attributed to biomass production and biological activity (Calvaruso et al., 2009), N-fixing crops such as alfalfa and to N fertilization specifically (Li et al., 2011; Tye et al., 2009). However, the acidification could not explain the lower vermiculite content in F1C2 compared to Alfa at the 0.2–0.3 m depth because there were no differences in Al, Si, and pH between them. The decreased vermiculite in F1C2 corresponded to its greater EK content, indicating the role of mineral transformation between vermiculite and illite through K^+ uptake in both acidic conditions as discussed earlier for the 2–5-μm particle size class. Therefore, vermiculite in the <2-μm particle size class formed preferentially through Al^{3+} dissolution and substitution of interlayer water in acidic and K-rich condition at the 0–0.2 m depth. At the deeper soils, where K^+ was less abundant and where the soil was less acidic, vermiculite in the <2-μm particle size class could also form through K^+ substitution in illite.

For the <2-μm particle size class at the 0.3–0.4 m depth, F0C0 contained more poorly and well-crystallized illite, but less vermiculite and smectite than F1C1, while F0C0 contained more well-crystallized illite than Alfa and less vermiculite and smectite than NatF (Fig. 3). These differences corresponded to the higher EK in F1C1 compared to F0C0 (Table 2), indicating the transformation from smectite and vermiculite into illite through adsorption of K^+ mobilized down to 0.4 m layer in F1C1. Despite no fertilization in F0C0, the greater content of smectite and vermiculite in NatF and Alfa than in F0C0 corresponded well to the lower EK in NatF and Alfa compared to F0C0. This greater content of smectite and vermiculite indicated transformation from illite into smectite and vermiculite due to K^+ release. This transformation could be explained by K^+ uptake in NatF and Alfa, presumably by the deeper roots down to 0.4 m depth in Alfa and NatF. The crop roots in F0C0 were shallower and could not take up K^+ from 0.3 to 0.4 m depth.

The transformation of poorly crystallized illite and mixed-layer minerals into vermiculite through deep root K^+ uptake and that of vermiculite into well-crystallized illite in the 2–5-μm particle size class is consistent with the transformation of smectite and vermiculite into illite in the <2-μm particle size class in the subsoils. These results also indicated that the content of illite and vermiculate became associated with smaller particle size during the mineral weathering that occurred in the surface soil. However, these phenomena were not controlled by soil acidification of the surface soil, but solely through isomorphic K^+ release and adsorption.

4.3 Effects of Soil Organic Matter on Clay Mineralogy

The repeated treatments using 30% H_2O_2 oxidized 65–90% soil organic matter (data not shown). The residual soil organic matter was likely the most strongly chemically bound organic matter with the minerals. The effect of SOM removal on the content of the clay minerals was detected (Figs. 2 and 3), and the content of the minerals was correlated to the residual SOC content, depending on mineral type (Fig. 6). The different correlations suggested different mechanisms of organic matter–mineral association as discussed later. Moreover, the H_2O_2 treatments did not affect the correlation between the K^+-filling minerals (illite and mixed-layer mineral) and non-K^+-filling minerals (vermiculite and smectite) (Fig. 7). These results suggested that soil organic matter association influenced the content of vermiculite or smectite detected, but not the transformation of minerals driven by K^+ substitution in both soil particle sizes.

The H_2O_2 oxidization treatment caused reduction or enhancement of diffraction peaks for different minerals compared with no H_2O_2 treatment

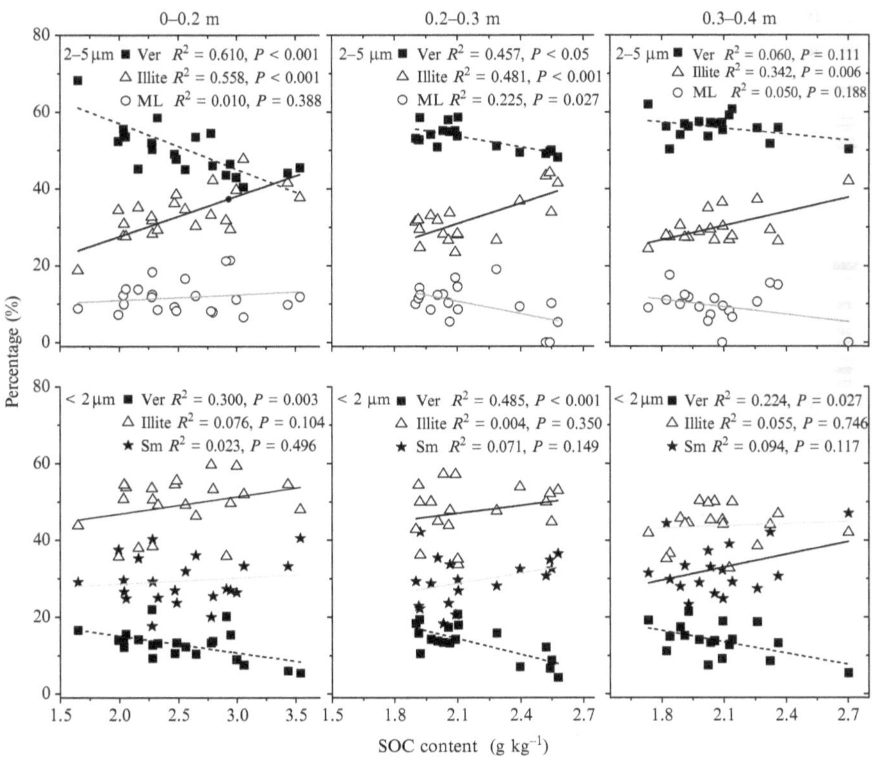

Fig. 7 Correlation of the content of clay minerals with the residual content of soil organic carbon (SOC) after the H_2O_2 treatment in all soils samples including the references (parent material and the matured Mollisol) for the 2- to 5- and <2-μm particle size classes at different soil depths.

(Fig. 1). The peak of kaolinite at 0.71 nm and the peak of illite at 1.0 nm in the Mg^{2+}–glycerol treatment were reduced (Fig. 1A), indicating decreased interlayer spacing after the H_2O_2 treatment. Although the H_2O_2 treatment could also oxidize Fe(II) or Mn(II) and thus change the interlayer charge and spacing, the increased interlayer spacing could be partly attributed to the removal of soil organic matter, which may be filled in the interlayers (Pernes-Debuyser et al., 2003). Kaolinite retains organic molecules between its layers as a host material due to its special assemblage, which is linked by van der Waals attraction and hydrogen bonds (Bhattacharyya and Gupta, 2008; Frost et al., 1998; Moretti et al., 2012). The peak of vermiculite at 1.42 nm for the 2- to 5-μm particle size class and the peak of smectite at 1.8 nm for the <2-μm particle size class in Mg^{2+}–glycerol treatment increased with H_2O_2 treatment (Fig. 1A and B), suggesting expansion of the interlayer spacing. In addition to the effect of oxidized Fe^{2+} or Mn^{2+}, the expansion of the interlay spacing could be partly attributed to the removal of organic matter, which is possibly coated on the surfaces of the minerals. Organic coatings may retard swelling of 2:1 minerals by inhibiting water access for interlayer hydration. The peak of vermiculite at 1.42 nm for the <2-μm particle size class did not change or decreased after the oxidization, particularly in F1C1 and F1C2. These results suggested no organic–vermiculite association or no organic filling in the interlayers of vermiculate for the <2-μm particle size class, possibly due to the acidic conditions such as in F1C1 and F1C2.

Soil organic matter removed from the PM and the young soils represents, respectively, old and newly formed organic matter during the 8-year restoration study. Therefore, the contrasting changes in the content of different mineral types after the oxidization (Figs. 2 and 3) provide information how old and young (years) organic matter is associated with soil minerals. The oxidization by H_2O_2 treatment decreased the content of kaolinite consistently for both the clay particle sizes of all the soil samples including PM (Figs. 2 and 3). These results suggest that either the old organic matter in the PM or the young organic matter in the soils was adsorbed on kaolinite weakly and occurred for both the particle size classes.

After the oxidization treatment, the changes in illite were small in the soils for both the clay particle size classes, but the magnitude of the changes were larger for the 2- to 5-μm particles in the PM than in the soils. These results suggested that the older organic matter was mainly associated with the interlayers of illite in the 2- to 5-μm particle size class and that there was little association of the young organic matter with the illite. Despite its occurrence as a 2:1 mineral, the oxidization caused less change in illite compared to the

1:1 mineral kaolinite, which was attributed to the smaller specific surface area of illite compared to kaolinite (Kaiser and Gugenberger, 2003). The oxidization caused dramatic increases in the content of vermiculite in both particle size classes for the PM and the content of smectite in the <2-μm particle size class for the soils (Figs. 2 and 3). When soil organic matter is low, it is more tightly associated with smectite and vermiculite due to their smaller size and larger specific surface area (Balabane and Plante, 2004; Wagai and Mayer, 2007) compared to illite and kaolinite. Kaiser and Gugenberger (2003) suggested that the organic association with smectite and vermiculite preferentially occurs at micropores and selectively stronger adsorption sites. These authors also propose that the remaining capacity of the surface of these minerals that is available for additional sorption is often reduced after organic association. The reduced CEC of the soils compared to PM (Table 2) suggests that the reduction of specific surface area could be attributed to organic–mineral association.

Following oxidization, the content of the mixed-layer minerals was detectable in the <2-μm particle size class, an occurrence where the relative content was reversed between PM and the young soils. This contrasting change was likely attributed to the composition of the mixed-layer minerals, indicating that the old organic matter was absorbed into the interlayers of the mixed-layer minerals, while the newly formed organic matter was adsorbed on the surfaces of the 2:1 minerals such as vermiculite or smectite. With the organic matter being associated more with vermiculite or smectite than with illite, the interlayer spacing would likely increase. Velde (2001) observed increased expansion of the illite–smectite mixed-layer minerals, which increased to a certain extent upon removal of soil organic matter. However, other studies observed little change in the interlayer spacing after oxidation of organic matter, which was explained by the contrasting expansion and shrinking of the interlayers during the addition and removal of organic matter (Pernes-Debuyser et al., 2003; Velde and Peck, 2002).

It was hypothesized that organic–mineral association may influence the mineral composition through colloidal transport from the surface soil horizons to the subsoil. Despite large variations in relative mineral contents, there were clear decreases in clay and silt in F0C0 and in silt in the three treatments for Alfa, F1C0, and F1C1, all with soil depth (Figs. 2 and 3), indicating downward vertical transport of clay and silt particles in these field treatments. The greatest downward transport in F0C0 suggested the important role of soil organic matter and soil tillage in transport processes. Despite the higher SOC content in Alfa than in the three treatments of NatF, F0C0, and F1C0, the leaching of silt particles in Alfa, but not in NatF, indicated an important role for soil acidification in vertical transport. Despite the relatively higher SOC content, the vertical transport of silt particles in F1C1 confirmed the importance of soil acidification on soil particle mobility. Therefore, soil tillage, soil acidification, and soil organic matter cumulatively can influence the mobilization and transport of silt and clay particles.

The vermiculite in both particle size classes was more abundant at the 0.3–0.4 m depth than at the 0–0.3 m depths, but there were no differences in the content of vermiculite in the 2- to 5-μm particle size class among the soils. These results indicated that changes in the vermiculite content in the <2-μm particle size class with soil depth may be related to silt- or clay-sized particle transport. The well-crystallized illite in the <2-μm particle size class increased in content with increasing soil depths in the three treatments of Alfa, F1C0, and F0C0 at the 0.3–0.4 m depth, indicating the transport of well-crystallized illite with the silt and/or clay particle size classes. The vermiculite and well-crystallized illite in the <2-μm particles had small capacity to form aggregates with increasing soil organic matter as only few organic matter would have been filled in the interlayers, particularly as coating on the surface of vermiculite in the 2- to 5-μm particles. The vermiculite and well-crystallized illite in the <2-μm particle size class had small sorption capacity to bind with organic matter and provide favorable particle interactions to form aggregates. Therefore, less organic matter would have been available to fill in the interlayers in the <2-μm particle size class in contrast to the vermiculite in the 2- to 5-μm particle size class. The latter could have been coated by more organic matter.

The smectite did not show high mobility, which was in contrast to the conclusions ofFrison et al. (2009) and Cornu et al. (2012) who attributed high mobility ofsmectite due to its small size in the coarse texture soils with low soil organic matter at the early stage of soil formation. In our study, the smectite could quickly transform into illite due to adsorption of K^+ uplifted through plant uptake and subsequent decomposition of plant litter. In addition, soil organic matter was quickly increased even for the nonfertilized soils (NatF, Alfa, and F0C0), and the smectite could be preferentially associated with organic matter due to its large specific surface area, resulting in a low mobility.

4.4 The Relationship Between Soil Depth and Clay Mineralogy

The principal component analysis demonstrated that soil depth had a stronger effect in relation to soil mineral composition than the field treatments and that the effects varied with particle size class and H_2O_2 treatment (Fig. 5). The stratifications by soil depth were largely attributed to the variation in the intensities of the soils processes modifying soil minerals and their complex interactions. These processes have been discussed in the previous sections and are summarized up in Fig. 8. The stratification by soil depth was mainly driven by contrasting changes in the (1) well-crystallized illite for

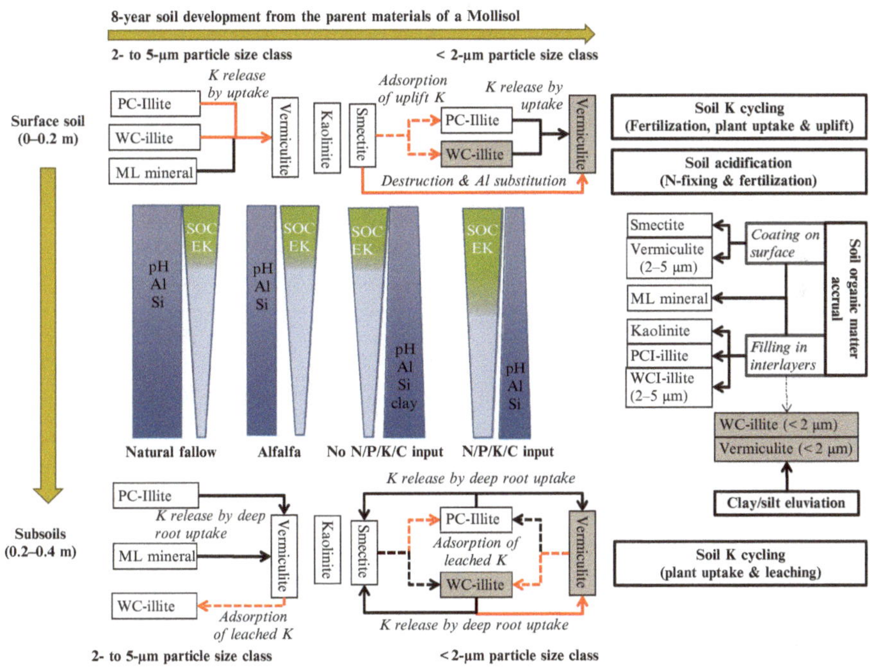

Fig. 8 Soil processes (in *bold boxes*) modifying soil minerals (in *light boxes*) through various mechanisms (in *italics*) in the 2- to 5- and <2-μm particle size classes from the surface soil (0–0.2 m) to the subsoils (0.2–0.4 m) over the 8-year soil development from the parent material of a Mollisol under different agricultural practices such as natural fallow with multiple grasses, N-fixing alfalfa and cropped soil with and without mineral (N, P, and K) fertilization, and with and without organic (C) amendment. Larger ends of wedges and deeper colors in the center illustrate greater values of soil pH and contents of exchangeable K, Al, and Si at the corresponding soil depths (0–0.2, 0.2–0.3, and 0.3–0.4 m). The *solid, dotted,* and *dashed lines* with *arrows* showing different mechanisms and transformation direction. The *dotted lighter lines* showed weaker organic–mineral association that the *bold thicker lines*. The *red lines* showed transformation of these clay minerals (illite and vermiculite) from the 2- to 5-μm particle size class into the <2-μm particle size class. The *bold boxes* with *gray background* showed these minerals (illite and vermiculite in the <2-μm particle size class) were affected by three transformation processes, such as K^+ substitution, organic–mineral association, and leaching.

both the 2- to 5- and <2-μm particle size class at the 0–0.2 m depth, (2) the vermiculite from the 2- to 5-μm particle size class, and (3) the smectite for the <2-μm particle size class at the 0.3–0.4 m depth.

This pattern with depth corresponded to the transformation of these minerals through K^+ transformations that were strongly differentiated with plant types. This differentiation with depth results from plant having different root depths, K fertilization, and ion transport as well as vertical transport of soil silt and clay particles. After removal of soil organic matter, the stratification by soil depth became much weaker for the 2- to 5-μm particle size class and no clear pattern among the soils for the <2-μm particle size class. The stratification was significant only for the soils such as Alfa, F1C1, and F1C2 that exhibited the strongest variations in SOC, EK, and pH with soil depth. The absence of a clear pattern among the soils suggested complex interactions between soil processes and the effects of the field treatments in the soil profile.

5. CONCLUSION

The quantitative decomposition of XRD spectra combined with the unique field treatments revealed rapid (within a decade) changes in soil clay minerals derived from the weathered PM, for agricultural soil restoration practices in the MO region of China. The clay mineral type and composition varied between the clay particle size classes due to their origin and the transformations that occurred, particularly for illite and vermiculate through K^+ release and adsorption in the interlayers. In this study, the following soil processes that modify the soil mineralogy were identified. (1) K^+ substitution in the interlayers for the transformation between illite and vermiculite or smectite. This process depends on soil K^+ cycling and is affected by the amount of K fertilizer, plant K^+ uptake, and uplift from different soil depths. (2) Structural destruction and Al^{3+} substitution which leads to the transformation between vermiculite and smectite. This process is affected by soil acidification from N-fixing crop and N fertilization. (3) Mineral–organic association influenced the translocation of vermiculite and illite in fine clay particles and the stratification of soil depth. Therefore, this process depends on soil organic matter accrual and, together with soil tillage and acidification, affects the mobility of silt and clay particles. (4) The interactions among the above soil processes.

The changes in soil mineralogy after the oxidization treatment showed that more soil organic matter was bound to vermiculite and smectite than to kaolinite and illite and that organic matter was bound to different minerals through different pathways. Soil organic matter was bound to vermiculite contained in the large clay particle size class and to smectite in the small clay particle size class through coating on their surfaces. However, soil organic matter was bound to kaolinite and illite through filling in the interlayers of illite and kaolinite assemblages. The soil organic–mineral association did not affect mineral transformation, but influenced mineral translocation. These results suggest that soil mineral composition of the more intensely weathered PM of the studied MO can change after agricultural restoration more quickly than previously known. The rate and extent of soil mineral transformations and translocation must be considered in agricultural practices that are used to manage soil functions and ecosystem services.

ACKNOWLEDGMENTS

The study was partially funded by the National Natural Science Foundation of China (41330855, 41201222) and EC-FP7 (SoilTrEC, Grant No. 244118). Y.F. Qiao, W.X. Zou, and J. Chen are acknowledged for maintaining the field experiment and for clay mineralogy analysis. B.Z. was supported by the Hundred Talents Program of the Chinese Academy of Sciences. N. Li was supported by the Youth Innovation Promotion Association CAS (2016211) and Excellent Young Talent Program of Northeast Institute of Geography and Agroecology, CAS.

REFERENCES

Bain, D.C., Griffen, D.T., 2002. Possible effects of land use on the clay mineralogy of a brown forest soil. Clay Miner. 37 (4), 663–670.

Bain, D.C., Mellor, A., Wilson, M.J., 1990. Nature and origin of an aluminous vermiculitic weathering product in acid soils from upland catchments in Scotland. Clay Miner. 25 (4), 467–475.

Balabane, M., Plante, A.F., 2004. Aggregation and carbon storage in silty soil using physical fractionation techniques. Eur. J. Soil Sci. 55, 415–427.

Barré, P., Velde, B., 2010. Clays developed under sequoia gigantia and prairie soils: 150 years of soil-plant interaction in the parks of French Châteaux. Clays Clay Miner. 6, 803–812.

Barré, P., Velde, B., Abbadie, L., 2007a. Dynamic role of 'illite-like' clay minerals in temperate soils: facts and hypotheses. Biogeochemistry 82, 77–88.

Barré, P., Velde, B., Catel, N., Abbadie, L., 2007b. Soil-plant potassium transfer: impact of plant activity on clay minerals as seen from X-ray diffraction. Plant and Soil 292, 137–146.

Barré, P., Velde, B., Fontaine, C., Catel, N., Abbadie, L., 2008. Which 2:1 clay minerals are involved in the soil potassium reservoir? Insights from potassium addition or removal experiments on three temperate grassland soil clay assemblages. Geoderma 146, 216–223.

Beck, L., Rousselière, H., Castaing, J., Duran, A., Lebon, M., Moignard, B., Plassard, F., 2014. First use of portable system coupling X-ray diffraction and X-ray fluorescence for in-situ analysis of prehistoric rock art. Talanta 129, 459–464.

Bhattacharyya, K.G., Gupta, S.S., 2008. Adsorption of a few heavy metals on natural and modified kaolinite and montmorillonite: a review. Adv. Colloid Interface Sci. 140, 114–131.

Birkeland, P.W., 1999. Soils and Geomorphology. Oxford University Press, Oxford.

Calvaruso, C., Mareschal, L., Turpault, M.P., Leclerc, E., 2009. Rapid clay weathering in the rhizosphere of Norway spruce and oak in an acid forest ecosystem. Soil Sci. Soc. Am. J. 73, 333–337.

Caner, L., Joussein, E., Salvador-Blanes, S., Hubert, F., Schlicht, J.F., 2010. Short-time clay-mineral evolution in a soil chronosequence in Oléron Island (France). J. Plant Nutr. Soil Sci. 173, 591–600.

Cébron, A., Beguiristain, T., Bongoua-Devisme, J., Denonfoux, J., Faure, P., Lorgeoux, C., Ouvrard, S., Parisot, N., Peyret, P., Leyval, C., 2015. Impact of clay mineral, wood sawdust or root organic matter on the bacterial and fungal community structures in two aged PAH-contaminated soils. Environ. Sci. Pollut. Res. Int. 22, 13724–13738.

Chazdon, R.L., 2008. Beyond deforestation: restoring forests and ecosystem services on degraded lands. Science 320, 1458–1460.

Claret, F., Sakharov, B.A., Drits, V.A., Velde, B., Meunier, A., Griffault, L., Lanson, B., 2004. Clay minerals in the meuse-Haute Marne underground laboratory (France): possible influence of organic matter on clay mineral evolution. Clays Clay Miner. 52, 515–532.

Colin-Belgrand, M., Dambrine, E., Bienaimé, S., Nys, C., Turpault, M.P., 2003. Influence of tree roots on nitrogen mineralization. Scand. J. For. Res. 18, 260–268.

Cornu, S., Besnault, A., Bermond, A., 2008. Soil podzolisation induced by reforestation as shown by sequential and kinetic extractions of Fe and Al. Eur. J. Soil Sci. 59, 222–323.

Cornu, S., Montagne, D., Hubert, F., Barré, P., Caner, L., 2012. Evidence of short-term clay evolution in soils under human impact. C. R. Geosci. 344, 747–757.

Denef, K., Six, J., 2005. Clay mineralogy determines the importance of biological versus abiotic processes for macroaggregate formation and stabilization. Eur. J. Soil Sci. 56, 287–303.

Dixon, J.B., Weed, S.B., 1989. Minerals in Soil Environments. American Society of Agronomy Press, Madison, WI.

Drever, J.I., Stillings, L.L., 1997. The role of organic acids in mineral weathering. Colloid Surf. A-Physicochem. Eng. Asp. 120, 167–181.

Dümig, A., Häusler, W., Steffens, M., Kögel-Knabner, I., 2012. Clay fractions from a soil chronosequence after glacier retreat reveal the initial evolution of organo-mineral associations. Geochim. Cosmochim. Acta 85, 1–18.

Egli, M., Mirabella, A., Fitze, P., 2001. Clay mineral transformations in soils affected by fluorine and depletion of organic matter within a time span of 24 years. Geoderma 103, 307–334.

Frison, A., Cousin, I., Montagne, D., Cornu, S., 2009. Soil hydraulic properties in relation to local rapid soil changes induced by field drainage: a case study. Eur. J. Soil Sci. 60, 662–670.

Fritsch, E., Balan, E., Nascimento, N.R., Allard, T., Bardy, M., Bueno, G., Derenne, S., Melfi, A.J., Calas, G., 2011. Deciphering the weathering processes using environmental mineralogy and geochemistry: towards a general model of laterite and podzol genesis in the upper Amazon basin. C. R. Geosci. 343, 188–198.

Frost, R.L., Kristof, J., Paroz, G.N., Tran, T.H., Kloprogge, J.T., 1998. The role of water in the intercalation of kaolinite with potassium acetate. J. Colloid Interface Sci. 204, 227–236.

Griffiths, B.S., Liu, Q., Wang, H.L., Zhang, B., Kuan, H.L., McKenzie, B.M., Hallett, P.D., Neilson, R., Daniell, T.J., 2008. Restoration of soil physical and biological stability are not coupled in response to plants and earthworms. Ecol. Restor. 26, 102–104.

Grim, R.E., 1953. Clay Mineralogy. McGraw-Hill, New York, p. 384.

Harris, J., 2009. Soil microbial communities and restoration ecology: facilitators or followers? Science 325, 573–574.

Haygartha, P.M., Ritz, K., 2009. The future of soils and land use in the UK: soil systems for the provision of land-based ecosystem services. Land Use Policy 26, S187–S197.

Huang, C.Q., Zhao, W., Liu, F., Tan, W.F., Koopal, L.K., 2011. Environmental significance of mineral weathering and pedogenesis of loess on the southernmost Loess Plateau, China. Geoderma 163, 219–226.

Jepson, W.B., 1984. Kaolins: their properties and uses. Philos. Trans. R. Soc. Lond. A 311 (1517), 411–432.

Jorgensen, S.E., Fath, B., 2008. Encyclopedia of Ecology. Elsevier Science Press, Amsterdam, pp. 3291–3295.

Kaiser, K., Gugenberger, G., 2003. Mineral surfaces and soil organic matter. Eur. J. Soil Sci. 54, 219–236.

Knops, J.M.H., Tilman, D., 2000. Dynamics of soil nitrogen and carbon accumulation for 61 years after agricultural abandonment. Ecology 81, 88–98.

Lanson, B., 1997. Decomposition of experimental X-ray diffraction patterns (profile fitting): a convenient way to study clay minerals. Clays Clay Miner. 45, 132–146.

Li, D., Velde, B., Li, F., Zhang, G., Zhao, M., Huang, L., 2011. Impact of long-term alfalfa cropping on soil potassium content and clay minerals in a semi-arid loess soil in china. Pedosphere 21 (4), 522–531.

Li, N., Yao, S.H., You, M.Y., Zhang, Y.L., Qiao, Y.F., Han, X.Z., Zhang, B., 2014. Contrasting development of soil microbial community structure under no-tilled perennial and tilled cropping during early pedogenesis of a Mollisol. Soil Biol. Biochem. 77, 221–232.

Li, N., Yao, S.H., Qiao, Y.F., Zou, W.X., You, M.Y., Han, X.Z., Zhang, B., 2015. Separation of soil microbial community structure by aggregate size to alarge extent under agricultural practices during early pedogenesis of a Mollisol. Appl. Soil Ecol. 88, 9–20.

Li, N., Pan, F.J., Han, X.Z., Zhang, B., 2016. Development of soil food web of microbes and nematodes under different agricultural practices during the early stage of pedogenesis of a Mollisol. Soil Biol. Biochem. 98, 208–216.

Liu, Y.L., Zhang, B., Li, Ch.L., Hu, F., Velde, B., 2008. Long-term fertilization influences on clay mineral composition and ammonium adsorption in a rice paddy soil. Soil Sci. Soc. Am. J. 72, 1580–1590.

Manhaes, R.S.T., Auler, L.T., Sthel, M.S., Alexandre, J., Massunaga, M.S.O., Carrió, J.G., dos Santos, D.R., da Silva, E.C., Garcia-Quiroz, A., Vargas, H., 2002. Soil characterisation using X-ray diffraction, photoacoustic spectroscopy and electron paramagnetic resonance. Appl. Clay Sci. 21, 303–311.

Margaret, S.T., Susan, E.T., Oliver, A.C., Peter, M.V., David, M.H., 1997. Mineral control of soil organic carbon storage and turnover. Nature 389, 170–173.

Mavris, C., Plötze, M., Mirabella, A., Giaccai, D., Valboa, G., Egli, M., 2011. Clay mineral evolution along a soil chronosequence in an Alpine proglacial area. Geoderma 165, 106–117.

McKeague, J.A., Brydon, J.E., Miles, N.M., 1971. Differentiation of forms of extractable iron and aluminium in soils. Soil Sci. Soc. Amer. Proc. 35, 33–38.

Menon, M., Rousseva, S., Nikolaidis, N.P., van Gaans, P., Panagos, P., de Souza, D.M., Banwart, S.A., 2014. SoilTrEC: a global initiative on critical zone research and integration. Environ. Sci. Pollut. Res. Int. 21 (4), 3191–3195.

Moore, D.E., Reynolds, R.C., 1997. X-Ray Diffraction and the Identification of Clay Minerals, second ed. Oxford University Press, Oxford.

Moretti, E., Storaro, L., Chessa, G., Talon, A., Callone, E., Mueller, K.J., Enrichi, F., Lenarda, M., 2012. Stepwise dansyl grafting on the kaolinite interlayer surface. J. Colloid Interface Sci. 375, 112–117.

Mouhamad, R., Alsaede, A., Iqbal, M., 2016. Behavior of potassium in soil: a mini review. Chem. Int. 2 (1), 47–58.

Muir, A., Anderson, B., Stephen, I., 1957. Characteristics of some Tanganyika soils. Eur. J. Soil Sci. 8 (1), 1–18.

Nelson, D.W., Sommers, L.E., 1996. Total carbon, organic carbon, and organic matter. In: Sparks, D.L. (Ed.), Methods of Soil Analysis. Part 3. Chemical Methods. Soil Science Society of America Book Series, vol. 5. ASA and SSSA, Madison, WI, pp. 961–1010.

Owliaie, H.R., Abtahi, A., Heck, R.J., 2006. Pedogenesis and clay mineralogical investigation of soils formed on gypsiferous and calcareous materials, on a transect, southwestern Iran. Geoderma 134, 62–81.

Paola, A., Barré, B., Cozzolino, V., Di-Meo, V., Velde, B., 2016. Short term clay mineral release and re-capture of potassium in a Zea mays field experiment. Geoderma 264, 54–60.

Pernes-Debuyser, A., Pernes, M., Velde, B., Tessier, D., 2003. Soil mineralogy evolution in the INRA 42 plots experiment (Versailles, France). Clays Clay Miner. 51, 577–584.

Ramirez-Perez, A.M., Paradelo, M., Novoa-Munoz, J.C., Arias-Estevez, M., Fernandez-Sanjurjo, M.J., Alvarez-Rodriguez, E., Nunez-Delgado, A., 2013. Heavy metal retention in copper mine soil treated with mussel shells: batch and column experiments. J. Hazard. Mater. 248–249, 122–130.

Rhoades, J.D., 1982. Cation exchange capacity. In: Page, A.L. (Ed.), Methods of Soil Analysis, Part 2, second ed. ASA and SSSA, Madison, pp. 149–157.

Righi, D., Huber, K., Keller, C., 1999. Clay formation and podzol development from post-glacial moraines in Switzerland. Clay Miner. 34 (2), 319–332.

Ruffell, A., Wiltshire, P., 2004. Conjunctive use of quantitative and qualitative X-ray diffraction analysis of soils and rocks for forensic analysis. Forensic Sci. Int. 145 (1), 13–23.

Sayen, S., Guillon, E., 2014. Aging effect on Zn retention on a calcareous soil: column experiments and synchrotron X-ray micro-spectroscopic investigation. Sci. Total Environ. 487, 545–556.

Survey Staff, Soil, 2010. Keys to Soil Taxonomy. United States Department of Agriculture, Natural Resources Conservation Service, Washington, DC, p. 328.

Szymański, W., Skiba, M., Nikorych, V.A., Kuligiewicz, A., 2014. Nature and formation of interlayer fillings in clay minerals in Albeluvisols from the Carpathian Foothills, Poland. Geoderma 235–236, 396–409.

Tsao, T.M., Wang, Y.N., Chen, Y.M., Chou, Y.M., Wang, M.K., 2013. Automated ultrafiltration device for environmental nanoparticle research and implications: a review. Micromachines 4 (2), 215–231.

Tye, A.M., Kemp, S.J., Poulton, P.R., 2009. Responses of soil clay mineralogy in the Rothamsted classical experiments in relation to management practice and changing land use. Geoderma 153, 136–146.

Velde, B., 2001. Clay minerals in the agricultural horizon of loams and silt loams in the Central United States. Clay Miner. 36, 277–294.

Velde, B., Barré, P., 2010. Soils, Plants and Clay Minerals. Springer-Verlag, Berlin Heidelberg.

Velde, B., Meunier, A., 2008. The Origin of Clay Minerals in Soils and Weathered Rocks. Springer-Verlag, Berlin.

Velde, B., Peck, T., 2002. Clay mineral changes in the morrow experimental plots, University of Illinois. Clays Clay Miner. 50, 364–370.

Velde, B., Goffé, B., Hoellard, A., 2003. Evolution of clay minerals in a chrono sequence of poldered sediments under the influence of a natural pasture development. Clays Clay Miner. 2, 205–217.

Viennet, J.C., Hubert, F., Ferrage, E., Tertre, E., Legout, A., Turpault, M.P., 2015. Investigation of clay mineralogy in a temperate acidic soil of a forest using X-ray diffraction profile modeling: beyond the HIS and HIV description. Geoderma 241, 75–86.

Wagai, R., Mayer, L., 2007. Sorptive stabilization of organic matter in soils by hydrous iron oxides. Geochim. Cosmochim. Acta 71 (1), 25–35.

Wagai, R., Mayer, L.M., Kitayama, K., 2009. Extent and nature of organic coverage of soil mineral surfaces assessed by a gas sorption approach. Geoderma 149, 152–160.

Wattel-Koekkoek, E.J.W., van Genuchten, P.P.L., Buurman, P., van Lagen, B., 2001. Amount and composition of clay-associated soil organic matter in a range of kaolinitic and smectitic soils. Geoderma 99, 27–49.

Wilkinson, H.T., Humphreys, G.S., 2005. Exploring pedogenesis via nuclide-based soil production rates and OSL-based bioturbation rates. Austr. J. Soil Res. 43, 767–779.

Wilson, M.J., 1999. The origin and formation of clay minerals in soils: past, present and future perspectives. Clay Miner. 34 (1), 7.

Wissing, L., Kölbl, A., Schad, P., Bräuer, T., Cao, Z.-H., Kögel-Knabner, I., 2014. Organic carbon accumulation on soil mineral surfaces in paddy soils derived from tidal wetlands. Geoderma 228–229, 90–103.

Xiong, Y., Li, Q.K., 1987. Soils in China. Science Press, Beijing.

Zanellia, R., Egli, M., Mirabellab, A., Giaccaib, D., Abdelmoula, M., 2007. Vegetation effects on pedogenetic forms of Fe, Al and Si and on clay minerals in soils in southern Switzerland and northern Italy. Geoderma 141, 119–129.

Zeller, B., Recous, S., Kunze, M., Moukoumi, J., Colin-Belgrand, M., Bienaimé, S., Ranger, J., Dambrine, E., 2007. Influence of tree species on gross and net N transformation in forest soils. Ann. For. Sci. 64, 151–158.

Zhang, W.Z., Chen, X.Q., Zhou, J.M., Liu, D.H., Wang, H.Y., Du, C.W., 2013. Influence of humic acid on interaction of ammonium and potassium ions on clay minerals. Pedosphere 23 (4), 493–502.

Zhang, Z.Y., Huang, L., Liu, F., Wang, M.K., Fu, Q.L., Zhu, J., 2016. Characteristics of clay minerals in soil particles of two Alfisols in China. Appl. Clay Sci. 120, 51–60.

CHAPTER SEVEN

Modeling Soil Aggregation at the Early Pedogenesis Stage From the Parent Material of a Mollisol Under Different Agricultural Practices

N. Li*[,1], M.-Y. You*[,1], B. Zhang*[,†,2], X.-Z. Han*[,2], S.K. Panakoulia[‡], Y.-R. Yuan[§], K. Liu*, Y.-F. Qiao*, W.-X. Zou*, N.P. Nikolaidis[¶], S.A. Banwart[||]

*National Observation Station of Hailun Agro-Ecology System, Key Laboratory of Mollisols Agroecology, Northeast Institute of Geography and Agroecology, Chinese Academy of Sciences, Harbin, PR China
[†]National Engineering Laboratory for Improving Fertility of Arable Soils, Institute of Agricultural Resources and Regional Planning, Chinese Academy of Agricultural Sciences, Beijing, China
[‡]School of Environmental Engineering, Technical University of Crete (TUC), Chania, Greece
[§]College of Geographical Science, Harbin Normal University, Harbin, PR China
[¶]School of Environmental Engineering, Technical University of Crete, University Campus, Chania, Greece
[||]School of Earth and Environment, University of Leeds, Leeds, United Kingdom
[2]Corresponding authors: e-mail address: zhangbin01@caas.cn; xzhan@iga.ac.cn

Contents

[1] These authors contribute equally to the work.

Advances in Agronomy, Volume 142
ISSN 0065-2113
http://dx.doi.org/10.1016/bs.agron.2016.10.007

Abstract

Soil aggregation and development of soil structure play a crucial role in determining soil functions and ecosystem services in Earth's Critical Zone. Soil organic matter (SOM) is the main constituent that binds mineral particles together into larger-sized aggregates. Many theoretical concepts have been proposed to explain soil aggregation and SOM accrual processes, but quantification of the processes remains lacking. We observed changes in aggregate size mass distribution and aggregate carbon content in 2 discrete years from an 8-year field experiment, which was conducted to determine how to speed up soil development and restoration with different agricultural practices. The 2-year data showed that the proportion of macroaggregates (>250 μm) increased with the decreasing proportions of microaggregates (53–250 μm) and soil texture units (<53 μm) and increasing proportion of particulate organic matter occluded in the macroaggregates. This aggregation process was more noticeable in the field treatments with higher organic carbon input. By using these data, we successfully calibrated the recently developed CAST mathematical model, which assumes that primary macroaggregates are first formed around plant-derived organic matter and primary microaggregates are then formed within the macroaggregates as the occluded organic materials are decomposed. The secondary microaggregates and macroaggregates are assumed to form after the breakdown of primary macroaggregates. The calibration processes separated the plant growth season from the frozen winter season. Some calibrated parameters were the same for all of the field treatments, and these parameters included the first-order rate constants of the fragmentation and decomposition of plant litter within the macroaggregates, the decomposition rates of organic matter incorporated and protected in microaggregates, and organic carbon associated with the free soil texture units. Other calibrated parameter values were affected by the field treatments, and the parameters were the first-order decay rate constants of the plant-derived organic matter in nonaggregated soils and in the secondary macroaggregates, the least carbon sources for the formation of the secondary macroaggregates, and the correction factor to adjust carbon with the mass flow of the soil texture units. The differences in the calibrated parameter values suggested that the rates of the primary macroaggregation and the secondary microaggregation were likely controlled by the intrinsic properties of the parent material rather than by the land use and agricultural practices, whereas the rates of the secondary macroaggregation process were likely affected by land use through controlling plant litter quality and quantity and by agricultural practices, such as soil tillage, and to a less extent by fertilizers that control organic input and then carbon and mass content of the soil texture units and microaggregates. The model results were consistent with measured changes over time of the aggregate mass distribution by size class, the total organic matter content, and the SOM bound in aggregates. However, the CAST model did not reproduce particularly well the mass of the soil texture units within different aggregate size classes nor the relative contribution of carbon stocks from different carbon pools within aggregates. The gap between observations and model results points the way forward to develop further the detailed process descriptions in the CAST modeling approach, for example, to consider primary microaggregation directly from basic soil texture units. Nevertheless, this study demonstrates that the CAST model is very powerful to predict the dynamics of soil aggregation and concurrent SOM change in response to changes in land use and soil management.

1. INTRODUCTION

Soils are highly complex materials, constituted from an extremely wide variety of mineral particles, decomposing organic matter, and living microbial organisms. These constituent fragments are formed into soil aggregates, which create soil structure by their close arrangement in space. Thus, soils are essentially porous materials, with remarkable soil properties, providing the physical architecture in and through which all soil processes occur (Ritz and Young, 2011). The development of soil structure determines soil functions and ecosystem services across the soil life cycle starting from initial soil formation, through stages of maximum production, to soil degradation which can occur through overly intensive land use (Banwart et al., 2011). Soil formation is classically considered as a very slow process under natural conditions, occurring, for example, over the timescales of interglacial maxima, until reaching a productive state where they may provide many ecosystem services to humans (Harrison and Strahm, 2008). Deterioration of soil structure and loss of its physical stability often lead to erosion losses of surface soils. In extreme cases, this leads to the surface exposure of parent materials (PMs), e.g., observed in the Southern and Northern China (Liu et al., 2011; Zhang et al., 2004) and in many locations around the world (Lal, 2003; Quinton et al., 2010). Once the soils are degraded, it is difficult to restore their previous productive states (Chazdon, 2008) and degradation may cause permanent loss of soil functions. Therefore, pressures from a burgeoning human population, land use changes, and intensifying human intervention raise significant concerns about whether soil formation can continue uninterrupted or be accelerated (Haygarth and Ritz, 2009; Wilkinson and Humphreys, 2005). Despite this interest, there have been remarkably few studies that examine the development of soil structure at the initial stage of soil formation (Andrianaki et al., 2017; Li et al., 2015; Schulz et al., 2013) compared to the overwhelming number of studies that examine well-developed soils undergoing soil degradation (Thomaz and Luiz, 2012; Zhang et al., 2004, 2013). Moreover, compared with the successful restoration of aboveground ecosystems, our understanding of soil formation, particularly the development of soil structure in the context of supporting vegetation restoration, is still very poor (Griffiths et al., 2008; Harris, 2009). A critical question is to what extent, and how quickly, agroecological practices can enhance soil structure formation at the initial stages of soil formation from PM, as a contribution to ecosystem restoration techniques.

Soil organic matter (SOM) is the main constituent that binds mineral particles together into larger-sized aggregates, particularly those which are physically stable when wetted. An aggregate hierarchy concept has been proposed by Tisdall and Oades (1982) and Oades (1984) to describe the mechanisms involving formation of microaggregates and macroaggregates. It has been postulated that different organic binding agents act at different hierarchical stages of soil aggregation (Tisdall and Oades, 1982) and microaggregates can form within macroaggregates (Oades, 1984). Land uses and soil management practices can either enhance or impede the development of soil structure mainly through their effects on plant litter inputs and SOM. The aggregate hierarchy concept and subsequent modifications have been tested using incubation experiments by comparing land uses and soil management practices (Bidisha et al., 2010; Elmholt et al., 2008; Kara and Baykara, 2014; Pagliai et al., 2004; Six et al., 2000). Elliott (1986) demonstrated direct links between (1) loss of SOM and decreased aggregation and (2) increased carbon content with aggregate sizes. Oades and Waters (1991) showed the decreased aggregate/particle size class mass distribution with increasing dispersive energy applied to soil for Alfisols and Mollisols, but not for Oxisols. With a series of incubation experiments, Golchin et al. (1994a,b, 1995) showed the dynamics of formation of microaggregates from mineral particles and their destruction into organic–mineral complexes. Angers et al. (1997) reported the redistribution of carbon from macroaggregates to microaggregates with time, indicating the formation of microaggregates within macroaggregates and after macroaggregate breakdown. Thereafter, several other studies proposed the role of particulate organic matter (POM) in the formation of microaggregates within macroaggregates and the importance of the process for the storage and stabilization of soil carbon (Jastrow, 1996; Jastrow et al., 2007). More recently, by comparing ecosystems with different intensities of soil disturbance such as tillage, Six et al. (1998, 2000) linked the loss of POM upon physical disturbance to the increased turnover of macroaggregates and demonstrated that a decreased macroaggregate turnover under no-tillage compared to conventional tillage led to stabilization of POM in microaggregates. These studies have significantly improved our understanding of the coupling between aggregates and SOM dynamics and the effects of physical disturbances on soil structure. This knowledge provides the basis for the formulation of the theory of physical protection as the main mechanism of SOM stabilization (Dungait et al.,

2012; Schmidt et al., 2011; Stockmann et al., 2013). Despite these advances, a quantitative methodology to interpret and predict the development over time of soil structure and the coupled SOM stabilization largely remains lacking. This knowledge gap is probably due to the multitude of interactions and feedbacks among different binding agents, environmental factors, and the natural heterogeneity of the soil as a three-dimensional porous system (Six et al., 2004; Stockmann et al., 2013). Moreover, the aggregation concepts have been developed based on the comparison of land uses using soils sampled from locations with known histories of land use change (Kara and Baykara, 2014; Six et al., 2004). However, there are limited studies with quantitative data on the temporal changes in soil aggregate mass distribution and aggregate organic carbon content under the same land use. The temporal data are vital to eliminate the differences in soil aggregation interactions that arise from different land uses, to better isolate and quantify the dominant mechanisms, and to provide an empirical basis to quantify rates of aggregation and soil structure development.

Traditional methods of quantitatively calculating soil structure are mostly indirect and empirical (Crawford, 1994; Giménez et al., 1997; Kuka et al., 2007). More recent modeling methods are based on various computing theories to reconstruct two- and three-dimensional images derived by optical microscopy applied on thin section and other advanced techniques such as X-ray computed tomography and nuclear magnetic resonance imaging (Papadopoulos et al., 2009; Peth et al., 2008; Wu and Zhang, 2011; Young and Crawford, 2004). Despite their increasing application, these methods are unable to distinguish microbes and fungi from soil minerals and other organic materials. In addition, there is a trade-off between sample size and resolution. Malamoud et al. (2009) first tried to mathematically model the effects of SOM on the development of soil structure (STRUC-C model), following the aggregate hierarchical concept (Tisdall and Oades, 1982). The STRUC-C model considered each aggregate type as a single carbon pool and did not consider the quality of input plant litter and the resulting POM in the aggregation after decomposition of the plant litter. Subsequently, Stamati et al. (2013) developed a new mathematical model (CAST) to calculate the coupled carbon, aggregation, and structure turnover in topsoil. To date the STRUC-C and CAST models have been validated with limited data by comparison of different land uses at distinct points in time (Malamoud et al., 2009; Panakoulia et al., 2017; Stamati et al., 2013).

We established a field experiment in 2004 to examine how to speed up soil development and restoration from the PM of a Mollisol under different agricultural practices. The changes in soil microbiology (Li et al., 2014, 2015, 2016) and soil mineralogy (Liu et al., 2017) have been reported. In this study, our general aims were to test the aggregate hierarchical concept and quantify the aggregate development with time. The specific objectives were (1) to determine changes in the mass distribution of aggregate size classes and their C content with time under the selected agricultural practices, (2) to calibrate the parameters of the CAST model using 2-year data sets in order to quantitatively predict the dynamics of soil structure and soil carbon, and (3) to identify the controlling factors responsible for the effect of agricultural practices on initial soil structure development from the Mollisol PM.

2. MATERIALS AND METHODS

2.1 Field Data Used for Model Evaluation

The experimental site was located at the National Observation Station of Hailun Agroecology System, Chinese Academy of Sciences, Heilongjiang province (47°26′N, 126°38′E), a central region of Mollisols in Northeast China. The region has a typical temperate continental monsoon climate. The mean annual temperature is 2.2°C, with the highest monthly temperature (35°C) in July and lowest in January (−38°C). The soil is frozen down to more than 1 m depth from late October to mid-April in the following year. The PM consists of soil texture with 420 g kg^{-1} clay and 356 g kg^{-1} silt and with dominant clay minerals of vermiculite, smectite, and illite.

The field experiment was established in 2004 to examine how quickly a degraded soil could be restored under different agricultural practices, with more information on experimental methods available elsewhere (Li et al., 2014, 2016). Briefly, the soil PM was excavated from the C horizon (>2 m) and used to replace the surface soil which had been removed to a depth of 0.8 m. The experiment was set up to study how to speed up soil development and restoration when the Mollisol was extremely degraded due to overly intensive use and exhibiting increasing gully erosion.

The field experiments were laid out following a randomized block design with four replicates. There were six treatments (Table 1): (1) natural fallow (NatF) with *Poa annua* L., *Equisetum arvense* L., and *Spodiopogon sibiricus* Trin.; (2) leguminous fallow (Alfa) with *Medicago sativa* L.; and (3)–(6) cropping rotation between soybean (*Glycine max* (Merrill.) L.) and maize (*Zea mays* L.) in alternative years under four organic and mineral input combinations with tilled soils. The tilled soils were applied either without

Table 1 Land Use, Plant Residue Management, Amounts of Organic C and Mineral Fertilizers Applied, and Soil Organic Carbon Content in 2008 and 2012 for the Different Field Treatments

Field Treatments	Land Use[a]	Plant Residue Management[b]	Tillage[c]	Organic C Input[d] ($mg\ ha^{-1}$)	Mineral Fertilizers[e] ($kg\ ha^{-1}$)			Soil Organic Carbon[f] ($g\ kg^{-1}$)	
					N	P	K	2008	2012
PM	Parent material							4.9 f	4.9 f
NatF	Natural fallow	Shoot and roots	No	0.95 d	0	0	0	5.58 e	7.64 cd
Alfa	Alfalfa	Shoot and roots	No	1.17 c	0	0	0	6.20 de	9.03 b
F0C0	Cropping	Roots only (shoot removed)	Yes	0.86 d	0	0	0	5.67 e	6.99 d
F1C0	Cropping	Roots only (shoot removed)	Yes	1.31 c	54§/192*	56	50	5.93 de	7.87 c
F1C1	Cropping	Roots and amended with soybean seeds and maize straw	Yes	2.84 b	54§/192*	56	50	8.49 bc	10.79 a
F1C2	Cropping	All crop shoot, seeds, and roots	Yes	3.70 a	54§/192*	56	50	8.28 bc	12.28 a

From Li, N., Pan, F.J., Han, X.Z., Zhang, B., 2016. Development of soil food web of microbes and nematodes under different agricultural practices during the early stage of pedogenesis of a Mollisol. Soil Biol. Biochem. 98, 208–216.

[a]The cropping system is an annual rotation in a sequence of soybean and maize in different years since 2004. The crop densities were 70,000 and 270,000 ha^{-1} for maize and soybean, respectively.

[b]The amendment rates in F1C1 were 2250 $kg\ ha^{-1}$ of baked soybean powder and 4500 $kg\ ha^{-1}$ of maize straw mixed homogeneously.

[c]Tillage was performed manually down to 0.2 m with ridges made in October after harvest.

[d]The amount of organic C input in 2012 was estimated based on the sources of organic incorporation. The average C concentrations were 428, 443, 442, and 431 $g\ kg^{-1}$ in aboveground biomass and root of the grasses, alfalfa, maize, and soybean, respectively. The root:shoot ratios were 2.0 and 0.4 in the grasses and alfalfa, and the root to yield ratios were 0.19 and 0.13 in maize and soybean, respectively. Different small letters in this column indicate significant differences among the field treatments.

[e]Diammonium phosphate and potassium sulfate were applied as source of N, P, and K. § is for soybean and * is for maize.

[f]Different small letters in this column indicate significant differences among the field treatments in 2 years.

chemical fertilization (F0) or with chemical fertilization (F1), and without the aboveground biomass incorporated (C0) or with the aboveground biomass partially incorporated (C1) or totally incorporated (C2) into the soil, resulting in a gradient in the annual rate of organic carbon inputs. The cropped soils were tilled to the 0.20-m depth using a spade before seeding. The crop was soybean in 2008 and maize in 2012. Following the practice of local farmers, the crops were generally sown in May and harvested in October. The amounts of mineral fertilization and organic C input are presented in Table 1.

Soil samples were collected from the 0–0.2-m depth from three replicated plots of each treatment after harvest of soybean in October 2008 and before sowing soybean in May 2012. Sampling was carried out in such a way so as not to break the aggregates and to minimize compression. In addition, four soil cores (100 cm^3) were taken to measure mass for determination of soil bulk density by drying at 105°C for 48 h. To simulate the evolution of soil carbon, soil aggregation, and soil structure from that of the PM under the different treatments of agricultural practices, the archive air-dried PM samples were also used for soil structure determination by aggregate mass fractionation by size class. The bulk soil samples were first generally broken down and then passed through an 8-mm sieve after manual elimination of visible plant litter and root residues. These soil samples were then used to determine aggregate size distribution following Elliott (1986) and to isolate the mass fractions of SOM in different aggregate size classes following Lichter et al. (2008) (Fig. 1). First, 100-g aliquots of soil samples that were dried at 40°C overnight were placed on a 250-μm sieve, submerged in deionized water for 5 min, and then moved up and down over 2 min with a stroke length of 3 cm for 30 strokes. Floating organic materials on the 250-μm sieve were removed after the vertical shaking and were not considered as aggregate-bound SOM. The soil and water passing through a 250-μm-sieve were poured into a 53-μm sieve, and the sieve was moved up and down for 10 strokes with a stroke length of 3 cm for 3 min. The soil remaining on each of the sieves and that remaining in the final water volume were collected and oven-dried at 40°C overnight and weighed to determine the mass distribution among the aggregate size classes, namely soil texture units (silt- and clay-sized particles) (<53 μm) (A1), microaggregates fraction (53–250 μm) (A2), and macroaggregates fraction (>250 μm) (A3).

The larger aggregate fractions (A3 and A2) (10 g) were further separated following the same procedures (Lichter et al., 2008) until the intraaggregate particulate organic matter was isolated from the >53-μm aggregates and the

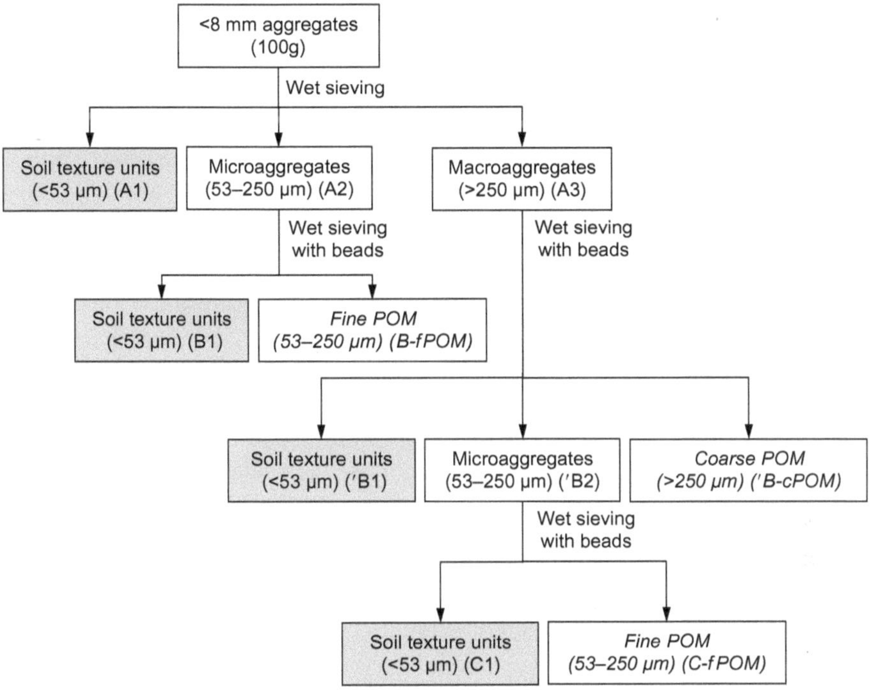

Fig. 1 Physical fractionation scheme by wet sieving without and with beads to determine aggregate size distribution and carbon content in aggregates and particulate organic matter (POM). A, B, 'B, and C indicate different hierarchy of aggregation as separated by wet sieving, and the subscripts, 1–3, represent the fractions of soil texture units (<53 μm), microaggregates (53–250 μm), and macroaggregates (>250 μm). The prescripts, f and c, represent the fine and coarse POM collected from the microaggregates (53–250 μm) and macroaggregates (>250 μm).

silt–sand fraction (Fig. 1). Briefly, aliquots of 10 g >53-μm aggregates were placed on top of a 250-μm mesh screen, immersed in deionized water, and gently shaken with 50 glass beads (4 mm in diameter) with continuous and steady water flow through the device to ensure that microaggregates were immediately flushed onto a 53-μm sieve and were not exposed to any further disruption by the beads (Six et al., 2000).

After all macroaggregates were broken up, the organic materials either on the 250-μm sieve or on the 53-μm sieve were collected and considered as POM originating, respectively, from the >250-μm intraaggregates (coarse POM, cPOM) and from the 53- to 250-μm intraaggregates (fine POM, fPOM). The fraction that passed through the 53-μm sieve was also collected and considered as silt–clay fraction. Although floating material was observed, the free POM and intraaggregate POM were not considered as separate mass fractions because free POM could be also from occluded POM and the proportion was often <1% for agricultural soils and its C-to-N ratio was often not different from that in occluded POM.

The aggregate size class mass was corrected using mineral sand content according to Denef et al. (2001), which may differ in different aggregate sizes. Therefore, aliquots of 5-g soil from the separated aggregates were dispersed with 0.5% sodium hexametaphosphate for 18 h on a rotary shaker at 190 rpm to pass through the sieves with corresponding aperture sizes. Sand-sized particles remaining on the sieves were collected, dried, and weighed. Sand-corrected aggregate weights were calculated according to Eq. (1):

$$\text{Sand corrected mass} = \text{aggregate mass} - [(\text{sand mass}/5\,\text{g}) \times \text{aggregate mass}] \quad (1)$$

All the POM pools and silt–clay fractions were oven-dried at 40°C, weighed, and used to determine the mass fraction of each size class and to measure the corresponding organic carbon mass content using a Vario EL CHN elemental analyzer (Heraeus Elementar Vario EL, Hanau, Germany). For initial soil state, the aggregate mass distribution by size class as well as the corresponding organic carbon mass contents were calculated by the size fraction mass multiplied by the carbon content in the PM. Similar data from soil samples taken during the field experiments were used for the calibration of the CAST model (Table 2; Fig. 1).

2.2 Model Description

The CAST model was developed based on the RothC mathematical model for soil carbon dynamics and was described in detail by Stamati et al. (2013). In general, the CAST defines three aggregate size classes, macroaggregates (>250 μm), microaggregates (53–250 μm), and soil texture units (<53 μm). The soil texture units, as organomineral complexes, are basic units to form into microaggregates and macroaggregates that then contain the organic matter associated with the textural units. The CAST model assumes that macroaggregates are primarily formed through the association of basic soil texture units with POM derived from fragmented and decomposed plant litter. Subsequently, microaggregates are primarily formed within macroaggregates due to decomposition of intraaggregate POM resulting in slaking (Balesdent et al., 2000; Helfrich et al., 2008; Plante and McGill, 2002; Puget et al., 2000; Six et al., 2002a,b, 2004). It is also assumed that secondary

Table 2 Calibrated Parameters of Plant Litter Composition (DPM to RMP Ratio), Decay Rate Constants of Fragmented Plant Materials Free in Soil and Within the Secondary Macroaggregates (A3), Carbon Sources Limiting the Formation of Secondary Macroaggregates, and Correction Factor to Adjust Carbon and Mass Flows Into Macroaggregates During the Plant Growth Season and Frozen Winter Season for the Different Field Treatments

		Decay Rate Constant ($Year^{-1}$) of Fragmented Plant Materials						Carbon Sources (%)				Correction Factor (cf) to Adjust Carbon and Mass Flows of Soil Texture Units		
	Plant Litter Composition	In No-Aggregated Soil				In Secondary A3		Of Secondary A3						
Field Treatment	DPM to RPM	cDPM	RPM	cRPM	fRPM	cRPM	cDPM	cRPM	cDPM	A1	A2	A1	A1 in A2	A1 in A3
Plant growth season														
NatF	0.15	0.5	0.005	0.005	0.005	0.006	0.0047	30	25	30	15	0.1	1.0	0.01
Alfa	0.25	0.5	0.005	0.005	0.005	0.006	0.0047	20	10	60	10	0.3	1.0	0.01
F0C0	0.25	0.5	0.005	0.005	0.005	0.006	0.0047	10	15	60	15	1.0	2.0	0.1
F1C0	1.0	1.0	0.005	0.01	0.01	0.006	0.0047	10	15	60	15	0.7	1.0	0.1
F1C1	1.2	1.0	0.005	0.01	0.01	0.025	0.025	15	20	60	5	0.7	1.0	0.1
F1C2	1.44	0.5	0.005	0.005	0.005	0.03	0.03	10	20	60	10	0.7	1.0	0.1
Frozen wither season														
NatF	0.15	0.05	0.0005	0.0005	0.0005	0.006	0.0047	20	20	30	30	0.1	1.0	0.01
Alfa	0.25	0.05	0.0005	0.0005	0.0005	0.006	0.0047	20	10	60	10	0.3	1.0	0.01
F0C0	0.25	0.25	0.0025	0.0025	0.0025	0.006	0.0047	10	15	60	15	1.0	2.0	0.05
F1C0	1.0	0.5	0.0025	0.005	0.005	0.006	0.0047	10	15	60	15	0.7	0.1	0.05
F1C1	1.2	0.5	0.0025	0.005	0.005	0.006	0.0047	15	20	60	5	0.7	0.1	0.05
F1C2	1.44	1.0	0.005	0.01	0.01	0.006	0.0047	10	20	60	10	0.7	0.1	0.05

A1, soil texture units (<53 μm); *A2*, microaggregates (53–250 μm); and *A3*, macroaggregates (>250 μm).
cDPM, coarse decomposable plant materials; *cRPM*, coarse resistant organic materials; *DPM*, decomposable plant materials; *fRPM*, fine resistant plant materials; *RPM*, resistant plant materials.

microaggregates and macroaggregates are formed during breakdown of aggregates. Therefore, with the aggregate turnover, different organic matter pools within the aggregates and associated with the soil texture units transform as associations within different aggregate size classes. The carbon pools in each size fraction exhibit distinct first-order rate constants (*k*) for carbon mineralization as described for bulk soil in RothC (Coleman and Jenkinson, 1999) (Fig. 2).

Plant litter is assumed to consist of decomposable plant material (DPM) and resistant plant material (RPM), which are transformed into cPOM and fPOM through fragmentation and decomposition prior to subsequent interaction with soil textural units. The POM is then incorporated in macroaggregates and microaggregates where it provides a source of carbon and energy for decomposition by microbes, with degraded organic matter finally protected by association with silt- and clay-size particles as organomineral complexes. The coarse and fine DPM and RPM existing in microaggregates and macroaggregates are considered to be components of POM, which are measured for calibration of the carbon dynamics rate constants. Microbial biomass (BIO), humified organic matter (HUM), and inert organic matter (IOM) are assumed to exist only in association with the soil texture units. IOM corresponds to biochar exhibiting turnover times in the range of centuries to millennia (Powlson et al., 2011) and was not included in this study. Each carbon pool decomposes following first-order kinetics with a specific rate constant. The decomposition rate constant (*k*) was corrected by the product (*a–c*) of three correction factors for the major factors to determine microbial activity ("*a*" for temperature, "*b*" for water deficit, and "*c*" for soil cover) as described for the RothC model (Coleman and Jenkinson, 1999). Atmospheric wet deposition and irrigation are combined as a single boundary condition for moisture inputs to soil. The carbon flow apportioned to the CO_2, BIO, and HUM carbon pools was determined by soil clay content. The model adjusted for soil texture by altering the partitioning between CO_2 evolved and (BIO + HUM) formed during decomposition, rather than by using a rate-modifying factor, such as that used for temperature (Coleman and Jenkinson, 2004). The ratio CO_2/(BIO + HUM) was calculated from the clay content of the soil using the following equation:

$$x = 1.67[1.85 + 1.60\exp(-0.0786\%\ \text{clay})] \quad (2)$$

where x is the ratio CO_2/(BIO + HUM), $x/(x+1)$ is evolved as CO_2, and $1/(x+1)$ is formed as BIO + HUM.

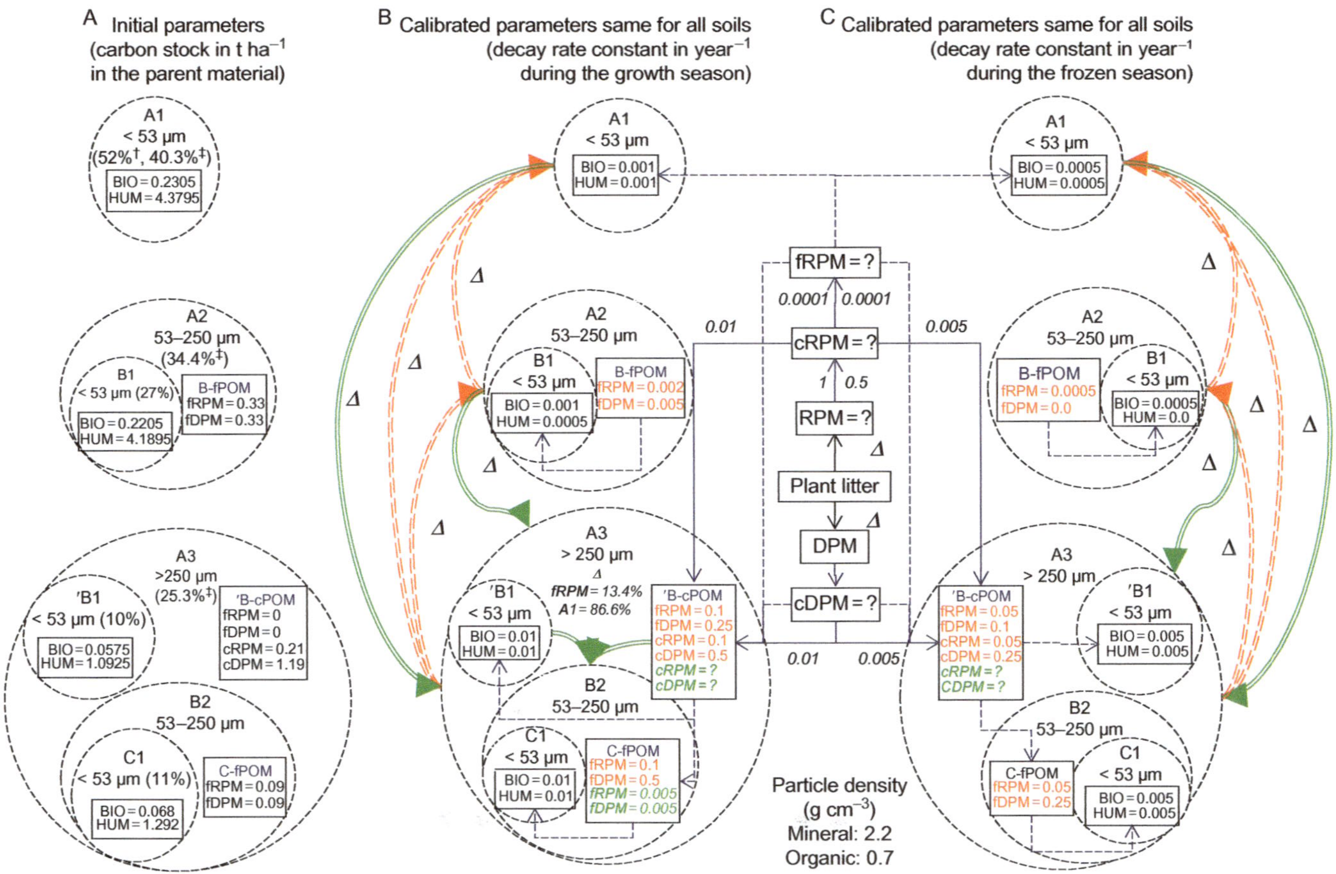

Fig. 2 See legend on next page.

The remaining organic matter pools associated with the soil texture units were set to 46% BIO and 54% HUM, which are the default values in RothC. The carbon saturation capacity of the soil texture units was not considered in this study. In each time step of the CAST model, the decomposition rate and mass loss of each organic pool were calculated; i.e., decomposition of each pool follows the calculation of aggregate breakdown, with the formation of new aggregates calculated after the updating the mass of the organic pools.

2.3 Model Parameterization and Calibration

To initiate the CAST model calculation, the initial carbon stock was estimated based on the measured carbon content in the soil texture unit and in the different POM fractions in the PM (Fig. 2A). To reduce the complexity of the input data set, the fPOM in free microaggregates (A2) and

Fig. 2 The conceptual structure of the CAST model, showing various organic matter pools in differently sized aggregates (*dotted circles*) and their transformation with aggregate turnover: (A) with the initialized parameters as for the parent material; (B) with the calibrated parameters insensitive to the field treatments during the plant growth season; and (C) during the frozen season. A, B (′B), and C indicate the first, second, and third hierarchy by which the silt–clay unit is isolated, with ′B from macroaggregates being distinguished from B from microaggregates. The subscripts 1, 2, and 3 indicate soil texture units (<53 μm), microaggregates (53–250 μm), and macroaggregates (>250 μm). *BIO*, microbial biomass; *DPM*, decomposable plant material; *HUM*, humified organic matter; *RPM*, resistant plant material; *POM*, particulate organic matter. The prescripts, f and c, represent the fine and coarse DPM or RPM, which exist as inter- and intramicroaggregates and aggregates and are components of the POM following the procedures described in Fig. 1. The *solid* and *dotted blue lines* indicate the directions of fragmentation and decomposition of plant materials, respectively. The *solid* and *dotted double lines* indicate the formation and breakdown of microaggregates and macroaggregates from and into silt–clay units, respectively. The pools of DPM and RPM in *black*, *red*, and *green* are those free in soil, within primary microaggregates and macroaggregates (formed during silt–clay unit and plant material association), and within secondary microaggregates and macroaggregates (reformed after aggregate breakdown). [†]The numbers following A1, B1, ′B1, and C1 were mass proportion of silt-clay unit in (A). [‡]The number following A1, A2, and A3 were aggregate size distribution in (A). In (B) and (C), the numbers following the organic matter pools are decay rate constants of fragmentation or decomposition (1/year), which are calibrated to be the same for all the field treatments. The symbol, Δ, indicates the carbon factor limiting formation of secondary microaggregates and macroaggregates with aggregate turnover, being the same for the formation of microaggregates (italic *Δ* in A3) with the decay rate constants (in *green* and *italic* in B2) different from those in the primary microaggregates in A3. The *question marks* (?) following the organic matter pools and the limiting carbon sources (Δ) are specific in the formation of the secondary macroaggregates in the field treatments as listed in Table 2.

microaggregates (B2) within the macroaggregates (A3) was assumed being equally apportioned to fine DPM and fine RPM (i.e., fDPM=fRPM). The fPOM in the macroaggregates (A3) and within the microaggregates (B2) was assumed to be zero. The cPOM in the macroaggregates (A3) was primarily attributed to cRPM, with a small fraction attributed to cDPM. The BIO was considered to be 5% of the total carbon in the soil texture units. The biochar (IOM) was considered to be zero due to no such sources.

To calibrate the CAST model, the input of plant carbon per month was calculated for each experimental treatment based on annual biomass production and management practices (tillage) and was introduced as input to the model (Table 1). Considering the large differences of the quality of the organic inputs, the DPM to RPM ratio was adjusted accordingly depending on their chemical structure measured using solid nuclear magnetic resonance (NMR) (data not shown) and refined by optimization to obtain the best fit to the measured organic matter pools. The decomposition rate constants of the different organic matter pools were calibrated by testing proportionally higher values for the unprotected and less protected pools and proportionally lowered for the most protected pools, assuming that decomposition occurred at a slower rate within macroaggregates as compared with nonaggregate-associated POM due to diffusion limitation of O_2 (g).

Considering the long-freezing period from mid-October to next April in the study area, when organic decomposition was very low, the CAST model was extended by changing the parameter values for the freezing period as compared to those used for the crop-growth period (Panakoulia et al., 2017). This new version of the CAST model provided the ability to simulate soil structure and carbon sequestration under extreme conditions where the decomposition rates of organic matter vary between seasons.

3. RESULTS

3.1 Mass and Carbon Distribution in Aggregates

The SOC content in the bulk soil increased during the 8-year field treatments, but the magnitude of the increases varied (Table 1). The SOC content was greater in the treatments of F1C1 and F1C2 than in other field treatments in 2008, and the differences among the treatments became significant among the treatments of NatF, Alfa, F0C0, and F1C0 in 2012 except for between treatments NatF and F0C0 (Table 1).

The aggregate size mass fraction and aggregate carbon content did not change consistently (Fig. 3). The mass fraction increased from 2008 to

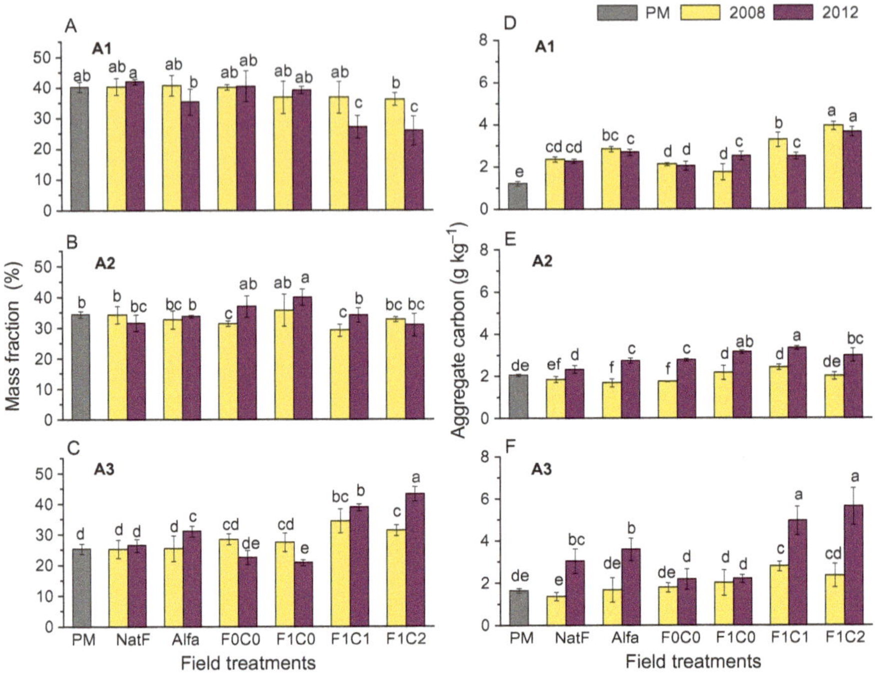

Fig. 3 Aggregate mass fraction (A–C) and aggregation carbon content (D–F) in soil texture units (<53 μm, A1), microaggregates (53–250 μm, A2), and macroaggregates (>250 μm, A3) measured by using wet sieving method for different field treatments.

2012 for the macroaggregate size class (A3) with the decreases in the fraction of soil texture units (A1) in treatments F1C1 and F1C2, while it increased from 2008 to 2012 for the microaggregate size class (A2) with the decreases in the fraction of soil texture units (A1) in treatments F0C0 and F1C0 (Fig. 3A–C). There was only a slight increase in the A3 fraction for treatment Alfa in 2012 and F1C1, F1C2 in 2008 and 2012. However, the SOC content increased from 2008 to 2012 for the A2 and A3 size classes, to a larger extent in the A3 size class than in the A2 size class (Fig. 3D–F). The SOC content in the A1 class did not change with time in any of the field treatments except for treatments F1C0 and F1C1. The SOC contents in all of the aggregate size classes were increased in both 2008 and 2012 compared to the PM.

The carbon content varied in different carbon pools, being largest in the soil texture units in the A2 size class (B1 in A2) (Fig. 4). The carbon content in the soil texture units increased with time only for a few treatments, e.g., the treatments of Alfa, F0C0, and F1C0 for those in the microaggregate size

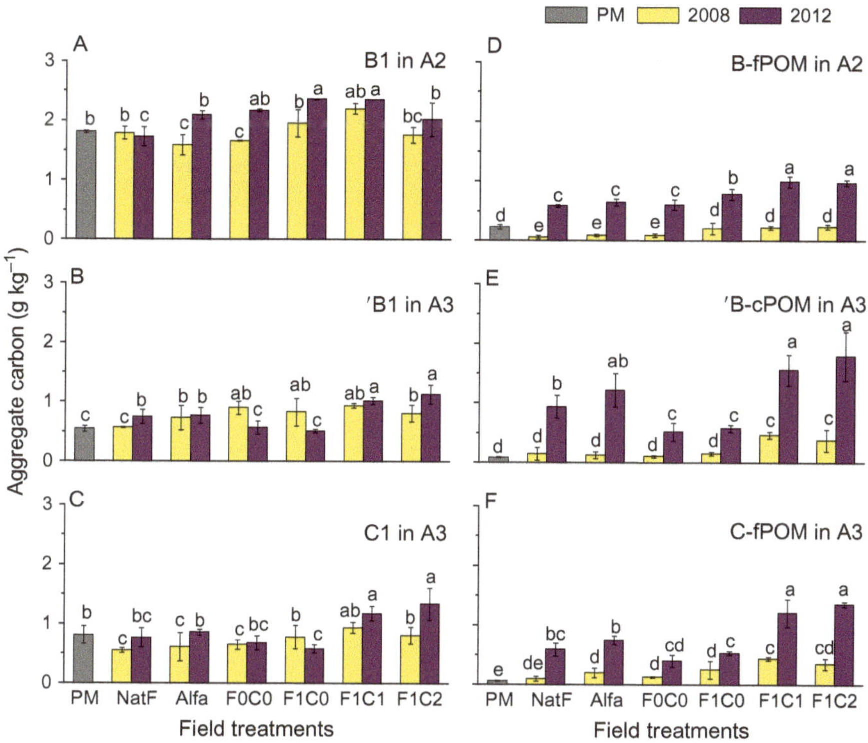

Fig. 4 Aggregate carbon content in the soil texture units free in the microaggregate size class (B1 in A2) (A), in the macroaggregate size class ('B1 in A3) (B), and in the microaggregate size class within the macroaggregate size class (C1 in A3) (C), and particulate organic matter free in the microaggregate size class (B1-fPOM in A2) (D), in the macroaggregate size class ('B-cPOM in A3) (E), and in the microaggregate size class within the macroaggregate size class (C-fPOM in A3) (F), measured by physical fractionation (Fig. 1) for different field treatments.

class (B1 in A2), and the treatment of F1C2 for those in the macroaggregate size classy (A2, 'B1, and C1 in A3) (Fig. 4A–C). The carbon content increased with time for all POM pools and the magnitudes of the increases varied among the field treatments to a larger extent for those in the A3 class than for the A2 class (Fig. 4E–G).

Utilizing the full data set acquired over 2 years, bivariate linear regression of the aggregate mass fractions of the A1 and A2 aggregate size classes was negatively and significantly correlated with that of the A3 aggregate size, with the slopes summing to one (respectively, with linear regression slopes of 0.42 and 0.58) and the determination coefficient larger for the A1 and A3 correlation than for the A2 and A3 correlation (0.87 vs 0.77) (regression calculations not shown, data plotted in Fig. 3A and B). There was no significant correlation in aggregate mass between the A2 and A1 size classes (Fig. 5C). The correlations in the aggregate carbon content were positive and significant among the three aggregate size classes (Fig. 5D–F). The slope and determination coefficient for the correlation in aggregate carbon content were 0.77 and 0.79 between the A3 and A2 size classes, 0.87 and 0.56

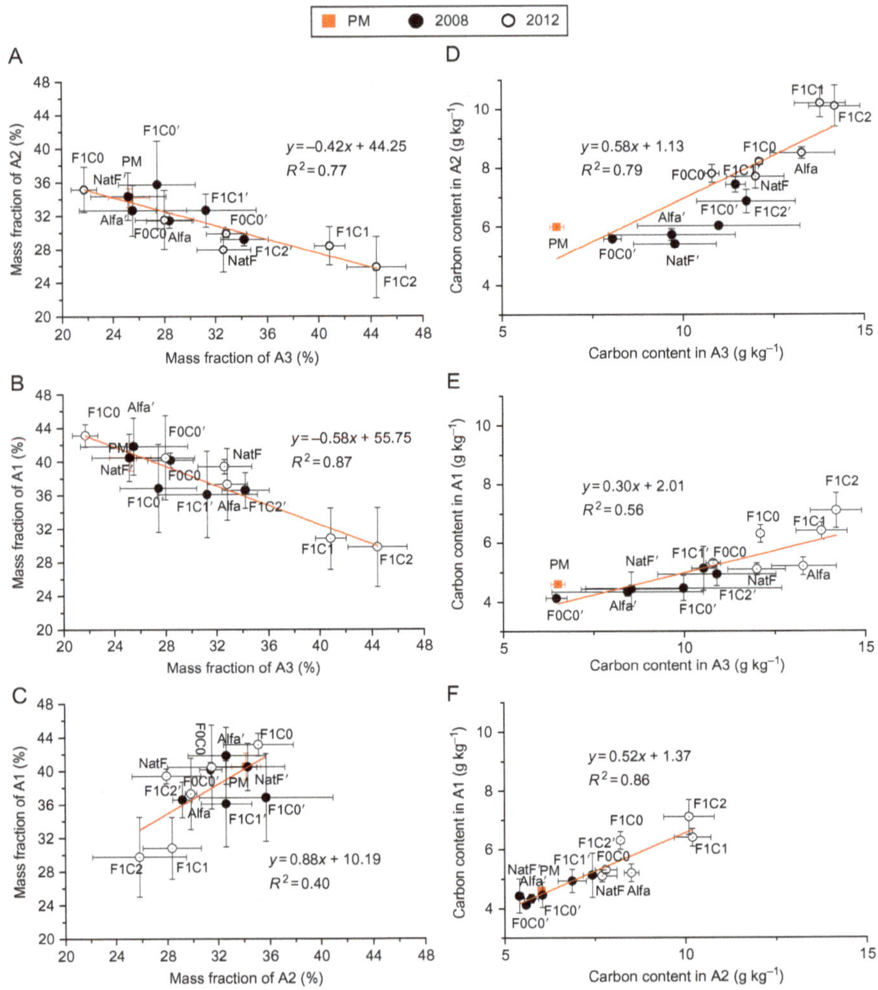

Fig. 5 Correlations in aggregate mass fraction (A–C) and aggregate carbon content (D–F) between the aggregate size classes and soil texture units using all the data sets acquired for all the field treatments in 2008 and 2012 and the parent material. A1, soil texture units (<53 μm), A2, microaggregates (53–250 μm), and A3, macroaggregates (>250 μm). The *bar* represents the standard deviation of means of three replicates.

between the A3 and A1 size classes, and 0.40 and 0.86 between the A2 and A1 size classes.

3.2 Model Calibration and Simulation

The CAST model was calibrated successfully with the measured data in 2008 and 2012 to capture over a 10-year period calculated changes in aggregate mass distribution and aggregate carbon stock for all the six field treatments (Fig. 6). The aggregate mass fraction increased substantially with time for the A3 size class and decreased for the A1 size class in the treatments NatF, Alfa, F1C1, and F1C2, with the magnitudes of the changes more profound for the treatments F1C1 and F1C2 than for the treatments NatF and Alfa. For the treatments F0C0 and F1C0, the mass fraction of the A3 size class decreased substantially with time, while the mass fractions of the A1 and A2 size classes increased slightly. The aggregate carbon stock substantially increased for the A3 aggregate size class for all the field treatments except for F0C0 and F1C1, which showed little change in aggregate carbon stock with time. All the field treatments showed little change in carbon stock for the A2 and A1 size classes.

The simulation captured less well the time evolution of the mass fraction of the soil texture units and the associated carbon stocks, as free soil texture units or incorporated into microaggregates and macroaggregates; this was observed for all field treatments (Fig. 7). Furthermore, the model was unable to distinguish the magnitude of the changes in the C content from different carbon pools within the macroaggregate size class (A3) (Fig. 8). The simulated changes in the mass fraction of the soil texture units were supported by the 2 years of measured data only for the treatment Alfa. The simulated changes in the total carbon content associated with the soil texture units were supported by the 2 years of measured data for all the field treatments. The magnitude of the simulated changes in the carbon stocks from different sources within the macroaggregate size fraction was generally not supported by the measured data. Among the carbon pools within the macroaggregate size class, only the simulated evolution of carbon stock in the C1 size class was supported by the measured data in all the field treatments except for F1C2.

3.3 Calibrated Parameters

The calibrated decay rate constants demonstrated that some parameters were the same for all the field treatments as presented in Fig. 2B and C and that others were different among the field treatments as listed in Table 2. Those

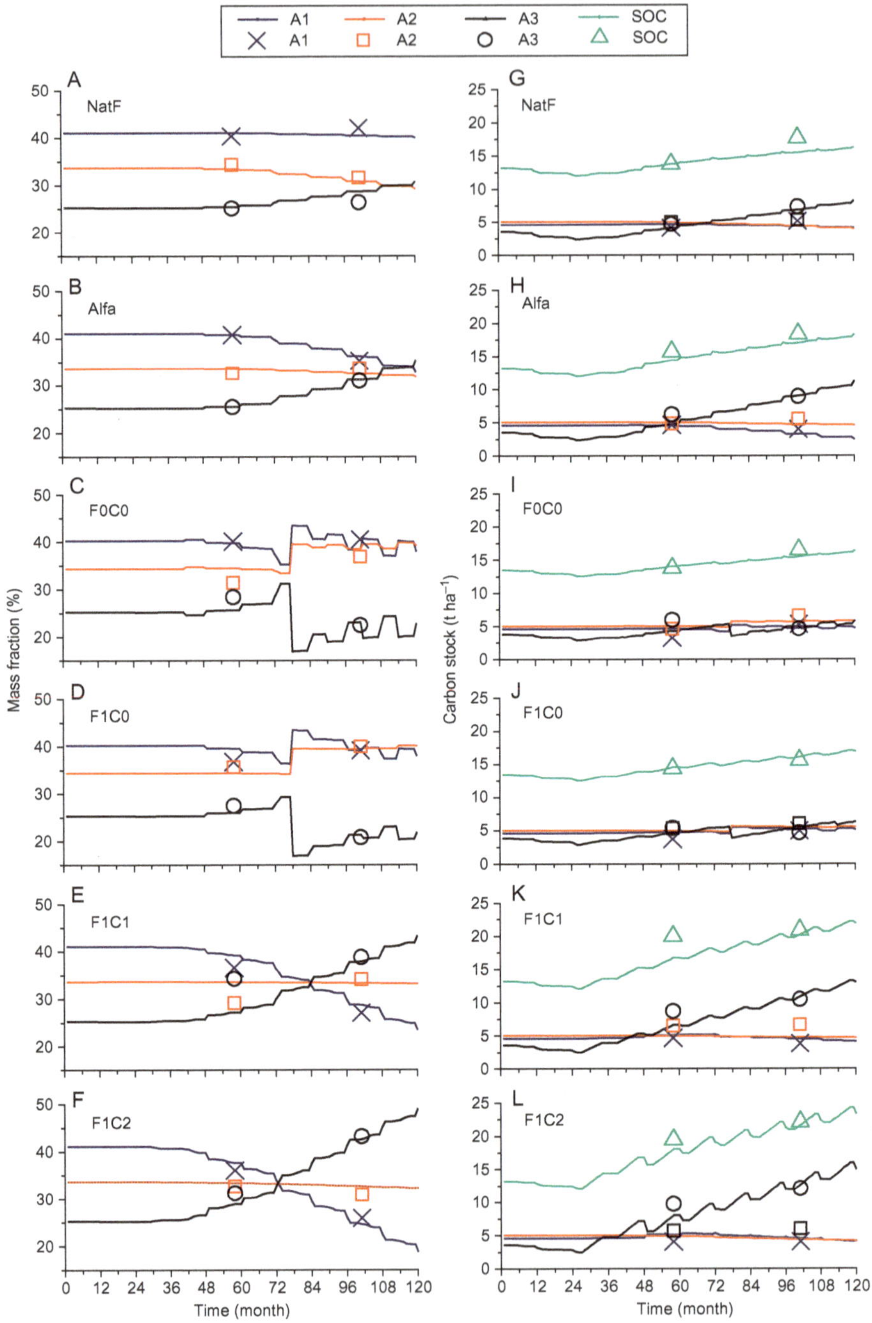

Fig. 6 Simulated evolution of aggregate mass fraction (A–F) and aggregate organic carbon stock (G–L) for the different field treatments.

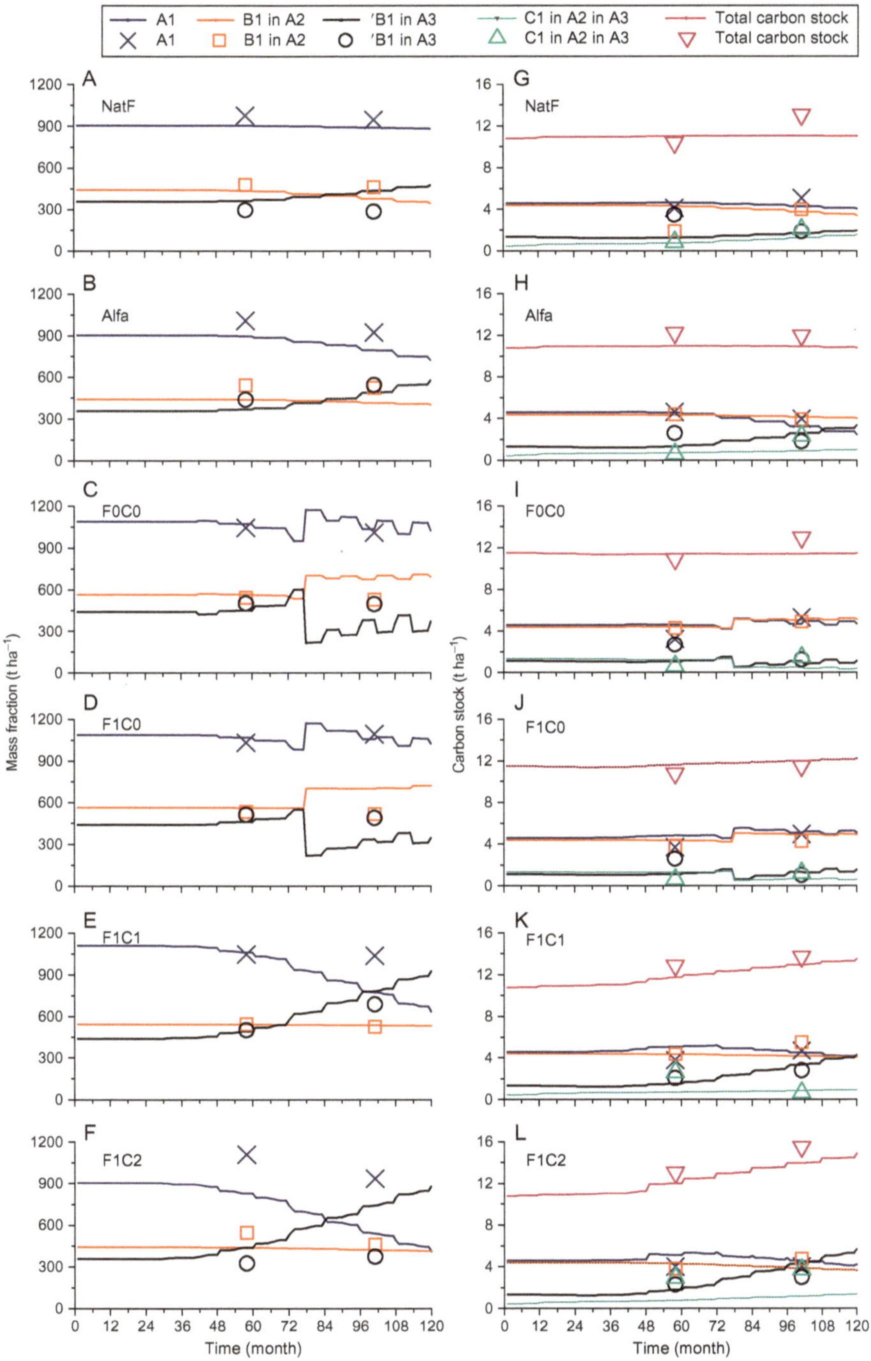

Fig. 7 Simulated evolution of mass fraction (A–F) and organic carbon stock (t ha^{-1}) (G–L) in soil texture units either in free status (A1) or within the microaggregate size class (A2) (B1 in A2) and the macroaggregate size class (A3) ('B1 in A3, C1 in A2 in A3), and in total (total) for the different field treatments.

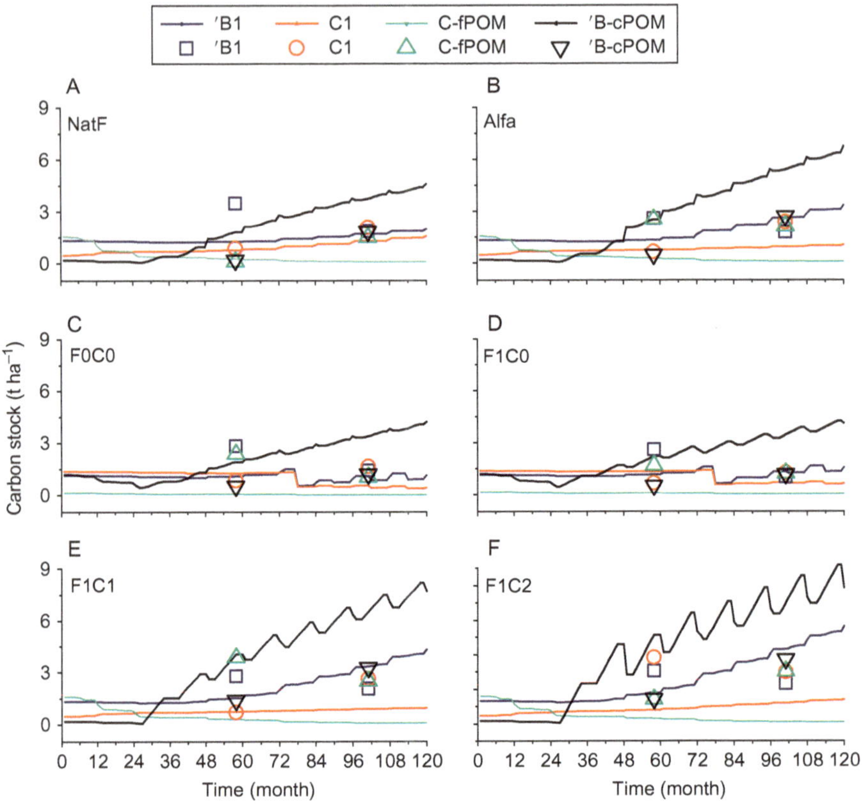

Fig. 8 Simulated evolutions of the carbon stock as in different carbon pools within the macroaggregate size class (A3), the free soil texture units ('B1), soil texture units in microaggregate size class (C1), fine particulate organic matter in the microaggregate size class (c-fPOM) and coarse particulate organic matter ('B-cPOM) for the different field treatments (A–F).

independent of field treatments were the rate constants of the fragmentation of plant materials (from RPM into cRPM and from cRPM into fRPM), the decomposition of DPM, fDPM, RPM, fRPM, HUM, and BIO carbon pools in the soil texture units in different sources (free or aggregate protected), and microaggregate and macroaggregate size classes (Fig. 2B and C). In addition, all of the field treatments had the same limiting carbon sources for the formation of secondary microaggregates within macroaggregates, which were 13.4% from fRPM carbon and 86.6% from the carbon associated with the soil texture units released from broken-down microaggregates or macroaggregates. All of these rate constants were larger during the plant growth season than during the frozen season. They also decreased from the fragmentation, and decomposition of plant litter-derived organic matter in the primary aggregates, through those in finer aggregates and soil texture units after breakdown of the primary aggregates and then to those in the secondary aggregates.

The calibrated decomposition rate constants were specific in the field treatments for the quality of plant litter (DPM/RPM ratio), the organic matter pools in nonaggregated soils (cDPM, RPM, cRPM, and fROM) and in the secondary macroaggregate size class (cRPM and cDPM) (Table 2). The DPM/RPM ratio was lowest in NatF, followed by Alfa and F0C0, F1C0, and F1C1, and greatest in F1C2. In the nonaggregated soil, the decomposition rate constants were greater in F1C0 and F1C1 compared to the other field treatments for cDPM, cRPM, and fROM during the plant growth period and greater in the fertilized soils compared to the fertilized soils for cDPM, RPM, cRPM, and fROM during the frozen season.

In addition, the carbon sources that were the least limiting factors for the formation of secondary macroaggregates were different among the field treatments. The carbon limiting factor demonstrated that the free soil texture units (A1) were the most important carbon sources, contributing 60% of carbon to form secondary macroaggregates in all the field treatments except for NatF during both plant growth and frozen seasons. In NatF, the proportion was 30%. The second important carbon sources were from cRPM in the treatments of NatF and Alfa, from cRPM and free microaggregate size class (A2) in the treatments of F0C0 and F1C0, and from cDPM in the treatments of F1C1 and F1C2 during the plant growth season and the frozen season except for treatment NatF. The carbon from the free microaggregate size class was the second important source in the treatment of NatF.

Moreover, the correction factor to adjust carbon flow with the mass flow of the soil texture units to format the secondary macroaggregates differed among the field treatments. It ranged from 1 to 2 from the microaggregate size class (A2) in the no-fertilized soils (NatF, Alfa, and F0C0) during the frozen season, from 0.1 to 1.0 from free soil texture units (A1), and from 0.01 to 0.1 in the aggregate-protected soil texture units ('B1) within the macroaggregate size (A3) during both the seasons in all treatments.

4. DISCUSSION

4.1 Soil Aggregation Processes at the Initial Stage of Soil Formation

The CAST model assumes that macroaggregates are first formed around POM and that microaggregates are then formed within the macroaggregates as the occluded POM decomposes (Stamati et al., 2013). Based on this conceptual model and the mathematical translation as the first-order kinetics for these processes, the aggregate mass distribution by size classes and the aggregate carbon content in the field treatments were well captured in 2 separate years (Fig. 6) when it was parameterized separately during the plant growth period and frozen period. The success of the CAST modeling demonstrated that the CAST model was constrained sufficiently by the data and that the primary macroaggregation process and the primary microaggregation within the macroaggregates occurred as well at the early stage of soil development for the studied Mollisol that has well-weathered PMs. These aggregation processes were also confirmed by the sum of the slopes being 1.0 for the linear correlations in mass fraction between the A3 size class and the A2 or A1 size class (Fig. 5).

The simulated changes in the mass of soil texture units and its associated carbon stock in different aggregate size classes (Fig. 7), and the carbon stock in different carbon pools within the macroaggregate size class (Fig. 8) were, however, not as well supported by the measured data. These results suggested that different or additional aggregation processes may occur at the initial soil formation stage from the PM of Mollisol, which may not be incorporated in the CAST model, or may occur differently than conceptualized in the model. For example, the positive correlations in mass fraction and in carbon content between the A2 and A1 size classes (Fig. 5C and F) suggested possible direct formation of microaggregates from basic soil texture units associating with organic matter at the early stage of soil development. The mineral surfaces of the soil texture units may provide substantial capacity to bind decomposing organic matter due to the initially low carbon content of the PM, with these texture units and associated surface-bound organic carbon then forming microaggregates. In an incubation experiment, Golchin et al. (1994a,b, 1995) demonstrated the dynamics of formation of microaggregates from mineral particles and their disaggregation into organic-mineral complexes.

There were some calibrated parameters that varied among the field treatments (Table 2), indicating that the secondary macroaggregation process was affected by land use and management practices as discussed later. The human intervention may interact with different aggregation processes, for example, influencing the availability of soil texture units and carbon availability during aggregate turnover processes. Understanding of these processes, such as direct formation of microaggregates from soil texture units, complex carbon association with soil texture units during aggregate turnovers, and incorporating these processes in the conceptual model of CAST, may improve the quantitative calculation of the mass fractions of soil texture units and their associated carbon stocks. The calibrated first-order rate constants for carbon decomposition varied with season and carbon pools associated with soil texture units and aggregates. These results indicated coupling processes between soil aggregation and organic matter incorporation in soil structure development. All the decay rate constants were much smaller for the frozen season than for the plant growth season. The lower rate constants during the frozen period correspond to inhibition of the biological processes for fragmentation of plant litter, its decomposition, and the incorporation of decomposition products into soil as humus and microbial biomass.

The decomposition rate constants of plant-derived organic matter were smaller for (1) carbon in aggregates than in nonaggregated soil, (2) aggregate-protected conditions than in free conditions, and (3) microaggregate-protected conditions than in macroaggregate-protected conditions (Fig. 2; Table 2). All of these differences in the first-order rate constants indicated that SOM received greater protection within aggregates, particularly in the microaggregates within macroaggregates, which was consistent with the processes described by the aggregate hierarchical theory (Six et al., 2000, 2004; Tisdall and Oades, 1982) that developed from observations on mature soils. Previous literature reported inhibition of soil aggregation due to loss of SOM as corroboration of the aggregate hierarchical theory (Tisdall and Oades, 1982) by comparing the effect of different land uses (Elliott, 1986; Kara and Baykara, 2014) and soil management practices (Bidisha et al., 2010; Elmholt et al., 2008; Six et al., 2000) on soil structure development. Our modeling corroborates the theory using time-course data under the same land uses as one of the first studies of this kind.

The modeling results also demonstrated that some calibrated parameters were independent of field treatments (Fig. 2B and C), but others were dependent on the field treatments (Table 2). The field treatment-independent parameters included the first-order rate constants of the fragmentation and decomposition of plant litter within the macroaggregates, the decomposition of organic matter incorporated and protected in microaggregates, and the carbon availability of fine POM and free soil texture units released from the breakdown of macroaggregates. Therefore, those same calibrated parameters suggest that the rates of the primary macroaggregation and the secondary microaggregation processes were likely controlled by the intrinsic properties of the PM rather than by land use and agricultural practices. In contrast, the calibrated parameters that differed among the field treatments included the first-order rate constants of the plant-derived organic matter in the nonaggregated soils and in the secondary macroaggregates, the limiting carbon sources in the secondary macroaggregates, and the carbon flow efficiency with the mass flow of the soil texture units. Thus, the differences in the parameters among the field treatments suggested that the secondary macroaggregation processes were controlled by the land use and agricultural practices as discussed later.

The same rates of the primary macroaggregation and the secondary microaggregation among the field treatments suggested that these processes were not inhibited, likely because of sufficient soil texture units and inputs of plant litter materials to provide mineral surface area and organic carbon for mineral–carbon association. The PMs of the Mollisol were well-weathered loess-like materials; physical weathering from large clods into fine soil texture units was very strong due to strong freezing and thawing processes. The breakdown of primary macroaggregates would provide sufficient free fine RPMs in all of the field treatments, to drive microaggregate formation, and this secondary microaggregation process may occur preferentially compared to the secondary macroaggregation process, which was not affected by the field treatments. Such rapid microaggregation processes have been reported elsewhere in the literature (Malamoud et al., 2009; Puget et al., 2000; Six et al., 2002a,b, 2004).

4.2 Effects of Agricultural Practices on Soil Aggregation and SOM

SOM dynamics is the balance between organic inputs from plant production and outputs through decomposition (Bronick and Lal, 2005; Johnston et al., 2009; von Lützow et al., 2006), and the decomposition rates are considered to be further governed by environmental and biological factors, rather than the recalcitrance of SOM (Dungait et al., 2012; Schmidt et al., 2011; Stockmann et al., 2013). The calibrated parameters demonstrated that the rates of the primary macroaggregation and microaggregation and the secondary microaggregation were the same for all the treatments. The similar magnitudes of the increases in POM carbon in all the field treatments (Fig. 4D) confirmed little effect of land use and agricultural practices through secondary microaggregation. Therefore, the differences in carbon content among the field treatments were attributed to the amount of organic carbon inputs and the factors controlling the secondary macroaggregation.

The calibrated parameters that differed among the field treatments included (1) the DPM to RPM ratio, (2) the decomposition rate constant of plant litter-derived organic matter in nonaggregated soil and in the secondary macroaggregates, (3) the carbon sources limiting the formation of secondary macroaggregates, and (4) the correction factor to adjust for the mass flow of soil structure units during aggregate turnover (Table 2). The difference in DPM to RPM ratio among the field treatments shows that the plant litter decomposed more slowly in the grasslands than in the cropped fields, in crop roots than in crop straw, both observations which are consistent with literature (Shiratio and Yokozawa, 2006; Woomer et al., 1994). Nitrogen availability may explain the higher DPM to RPM ratio in Alfa than in NatF, in F1C0 than in F0C0 (Finn et al., 2015; Kochsiek and Knops, 2013), and the high protein in soybean applied to the soil may be another reason for the lower DPM to RPM ratio in F1C1 than in F1C2. The plant litter quality further influenced the decomposition of plant litter-derived organic matter (cDPM, cRPM, and fRPM) in nonaggregated soils and in the secondary macroaggregates (A3). The plant litter quality further influenced carbon from POM, and the soil texture unites as indicated by the correction factor to adjust the mass flow of the soil texture units to form the secondary macroaggregates.

The decomposition rate constants of cRPM and cDPM in the A3 size class were larger for the F1C1 and F1C2 treatments, compared to the other field treatments, demonstrating that the macroaggregate turnover was much faster for the F1C1 and F1C2 treatments (4.8–5.7 vs 23.7–30.4 years), likely due to their high decomposability as well as the larger amount of organic inputs. The higher the decomposability and the larger the amount of organic input, the more active the microbial communities (Li et al., 2014) and, plausibly, the more rapid and extensive the aggregation (Kong et al., 2005; Trivedi et al., 2015). The POM pools were the largest liming carbon source for the formation of the secondary macroaggregates in treatment NatF, which was also likely attributed to its relatively low decomposition rate. In addition, the treatments with lower organic carbon input (F0C0 and F1C0) were limited more by the carbon from the microaggregate size class (A2) and less by that from cDPM and/or cRPM in the formation of the secondary macroaggregates compared to the treatments with higher organic carbon input (F1C1 and F1C2). These results suggested faster breakup of the macroaggregates as indicated by the larger decomposition decay rate constants of cRPM and cDPM in F1C1 and F1C2.

The correction factor describes whether mineral surface area, or the amount of available organic matter for binding to the mineral surface, is the limiting factor for the formation of secondary macroaggregates

(Stamati et al., 2013). A correction factor value close to one means that the mass flow of soil texture units is linearly proportional to the organic carbon flow. A value significantly higher than one suggests that substantial parts of the mineral surfaces are not covered by organic matter. A value significantly lower than one indicates an excess of organic matter with high organic C concentration, where the microaggregation is induced. Therefore, the lower correction factors, suggested higher carbon protection efficiency or more carbon bound to the mineral surface and protected during the formation of the secondary macroaggregates. Therefore, the quality and quantity of plant litter play a central role in determining the carbon availability and the carbon protection efficiency for secondary macroaggregation. This concept is consistent with traditional concepts that SOM is stabilized through selective decomposition (Sollins et al., 2007; von Lützow et al., 2006), but different from those recently proposed concepts that plant litter recalcitrance is less important for SOM stabilization (Dungait et al., 2012; Schmidt et al., 2011; Stockmann et al., 2013).

The differences in the calibrated parameters among the field treatment also suggest the direct role of tillage on the secondary aggregation. The tilled soils had the largest carbon contribution from soil texture units and lower carbon contributions from POM particularly in the F0C0 and F1C0 treatments compared to the nontilled treatments of NatF and Alfa, though they had similar amount of organic carbon inputs. Soil tillage mechanically abrades PM and primary macroaggregates, releasing soil texture units and POM. However, the decomposition rate constant and the correction factor to adjust carbon and mass flow of soil texture units demonstrated that the POM decomposition was much faster in the nonaggregated soils and the soil texture unit-associated carbon was much lower in the tilled soils than in the nontilled soils. Therefore, soil tillage increased the amount of free soil texture units, but decreased carbon protection efficiency, which likely impeded secondary macroaggregation at low organic carbon input. Despite no-tillage, the carbon contribution from the soil texture unit in Alfa was as high as in the tilled soils, and the high carbon availability from the soil texture units was likely attributed to the lower soil pH under N-fixing alfalfa. The soil carbon availability from the soil texture units is high at low soil pH, when bacteria are dominant over fungi, resulting in low aggregate stability (Leifeld et al., 2008; Nicolardot et al., 2001). Although the treatment F0C0 exhibited a low DPM to RPM ratio as did the treatment Alfa, it had the highest values for the correction factor (Table 2), indicating that it had the lowest carbon protection efficiency when the soil texture units were incorporated into the secondary macroaggregation due to soil tillage.

Therefore, the land use and agricultural practices influenced secondary macroaggregation by determining the quality and quantity of plant litter that controlled the decomposition rate of the plant-derived organic matter after the breakdown of primary macroaggregates, and by soil tillage that influenced the carbon sources and their availability from the soil texture units and also affected the decomposition rate of protected POM after it was released from the breakdown of the primary macroaggregates.

5. CONCLUSION

The CAST model captured well the dynamics of soil structure development and carbon sequestration observed in 2 separate years for the different field treatments emulating the early stage of soil development from the PM of a Mollisol under different land use and agricultural practices. Some calibrated parameters were independent of the field treatments, indicating that the rates of primary macroaggregation and microaggregation and the formation of secondary microaggregates were similar among the field treatments despite differing in the plant growth season and in the frozen winter season. Other calibrated parameters varied between the field treatments, indicating that the rates and extent of secondary macroaggregate formation were dependent on the quality and quantity of plant litter input and soil tillage practices. For these cases, the variations in soil structure and carbon stocks among the field treatments were mainly caused during the secondary macroaggregation, but not during the primary macroaggregation. However, the CAST model did not capture particularly well the mass of the soil texture units within different aggregates, and it did not capture the magnitude of the trends in the carbon stocks of different carbon pools within aggregates. This suggest that processes affecting these results may be improved by incorporating additional information, such as considering primary microaggregation process directly from basic soil texture units. This hypothesis needs further studies using different soil types and time series data so as to incorporate it into dynamic models such as the CAST model. Nevertheless, this study demonstrates that the CAST model is a powerful tool to analyze and to predict the dynamics of soil aggregation and concurrent SOM changes in response to changes in land use and soil management.

ACKNOWLEDGMENTS

The work was partially funded by the National Natural Science Foundation of China (41330855, 41571219) and FP-7 (Soil TrEC, 244118). B.Z. was supported by the Hundred Talents Program of the Chinese Academy of Sciences. N.L. was supported by the Youth Innovation Promotion Association CAS (2016211) and Excellent Young Talent Program of Northeast Institute of Geography and Agroecology, CAS (DLSYQ15003).

REFERENCES

Andrianaki, M., Bernasconi, S.M., Nikolaidis, N.P., 2017. Quantifying the incipient development of soil structure and functions within a glacial forefield chronosequence. Adv. Agron. 142, 215–239.

Angers, D.A., Bolinder, M.A., Carter, M.R., Gregorich, E.G., Drury, C.F., Liang, B.C., Voroney, R.P., Simard, R.R., Donald, R.G., Beyaert, R.P., Martel, J., 1997. Impact of tillage practices on organic carbon and nitrogen storage in cool, humid soils of eastern Canada. Soil Tillage Res. 41 (3–4), 191–201.

Balesdent, J., Chenu, C., Balabane, M., 2000. Relationship of soil organic matter dynamics to physical protection and tillage. Soil Tillage Res. 53, 215–230.

Banwart, S., Bernasconi, S.M., Bloem, J., Blum, W., Brandao, M., Brantley, S., Chabaux, F., Duffy, C., Kram, P., Lair, G., Lundin, L., Nikolaidis, N., Novak, M., Panagos, P., Ragnarsdottir, K.V., Reynolds, B., Rousseva, S., de Ruiter, P., van Gaans, P., van Riemsdijk, W., White, T., Zhang, B., 2011. Soil processes and functions in Critical Zone Observatories: hypotheses and experimental design. Soil Sci. Soc. Am. J. 10, 974–987.

Bidisha, M., Joerg, R., Yakov, K., 2010. Effects of aggregation processes on distribution of aggregate size fractions and organic C content of a long-term fertilized soil. Eur. J. Soil Biol. 46 (6), 365–370.

Bronick, C.J., Lal, R., 2005. Soil structure and management: a review. Geoderma 124, 3–22.

Chazdon, R.L., 2008. Beyond deforestation: restoration forests and ecosystem services on degraded lands. Science 320, 1458–1460.

Coleman, K., Jenkinson, D.S., 1999. RothC-26.3—A Model for the Turnover of Carbon in Soil: Model description and Windows Users Guide. November 1999 issue, Lawes Agricultural Trust, Harpenden, UK. ISBN: 0 951 4456 8 5.

Coleman, K., Jenkinson, D.S., 2004. RothC—A Model for the Turnover of Carbon in Soil: Model Description and User's Guide. Rothamsted Research, Harpenden, Herts.

Crawford, J.W., 1994. The relationship between structure and the hydraulic conductivity of soil. Eur. J. Soil Sci. 45 (4), 493–502.

Denef, K., Six, J., Bossuyt, H., Frey, S.D., Elliott, E.T., Merckx, R., Panstian, K., 2001. Influence of dry-wet cycles on the interrelationship between aggregate, particulate organic matter, and microbial community dynamics. Soil Biol. Biochem. 33, 1599–1611.

Dungait, J.A.J., Hopkins, D.W., Gregory, A.S., Whitmore, A.P., 2012. Soil organic matter turnover is governed by accessibility not recalcitrance. Glob. Chang. Biol. 18, 1761–1796.

Elliott, E.T., 1986. Aggregate structure and carbon, nitrogen, and phosphorus in native and cultivated soils. Soil Sci. Soc. Am. J. 50, 627–633.

Elmholt, S., Schjønning, P., Munkholm, L.J., Debosz, K., 2008. Soil management effects on aggregate stability and biological binding. Geoderma 144, 455–467.

Finn, D., Page, K., Catton, K., Strounina, E., Kienzle, M., Robertson, F., Armstrong, Roger, Dalal, R., 2015. Effect of added nitrogen on plant litter

decomposition depends on initial soil carbon and nitrogen stoichiometry. Soil Biol. Biochem. 91, 160–168.
Giménez, D., Allmaras, R.R., Nater, E.A., Huggins, D.R., 1997. Fractal dimensions for volume and surface of interaggregate pores-scale effects. Geoderma 77, 19–38.
Golchin, A., Oades, J.M., Skjemstad, J.O., Clarke, P., 1994a. Soil structure and carbon cycling. Aust. J. Soil Res. 32, 1043–1068.
Golchin, A., Oades, J.M., Skjemstad, J.O., Clarke, P., 1994b. Study of free and occluded particulate organic matter in soils by solid state ^{13}C CP/MAS NMR spectroscopy and scanning electron microscopy. Aust. J. Soil Res. 32, 285–309.
Golchin, A., Clark, P., Oades, J.M., Skjemstad, J.O., 1995. The effects of cultivation on the composition of organic matter and structural stability of soils. Aust. J. Soil Res. 33, 975–993.
Griffiths, B.S., Liu, M.Q., Wang, H.L., Zhang, B., Kuan, H.L., McKenzie, B.M., Hallett, P.D., Neilson, R., Daniell, T.J., 2008. Restoration of soil physical and biological stability are not coupled in response to plants and earthworms. Ecol. Restor. 26, 102–104.
Harris, J., 2009. Soil microbial communities and restoration ecology: facilitators or followers? Science 325, 573–574.
Harrison, R.B., Strahm, B., 2008. Soil formation. In: Jorgensen, S.E., Fath, B. (Eds.), Encyclopedia of Ecology. Elsevier, pp. 3291–3295.
Haygarth, P.M., Ritz, K., 2009. The future of soils and land use in the UK: soil systems for the provision of land-based ecosystem services. Land Use Policy 26, S187–S197.
Helfrich, M., Ludwig, B., Potthoff, M., Flessa, H., 2008. Effect of litter quality and soil fungi on macroaggregate dynamics and associated partitioning of litter carbon and nitrogen. Soil Biol. Biochem. 40, 1823–1835.
Jastrow, J.D., 1996. Soil aggregation formation and the accrual of particulate and mineral-associated organic matter. Soil Biol. Biochem. 28, 665–676.
Jastrow, J.D., Amonette, J.E., Bailey, V.L., 2007. Mechanisms controlling soil carbon turnover and their potential application for enhancing carbon sequestration. Climat. Change 80, 5–23.
Johnston, A.E., Poulton, P.R., Coleman, K., 2009. Soil organic matter: its importance in sustainable agriculture and carbon dioxide fluxes. Adv. Agron. 101, 1–57.
Kara, O., Baykara, M., 2014. Changes in soil microbial biomass and aggregate stability under different land uses in the northeastern Turkey. Environ. Monit. Assess. 186 (6), 3801–3808.
Kochsiek, A.E., Knops, J.M.H., 2013. Effects of nitrogen availability on the fate of litter-carbon and soil organic matter decomposition. Br. J. Environ. Clim. Change 3 (1), 24–43.
Kong, A.Y.Y., Six, J., Bryant, D.C., Denison, R.F., van Kessel, C., 2005. The relationship between carbon input, aggregation, and soil organic carbon stabilization in sustainable cropping systems. Soil Sci. Soc. Am. J. 69 (4), 1078–1085.
Kuka, K., Franko, U., Rühlmann, J., 2007. Modelling the impact of pore space distribution on carbon turnover. Ecol. Model. 208, 295–306.
Lal, R., 2003. Soil erosion and the global carbon budget. Environ. Int. 29, 437–450.
Leifeld, J., Zimmermann, M., Fuhr, J., 2008. Simulating decomposition of labile soil organic carbon: effects of pH. Soil Biol. Biochem. 40 (12), 2948–2951.
Li, N., Yao, S.H., You, M.Y., Zhang, Y.L., Qiao, Y.F., Zou, W.X., Han, X.Z., Zhang, B., 2014. Contrasting development of soil microbial community structure under nottilled perennial and tilled cropping during early pedogenesis of a Mollisoll. Soil Biol. Biochem. 77, 221–232.
Li, N., Yao, S.H., Qiao, Y.F., Zou, W.X., You, M.Y., Han, X.Z., Zhang, B., 2015. Separation of soil microbial community structure by aggregate size to a large extent under agricultural practices during early pedogenesis of a Mollisol. Appl. Soil Ecol. 88, 9–20.

Li, N., Pan, F.J., Han, X.Z., Zhang, B., 2016. Development of soil food web of microbes and nematodes under different agricultural practices during the early stage of pedogenesis of a Mollisol. Soil Biol. Biochem. 98, 208–216.

Lichter, K., Govaerts, B., Six, J., Sayre, K.D., Deckers, J., Dendooven, L., 2008. Aggregation and C and N contents of soil organic matter fractions in a permanent raised-bed planting system in the highlands of Central Mexico. Plant Soil 305, 237–252.

Liu, X.B., Zhang, S.L., Zhang, X.Y., Ding, G.W., Cruse, R.M., 2011. Soil erosion control practices in Northeast China: a mini-review. Soil Tillage Res. 117, 44–48.

Liu, Y.-L., Yao, S.-H., Li, N., Qiao, Y.F., Zou, W.X., Han, X.Z., Zhang, B., Banwart, S.A., 2017. Soil mineralogy changes with different agricultural practices during 8-year soil development from the parent material of a Mollisol. Adv. Agron. 142, 143–179.

Malamoud, K., McBratney, A.B., Minasny, B., Field, D.J., 2009. Modeling how carbon affects soil structure. Geoderma 149, 19–26.

Nicolardot, B., Recous, S., Mary, B., 2001. Simulation of C and N mineralisation during crop residue decomposition: a simple dynamic model based on the C:N ratio of the residues. Plant Soil 228, 83–103.

Oades, J.M., 1984. Soil organic matter and structural stability: mechanisms and implications for management. Plant Soil 76 (1–3), 319–337.

Oades, J.M., Waters, A.G., 1991. Aggregate hierarchy in soils. Aust. J. Soil Res. 29, 815–828.

Pagliai, M., Vignozzi, N., Pellegrini, S., 2004. Soil structure and the effect of management practices. Soil Tillage Res. 79, 131–143.

Panakoulia, S.K., Nikolaidis, N.P., Paranychianakis, N.V., Menon, M., Schiefer, J., Lair, G.J., Krám, P., Banwart, S.A., 2017. Factors controlling soil structure dynamics and carbon sequestration across different climatic and lithological conditions. Adv. Agron. 142, 241–276.

Papadopoulos, A., Bird, N.R.A., Whitmore, A.P., Mooney, S.J., 2009. Investigating the effects of organic and conventional management on soil aggregate stability using X-ray computed tomography. Eur. J. Soil Sci. 60, 360–368.

Peth, S., Horn, R., Beckmann, F., Donath, T., Fischer, J., Smucker, A., 2008. Three-dimensional quantification of intra-aggregate pore-space features using synchrotron-radiation-based microtomography. Soil Sci. Soc. Am. J. 72, 897–907.

Plante, A.F., McGill, W.B., 2002. Soil aggregate dynamics and the retention of organic matter in laboratory-incubated soil with differing simulated tillage frequencies. Soil Tillage Res. 66, 79–92.

Powlson, D.S., Whitmore, A.P., Goulding, K.W.T., 2011. Soil carbon sequestration to mitigate climate change: a critical re-examination to identify the true and the false. Eur. J. Soil Sci. 62, 42–55.

Puget, P., Chenu, C., Balesdent, J., 2000. Dynamics of soil organic matter associated with particle-size fractions of water-stable aggregates. Eur. J. Soil Sci. 51, 595–605.

Quinton, J.N., Govers, G., Van Oost, K., Bardgett, R.D., 2010. The impact of agricultural soil erosion on biogeochemical cycling. Nat. Geosci. 3, 311–314.

Ritz, K., Young, L., 2011. The Architecture and Biology of Soils: Life in Inner Space. CABI, Wallingford, Oxfordshire, UK.

Schmidt, M.W.I., Torn, M.S., Abiven, S., Dittmar, T., Guggenberger, G., Janssens, I.A., Kleber, K., Kögel-Knabner, I., Lehmann, J., Manning, D.A.C., Nannipieri, P., Rasse, D.P., Weiner, S., Trumbore, S.E., 2011. Persistence of soil organic matter as an ecosystem property. Nature 478, 49–56.

Schulz, S., Brankatschk, R., Dümig, A., Kögel-Knabner, I., Schloter, M., Zeyer, J., 2013. The role of microorganisms at different stages of ecosystem development for soil formation. Biogeosciences 10, 3983–3996.

Shiratio, Y., Yokozawa, M., 2006. Acid hydrolysis to partition plant material into decomposable and resistant fractions for use in the Rothamsted carbon model. Soil Biol. Biochem. 38, 812–816.
Six, J., Elliott, E.T., Paustian, K., Doran, J.W., 1998. Aggregation and soil organic matter accumulation in cultivated and native grassland soils. Soil Sci. Soc. Am. J. 62, 1367–1377.
Six, J., Elliott, E.T., Paustian, K., 2000. Soil macroaggregate turnover and microaggregate formation: a mechanism for C sequestration under no-tillage agriculture. Soil Biol. Biochem. 32, 2099–2103.
Six, J., Conant, R.T., Paul, E.A., Paustian, K., 2002a. Stabilization mechanisms of soil organic matter: implications for C-saturation of soils. Plant Soil 241, 155–176.
Six, J., Feller, C., Denef, K., Ogle, S.M., de Moraes Sa, J.C., Albrecht, A., 2002b. Soil organic matter, biota and aggregation in temperate and tropical soils—effects of no-tillage. Agronomy 22, 755–775.
Six, J., Bossuyt, H., Degryze, S., Denef, K., 2004. A history of research on the link between (micro) aggregates soil biota, and soil organic matter dynamics. Soil Tillage Res. 79, 7–31.
Sollins, P., Swanston, C., Kramer, M., 2007. Stabilization and destabilization of soil organic matter—a new focus. Biogeochemistry 85, 1–7.
Stamati, F.E., Nikolaidis, N.P., Banwart, S., Blum, W.E.H., 2013. A coupled carbon, aggregation, and structure turnover (CAST) model for topsoils. Geoderma 211–212, 51–64.
Stockmann, U., Adams, M.A., Crawford, J.W., Field, D.J., Henakaarchchi, N., Jenkins, M., Minasny, B., McBratney, A.B., Courcelles, V.D.R., Singh, K., Wheeler, I., Abbott, L., Angers, D.A., Baldock, J., Bird, M., Brookes, P.C., Chenu, C., Jastrow, J.D., Lal, R., Lehmann, J., O'Donnell, A.G., Parton, W.J., Whitehead, D., Zimmermann, M., 2013. The knowns, known unknowns and unknowns of sequestration of soil organic carbon. Agric. Ecosyst. Environ. 164, 80–99.
Thomaz, E.L., Luiz, J.C., 2012. Soil loss, soil degradation and rehabilitation in a degraded land area in Guarapuava (Brazil). Land Degrad. Dev. 23 (1), 72–81.
Tisdall, J.M., Oades, J.M., 1982. Organic matter and water stable aggregates in soils. J. Soil Sci. 33, 141–163.
Trivedi, P., Rochester, I.J., Trivedi, C., Van Nostrand, J.D., Zhou, J.Z., Karunartne, S., Andserson, I.C., Singh, B.K., 2015. Soil aggregate size mediates the impacts of cropping regimes on soil carbon and microbial communities. Soil Biol. Biochem. 91, 169–181.
von Lützow, M., Köl-Knabner, I., Ekschmitt, K., Matzner, E., Guggenberger, G., Marschner, B., Flessa, H., 2006. Stabilization of organic matter in temperate soils: mechanisms and their relevance under different soil conditions—a review. Eur. J. Soil Sci. 57, 426–445.
Wilkinson, M.T., Humphreys, G.S., 2005. Exploring pedogenesis via nuclide-based soil production rates and OSL-based bioturbation rates. Aust. J. Soil Res. 43, 767–779.
Woomer, P.L., Martin, A., Albrecht, A., Resck, D.V.S., Scharpenseel, H.W., 1994. The importance and management of soil organic matter in the tropics. In: Woomer, P.L., Swift, M.J. (Eds.), The Biological Management of Tropical Soil Fertility. Wiley, Chichester, UK, pp. 47–80.
Wu, K.J., Zhang, X.X., 2011. Modelling soil structure and processes. In: Ritz, K., Young, I. (Eds.), The Architecture and Biology of Soils: Life in Inner Space. CPI Group (UK) Ltd., Croydon, pp. 13–36.
Young, I., Crawford, J., 2004. Interactions and self-organization in the soil-microbe complex. Science 304 (5677), 1634–1637.

Zhang, B., Yang, Y.S., Zepp, H., 2004. Effect of vegetation restoration on soil and water erosion and nutrient losses of a severely eroded clayey Plinthudult in Southeastern China. Catena 57, 77–90.
Zhang, B., Wang, H.L., Yao, S.H., Bi, L.D., 2013. Litter quantity confers soil functional resilience through mediating soil biophysical habitat and microbial community structure on an eroded bare land restored with mono *Pinus massoniana*. Soil Biol. Biochem. 57, 556–567.

CHAPTER EIGHT

Quantifying the Incipient Development of Soil Structure and Functions Within a Glacial Forefield Chronosequence

M. Andrianaki*[,1], S.M. Bernasconi*, N.P. Nikolaidis[†]
*Geological Institute, ETH Zurich, Zürich, Switzerland
[†]School of Environmental Engineering, Technical University of Crete, University Campus, Chania, Greece
[1]Corresponding author: e-mail address: maria.andrianaki@erdw.ethz.ch

Contents

Abstract

In this study, we used the newly developed mathematical model CAST (carbon, aggregation, and structure turnover) to simulate the accumulation of soil organic carbon and the development of soil structure along the 150 years soil chronosequence of the Damma glacier Critical Zone Observatory (CZO). As first step in the modeling procedure, an adaptation of the widely used multipool Rothamsted carbon model RothC was tested. The calibration of the model is based on the extensive dataset available for the Damma glacier CZO and a published reconstruction of climate back to 1867. The adapted RothC model simulates effectively the dynamics of total organic

Advances in Agronomy, Volume 142
ISSN 0065-2113
http://dx.doi.org/10.1016/bs.agron.2016.10.013

carbon (TOC), microbial biomass (BIO), and humified organic matter (HUM) pools, while decomposable plant material (DPM) and resistant plant material (RPM) pools which are calculated by the model are slightly overestimated compared to measured values. In a second set of simulations, the CAST model was applied and captured effectively the carbon content of the three aggregate classes which are considered by the model (macroaggregates: >250 μm (AC3), microaggregates: 53–250 μm (AC2), silt–clay-sized aggregates: <53 μm (AC1)). The major percentage of aggregates belongs to the AC3 class, which consists of the particulate organic matter primarily loosely bound with the mineral particles. The amount of aggregates within the AC1 aggregate class is low, while the formation of the AC2 aggregates has not been initiated at this stage of soil development. One of the most important finding of this study is that the CAST model realistically calculates the processes during the initial stages of soil structure development, even though the soils of the Damma glacier CZO soil chronosequence are characterized by very poor soil structure and low clay content. The main process for soil development at this early stage is identified to be the macroaggregation through the addition of particulate organic matter. Microaggregation of smaller-sized particles and aggregates is not found to be a major contribution to soil structure development at this stage and it is only observed at the end of the chronosequence. Finally, despite the extreme climatic conditions of the alpine environment, the young nature of the soils and the short growing season, both models performed very well after calibration of the carbon turnover rates.

1. INTRODUCTION

1.1 Soil Organic Carbon Modeling

The dynamics of soil organic carbon (SOC) are fundamental controls on the development and maintenance of soil functions. A better understanding of SOC dynamics, therefore, is a prerequisite for predicting the response of soil systems to changes in land-use, climate, and/or management practices. Many models have been developed to simulate the storage and turnover of SOC and most of them are based on the approach that SOC is distributed in mass pools with different turnover rates (Coleman and Jenkinson, 1999; Parton et al., 1988). The number of the pools is between two to six and the characteristic time scale of carbon turnover for the different pools can vary from a few weeks to thousands of years (Smith et al., 1997). Each of these pools represents a real contribution to the total mass SOC such as, for example, plant litter, stable or humified organic matter, and different forms of stored organic matter bound to mineral particles or within soil aggregates. In order to account for environmental conditions that affect soil carbon mineralization, soil moisture, or plant growth, the turnover rates are modified using empirical factors or the models are coupled with simpler submodels that modify the rate constants for soil carbon decomposition.

Even though SOC modeling has advanced significantly, there are still limitations on our ability to predict present but also future SOC dynamics. Smith et al. (1997) compared the performance of nine models using data from seven long-term experiments. They concluded that the increased uncertainty due to changes in environmental conditions can be addressed by the coupling of SOC models with other submodels, despite the fact that it increases the complexity of the modeling procedure and the data requirements. Furthermore, another important limitation of the SOC models is that they fail to include successfully physical protection of organic carbon that is associated with mineral particles or soil aggregates. Many recent studies showed that the most important mechanisms for the stabilization of SOC are (1) interaction/adsorption of organic matter on silt- and clay-sized mineral particles and (2) the protection of SOC that is occluded within macro- or microaggregates (Six et al., 2002). The amount of SOC in the different pools is controlled mainly by these mechanisms and less by the initial plant input and therefore it is crucial that these processes are described by the models. The most recent and complete modeling attempt, with the addition of the physical protection mechanisms, is the development of the carbon, aggregation, and structure turnover (CAST) model by Stamati et al. (2013), which is a combination of the STRUC-C model, which simulates soil aggregate stability/turnover and soil structure (Malamoud et al., 2009), and the RothC model for soil carbon dynamics.

In addition, it is essential to address the fact that most of the available models have been developed for agricultural or forested soils with site-specific calibration. Subsequently, there is the need to test their performance under different types of soils and climatic conditions. For this reason, here we used an adapted version of the multipool Rothamsted carbon model, ROTHC (Coleman and Jenkinson, 1999) and the CAST model (Stamati et al., 2013), to simulate carbon build-up in the soil chronosequence of the Damma glacier Critical Zone Observatory (CZO) forefield, which is characterized by extreme climatic conditions and special soil types.

1.2 The RothC Model

The RothC model was developed to simulate SOC stocks in the Rothamsted long-term field experiments based in the United Kingdom.

It has been applied to assess soil carbon dynamics in agricultural land, grassland, and forested land in different climates (Falloon and Smith, 2002; Ludwig et al., 2010; Sainju et al., 2007; Todorovic et al., 2010). SOC is distributed into five different pools. The main carbon mass input described by the model is the plant material that reaches the soil surface together with root exudates, and is separated conceptually into decomposable (DPM) and resistant (RPM) plant material with an empirical DPM:RPM ratio. DPM and RPM decompose to form CO_2, microbial biomass (BIO), and humified organic matter (HUM) following a first-order reaction. The BIO and HUM pools are further decomposed to CO_2, BIO, and HUM. Decomposition rates are corrected for temperature with the factor (a); for water deficit with the factor (b); and for the soil cover with the factor (c) (Coleman and Jenkinson, 1999). Finally, inert organic matter (IOM) with a nominal radiocarbon age of 50,000 years is stored in the IOM pool and can be estimated using the equation proposed by Falloon et al. (1998). Despite its simplicity, RothC model shows a good performance under a range of land uses and climate conditions (Smith et al., 1997). In this study, an adapted version of RothC model was used, as a first step in the calibration procedure, in order to confirm the applicability of both RothC and CAST on a glacier forefield chronosequence and to define the values of common model parameters. The RothC model was chosen because it is one of the two models that were used as a basis for the development of the CAST model.

1.3 The CAST Model

The CAST model is one of the carbon models that incorporates the physical protection of SOC within the soil aggregates. It is based on the theory that macroaggregates are formed in the presence of particulate organic matter (POM), after colonization of the latter by communities of decomposer microbes and encrustation of POM by interaction with clay particles and microaggregates to form macroaggregates. The decomposition of the POM fragment results in disruption of aggregated particles and the release of microaggregates (Six et al., 2002; Stamati et al., 2013). The aggregate types considered in the model are (i) the silt–clay-sized aggregates which have a particle size <53 μm, named AC1; (ii) the microaggregates of the size fraction 53–250 μm, named AC2; and (iii) the macroaggregates, which are >250 μm, named AC3. SOC pools in the model are the same as those in RothC, which are distributed in the different types of aggregates.

SOC, in the AC1 type aggregates, is distributed within BIO, HUM, and IOM pools, in the AC2 aggregate type is distributed within BIO, HUM, IOM, and fine DPM and RPM pools and finally in the AC3 aggregate type SOC is distributed within BIO, HUM, IOM, and fine and coarse DPM and RPM pools. Each carbon pool in the different aggregate classes, apart from the IOM, is decomposed producing CO_2, BIO, and HUM. Carbon decomposition or accumulation follows first-order kinetics. The decomposition rate constants (k) are different for each pool in each aggregate type and are corrected for pedoclimatic conditions of moisture and temperature and the soil cover using the (a), (b), and (c) factors described earlier. Macroaggregation, which is one of the processes taken into account in the CAST model, is induced by plant input and specifically the decomposable plant material. On the other hand, microaggregation depends on the rate constants for aggregation, the availability of plant material, and the availability of AC1 particles. Both processes are considered to follow first-order kinetics. Furthermore, macroaggregates are destroyed when slaking occurs, releasing stabilized microaggregates and silt–clay-sized aggregates and highly decomposed residual particulate organic matter (Stamati et al., 2013).

1.4 The Damma Glacier CZO

Since the end of last little ice age, about 1850 AD, alpine glaciers have been retreating, exposing barren rock and debris, which within only a few years, become colonized by soil biota, lichens, and plants. This leads to the development of initial ecosystems and soil chronosequences that represent the succession of different soil development stages and vegetation. The very early stage of soil development is characterized by low plant diversity, low contents of soil carbon and nitrogen, and of high richness of bacterial first colonizers (Brankatschk et al., 2011; Kabala and Zapart, 2012; Zumsteg et al., 2012). It has been observed that organic carbon and nitrogen accumulate very rapidly in glacier forefields until the soil reaches the age of approximately 200 years when the accumulation rate decreases significantly (Egli et al., 2010; He and Tang, 2008). Carbon dynamics are considerably controlled by the soil structure and the formation and destruction of soil aggregates (Six et al., 2004). Two of the most important mechanisms for the stabilization of organic matter in soils are its protection within soil aggregates, which exhibit pores which are small enough to prevent diffusion access by enzymes and other degradative substances and the occurrence of organic matter adsorption on mineral surfaces (Mayer, 1994; Schmidt et al., 2011; Six et al., 2002; von Lutzow et al., 2006). However, to the best of our knowledge, only the study of Dumig et al. (2012) investigates the evolution of soil structure in the very early stages of soil development and in relatively extreme physical environments such as glacier forefields. They found that the formation of organomineral associations start with the adsorption of organic matter on mineral surfaces, mainly ferrihydrite. In the initial stages of soil formation, organic matter accumulation happens in distinct zones or layers as it can be shown by the large loadings of organic carbon (Dumig et al., 2012). In this study, we investigated the ability of the CAST model to simulate the development of the soil structure and carbon accumulation in the Damma glacier forefield in the 150-years-old soil chronosequence.

2. SITE DESCRIPTION

The Damma glacier watershed is one of the CZOs established within the European project SoilTrEC (Banwart et al., 2011) building on the interdisciplinary project BigLink (Bernasconi et al., 2008, 2011). It is located in the central Alps in Switzerland, Canton of Uri, and is characterized by an alpine climate with an average yearly temperature of 1°C and precipitation 2400 mm $year^{-1}$ (Kobierska et al., 2013). The altitude of the watershed is between 1790 and 3200 m above sea level, the geology is uniform across the whole watershed, and the bedrock is composed of coarse-grained granite of the Aare massif, which was metamorphosed under green schist faces conditions and is composed of quartz, plagioclase, potassium feldspar, biotite, and muscovite (Schaltegger, 1990). A full description of the site can be found in Bernasconi et al. (2011). Damma glacier which covers 50% of the watershed has retreated at an average rate of 10 m per year since 1921, when systematic annual measurements started (Kobierska et al., 2011). However, the recession of the glacier was reversed twice in the past, during 1920–1928 and 1970–1992, resulting in two small terminal moraines. Due to glacier readvance, the soil chronology is not continuous but consists of three groups of soil ages. The youngest sites include soils from 6 to 14 years old, the intermediate group comprises of locations freed from the ice between 1930 and 1950 with ages of 57–79 years old, and the oldest group includes soils that started to form between 1870 and 1897 with ages from 108 to 140 years, as seen in Fig. 1. The soils located between these two zones of the chronosequence have been eroded during glacial advances. The soil chronosequence is developing at the bottom of the glacier valley,

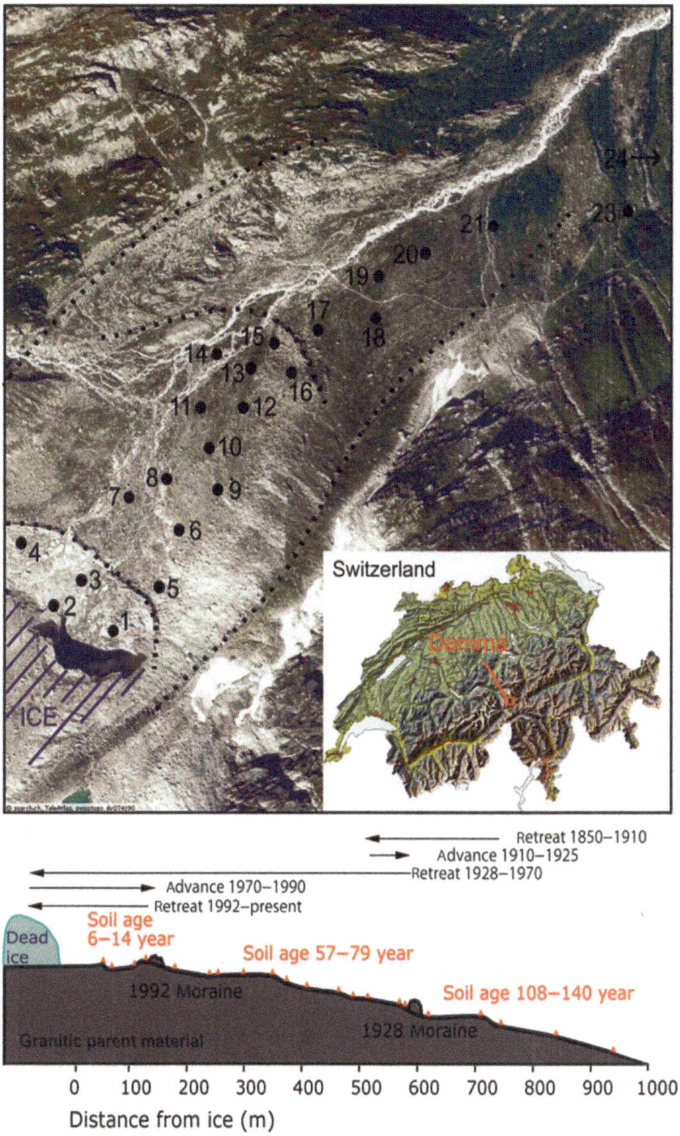

Fig. 1 Location and aerial photograph of the Damma glacier CZO showing all the sampling sites. The sketch of the forefield shows the age of the sampling sites and the glacier movements based on the glacier retreat history. *After Bernasconi, S.M., Bauder, A., Bourdon, B., Brunner, I., Bunemann, E., Christl, I., Derungs, N., Edwards, P., Farinotti, D., Frey, B., Frossard, E., Furrer, G., Gierga, M., Goransson, H., Gulland, K., Hagedorn, F., Hajdas, I., Hindshaw, R., Ivy-Ochs, S., Jansa, J., Jonas, T., Kiczka, M., Kretzschmar, R., Lemarchand, E., Luster, J., Magnusson, J., Mitchell, E.A.D., Venterink, H.O., Plotze, M., Reynolds, B., Smittenberg, R.H., Stahli, M., Tamburini, F., Tipper, E.T., Wacker, L., Welc, M., Wiederhold, J.G., Zeyer, J., Zimmermann, S., Zumsteg, A., 2011. Chemical and biological gradients along the Damma glacier soil chronosequence, Switzerland. Vadose Zone J. 10 (3), 867–883.*

has a total length of 1 km and is situated at an elevation range from 2150 to 1900 m (Bernasconi et al., 2011; Smittenberg et al., 2012).

3. SITE SAMPLING AND ANALYTICAL METHODS

The first soil sampling campaign was conducted in September 2007 within the framework of the BigLink project. The site selection, sampling strategy, sample preparation, analyzes, and results have been described extensively in Bernasconi et al. (2011). Twenty-three sites were sampled in total, 21 of which were situated in the glacier forefield with ages from <10 to 140 years and two additional samples from reference sites located outside the forefield which were ice-free at least for 10,000 years (Fig. 1). In addition, soil samples were collected in September 2012 and analyzed, according to Elliott (1986), for the following water stable aggregate (WSA) fractions: (i) macroaggregates (>250 μm), (ii) microaggregates (53–250 μm), and (iii) silt–clay-sized microaggregates and minerals (<53 μm). Total organic carbon (TOC) and total nitrogen (TN) were measured for each aggregate fraction.

In the framework of the SoilTreC project, another soil sampling campaign was conducted in the summer of 2010. Four soil profiles, one for each soil group in the forefield and one reference site, were fully characterized and samples were collected and analyzed for a large number of parameters. The results for TOC, TN, and microbial carbon were used for the model validation.

4. INPUT PARAMETERS

4.1 Modified RothC Model

In this study, the RothC version 26.3, modified for Excel (Todorovic et al., 2010), and further modified for the purposes of this study, was used. The model input requirements are (i) meteorological data (average monthly mean temperatures, precipitation, and open-pan evaporation), (ii) monthly plant input and soil cover, and (iii) clay content and sampling depth. For the calibration of the model, the results of the model for TOC and the carbon mass stocks in the different pools were compared, respectively, with the measured values of total SOC and carbon in the different soil fractions.

Meteorological data: The rate constants of the model are corrected for temperature, water deficit, and soil cover through application of the respective correction factors (a), (b), and (c) described earlier, which are calculated as described by Coleman and Jenkinson (1999). In this study, the Excel version of RothC model was modified in order to reconstruct the time sequence of historical conditions and calculate new (a), (b), and (c) factors for each simulation month. Mean monthly precipitation data for the period 1981–2010 were taken from the meteorological station of the ANETZ network, located at Gütsch (2283 m a.s.l.), which correlates better with the weather station located in the middle of Damma watershed (Magnusson et al., 2011). Monthly soil temperatures, previously published in Guelland et al. (2013), were used for the simulation until 1992. For the period from 1867 to 1992, temperature was reconstructed using a model based on observed and past data (Farinotti et al., 2010). For the years of glacier readvance soils were covered by ice and therefore soil temperature and precipitation were set to lower values.

Plant input: Extensive studies in the Damma glacier forefield showed that the vegetation of the very young sites consists of pioneer forbs, while the intermediate sites are dominated by grasses and the older sites by woody shrubs (Bernasconi et al., 2011). The total biomass and subsequently the plant inputs increase significantly with soil age along the chronosequence, from low values in the young sites to higher values in the intermediate sites, reaching steady state at the older sites (Bernasconi et al., 2011). The measurements of biomass, the plant litter, and moss layer along the whole chronosequence (Bernasconi et al., 2011; Smittenberg et al., 2012) helped us to estimate accurately the plant input, which proved to be of high significance for the modification and calibration of the model. The plant input alterations along the chronosequence were incorporated in the model by modifying the excel version, in order to reconstruct the historical time series of plant inputs by specifying the value of plant input for each month of simulation.

During the two periods of glacier readvance, plant input reaches low values, since biomass is either destroyed or redistributed at the area affected by the movement of the glacier. Consequently, it was assumed that the plant input for the model during the years of glacier movement is the total biomass of the last year before the readvancing, distributed evenly throughout this period.

Clay content and soil depth: Both soil depth and clay content increase along the soil chronosequence with the development of the soil profile, but only to a small extent (Bernasconi et al., 2011). Therefore, an average value of 3.14% was used, while the soil depth was set to 5 cm.

Soil organic carbon: Soil TOC increases from 0.1–0.4% in the very young sites, to 2–4% at the older sites and 8–9% at the reference sites (Bernasconi et al., 2011; Smittenberg et al., 2012). Within the two young soil groups, TOC increases exponentially as the age of the soil increases. In the oldest soil age group, the increase of TOC values is not exponential and at the end of the chronosequence it can be observed that this increase is much smaller in comparison to the first two soil groups, indicating that TOC tends to reach a steady value. TOC contents decrease significantly when Damma glacier readvanced, and at the sites just below the moraines they are much lower than expected.

For the partitioning of SOC in the different soil carbon pools, the density fractionation data of Smittenberg et al. (2012) and the dissolved organic carbon (DOC) measurements of Guelland et al. (2013) were used. The partitioning scheme was based on the studies of Zimmermann et al. (2007) and Stamati et al. (2013). IOM was considered to be equal to zero. Free light organic carbon, light organic carbon within the soil particles size fraction 63–2000 μm, and DOC were considered to be the sum of DPM and RPM pools since these fractions represent the amount of organic carbon that derives from the plant debris. The carbon contained in the most resistant fraction, which is the part of the soil with particle size fraction of 63–2000 μm, together with the carbon content within the size fraction smaller than 63 μm, was accounted for as the BIO and HUM pools. Subsequently, the sums of the DPM and RPM as well as the BIO and HUM pools were divided in the different pools according to the DPM/RPM ratio set to 1.44, which is the standard value of the model for grassland and was also used by Zimmermann et al. (2007) for alpine pastures. Furthermore, the percentage of the decomposed material that goes to the BIO pool was considered to be 60%.

4.2 CAST Model

The model input requirements of the CAST model are (i) meteorological data (average monthly mean temperatures, precipitation, and open-pan evaporation), (ii) monthly plant input and soil cover, (iii) clay content and sampling depth, and (iv) the distribution of carbon content in the different aggregate fractions. The meteorological data and the plant input used for the CAST model are the same as those for the modified version of RothC that is described earlier.

The CAST model not only simulates the accumulation of carbon but also the development of soil structure (Stamati et al., 2013). Therefore, the silt and clay content of the soil are important additional input parameters. However, in the Damma forefield, silt and clay content increases with soil age and therefore, in order to achieve better representation of soil texture for the model input, each soil age group was simulated separately, using the average of the soil physical characteristics of each group. For the initialization of the model, the carbon content in the carbon pools was set to be the carbon content of the youngest soil. The IOM pool is considered to be zero.

5. MODEL CALIBRATION

The modified RothC model was initialized using the measured carbon contents of the youngest soil. Each carbon pool in the model decomposes following a first-order reaction with a characteristic rate constant. The default rate constants of the model are defined by its developers based on the Rothamsted long-term experiment data, presented in Table 1. However, the Damma forefield has very different climate, altitude, and growing season compared to the Rothamsted experiment and the application of the model with the default rates led to model results that did not match the measured values, as presented in Fig. 2. For this reason, it was essential to modify the model by changing the rates of each pool during calibration, using the accurate plant input measurements described earlier. It is essential to point out the fact that the modification of the model and the alteration of the rates were made possible by the availability of the plant input data. The performance of the model was evaluated by the root mean square error (RMSE) and the modeling efficiency EF

Table 1 The Default and Modified in This Study, Decomposition Rate Constants (1/year) of the RothC Model

	Default	Damma
DPM	10.00	5.00
RPM	0.30	1.20
BIO	0.66	1.50
HUM	0.02	1.00

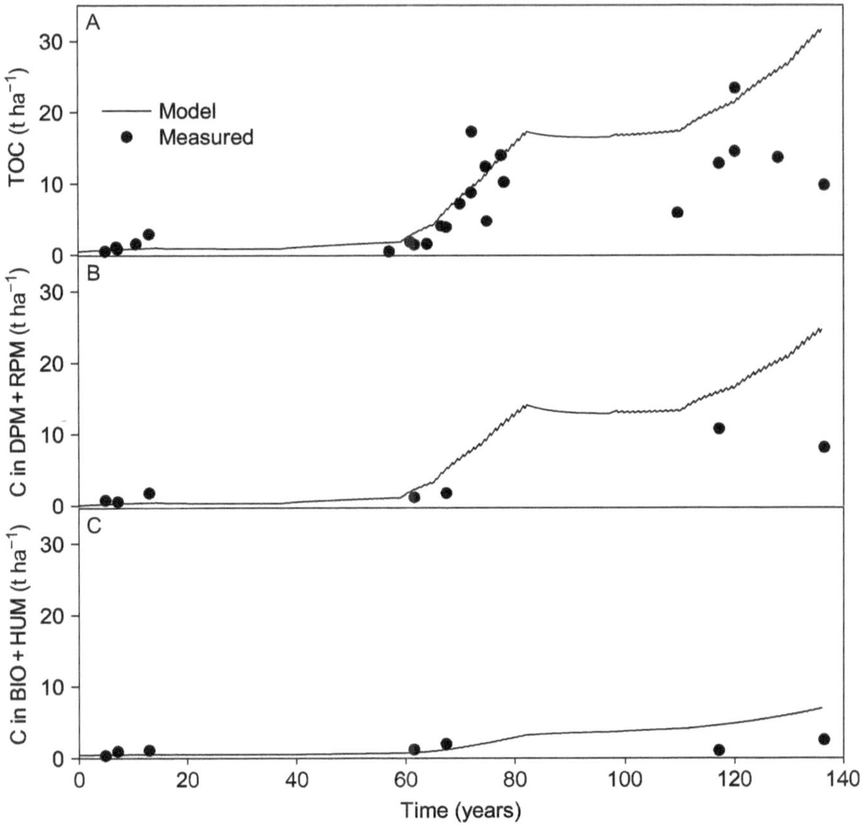

Fig. 2 Results of the noncalibrated RothC model in comparison to measured values of the TOC (A), the carbon content of the DPM plus RPM pools (B), and the BIO plus HUM pools (C).

(Smith et al., 1997). The best performance was defined as when the lowest value of RMSE was achieved and EF was positive and closest to 1 (Ludwig et al., 2007).

The new rate constants in comparison to the default rate constants of the model are presented in Table 1. Due to the short growing season and the alpine climate in the Damma forefield, the life cycle of the soil microbes is short and the turnover rate of the microbial carbon is fast (Smittenberg et al., 2012). This is a fact that explains the higher value of the BIO rate constant, 1.5 year^{-1}, than the default one, 0.66 year^{-1}. In addition, most of the organic material in Damma soils is rapidly mineralized and for this reason, the rate constant of the HUM pool is also set to 1 year^{-1}, a value that is higher than the original one (0.02 year^{-1}). Furthermore, although most of the SOC originates from fresh plant input produced mainly by grasses (Smittenberg et al., 2012), its turnover rate is not expected to be faster than that of the Rothamsted experiment, due to the colder climate and the different type of vegetation. Furthermore, the amount of the resistant plant material is lower in comparison to, i.e., agricultural land (Smittenberg et al., 2012). For this reason, the rate constant for decomposition of the DPM pool was set to 5 $year^{-1}$, half the value of the default one. However, the constant for the RPM pool was changed to the higher value of 1.2 $year^{-1}$ instead of the default value of 0.3 $year^{-1}$.

The CAST model was applied and calibrated for each soil age group separately, due to the different soil physical parameters. The calibration of the model was initiated using the same constants for DPM and RPM pools that were used in the calibration of the RothC model and the calibration was continued by considering that carbon within the most protected pools, e.g., the HUM pool of the AC1 type of aggregates within the AC3 type of aggregates, has longer residence time and slower decomposition rates. The most important rates for the calibration of the model and for the different soil age groups increase as the soil age increases (Table 2). That means that in the very young soils, where there is almost no distinct soil structure and the TOC values are low, the turnover rates are required to be set to lower values than the original values in order for the model to successfully simulate these conditions. On the other hand, the older soils are characterized by higher TOC values, slightly higher clay content and a primitive soil structure. In this case, the turnover rates for the microbial carbon and plant litter are rapid and therefore the rates of the model increase. The comparison of the model rates with the rates applied in different study areas is the subject of a separate paper.

Table 2 Comparison of the Default and the Calibrated Values of the Rate Constants (1/year) of the CAST Model, for the Young, Intermediated, and Old Soil Age Groups

	Default	Young	Intermediate	Old
Fragmentation				
RPM to RPMc	15	1	20	14
RPMc to RPMf	0	0.0001	0.001	0.001
RPMc(AC3) to RPMf(AC3)	0.1	0.01	0.1	0.1
DPMc(AC3) to DPMf(AC3)	0.5	0.01	0.5	0.5

Continued

Table 2 Comparison of the Default and the Calibrated Values of the Rate Constants (1/year) of the CAST Model, for the Young, Intermediated, and Old Soil Age Groups—cont'd

	Default	Young	Intermediate	Old
Macroaggregation				
RPMc	0.6	0.1	0.1	0.1
DPMc	0.47	0.1	0.1	0.5
Decomposition				
Fresh plant input				
DPM	10.45	3	7	10
RPM	0.305	0.005	0.1	0.2
RPMc	0.305	0.01	0.1	0.2
RPMf	0.305	0.01	0.1	0.2
AC3 aggregate type				
BIO(AC1) within AC3	0.6	0.01	0.01	0.1
HUM(AC1) within AC3	0.0031	0.01	0.01	0.1
BIO(AC2) within AC3	0.6	0.01	0.01	0.1
HUM(AC2) within AC3	0.0021	0.01	0.01	0.1
AC2 aggregate type				
BIO(AC2)	0.6	0.001	0.0001	0.1
HUM(AC2)	0.0051	0.0051	0.00051	0.1
RPMf(AC2)	0.2069	0.001	0.001	0.01
DPMf(AC2)	1.5	0.005	0.0005	0.005
AC2 aggregate type				
BIO(AC1)	0.6	0.001	0.01	0.25
HUM(AC1)	0.0051	0.001	0.01	0.25

6. RESULTS

6.1 Modified RothC Results

Fig. 3A–C presents the results of the RothC model for TOC, DPM + RPM, and BIO + HUM pools, respectively, in comparison to the measured data. The first years of simulation represent the carbon accumulation in the young

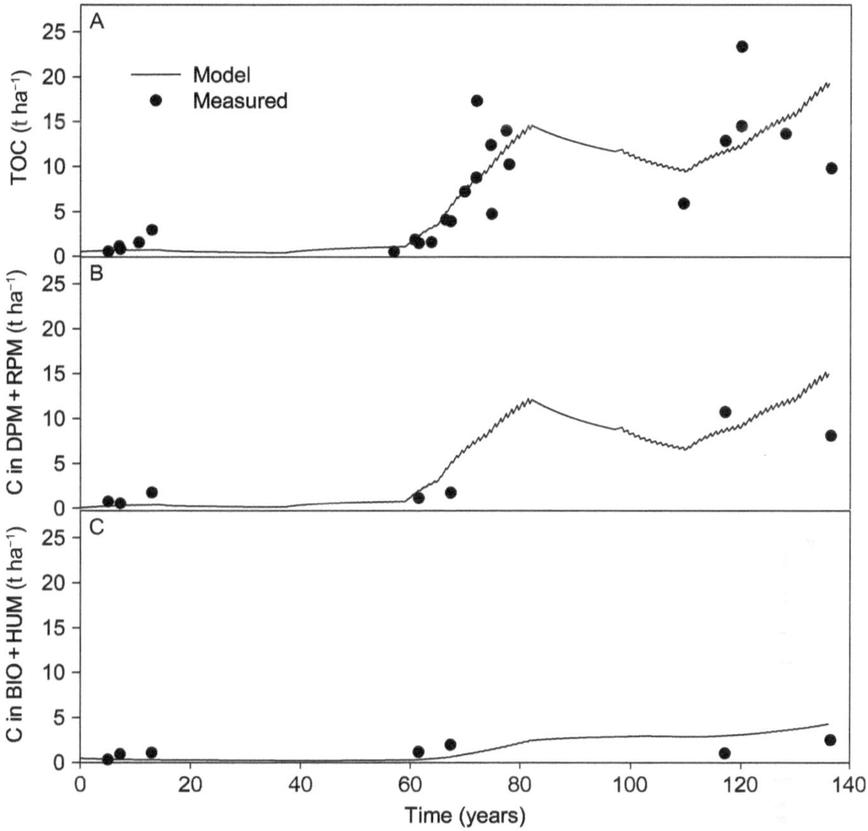

Fig. 3 Results of the modified version of the RothC model in comparison to the TOC (A), the carbon content of the DPM plus RPM pools (B), and the BIO plus HUM pools (C).

soil group. It can be observed that the simulated values of TOC during this period are lower than the observed data, with a difference of 2 t ha^{-1} at the last sampling point of the young soil age group. During the early stages of soil development in the Damma forefield, the vegetation cover is scarce and the carbon input is low. As a result the main role in carbon dynamics is held by the microbial community, a fact that might not be adequately modeled. In addition, an important source of carbon in the forefield is the dust input from the wind or the snow, which was measured to be approximately 7.5 kg ha^{-1} $year^{-1}$ (Brankatschk et al., 2011), and the organic material that enters the forefield through snow avalanches and surface runoff from the entire watershed. For these reasons, which are not accounted for in the model, the RothC model slightly underestimates the carbon accumulation for the very young soils.

From 1970 to 1990, i.e., from the 15th to 37th year of simulation, the Damma glacier advances due to a short period of cooler climate. During this period, plant input is low, 0.1 t ha^{-1} $year^{-1}$, resulting in the low observed values for TOC (Bernasconi et al., 2011), which is however well captured by the model.

Model results for the intermediate soil age group are shown from the 38th to 82nd year of simulation. The efficiency of the model in simulating TOC and HUM plus BIO pools is very satisfactory. However, measured DPM and RPM values are lower than those estimated by the model. It is possible that the fractionation procedure that was used to relate the measured soil carbon with the DPM and RPM pools of the model could not capture the full amount of the organic matter that originates from plant litter. This observation agrees with results for alpine pastures from Zimmermann et al. (2007), who also found differences between the modeled and measured DPM and RPM pools.

The 83rd and 97th year of simulation show the first glacier readvance, which occurred between 1910 and 1925, due to which the 1928 moraine was formed (Fig. 1). The TOC measurements showed that the carbon accumulation was affected by the movement of the glacier and the TOC decreased as a result, as is also captured by the model. Finally, the last part of the simulation represents the older soils, with the age of 98 to 136 years. It can be noted that the measured values of TOC are quite variable because of the strong heterogeneity of the field site (Bernasconi et al., 2011). Even so the clearly increasing trend of TOC could be satisfactorily modeled by RothC. Moreover, BIO plus HUM pools as well as DPM plus RPM pools are also well simulated.

In order to further assess the performance of the model, the CO_2 production simulated by the RothC model was compared with the measured carbon fluxes of Guelland et al. (2013). The measured values of the soil CO_2-efflux are 8.6 ± 1 g m^{-2} $year^{-1}$ for the young soils, 67 ± 17 g m^{-2} $year^{-1}$ for the intermediate soil age group, and 161 ± 67 g m^{-2} $year^{-1}$ for the older soils while the simulated values are 5.4, 60.2, and 93.1 g m^{-2} $year^{-1}$. It should also be noted that the measurements of the CO_2 effluxes revealed that the CO_2 production is ongoing even during the winter when the Damma forefield is completely covered by snow and the soil temperatures are close to zero (Guelland et al., 2013). However, there is a very good agreement of the model results with the measured values for the intermediate and older soil age group. The overall performance of the model is considered to be very good in reproducing the measured values.

The soil sampling sites immediately below the two moraines in Damma glacier forefield are characterized by lower values of TOC than the ones expected from their age, e.g., 0.5, 1.4, and 1.5 t ha^{-1} at the 57-, 61-, and 64-year-old soils, respectively, and 6 t ha^{-1} at the 110-years-old soil. This indicates that these sites are probably disturbed by the glacier and sediment movements and by changes in the riverbed of the Dammareuss stream. In order to understand the reasons of disturbance and the low content of TOC in these sites, two scenarios were simulated with the modified version of RothC. For the first scenario (Scenario 1), the plant input was regarded as not being affected by the glacier movement and it remained constant and equal to the year before the readvance. For the second scenario (Scenario 2), the plant input during glacier readvance was considered to be affected by the glacier movement and the vegetation in the areas where the glacier

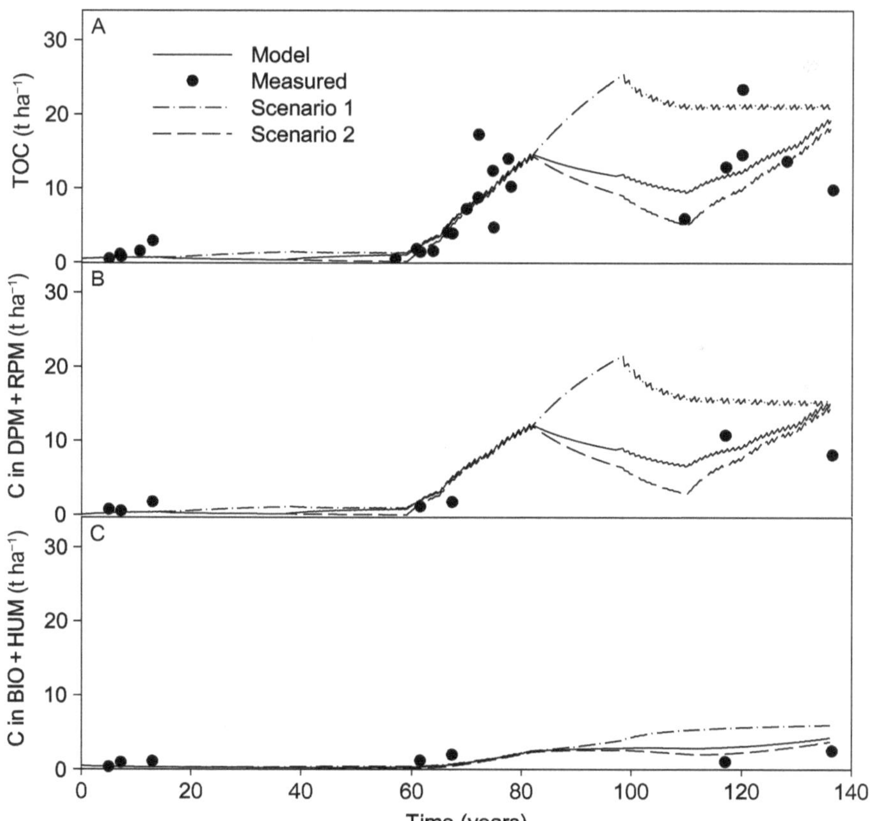

Fig. 4 Comparison of the modified RothC results for Scenarios 1 and 2 with the original values of the model and the measured results for the TOC (A), DPM plus RPM pools (B), and BIO plus HUM pools (C).

readvanced was considered to be completely destroyed so that the plant input is zero. The model was not recalibrated for the scenarios and the rates used are the ones of the calibrated model.

The results for Scenarios 1 and 2 described earlier, in comparison to the observed data and the originally calibrated model are presented in Fig. 4A–C for TOC, DPM–RPM, and BIO–HUM, respectively. The values of TOC for Scenario 1, for which no disturbance in the plant input is assumed, are significantly higher than the ones of the initial version of the model, especially for the older soil age group. This is a strong indication that the movements of the glacier disturbed the development of soils as well as the vegetation and carbon sequestration, especially in the areas that were covered by the glacier for a second time. Results for Scenario 2, for which zero plant input during the glacier readvance was considered, show that the sites close to the two moraines, with ages of 57, 61, 62, 64, and 110 years are better described by this scenario, since the model results agree more with observed values. This outcome verifies that these sites are the ones mostly disturbed. This observation supports the findings of Smittenberg et al. (2012), who suggested that the disturbance is caused by a decrease in plant input due to physical perturbation and/or addition of TOC-free sand to existing soil.

6.2 Simulations With Cast

The results for the CAST model compared with the measured values for TOC, the carbon pools within the aggregates AC1 and AC2 and the pool within AC3 aggregates together with the free organic matter, are presented in Fig. 5. Similarly to the concept of the simulation with ROTHC, the first 14 years show the results of the very young soils. The most recent glacier readvance is presented from the 15th to the 37th year, while the intermediate soil age group is simulated from the 38th to 82nd year of simulation. The first glacier readvance is represented in the 83rd and 97th year and finally the older soils are simulated in the 98th to the 136th year of the model.

The content of TOC in the CAST model is determined not only by the decomposition of the plant input but also by the processes that occur within each aggregate class. However, despite the complexity of these processes the CAST model simulates the TOC effectively. As explained extensively earlier, the two glacier readvances caused disturbance in the values of the TOC, a fact that explains the drop of the observed TOC to

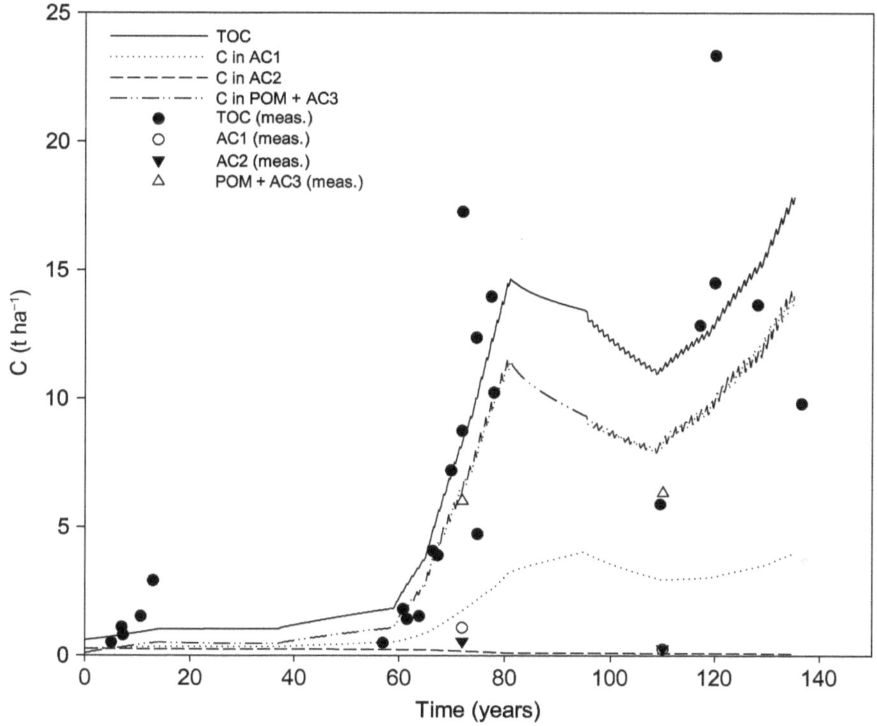

Fig. 5 Results of the CAST model simulation of the evolution of the carbon content of the soil TOC, aggregate type AC1, aggregate type AC2, and the aggregate type AC3 plus the coarse POM of the nonaggregated pools (cPOM). *Points* indicate the field measurements of the same pools.

the value of 0.5 t ha^{-1} from the 15th to 37th year of simulation and to the value of 6 t ha^{-1} from the 83rd and 97th year of soil age. This decrease is also depicted in the results of the CAST model.

In the Damma glacier forefield, most of the soil organic matter comprises of free organic matter and only a small fraction, 0.5% of TOC at the older soil age group, is observed to be strongly bound to the sand and silt–clay mineral fraction (Smittenberg et al., 2012). Due to the low amount of mineral bound carbon, the carbon content within the AC1 aggregate type is also low and equals 1.1 ± 0.2 t ha^{-1} for the 72-years-old soil and 0.3 ± 0.2 t ha^{-1} for the 110-years-old soil (Fig. 5). Furthermore, since most of the soil organic material is relatively fresh plant litter in particulate form and considering that the soil processes in Damma forefield are at an initial stage of soil development, the AC3 aggregates consist of the particulate organic matter primarily loosely bound with the mineral particles and comprise

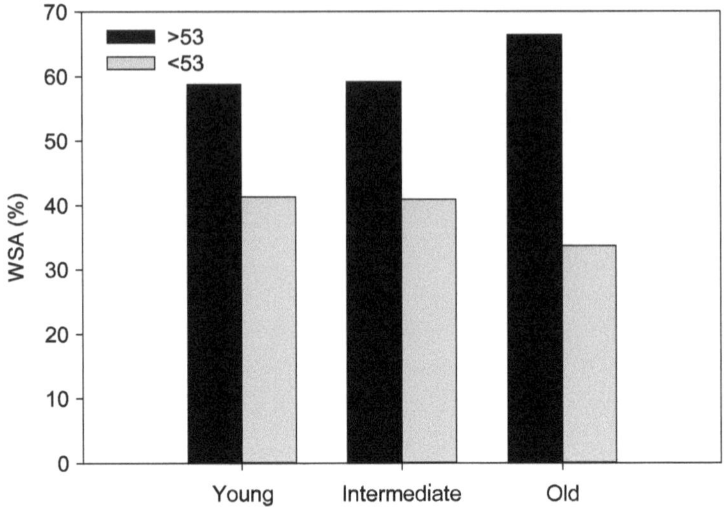

Fig. 6 Percentage of the WSA fractions above and below 53 μm in the young, intermediate, and old soil age group.

the major percentage of aggregates measured in these soils (Fig. 6). Therefore, the carbon content of the AC3 aggregate type is 4.8 ± 0.6 t ha^{-1} for the 72-years-old soil and 5.1 ± 0.5 t ha^{-1} for the 110-years-old soil. On the other hand, the formation of the AC2 aggregates has not been initiated at this stage of soil development, leading to low carbon content of AC2 aggregates, which does not exceed the value of 0.6 ± 0.2 t ha^{-1}. In spite of these unusual conditions of the soils in the Damma forefield, the CAST model performed satisfactorily and proved to be an effective model representation for the dominant soil processes that control the carbon pools within the different aggregate fractions as well as the free organic matter.

Fig. 6 shows the percentage of the measured aggregate fractions below and above the 53 μm size from the total aggregates in the three different soil age groups. What is observed is a small increasing trend from the young and intermediate soil group to the older group in the fraction above 53 μm while, subsequently, the fraction below 53 μm slightly decreases. This trend could mean that as the soils get older, the aggregates within the small fraction bind together with particulate organic matter forming larger microaggregates. This trend was also depicted in the CAST model since the only aggregation process for the young and intermediate soil group is the macroaggregation determined by particulate organic matter, while microaggregation appeared as a process only for the older soil age group.

However, it should be noted that due to the unusual nature of the measured aggregates even at the older soil age group, any final conclusions are difficult to be made.

Finally, in order to further assess the performance of the model, the emissions of CO_2 and the results for dissolved carbon leaching as calculated by the CAST model were compared with the measured carbon fluxes and DOC values of Guelland et al. (2013). The measured values of the soil CO_2-efflux are 8.6 ± 1 g m^{-2} year^{-1} for the young soils, 67 ± 17 g m^{-2} year^{-1} for the intermediate soil age group, and 161 ± 67 g m^{-2} year^{-1} for the older soil age group while the simulated values are 3.5, 62.7, and 98.6 g m^{-2} year^{-1}. Similarly to RothC, model results for CO_2 production agree relatively well with the measured values for the older and intermediate soils but for the young soils they are slightly underestimated. The total carbon stocks for dissolved carbon leaching of the CAST model are very close to zero which agree very well with the low measured values of 0.41 ± 0.02 g m^{-2} year^{-1} for the young soils, 1.6 ± 0.5 g m^{-2} year^{-1} for the intermediate soils, and 7.4 ± 1.6 g m^{-2} year^{-1} for the old soils.

7. CONCLUSIONS

This work is the first attempt, as far as we know, to simulate carbon accumulation and incipient soil development in a soil chronosequence of a glacier forefield. For this reason, we used a modified version of the widely used soil carbon dynamics model, RothC, and a recently developed model for carbon accumulation and soil structure dynamics, CAST.

Initially, the RothC model was tested using its default rates, which, however, lead to unsatisfactory results. These unsatisfactory results can be explained by the fact that Damma glacier forefield is characterized by the young nature of its soils, the short growing season, and the extreme climatic conditions, and is very different from the site where RothC developed and from the type of sites for which the model is generally applied. For this reason, we decided to alter the model by changing the turnover rate constants of the different carbon pools. This modification was made possible by the existing measurements of the plant input and biomass along the soil chronosequence. It was found that despite the two glacier readvances, the modified model generally simulates successfully the TOC, BIO, and HUM pools, while DPM and RPM pools are slightly overestimated. Furthermore, the CO_2 emissions calculated by the model are consistent with the measured values.

The newly developed CAST model was calibrated successfully and it captures well the accumulation of TOC. The soil processes at the Damma glacier forefield are at an initial stage of development. It is likely that the aggregation processes that develop soil structure, as well as the carbon accumulation, have not reached steady state. The WSA content of this soil system consists mainly of silt- and sand-sized particles loosely bound with particulate organic matter. This observation is also described by the results of the calibrated CAST model since the dominant aggregation process that is shown to be significant from the results of the model calibration is the macroaggregation determined by particulate organic matter and microaggregation is only present during the last years of simulation. Furthermore, the carbon content of the different aggregate types: (macroaggregates: >250 μm (AC3), microaggregates: 53–250 μm (AC2), silt–clay-sized aggregates: <53 μm (AC1)) and the free organic matter, even if it is considerably low in comparison to other types of soils, is well simulated. Consequently, despite the fact that the soil chronosequence of the Damma glacier forefield comprises of very young soils, with very low clay content, and in an alpine environment, the CAST model works effectively.

The accurate prediction of SOC dynamics and soil structure development for various soil types in different climatic conditions and land uses is essential for the assessment of the impact of climate change on SOC and the subsequent feedback effects. A crucial step toward this direction is the application and calibration of SOC and soil structure models in sites with diverse characteristics. In this context, this study is a unique contribution to the SOC modeling research, since it is a first attempt to test SOC models on a glacier forefield or soil chronosequence of the very earliest stages of soil formation—in this case under extreme alpine conditions. Both models, despite the complexity of the study site, gave satisfactory results after calibration. Furthermore, a significant finding of this study is that a relatively complex set of soil processes as represented by CAST, which includes the important function of physical protection of the SOC through aggregate formation, managed to simulate efficiently the very young soils of the Damma glacier forefield. A next step for future research would be to test whether the calibrated rates derived from this study would produce equally good results in similar sites.

Finally, we would like to point out the importance of the existence of the CZOs, such as the Damma glacier forefield. In this study, the application and the validation of both models were made possible by using the abundance of already existing data for the Damma glacier forefield and by the support provided by other European CZOs such as the Koiliaris CZO, where the CAST model was initially developed, and the Fuchsenbig-Marchfeld, Austria CZO, which provided additional datasets to support CAST model development and testing. Therefore, we concluded that the CZOs are ideal study sites for the interdisciplinary research of the many processes occurring at the biosphere–hydrosphere–geosphere interface and sites where new research methods and models, such as CAST, can be developed and tested.

ACKNOWLEDGMENTS

This study was supported by the European Commission FP 7 Collaborative Project Soil Transformations in European Catchments (SoilTrEC) (Grant Agreement No. 244118) and by the Project BigLink funded by the Competence Center Environment and Sustainability of the ETH Domain. We would like to thank F.E. Stamati, M. Kotronakis, and all the other staff of the Laboratory of Hydrochemical Engineering and Soil Remediation at the Technical University of Crete for their support with the model CAST and Dr. Juna Shrestha for her comments on the manuscript. Finally, we would like to thank all the reviewers.

REFERENCES

Banwart, S., Bernasconi, S.M., Bloem, J., Blum, W., Brandao, M., Brantley, S., Chabaux, F., Duffy, C., Kram, P., Lair, G., Lundin, L., Nikolaidis, N., Novak, M., Panagos, P., Ragnarsdottir, K.V., Reynolds, B., Rousseva, S., de Ruiter, P., van Gaans, P., van Riemsdijk, W., White, T., Zhang, B., 2011. Soil processes and functions in critical zone observatories: hypotheses and experimental design. Vadose Zone J. 10 (3), 974–987.

Bernasconi, S.M., Christl, I., Hajdas, I., Zimmermann, S., Hagedorn, F., Smittenberg, R.H., Furrer, G., Zeyer, J., Brunner, I., Frey, B., Plotze, M., Lapanje, A., Edwards, P., Venterink, H.O., Goransson, H., Frossard, E., Bunemann, E., Jansa, J., Tamburini, F., Welc, M., Mitchell, E., Bourdon, B., Kretzschmar, R., Reynolds, B., Lemarchand, E., Wiederhold, J., Tipper, E., Kiczka, M., Hindshaw, R., Stahli, M., Jonas, T., Magnusson, J., Bauder, A., Farinotti, D., Huss, M., Wacker, L., Abbaspour, K., Project, Biglink, 2008. Weathering, soil formation and initial ecosystem evolution on a glacier forefield: a case study from the Damma Glacier, Switzerland. Mineral. Mag. 72 (1), 19–22.

Bernasconi, S.M., Bauder, A., Bourdon, B., Brunner, I., Bunemann, E., Christl, I., Derungs, N., Edwards, P., Farinotti, D., Frey, B., Frossard, E., Furrer, G., Gierga, M., Goransson, H., Gulland, K., Hagedorn, F., Hajdas, I., Hindshaw, R., Ivy-Ochs, S., Jansa, J., Jonas, T., Kiczka, M., Kretzschmar, R., Lemarchand, E., Luster, J., Magnusson, J., Mitchell, E.A.D., Venterink, H.O., Plotze, M., Reynolds, B., Smittenberg, R.H., Stahli, M., Tamburini, F., Tipper, E.T., Wacker, L., Welc, M., Wiederhold, J.G., Zeyer, J., Zimmermann, S., Zumsteg, A., 2011. Chemical and biological gradients along the Damma glacier soil chronosequence, Switzerland. Vadose Zone J. 10 (3), 867–883.

Brankatschk, R., Towe, S., Kleineidam, K., Schloter, M., Zeyer, J., 2011. Abundances and potential activities of nitrogen cycling microbial communities along a chronosequence of a glacier forefield. ISME J. 5 (6), 1025–1037.

Coleman, K., Jenkinson, D.S., 1999. A model for the turnover of carbon in soil. Model Description and Windows Users Guide. Lawes Agricultural Trust, Harpenden, UK. November 1999 issue (modified August 2008).

Dumig, A., Hausler, W., Steffens, M., Kogel-Knabner, I., 2012. Clay fractions from a soil chronosequence after glacier retreat reveal the initial evolution of organo-mineral associations. Geochim. Cosmochim. Acta 85, 1–18.

Egli, M., Mavris, C., Mirabella, A., Giaccai, D., 2010. Soil organic matter formation along a chronosequence in the Morteratsch proglacial area (Upper Engadine, Switzerland). Catena 82 (2), 61–69.

Elliott, E.T., 1986. Aggregate structure and carbon, nitrogen, and phosphorus in native and cultivated soils. Soil Sci. Soc. Am. J. 50 (3), 627–633.

Falloon, P., Smith, P., 2002. Simulating SOC changes in long-term experiments with RothC and CENTURY: model evaluation for a regional scale application. Soil Use Manage. 18 (2), 101–111.

Falloon, P., Smith, P., Coleman, K., Marshall, S., 1998. Estimating the size of the inert organic matter pool from total soil organic carbon content for use in the Rothamsted carbon model. Soil Biol. Biochem. 30 (8–9), 1207–1211.

Farinotti, D., Magnusson, J., Huss, M., Bauder, A., 2010. Snow accumulation distribution inferred from time-lapse photography and simple modelling. Hydrol. Process. 24 (15), 2087–2097.

Guelland, K., Hagedorn, F., Smittenberg, R.H., Göransson, H., Bernasconi, S.M., Hajdas, I., Kretzschmar, R., 2013. Evolution of carbon fluxes during initial soil formation along the forefield of Damma glacier, Switzerland. Biogeochemistry 113, 545–561.

He, L., Tang, Y., 2008. Soil development along primary succession sequences on moraines of Hailuogou Glacier, Gongga Mountain, Sichuan, China. Catena 72 (2), 259–269.

Kabala, C., Zapart, J., 2012. Initial soil development and carbon accumulation on moraines of the rapidly retreating Werenskiold Glacier, SW Spitsbergen, Svalbard archipelago. Geoderma 175, 9–20.

Kobierska, F., Jonas, T., Magnusson, J., Zappa, M., Bavay, M., Bosshard, T., Paul, F., Bernasconi, S.M., 2011. Climate change effects on snow melt and discharge of a partly glacierized watershed in Central Switzerland (SoilTrec Critical Zone Observatory). Appl. Geochem. 26 (Suppl. 0), S60–S62.

Kobierska, F., Jonas, T., Zappa, M., Bavay, M., Magnusson, J., Bernasconi, S.M., 2013. Future runoff from a partly glacierized watershed in Central Switzerland: a two-model approach. Adv. Water Resour. 55, 204–214.

Ludwig, B., Schulz, E., Rethemeyer, J., Merbach, I., Flessa, H., 2007. Predictive modelling of C dynamics in the long-term fertilization experiment at Bad Lauchstadt with the Rothamsted Carbon Model. Eur. J. Soil Sci. 58 (5), 1155–1163.

Ludwig, B., Hu, K.L., Niu, L.A., Liu, X.J., 2010. Modelling the dynamics of organic carbon in fertilization and tillage experiments in the North China Plain using the Rothamsted Carbon Model-initialization and calculation of C inputs. Plant and Soil 332 (1–2), 193–206.

Magnusson, J., Farinotti, D., Jonas, T., Bavay, M., 2011. Quantitative evaluation of different hydrological modelling approaches in a partly glacierized Swiss watershed. Hydrol. Process. 25 (13), 2071–2084.

Malamoud, K., McBratney, A.B., Minasny, B., Field, D.J., 2009. Modeling how carbon affects soil structure. Geoderma 149, 19–26.

Mayer, L.M., 1994. Surface-area control of organic-carbon accumulation in continental-shelf sediments. Geochim. Cosmochim. Acta 58 (4), 1271–1284.

Parton, W.J., Stewart, J.W.B., Cole, C.V., 1988. Dynamics of C, N, P and S in grassland soils: a model. Biogeochemistry 5 (1), 109–131.

Sainju, U.M., Schomberg, H.H., Singh, B.P., Whitehead, W.F., Tillman, P.G., Lachnicht-Weyers, S.L., 2007. Cover crop effect on soil carbon fractions under conservation tillage cotton. Soil Tillage Res. 96 (1–2), 205–218.

Schaltegger, U., 1990. The central Aar Granite—highgly differentiated calc-alkalinemagmatism in the Aar Massif (Central Alps, Switzerland). Eur. J. Mineral. 2 (2), 245–259.

Schmidt, M.W.I., Torn, M.S., Abiven, S., Dittmar, T., Guggenberger, G., Janssens, I.A., Kleber, M., Kogel-Knabner, I., Lehmann, J., Manning, D.A.C., Nannipieri, P., Rasse, D.P., Weiner, S., Trumbore, S.E., 2011. Persistence of soil organic matter as an ecosystem property. Nature 478 (7367), 49–56.

Six, J., Conant, R.T., Paul, E.A., Paustian, K., 2002. Stabilization mechanisms of soil organic matter: implications for C-saturation of soils. Plant and Soil 241 (2), 155–176.

Six, J., Bossuyt, H., Degryze, S., Denef, K., 2004. A history of research on the link between (micro)aggregates, soil biota, and soil organic matter dynamics. Soil Tillage Res. 79 (1), 7–31.

Smith, P., Smith, J.U., Powlson, D.S., McGill, W.B., Arah, J.R.M., Chertov, O.G., Coleman, K., Franko, U., Frolking, S., Jenkinson, D.S., Jensen, L.S., Kelly, R.H., Klein-Gunnewiek, H., Komarov, A.S., Li, C., Molina, J.A.E., Mueller, T., Parton, W.J., Thornley, J.H.M., Whitmore, A.P., 1997. A comparison of the performance of nine soil organic matter models using datasets from seven long-term experiments. Geoderma 81 (1–2), 153–225.

Smittenberg, R.H., Gierga, M., Goransson, H., Christl, I., Farinotti, D., Bernasconi, S.M., 2012. Climate-sensitive ecosystem carbon dynamics along the soil chronosequence of the Damma glacier forefield, Switzerland. Glob. Chang. Biol. 18 (6), 1941–1955.

Stamati, F.E., Nikolaidis, N.P., Banwart, S., Blum, W.E.H., 2013. A coupled carbon, aggregation, and structure turnover (CAST) model for topsoils. Geoderma 211–212, 51–64.

Todorovic, G.R., Stemmer, M., Tatzber, M., Katzlberger, C., Spiegel, H., Zehetner, F., Gerzabek, M.H., 2010. Soil-carbon turnover under different crop management: evaluation of RothC-model predictions under Pannonian climate conditions. J. Plant Nutr. Soil Sci. 173 (5), 662–670.

von Lutzow, M., Kogel-Knabner, I., Ekschmitt, K., Matzner, E., Guggenberger, G., Marschner, B., Flessa, H., 2006. Stabilization of organic matter in temperate soils: mechanisms and their relevance under different soil conditions—a review. Eur. J. Soil Sci. 57 (4), 426–445.

Zimmermann, M., Leifeld, J., Schmidt, M.W.I., Smith, P., Fuhrer, J., 2007. Measured soil organic matter fractions can be related to pools in the RothC model. Eur. J. Soil Sci. 58 (3), 658–667.

Zumsteg, A., Luster, J., Goeransson, H., Smittenberg, R.H., Brunner, I., Bernasconi, S.M., Zeyer, J., Frey, B., 2012. Bacterial, archaeal and fungal succession in the forefield of a receding glacier. Microb. Ecol. 63 (3), 552–564.

CHAPTER NINE

Factors Controlling Soil Structure Dynamics and Carbon Sequestration Across Different Climatic and Lithological Conditions

S.K. Panakoulia*[,1], N.P. Nikolaidis[†], N.V. Paranychianakis[†], M. Menon[‡], J. Schiefer[§], G.J. Lair[¶], P. Krám[||], S.A. Banwart[#]

*School of Environmental Engineering, Technical University of Crete (TUC), Chania, Greece
[†]School of Environmental Engineering, Technical University of Crete, University Campus, Chania, Greece
[‡]The University of Sheffield, Sheffield, United Kingdom
[§]University of Natural Resources and Life Sciences (BOKU), Vienna, Austria
[¶]Institute of Soil Research, University of Natural Resources and Life Sciences (BOKU), Vienna, Austria
[||]Czech Geological Survey, Prague, Czech Republic
[#]School of Earth and Environment, University of Leeds, Leeds, United Kingdom
[1]Corresponding author: e-mail address: spanakoulia@gmail.com

Contents

Abstract

Soil organic carbon (SOC) is a strong determinant of soil fertility through its positive effects on soil structure and soil chemical and biological properties which in turn stimulate primary production. The objective of this work was to simulate field sites that represent different land uses and management practices in three continents, in order to identify the most important factors controlling soil structure dynamics and C sequestration across different climatic and lithological conditions as well as to quantify

Advances in Agronomy, Volume 142
ISSN 0065-2113
http://dx.doi.org/10.1016/bs.agron.2016.10.008

the rates of the aforementioned processes. The carbon, aggregation, and structure turnover (CAST) model was used to simulate SOC sequestration, aggregate formation, and structure dynamics in three field sites including nontilled soils of natural ecosystems and tilled soils of agricultural fields in Europe (Critical Zone Observatories (CZO) of the SoilTrEC network) and one site in North America. Derived data from the simulations' results of SOC stocks and water-stable aggregate (WSA) particle size distribution, together with the respective results of three additional sites (Damma Glacier CZO, Milia (Greece), and Heilongjiang Mollisols (China)), were statistically analyzed in order to determine the factors affecting SOC sequestration and soil structure development. The natural ecosystems include nontilled soils covered with natural local vegetation, while the agricultural sites include cultivated and tilled soils covered with crops. The natural ecosystems were represented by Damma Glacier CZO (Switzerland), Heilongjiang Mollisols (China), Koiliaris CZO (Greece), Clear Creek (United States), and the Slavkov Forrest CZO (Czech Republic), whereas the agricultural field sites were located at Heilongjiang Mollisols (China), Koiliaris CZO (Greece), Clear Creek (United States), Marchfeld CZO (Austria), and Milia (Greece). Principal component analysis (PCA) identified clay content, bulk density, climatic conditions (precipitation and evapotranspiration), organic matter (OM), and its decomposition rates as the most important factors that controlled soil structure development. The relative importance of each of these factors differs under differing climatic and lithological conditions and differing stages of soil development. Overall, the modeling results for both natural ecosystems and agricultural fields were consistent with the field data. The model reliably simulated C and soil structure dynamics in various land uses, climatic conditions, and soil properties providing support for the underlying conceptual and mathematical modeling and evidence that the CAST model is a reliable tool to interpret soil structure formation processes and to aid the design of sustainable soil management practices.

1. INTRODUCTION

1.1 Soil Threats

The adoption of intensive agricultural management practices, deforestation, and livestock overgrazing are leading to accelerated soil losses rates, which exceed soil formation rates by approximately two orders of magnitude (Brantley et al., 2007) with respective consequences on the role of soil in vegetation productivity worldwide (CEC, 2006). The EU as part of the 2006 Environment Policy Review previously published the Thematic Strategy for Soil Protection (CEC, 2006) as one of seven thematic priorities. On January 1, 2016, the 17 Sustainable Development Goals of the 2030 Agenda for Sustainable Development, adopted by world leaders in September 2015, officially came into force. One of the "17 Goals to transform our World" by the United Nations aims to protect, restore, and promote sustainable use of terrestrial ecosystems, sustainably manage forests, combat desertification, and halt and reverse land degradation, and halt biodiversity loss (General Assembly, United Nations, 2015). Forests provide shelter to more than 80% of all terrestrial species; 2.6 billion people depend directly on agriculture while 12 million hectares are lost every year due to drought and desertification, where 20 million tons of grain could have been grown. Over 80% of the human diet is provided by plants and 60% of the energy intake is provided only by three cereals. As a result 74% of the poor are directly affected by land degradation (General Assembly, United Nations, 2015). Despite its critical role in the sustenance of the biosphere and meeting the food requirements of more than 7 billion global inhabitants, our knowledge of the soil functions and their response to human activity is far from complete. Thus, a great challenge in the soil science is to improve our understanding of soil processes, particularly for the critical soil functions as they are defined by EU Soil Thematic Strategy, and to develop suitable tools that will allow us to simulate and quantitatively evaluate the impacts of currently applied management practices, the potential impacts of future management practices, or the shifts in environmental conditions including changing land use and climate.

1.2 Soil Organic Matter and Fertility

Soil fertility strongly depends on soil organic matter (SOM) (Lal, 2004, 2015; Tiessen et al., 1994) by improving soil physical, chemical, and biological properties that support primary production by vegetation. The formation of water-stable aggregate (WSA) favors the sequestration of C in the soils by protecting it from decomposition (Tisdall and Oades, 1982), reducing in this way CO_2 emissions to the atmosphere (Lal, 2004). The distribution of aggregates between the different size classes has been related to soil structure, SOM, and biological activity and represents soil's ability to resist disintegration by disruptive forces (Six et al., 2000). The formation of WSA, due to organic matter (OM) addition, improves soil structure and the hydraulic properties of bulk soil through the formation of larger connected pores resulting in increased bulk permeability to fluid flow and improved drainage, while increasing the water holding capacity within the microscopic pores of the larger aggregates. WSA formation is directly related to protection of SOM. SOM can chemically bind to soil mineral particles, making SOM less bioavailable in sorbed form, and also altering the surface properties of soil mineral to favor particle–particle binding and aggregation (Nikolaidis and Bidoglio, 2013). Formation of larger aggregates also provides a redox barrier by holding water within the microscopic pores of the aggregates, which creates a diffusion barrier to O_2 and protects the SOM against oxidative microbial degradation. On the other hand, soil structure is affected adversely by the mechanical shearing created by tillage, by freezing and thawing that result from extreme temperature fluctuations, and by the compaction that is due to the use of heavy machinery (Ross et al., 2015).

Restoration of soil fertility through soil management can be achieved by applying appropriate agroecological practices (Lal, 2013). Crop rotation, organic C addition as SOM with associated nutrient elements, reduced tillage, use of cover crops during fallow periods, and controlled grazing are among the most commonly applied practices for sustainable agriculture (Milne et al., 2015). Incorporation of plant residues and organic amendments, such as compost and manure, combined with reduced tilling enhances C sequestration (Lal, 2015) and plant production (Li et al., 2016; Liao et al., 2015) as well as soil hydraulic properties (hydraulic conductivity, water holding capacity, aeration, and porosity) (Udom et al., 2016). Field experiments with organic amendments have shown that manure improves soil structure and protection of aggregate-associated SOM much more than plant compost in terms of aggregate size and C content (Udom et al., 2016), while evidence has been provided that a mixture of compost (e.g., municipal solid waste derived) and manure (of a rate 70/30) is equally beneficial, resulting in the increase of the large ($>$250 μm) WSA mass fraction (Kotronakis et al., 2016; Udom et al., 2016). The effectiveness of these practices depends strongly on OM application rate and composition, climate, and other management practices.

1.3 Modeling Soil Carbon and Structure Dynamics

Mathematical models have been developed to simulate and predict soil structure development and C sequestration. These soil properties, as measured by WSA and soil C stocks and their rates of formation, provide proxy measures for soil fertility, defined here as the capacity of soil to support the rate of primary production by vegetation (e.g., g C fixed year^{-1} kg soil^{-1}). Over the past decades, several mathematical models were developed to simulate soil organic carbon (SOC) dynamics. Mathematical models such as CENTURY (Parton et al., 1987), RothC (Coleman and Jenkinson, 1999; Jenkinson, 1990), DNDC (Gilhespy et al., 2014; Li et al., 2005), and APSIM (Keating et al., 2003) have been used (Álvaro-Fuentes and Paustian, 2010; Andrianaki et al., 2016; Carvalho Leite et al., 2004;

Dou et al., 2014; Galdos et al., 2009; Gao et al., 2008; Goglio et al., 2014; Luo et al., 2013; Poeplau and Don, 2015; Stamati et al., 2013b; Zhang et al., 2016) to assess the impact of land management on SOM stocks and to investigate the ability of these models to simulate long-term (decades–centuries) SOM dynamics across different ecosystems (Smith et al., 1997). However, these models do not consider the effect of structure and its feedbacks on organic C dynamics. Several studies have revealed a strong effect of structure on SOC turnover (Jastrow et al., 2007), identifying the link between OM decomposition and aggregate stability (Abiven et al., 2009) and summarizing the dynamics of SOC turnover with changes in soil structure (Nikolaidis and Bidoglio, 2013). Existing models have been updated (Coleman and Jenkinson, 1999) or further modified (Jenkinson and Coleman, 2008; Nadeu et al., 2015) to incorporate the contribution of factors affecting SOC dynamics and aggregate formation. Over the last decade new models have been developed, and new conceptual frameworks on modeling of soil functions have been proposed. Abiven et al. (2008) developed the Pouloud model to predict the impact of organic residues incorporation on aggregate stability, under field conditions. The Struc-C model (Malamoud et al., 2009) was based on the RothC-26.3 model and linked SOM dynamics with soil aggregation and soil structure. The InVEST model (Nelson et al., 2009) was structured to predict changes in ecosystem services under different land use/land cover change scenarios, incorporating subroutines assessing water service, soil conservation, C sequestration, biodiversity conservation, and commodity production value. The SoilGen2 model (Finke, 2012), which is a further development of the SoilGen1 model (Finke and Hutson, 2008), is a 1D model that simulates the pedogenesis of various parent materials and includes clay formation, and has been used to successfully simulate soil formation. Segoli et al. (2013) developed the AggModel which is a combination of an aggregate dynamics model and a SOM dynamics model, where the C pools are not conceptual but directly measured. Finally, Stamati et al. (2013a) developed the carbon, aggregation, and structure turnover (CAST) model using an aggregation mechanism similar to the Struc-C model approach and modeling the carbon sequestration and turnover rates in each aggregate size.

The objective of this work is to use the results of the CAST model simulations of SOC and WSA dynamics for seven sites around the world with different land use management practices, including natural and agricultural ecosystems, in order to characterize the process rates and the factors controlling soil structure dynamics and C sequestration across different climatic and lithological conditions. A meta-modeling PCA integrates the results, clusters the sites by dominant factors influencing SOC and structure dynamics, and identifies the principal factors controlling C sequestration within the clustered sites.

2. METHODOLOGY AND METHODS

The CAST model (Stamati et al., 2013a) was used to simulate soil structure dynamics and C sequestration in seven sites across the world. The sites were the Koiliaris CZO, Slavkov Forest CZO, Marchfeld CZO, Clear Creek (Iowa, United States), Damma Glacier CZO, Heilongjiang Mollisols (China), and Milia (Greece), representing various climates, land management practices, soil properties, and histories. Table 1 presents the list of the natural ecosystems and agricultural fields of each study site.

The simulations of the CAST model from the Marchfeld CZO, Slavkov Forest CZO, and the agricultural fields of Koiliaris CZO and Clear Creek are presented, while the simulations of Damma Glacier CZO, Heilongjiang

Table 1 Natural Ecosystems (Nontilled Soils) and Agricultural Fields (Tilled Soils) of the Study Sites

Natural Ecosystems	Agricultural Fields
Koiliaris CZO—Calcaric Regosols—set aside field	Koiliaris CZO—Calcaric Regosols—cropland
Marchfeld CZO—Chernozems—forest development	Marchfeld CZO—Chernozems—land use conversion from forest to cropland
Marchfeld CZO—Chernozems—land use conversion from forest to grassland	
Slavkov Forest CZO—Podzols—forestry	
Damma Glacier CZO—Lithic Leptosols	
Heilongjiang Mollisols: two fields with natural fallow and alfalfa	Heilongjiang Mollisols: four fields with soybean–maize rotation
Clear Creek—Millisols—set aside field	Clear Creek—Mollisols—cropland
	Milia—Eutric Lithosols: three fields with different practices of compost application—tilling

Fig 1 Geographic distribution of the seven simulated sites (Google Earth, 2016).

Mollisols, Milia, and the natural ecosystems of Koiliaris CZO and Clear Creek are described previously (Andrianaki, 2016; Li et al., 2017; Stamati et al., 2013a; Vavlas et al., 2014). The geographic distribution of the sites evaluated in this work is shown in Fig. 1.

All model simulations were performed using consistent guidelines. The simulations were compared in terms of the stocks and flows of C taking into account C sequestration, microbial biomass, and the CO_2 flux as well as the changes in the mass distribution of WSA size classes. A comparison of the sites was conducted regarding the rate constants related to the processes of plant biomass fragmentation, formation of micro- and macroaggregates, SOC decomposition, and aggregate disruption. Finally, PCA was performed in order to identify the most significant model input, output, and calibration parameters as the factors controlling SOC and soil structure dynamics and to identify sites across climatic and lithological gradients with similar responses to SOC sequestration. The model parameters that were included in the PCA were selected based on a parameter sensitivity analysis conducted during the calibration process. The most sensitive (i.e., affecting the simulation results) input, output, and model parameters were included in the PCA. The PCA was performed using the MiniTab 17 statistical software.

2.1 Methods

2.1.1 Model Description

The CAST model simulates the mechanisms of aggregation assuming three size classes of aggregates: the silt–clay-sized aggregates (AC1, <53 μm), the microaggregates (AC2; 53–250 μm), and the macroaggregates (AC3, >250 μm). Fig. 2 presents a schematic representation of the concept of WSA formation modified from Stamati et al. (2013a). The model assumes that macroaggregates are formed around large particulate organic matter (POM), followed by the inclusion of microaggregates within the macroaggregates. Microbial decomposers of plant residues provide the extracellular polymeric "glue" by which mineral particles and small aggregates bind to form macroaggregates around the POM (Phase I). Clay-sized mineral particles with relatively larger specific surface area provide complexation capacity to chemically bind constituent molecular components of SOM to the mineral surfaces, which protects the OM and favors inclusion of the clay-size fraction in micro- and macroaggregate formation (Nikolaidis and Bidoglio, 2013). The macroaggregate POM is further decomposed and the resulting finely fragmented POM is encrusted with silt–clay-sized aggregates leading to the formation of microaggregates

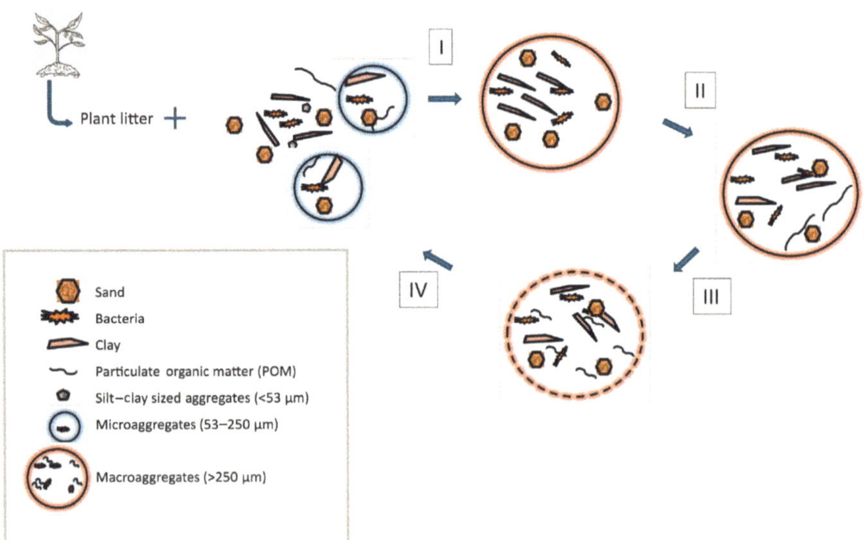

Fig. 2 Schematic representation of WSA formation modified from Stamati et al. (2013a). The particulate components that make up aggregates are not drawn to relative scale of their physical size.

within macroaggregates (AC2 in AC3) (Phase II). Decreased microbial activity inside the macroaggregates due to decreased availability of C and energy following biodegradation of POM reduces the supply of microbial polymers and aggregate disruption occurs (Phase III) with instant release of stable AC2 and AC1 aggregates and the fragmented POM becomes unprotected (Phase IV). When fresh plant residues enter the soil, new macroaggregates form and the cycle of aggregation and disaggregation continues.

Fig. 3 presents a schematic representation of the aggregation process of the CAST model, its aggregate fractions, and the C pools sequestered in them. Each arrow represents a mass transformation rate that is translated mathematically by the law of mass action into a linear rate equation with a first-order rate constant that defines the rate as proportional to the mass of the reactive material that is being transformed. The figure shows the phases of WSA formation together with the C pools and fluxes.

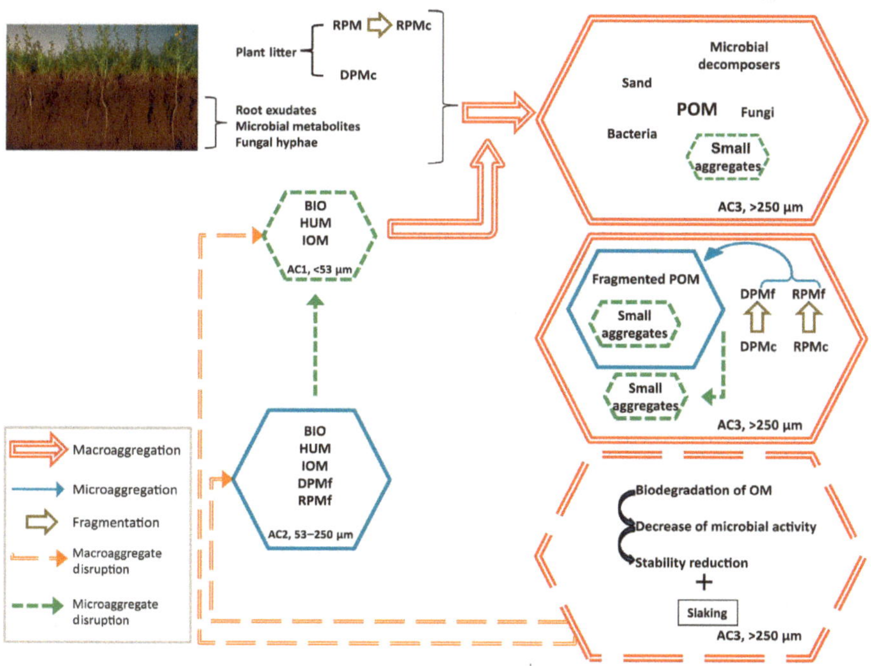

Fig. 3 Schematic representation of aggregation process, aggregate fractions, and C pools included in the CAST model. The three large hexagons representing the AC1 macroaggregate pool show in sequence, from top to bottom, three stages of macroaggregate transformation from initial POM entering the soil, until it is decomposed resulting in breakup and release of AC2 and AC3 aggregate fractions.

Fresh plant residue is characterized by the decomposable (DPM) and resistant plant material (RPM). Both DPM and RPM are fragmented and comprise the coarse fraction of DPMc and RPMc which further break down to fine fractions DPMf and RPMf. In the initial phase of aggregation, the AC3 aggregates are comprised of POM, AC1 aggregates, sand, bacteria, and fungi. At the second stage, coarse plant materials, DPMc and RPMc, are decomposed within the AC3 into fine, fragmented DPMf and RPMf which further facilitates the formation of the AC2 microaggregates within the AC3. Further biodegradation of the OM decreases the microbial activity and the stability of the macroaggregates which eventually break down into aggregates of the AC1 and AC2 size fractions.

The CAST model has been further modified to incorporate the effect of tilling and the impact of frozen soil on the aggregation/disaggregation mechanisms in order to improve the model versatility and extend the conditions that can be represented by simulation of the dynamics of SOM. Tilling was incorporated in the model by making the WSA destruction parameters time-variable. When there is tilling, the destruction rate parameter can be changed to a higher value, depending on the tilling intensity, which upon cessation of tilling subsequently reverts to its initial value. Similarly, frozen soil was incorporated into the model by allowing the rate constants of aggregation and decomposition of OM to vary with temperature. When the ground was defined as frozen, the rate constants of aggregation and OM decomposition can be changed to lower values which, over time, can then revert to their normal values.

The most important input parameters of the CAST model include climatic data such as temperature, precipitation and evapotranspiration (ET), soil properties (silt and clay content and bulk density), and WSA distribution and SOC stock distribution within each of the defined fractions of WSA. More specifically, samples from the topsoil (0–10 cm) for each field site were analyzed for the AC1, AC2, and AC3 fractions of WSA according to the methodology developed by Elliott (1986). AC3 aggregates were further separated to coarse POM (POMc), sand, easily dispersed silt–clay fraction, and AC2 aggregates according to Lichter et al. (2008). The AC2 aggregates within the AC3 aggregates were further separated into fine POM (POMf), sand, and the silt–clay fractions. The free AC2 aggregates were also separated into POMf and silt–clay fractions. These measurements are performed for every soil sample in order to obtain parameter values both for the initialization and calibration of the model.

2.1.2 Site Description

The data and experimental results used in the simulations represent conditions of natural ecosystems and agricultural practices. Table 2 presents a summary of the sites and their respective management. A description of the site details follows.

2.1.2.1 Damma Glacier CZO—Switzerland

Damma Glacier CZO is located at the central Alps, Canton Uri, Switzerland, and is a 9.9 km^2 catchment with an elevation range between 1940 and 3630 m above sea level. The glacier has been retreating since 1850, forming a soil chronosequence on a relatively flat area of about 1 km length at an altitude between 1950 and 2050 m. The glacier recession was reversed two times, during 1920–1928 and 1970–1992, which resulted in two small terminal moraines. Therefore, the chronosequence consists of three groups of soil ages. The youngest sites include soils from 6 to 14 years old, the intermediate group comprises of soils developed between 1930 and 1950, and the oldest group includes soils that started to evolve during 1870–1897. The soils have been classified as Lithic Leptosols. The soils at locations in the chronosequence, between those representing these stages of soil formation, have been eroded during glacial advances (Banwart et al., 2011; Bernasconi et al., 2011). The CAST model was used to simulate the accumulation of SOC and the development of soil structure along the chronosequence (Andrianaki et al., 2016). As the chronosequence is not continuous there were five simulations, three of the different soil ages (young soils, intermediate soils, and old soils) and two of the readmissions of the glacier. The calibration of the models is based on the extensive dataset available for the Damma Glacier CZO and a climate reconstruction back to 1867 (Smittenberg et al., 2012).

2.1.2.2 Heilongjiang Mollisols—China

The Heilongjiang Mollisols experimental field site is located in the central region of the Mollisols in Northeast China. The experimental site was established in 2004 at the State Key Experimental Station of Agroecology, Chinese Academy of Sciences, Hailun, Heilongjiang province. The region has a typical temperate continental monsoon climate with a hot summer and cold winter. The soils have evolved from sedimentary materials of loamy loess. Parent material was removed from the C horizon (>2 m) and replaced the surface soil down to 0.8 m. The experiment was set up to study soil development and restoration from an extremely degraded soil

Table 2 Summary of Sites Simulated With a Description of Soil Management

Site—Treatment	Description
Heilongjiang Mollisols—NatF	Natural fallow—no fertilization, organic input, no tilling
Heilongjiang Mollisols—Alfa	Alfalfa—no fertilization, organic input, no tilling
Heilongjiang Mollisols—F0C0	Soybean–maize rotation—no fertilization, no organic input, tilling
Heilongjiang Mollisols—F1C0	Soybean–maize rotation—fertilization, no organic input, tilling
Heilongjiang Mollisols—F1C1	Soybean–maize rotation—fertilization, low organic input, tilling
Heilongjiang Mollisols—F1C2	Soybean–maize rotation—fertilization, high organic input, tilling
Koiliaris natural	Koiliaris set aside
Koiliaris agricultural	Koiliaris agricultural management—green vegetable, light tilling
Clear Creek natural	Clear Creek set aside
Clear Creek agricultural	Clear Creek agricultural management—corn and soybeans, tilling
Milia1	Milia terrace 1—compost application for 10 years, every year
Milia2	Milia terrace 2—compost application for 8 years, every year and then fallow for 2 years
Milia3	Milia terrace 3—compost application for 10 years, every third year
Damma young	Young soils of Damma Glacier CZO—ages from 6 to 14 years old
Damma intermediate	Intermediate soils of Damma Glacier CZO—ages from 57 to 79 years old
Damma old	Old soils of Damma Glacier CZO—ages from 108 to 140 years old
Marchfeld forest	Marchfeld forest development (0–400 years forest)
Marchfeld grassland	Marchfeld land use conversion from forest to grassland (0–200 year forest—200–400 year grassland)
Marchfeld agricultural	Marchfeld land use conversion from forest to cropland (0–200 year forest—200–400 year cropland)
Slavkov forest	Slavkov Forest CZO—forestry

(Li et al., 2017). The field experiment included six treatments, two natural ecosystems and four agricultural fields, in order to compare: (a) nontilled soils under fallow and soil planted with alfalfa and (b) tilled soils with rotation of soya and maize in alternate years and different combinations of mineral fertilization (F) and organic (C) amendments. More specifically the tilled soils were managed (i) without fertilization and organic amendment (F0C0) after the aboveground biomass was removed, (ii) with fertilization and no organic C amendment (F1C0), (iii) with fertilization and low amount of organic C input (F1C1) after only partial aboveground biomass of two crops was incorporated into soil, and (iv) with fertilization and high amount of organic C input (F1C2) after all the aboveground biomass of the seasonal crop was incorporated into the soil (Li et al., 2017).

2.1.2.3 Koiliaris CZO—Greece

Koiliaris River catchment is a CZO that represents severely degraded soils due to intense agricultural practices applied for many centuries. It represents typical Mediterranean dry-lands soils evolving under imminent threat of desertification. The Koiliaris CZO is located 25 km east of the city of Chania, Crete, Greece. The total watershed area is approximately 130 km^2 and the main supply of water originates from the White Mountains. An additional area of 50 km^2 outside of the Koiliaris CZO is hydrogeologically connected due to limestone bedrock–karst terrain. Water erosion is recognized as the most important soil degrading threat due to the clearing of forests and natural vegetation, the livestock overgrazing, and the tilling of crops (Stamati et al., 2013a,b). Simulations performed for the Koiliaris CZO included both natural ecosystem and agricultural management scenarios, and the soil type is Calcaric Regosol.

2.1.2.4 Clear Creek—Iowa, United States

The Clear Creek site is located on the outskirts of Iowa City, United States. It is representative of humid continental climates with coarse-textured (sandy loam) Mollisols. The cropping period lasts from May to September, while during the winter period the soils are covered by snow. The site has been selected to study the impacts of land use conversion, from agricultural use to natural vegetation, on soil functions (Stamati et al., 2013a,b).

2.1.2.5 Slavkov Forest CZO—Czech Republic

Slavkov Forest CZO is a Protected Landscape Area located in the northwestern Czech Republic, 120 km west of Prague. The catchment area is

0.273 km^2, with elevations in the range of 829–949 m above sea level. There is intense silviculture with rapidly aggrading Norway spruce monoculture stands since 1850 on nutrient-depleted soils, mostly Podzols developed on granite. The mean stand age is about 40 years, and closed canopy forest covers 82% of the CZO, while clearings with young seedlings cover 18% of the catchment. The most important threats to the soil include nutrient leaching, acidity, metal toxicity, harvest erosion and compaction, low biodiversity due to monoculture, C loss due to elevated organic C export, and atmospheric deposition of anthropogenic pollutants (Banwart et al., 2011).

2.1.2.6 Marchfeld CZO—Austria

The study area is located in the Danube River flood plain downstream of Vienna in the "Marchfeld," with little variation in topography and climate. During alpine glaciations, the Danube continuously incised into the uplifting Tertiary basin fill and accumulated meltwater terraces. The floodplain is morphologically subdivided into two units: the recent floodplain sensu stricto and a slightly elevated area covered by older fluvial deposits. The soils in the Marchfeld CZO are Chernozems and create a chronosequence of soil development covering thousands of years, which allows the study of temporal soil development and also the effects of various land uses. The data from Marchfeld CZO were used to perform three simulations of the land use conversion (Rampazzo Todorovic et al., 2014). The initial conditions for each case are freshly deposited sediments with specified soil texture but without structure. Every simulation represents the evolution of C sequestration and soil structure development. The first 200 years of each simulation represent the forest development. The term "Forest" refers to the simulation of subsequent steady-state climax forest for 400 years. The term "Cropland" refers to the simulation of land use conversion from forest (initial 200 years) to cropland (total 400 years), and the term "Grassland" refers to the simulation of land use conversion from forest (initial 200 years) to grassland (total 400 years). The lengths of these periods were inferred from land use history.

2.1.2.7 Milia—Greece

Milia represents a strongly eroded Eutric Lithosol soil in which restoration practices including terraces formation and incorporation of organic amendments have been applied. The elevation of the study area is 500 m above sea level. Soil samples were collected from three terraces subjected to cultivation with varying compost application practices. Simulations were performed to represent a 10-year period of farming operations (Vavlas et al., 2014). The frequency of compost application was twice per year, through tilling, corresponding to a total application of 8 t C/ha. For terrace 1 (Milia 1), the compost was applied annually for 10 years, while for terrace 2 (Milia 2) it was applied for 8 years followed by 2 years fallow. Finally, for terrace 3 (Milia 3), the organic amendment was applied once every 3 years.

3. RESULTS AND DISCUSSION

The interpretation of soil properties, climatic conditions, and land use and management practices in the study areas reveals great variability. The sites are characterized by variable climatic conditions ranging from 0–4°C of mean annual temperature of Damma Glacier CZO to 17.6–18.1°C of the Greek sites. Precipitation also varies from 1898 mm/year of Damma Glacier CZO to 510 mm/year of the Heilongjiang Mollisols. In addition, the sites cover a wide variety of soil types at different stages of soil development, from Lithic Leptosol of Damma Glacier to Eutric Lithosols of Milia, to Calcaric Regosols of Koiliaris, and then to more developed Podzols, Chernozems, and Mollisols of Slavkov Forest, Marchfeld, and China/United States, respectively. Organic carbon input to the soils at the natural ecosystem sites was approximately proportional to the temperature gradient as it is shown in the site data comparison of Table 3. The lowest C inputs occurred at the Damma Glacier CZO and the Heilongjiang Mollisols and approached 1 t C/ha/year as a consequence of the short growing season imposed by the low temperatures. As the annual average temperature increases to 5.3°C in Slavkov Forest CZO, the C input also increased to 2.75 t C/ha/year. For higher temperatures, such as that of Clear Creek with 10.3°C and of Koiliaris CZO 18.1°C, the amount of C inputs was strongly mediated by the precipitation level and its seasonal distribution. For instance, the C input of Clear Creek was 5.8 t C/ha/year due to the availability of precipitation during the summer (923 mm/year), whereas in Koiliaris CZO is 3.8 t C/ha/year due to semiarid climate (652 mm/year). Organic carbon input to soils at the agricultural sites depends on the applied management practices, e.g., tilling intensity, below ground biomass of crop production, and the amount of aboveground biomass incorporated to the soil. At these sites, annual C input varies from 0.205 t C/ha at Marchfeld to 8 t C/ha at Milia.

Table 3 Comparison of the Site Data: SOC, Soil Characteristics, and Meteorological Data

Site	Soil Characteristics			SOC Characteristics		Meteorological Data		
	Bulk Density (g/cm^3)	Silt–Clay (%)	Clay (%)	DPM to RPM Ratio	Initial SOC Mass (t C/ha/year)	Mean Temperature (°C)	Mean Precipitation (mm)	Mean Pan Evaporation (mm)
Heilongjiang Mollisols Alfa	1.35	77.6	42.0	0.25	13.77	2.5	510.6	514
Heilongjiang Mollisols NatF	"	"	"	0.15	"	"	"	"
Heilongjiang Mollisols F0C0	"	"	"	0.40	"	"	"	"
Heilongjiang Mollisols F1C0	"	"	"	1.00	"	"	"	"
Heilongjiang Mollisols F1C1	"	"	"	1.20	"	"	"	"
Heilongjiang Mollisols F1C2	"	"	"	1.44	"	"	"	"
Koiliaris CZO natural	1.18	67.0	30.0	0.67	34.92	18.1	651.9	1916
Koiliaris CZO agricultural	1.11	"	"	1.44	58.55	"	"	"
Milia 1	1.00	34.0	3.30	0.43	33.90	17.6	1494.6	1601.7
Milia 2	"	"	"	"	"	"	"	"
Milia 3	"	"	"	"	"	"	"	"
Clear Creek natural	1.11	37.0	7.00	1.51	18.52	10.3	923.0	1413.6

Clear Creek agricultural	"	"	"	2.00	32.38	"	"	"
Slavkov Forest CZO	1.30	47.0	11.00	0.25	55.36	5.3	1049.0	442.7
Marchfeld CZO forest	2.00	73.6	16.36	0.25	20.00	9.1	687.0	727.7
Marchfeld CZO grassland	1.26	"	"	0.70	27.11	"	"	"
Marchfeld CZO cropland	1.26	"	"	1.44	"	"	"	"
Damma Glacier CZO young soils	1.50	35.0	3.10	1.44	0.60	4.0	1898.3	242.7
Damma Glacier CZO intermediate soils	1.50	"	"	"	1.05	2.9	"	"
Damma Glacier CZO old soils	1.00	"	"	"	13.5	2.7	"	"

", the same value as the above cell.
DPM, decomposable plant material; *RPM*, resistant plant material.

A schematic diagram of the study sites placed along a temperature gradient representing different stages of soil evolution is illustrated in Fig. 4. The sites are placed according to the annual mean temperature, beginning from the Heilongjiang Mollisols and Damma Glacier CZO, to Slavkov Forest CZO, Marchfeld CZO, Clear Creek ending to Milia and Koiliaris CZO. The initial soil conditions for the model simulations of each site, in terms of clay and macroaggregate content, and the rates of macroaggregates

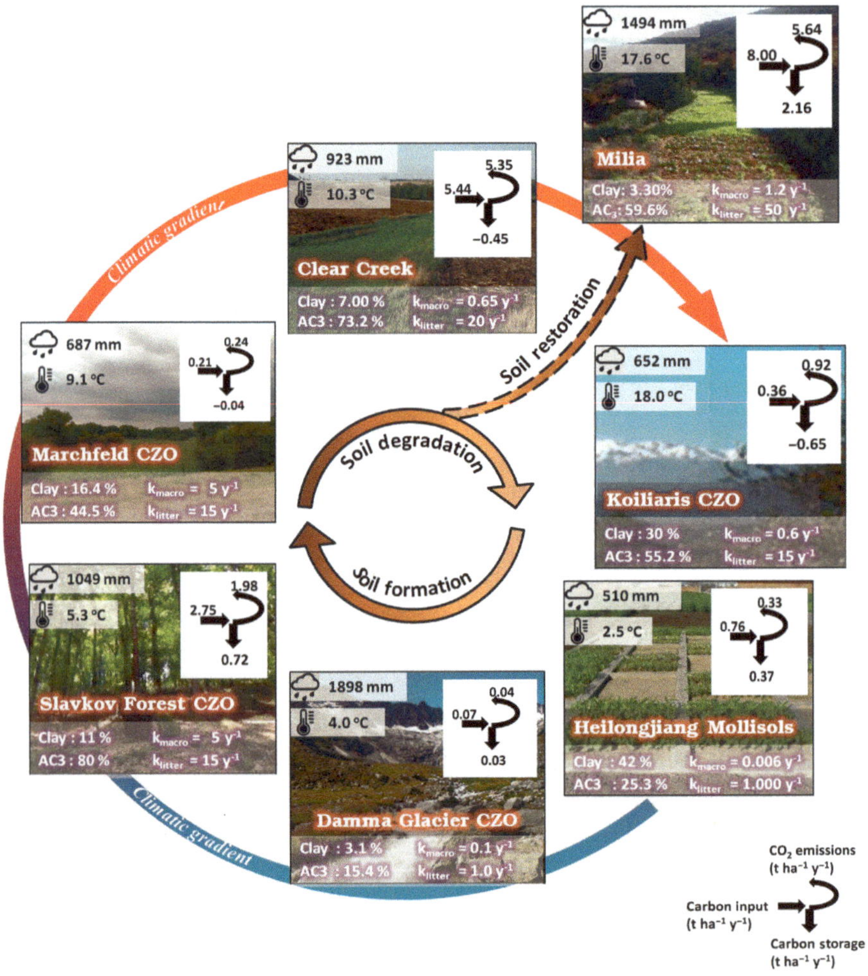

Fig. 4 Study sites placed on a temperature gradient representing different stages of soil evolution from soil formation to soil degradation, including soil restoration. The ordering of sites according to soil formation/degradation pathway also coincided with the temperature gradient.

formation and decomposition of plant litter are also illustrated in Fig. 4. The insert in each site are the C flux balances which include C input, C storage, and CO_2 emissions. The ordering of sites according to temperature also coincides with the different stages of soil evolution starting from soil formation due to weathering in Damma Glacier CZO and the parent material of Heilongjiang Mollisols, moving to soils used for forestry at Slavkov Forest CZO and arable land at Marchfeld CZO, ending at the relatively degraded soils of Clear Creek and Koiliaris CZO due to permanent cropping and overgrazing. The C flux balances show soil degradation due to C losses represented by the negative values of C storage. The site of Milia is placed outside the circle of soil formation and soil degradation because it represents soil restoration conditions due to organic carbon addition.

3.1 Calibration results

The simulations for the Koiliaris CZO, Clear Creek, Marchfeld CZO, and Slavkov Forest CZO sites with regard to the distribution of SOC stocks and WSA, and their interpretation with the field data used for model calibration, are shown in Figs. 5 and 6, respectively. The calibration parameters of the CAST model and their description are summarized in Table 4. The descriptions of the parameters indicating soil initial conditions are presented in Table S1 (Supplementary Information: http://dx.doi.org/10.1016/bs.agron.2016.10.008).

A comparison of the CO_2 fluxes, C stocks, and microbial biomass is presented in Table 5. The table presents the initial C content of each site that ranges from 0.6–14.8 t C/ha at Damma Glacier CZO to 13.5 t C/ha for the Heilongjiang Mollisols to 55.4 t C/ha for Slavkov Forest, to 18.5 t C/ha for Clear Creek, and to 34.9 t C/ha for Koiliaris CZO soil, for the natural sites. Regarding the agricultural sites, the initial SOC mass ranges from 13.77 t C/ha for the Heilongjiang Mollisols to 58.55 t C/ha for Koiliaris CZO soil. Milia had an initial SOC mass of 33.9 t C/ha, Clear Creek 32.38 t C/ha, and Marchfeld CZO 27.11 t C/ha. The values of initial SOC mass of Damma Glacier CZO and Heilongjiang Mollisols can be explained by the fact that Damma Glacier CZO is a very new, 150-year-old soil, while the Heilongjiang Mollisols were C horizon soil brought to the soil surface and placed under cultivation. The high value for Slavkov Forest CZO is due to the fact that it is a forested site relatively undisturbed for the past 70 years, while the low value of Clear Creek is due to intense cultivation

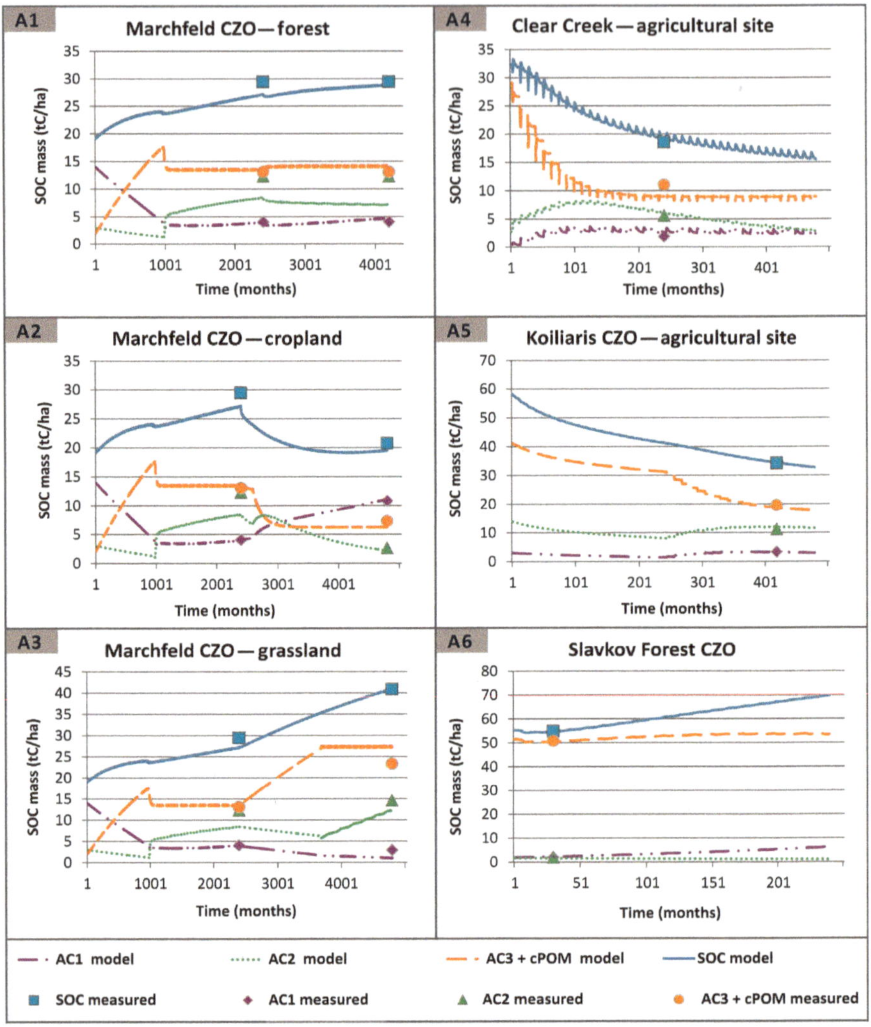

Fig. 5 Calibration results of Koiliaris CZO (agricultural—A5), Clear Creek (agricultural—A4), Marchfeld CZO (all managements—A1, A2, A3), and Slavkov Forest (A6) simulations for the SOC stock distribution together with the field (measured) data used for the calibration. Regarding the goodness of fit of the calibration of the model, the mean RMSE for the SOC is 0.55, which corresponds to 0.55 t C/ha.

of the Iowa, United States soils and the sandy/low clay content nature of the soil that offers limited mineral surface for binding and forming organo-mineral complexes that protect and sequester OM. Finally, the initial SOC mass of Koiliaris CZO is indicative of its relatively high clay content and lower intensity of agricultural practices.

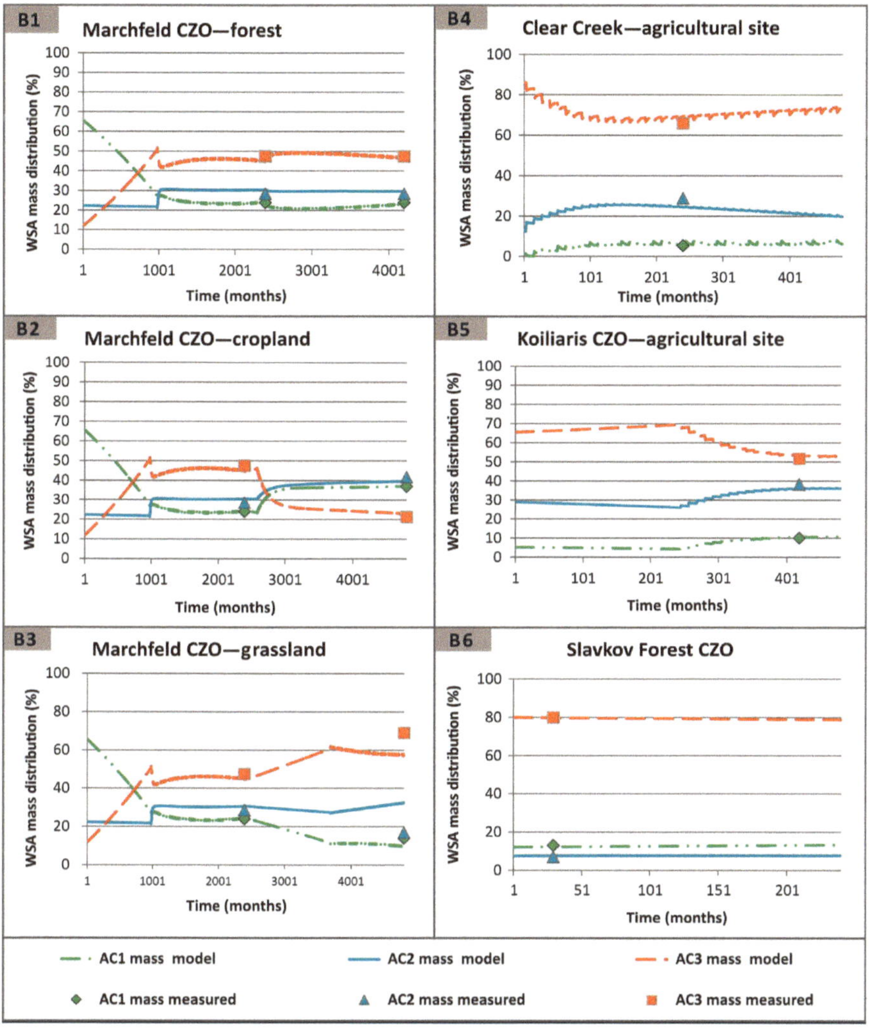

Fig. 6 Calibration results of Koiliaris CZO (agricultural—B5), Clear Creek (agricultural—B4), Marchfeld CZO (all managements—B1, B2, B3), and Slavkov Forest (B6) simulations for the WSA distribution together with the field data used for the calibration. Regarding the goodness of fit of the calibration of the model, the mean RMSE for the WSA is 1.29.

Regarding the natural ecosystems, the young soils of Damma Glacier CZO and Heilongjiang Mollisols sequester higher amounts of the C input, up to 0.31 (28%) and 0.47 t C/h/year (40%), respectively, while the older soils of the Clear Creek and Koiliaris CZO sequester significant lower proportions, 5% and 13%, respectively, even though the annual

Table 4 Description of the Calibration Parameters of the CAST Model

Main CAST Model Calibration Parameters		Description
Fragmentation	RPM to RPMc	Rate constant of fragmentation of resistant plant material to coarse resistant plant material
	RPMc to RPMf	Rate constant of fragmentation of coarse resistant plant material to fine resistant plant material
	RPMc(AC3) to RPMf(AC3)	Rate constant of fragmentation of coarse resistant plant material to fine resistant plant material within macroaggregates
	DPMc(AC3) to DPMf(AC3)	Rate constant of fragmentation of coarse decomposable plant material to fine decomposable plant material within macroaggregates
Macroaggregation	RPMc	Rate constant of coarse resistant plant material mass transfer for macroaggregate formation
	DPMc	Rate constant of coarse decomposable plant material mass transfer for macroaggregate formation
Microaggregation	RPMf(AC2inAC3)	Rate constant of fine resistant plant material mass transfer for microaggregate formation within the macroaggregates
	DPMf(AC2inAC3)	Rate constant of fine decomposable plant material mass transfer for microaggregate formation within the macroaggregates
Decomposition	Fresh plant input (DPM)	Rate constant of decomposition of decomposable plant material from the plant litter pool
	BIO(AC1) within AC3	Rate constant of decomposition of biomass carbon pools in aggregate type 1 within aggregate type 3 ($AC1_{withinAC3}$)
	HUM(AC1) within AC3	Rate constant of decomposition of humus in aggregate type 1 within aggregate type 3 ($AC1_{withinAC3}$)

Table 4 Description of the Calibration Parameters of the CAST Model—cont'd

Main CAST Model Calibration Parameters		**Description**
	BIO(AC2) within AC3	Rate constant of decomposition of biomass carbon pools in aggregate type 2 within aggregate type 3 ($AC2_{withinAC3}$)
	HUM(AC2) within AC3	Rate constant of decomposition of humus in aggregate type 2 within aggregate type 3 ($AC2_{withinAC3}$)
	BIO(AC2)	Rate constant of decomposition of biomass carbon pools in aggregate type 2
	HUM(AC2)	Rate constant of decomposition of humus in aggregate type 2
	BIO(AC1)	Rate constant of decomposition of biomass carbon pools in aggregate type 1
	HUM(AC1)	Rate constant of decomposition of humus in aggregate type 1
Contribution in macroaggregation	RPMc	Percent composition of macroaggregates (AC3) by coarse resistant plant material
	DPMc	Percent composition of macroaggregates (AC3) by coarse decomposable plant material
	AC1	Percent composition of macroaggregates (AC3) by silt–clay-sized aggregates (AC1)
	AC2	Percent composition of macroaggregates (AC3) by microaggregates (AC2)
Contribution in microaggregation	RPMf within AC3	Percent composition of microaggregates (AC2) by fine resistant plant material within macroaggregates (AC3)
	DPMf within AC3	Percent composition of microaggregates (AC2) by fine decomposable plant material within macroaggregates (AC3)
	AC1 within AC3	Percent composition of microaggregates (AC2) by silt–clay-sized aggregates (AC1) within macroaggregates (AC3)

Continued

Table 4 Description of the Calibration Parameters of the CAST Model—cont'd

Main CAST Model Calibration Parameters		Description
Disruption	$DPM_f + DPM_c$ within AC3	Fine and coarse DPM pool contents of the AC3 aggregate type, below which macroaggregates are considered unstable
	$DPM_f + DPM_c$ AC2 within AC3	Fine and coarse DPM pool contents of the AC2 aggregate type within AC3 aggregate type, below which microaggregates within macroaggregates are considered unstable
	$DPM_f + DPM_c$ within AC2	Fine and coarse DPM pool contents of the AC2 aggregate type, below which microaggregates are considered unstable

storage is maintained close to that of the young soils (0.27 t C/ha/year for Clear Creek and 0.48 t C/ha/year for Koiliaris CZO). These differences can be explained by climatic conditions, soil structure, and land use management. Slavkov Forest CZO presents intermediate values with 26% (0.72 t C/ha/year) of the C input to be sequestered annually. The CO_2 emissions follow the opposite trend to that observed for C storage with the young soils having lower emissions compared to the older soils. As for the cultivated sites, the initial SOC mass (Table 3) ranged from 13.77 t C/ha for the Heilongjiang Mollisols to 58.55 t C/ha for Koiliaris CZO soil. Milia had an initial SOC mass of 33.9 t C/ha, Clear Creek 32.38 t C/ha, and Marchfeld CZO 27.11 t C/ha. The values of initial SOC mass of Heilongjiang Mollisols can be explained by the fact that the Heilongjiang Mollisols had a well-developed C horizon when the cultivation started. C input ranged from 0.2 t C/ha at Marchfeld to 8 t C/ha in Milia. The data from the Heilongjiang Mollisols showed that the higher the C input, the more the C storage, CO_2 emissions, and bacterial biomass. A similar trend was found for the Milia soils. Comparison of sites with similar C input such as the Koiliaris CZO and Marchfeld CZO revealed that climate (i.e., temperature, rainfall) and water availability through irrigation played an important role. The warmer site (Koiliaris CZO) showed higher CO_2 emissions and loss of organic C despite the lower bacterial stock. Similar emissions of CO_2 to those observed in the Koiliaris CZO were observed in the Heilongjiang Mollisols which could be attributed to the higher C input.

Table 5 Comparison of Simulation Results With Regard to: Annual C Stock and Fluxes

Site	Simulation period (y)	Annual C Input (t C/ha/year)			Annual C Storage (t C/ha/year)			Annual CO_2 emissions (t C/ha/year)			Bacterial Stock (BIO—t C/ha/year)	
		Average	Min	Max	Average	Min	Max	Average	Min	Max	Average	% of Total Stock
Heilongjiang Mollisols—NatF	10	0.76	0.00	0.95	0.37	−0.61	1.12	0.33	0.08	0.49	8.00	4.49
Heilongjiang Mollisols—Alfa	"	0.94	"	1.17	0.47	"	1.31	0.41	0.11	0.57	8.35	4.52
Heilongjiang Mollisols—F0C0	"	0.69	"	0.86	0.26	−0.54	0.65	0.38	0.11	0.6	8.51	4.92
Heilongjiang Mollisols—F1C0	"	1.05	"	1.31	0.32	"	0.71	0.67	0.26	1.15	10.17	5.72
Heilongjiang Mollisols—F1C1	"	2.27	"	2.84	0.84	"	1.57	1.36	0.26	2.43	13.6	6.51
Heilongjiang Mollisols—F1C2	"	2.96	"	3.70	0.99	"	2.42	1.89	0.26	3.69	16.18	7.30
Koiliaris CZO natural	100	3.80	3.80	3.80	0.48	−0.8	1.07	3.29	2.73	4.47	15.8	2.10
Koiliaris CZO agricultural	40	0.36	0.36	0.36	−0.65	−2.34	−0.31	0.92	0.57	2.58	11.60	2.30
Milia 1	20	8.00	8.00	8.00	2.16	0.28	4.29	5.64	2.61	7.72	39.80	5.54
Milia 2	"	3.20	0.00	"	−0.21	−5.59	4.29	3.2	1.23	5.59	30.76	5.98
Milia 3	"	1.20	0.00	"	−0.65	−5.58	5.72	1.65	0.66	3.09	18.64	5.39
Clear Creek natural	100	5.60	5.60	5.60	0.27	0.03	2.00	5.29	2.79	5.57	36.35	7.50
Clear Creek agricultural	40	5.44	5.44	5.44	−0.42	−1.15	−0.12	5.35	5.07	6.19	7.90	3.08

Continued

Table 5 Comparison of Simulation Results With Regard to: Annual C Stock and Fluxes—cont'd

Site	Simulation period (y)	Annual C Input (t C/ha/year)			Annual C Storage (t C/ha/year)			Annual CO_2 emissions (t C/ha/year)			Bacterial Stock (BIO—t C/ha/year)	
		Average	Min	Max	Average	Min	Max	Average	Min	Max	Average	% of Total Stock
Slavkov Forest CZO	20	2.75	2.75	2.75	0.72	−0.97	0.93	1.98	1.81	2.78	24.08	3.22
Marchfeld CZO initial forest	200	0.22	0.22	0.22	0.04	0.04	0.04	0.18	0.08	0.29	17.48	6.01
Marchfeld CZO grassland	"	0.15	0.15	0.15	0.07	0.05	0.09	0.08	0.06	0.09	28.16	6.79
Marchfeld CZO cropland	"	0.21	0.21	0.21	−0.04	−1.03	0.01	0.24	0.19	1.23	32.56	13.22
Damma Glacier CZO—young soils	14	0.07	0.07	0.07	0.03	0.02	0.03	0.04	0.02	0.04	3.86	36.56
Damma Glacier CZO—intermediate soils	44	0.72	0.10	1.80	0.31	0.03	0.82	0.4	0.04	1.11	4.72	12.40
Damma Glacier CZO—old soils	40	1.12	0.20	2.00	0.24	−0.10	0.60	0.81	0.28	1.45	12.11	14.08

", the same value as the above cell.

The poor soil structure of the Heilongjiang Mollisols and the Damma Glacier CZO seems to improve with time. This is evident in the Damma Glacier CZO chronosequence where the macroaggregate AC3 fraction increases from 15% to 46%. In the Heilongjiang Mollisols the macroaggregate fraction ranged between 29% and 35%. Even greater proportions of macroaggregates were found in the Clear Creek and Slavkov Forest CZO soils that approached 79% and 78%, respectively. The corresponding proportion for the Koiliaris CZO scrublands was 47%. Finally, in the forests and grasslands of the Marchfeld CZO the macroaggregate fraction was 45% and 57%, respectively.

Regarding the cultivated sites, the Heilongjiang Mollisols macroaggregate fraction ranged from 21% to 48% depending on the management practice. Milia macroaggregate fractions ranged from 45% to 70% with the highest proportions corresponding to the treatment of the annual application of OM amendment. Finally, the Koiliaris CZO macroaggregate fraction was 53%, Clear Creek was 74%, and Marchfeld CZO was 23%, again reflecting differences regarding climate, agricultural practices, and soil properties. A decrease in the macroaggregate fractionation was observed in Koiliaris CZO natural ecosystem, which was accompanied by an increase in the microaggregate fraction AC2. The chronosequence at the Marchfeld CZO site, showed that the distribution of the large fraction of WSA, the AC3 fraction, was increased significantly during the first 200 years of the forest development. The soil initially was consisted of sediment deposition and the natural vegetation led to aggregate formation and soil structure development.

The values for the calibration parameters of the CAST model for the simulations of natural ecosystems and agricultural sites are summarized in Table S2 (http://dx.doi.org/10.1016/bs.agron.2016.10.008) and Table S3 (http://dx.doi.org/10.1016/bs.agron.2016.10.008), respectively, while a range of these values is presented in Table S4 (http://dx.doi.org/10.1016/bs.agron.2016.10.008) in order to provide an initial database for future modeling activities. Overall, the modeling results of the CAST model for both the natural ecosystems and agricultural management sites were consistent with the field data. The model has been able to describe SOC and soil structure dynamics in a wide variety of natural ecosystem and agricultural sites around the world. A broad comparison between the process rates constants of the agricultural management sites and the natural ecosystems shows that the rates of fragmentation, microaggregation, and disruption are higher for the agricultural sites, while the rates for macroaggregation and decomposition of OM are higher for the natural ecosystems.

3.2 Factors Controlling Soil Structure Dynamics and Carbon Sequestration

Our findings confirmed those of earlier studies that the factors regulating the sequestration of C in soils are driven by complex interactions within the soil–plant–water system. A PCA was employed to classify the soils studied in the present work in terms of their status considering climatic parameters, basic soil properties, SOC and macroaggregate content at the initial and final stages of simulation and their annual rates of change, C input rates to soil through litter fall and organic amendments, the decomposition rates of the soil C pools, and the magnitude of disruption of the aggregates including the impact of tilling.

The score and the loading plots of the components of the PCA are presented in Figs. 7 and 8 in order to illustrate the clustering of the sites and the significance of each parameter in this clustering. The 35 parameters used in the PCA together with the eigenvalues for the first two principal components are summarized in Table S5 (http://dx.doi.org/10.1016/bs.agron.2016.10.008). The most important parameters that explained the variability in the first component are the initial macroaggregate fraction as percentage of total mass, the initial value of the SOC normalized to the mass content of the silt and clay, the temperature, the bulk density, and the ET. Similarly, the most important parameters included in the second component are the percent mass fraction of macroaggregates, the percent mass fraction of silt-sized aggregates, the decomposition rates of humified and bacterial organic C pools, and the fragmentation rate of RPM by soil fauna. PC1 could explain 32.1% of the variability of the samples, while together with PC2 49.7% of the variability.

Fig. 7 presents the clustering of the sites as a result of the PCA. A plot of PC1 vs PC2 of all the sites resulted in the development of four clusters. The first cluster includes the sites of Heilongjiang Mollisols (all management practices) and Damma Glacier CZO (new and intermediate soils) which represents soils at their early stage of evolution. For the Heilongjiang Mollisols, this reflects the introduction of C horizon parent material to the soil surface, hence exhibiting characteristics of a young soil. The second cluster includes all the sites of Marchfeld CZO, Slavkov Forest CZO, Damma Glacier CZO (older soils), and the natural ecosystems of Koiliaris CZO. These sites are either natural or set aside sites or flood plain soils with a high degree of fertility. The third cluster includes the agricultural sites of the Koiliaris CZO and all the Clear Creek studies which represent

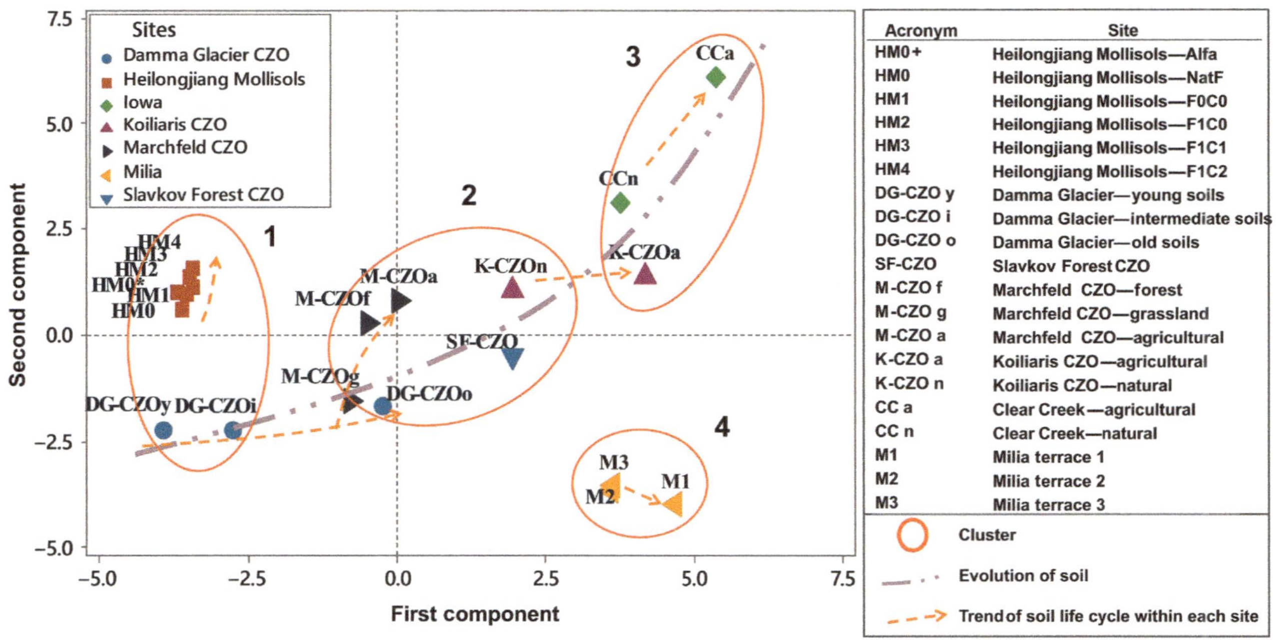

Fig. 7 Score plot of the components PC1 (first component) and PC2 (second component) of the principal component analysis.

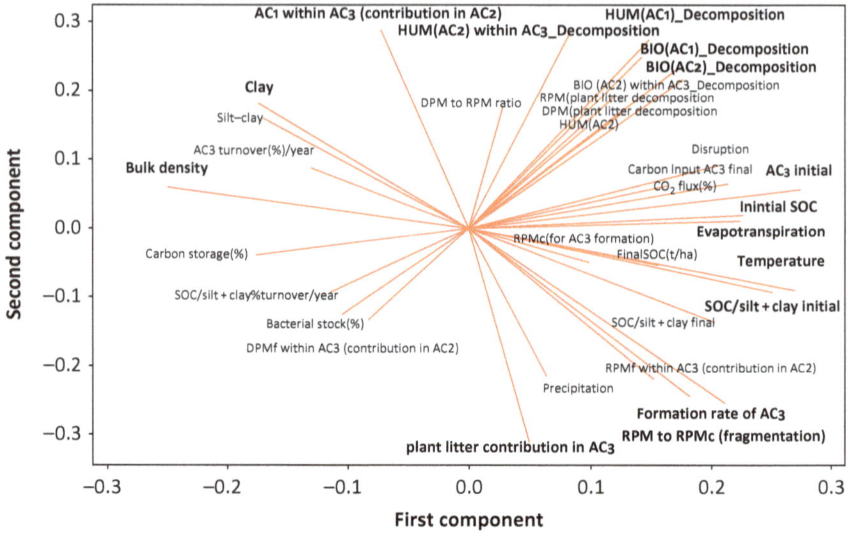

Fig. 8 Loading plot of the components PC1 (first component) and PC2 (second component) of the principal component analysis.

intensively managed soils. Finally, the fourth cluster includes the site of Milia (all management practices) where intensive C sequestration occurs due to large C additions. As it can be seen from Fig. 7, PC1 differentiates the sites starting from the left with soils at an early stage of evolution, moving to the second cluster in the middle with mature soils and moving to the upper right of the graph with the intensively managed soils. PC2 differentiates the sites based on the degree of intensification vs natural ecosystem sites but with C addition. Cluster 4 accounts for the Milia site which diverges from the main trend of clusters 1 through 3 due to extensive C addition. Similarly, the sites within Cluster 1 identify a trend from the early diagenesis sites of Damma Glacier CZO to the cultivated C horizon sites of the Heilongjiang Mollisols. It is interesting to note that PC2 differentiates the Chinese Heilongjiang Mollisols based on the amount of C added.

In terms of soil carbon evolution, the four clusters of Fig. 7 represent the sites dominated by carbon fixation (Cluster 1), the sites with stable SOC content (Cluster 2), the sites dominated by SOC mineralization and net loss (Cluster 3), and the sites with large artificial SOC addition (Cluster 4).

Fig. 8 presents the loading plot of the PCA in order to identify the most important parameters contributing to the differentiation of the clusters. The factors contributing to the differentiation of Cluster 1 include bacterial biomass as a proxy for microbial decomposer activity, SOC storage, the increase of the fertility factor SOC/(silt + clay), and the increase of the macroaggregate fraction. Cluster 2 represents older soils which have probably reached a steady state. This cluster is placed in the middle of the diagram where all the factors play a seemingly equal role. Cluster 3 represents degraded soils where the decomposition rates of the C pools are higher, the DPM outweighs the RPM, and the disruption of macroaggregates is increased. Cluster 4 represents soils restored by beneficial intervention practices where the aggregation rates are higher, and the plant material is decomposed by fauna and contributes to aggregation. The trend presented in the PCA in Fig. 7 is related to the evolutionary nature of the soils beginning from the young soils and parent material, proceeding to the older soils, forests, and arable lands, ending either at soil degradation through continuous cultivation or to restoration through beneficial practices of previously intensively managed soils. In broad terms the evolution of soil structure and increasing intensity of human exploitation of the resulting soil functions correlates with the trajectory shown in Fig. 7. This trajectory starts from young soils at the left, to productive soils in the center, to heavily impacted and very intensively managed soils at the upper right. The dotted arrows show the trend of soil evolution within each CZO. For the Heilongjiang Mollisols, the trend starts at the natural ecosystems ending at the cultivation practices. For the Damma Glacier CZO soils, it begins at the young soils, ending at the older soils. For the Marchfeld CZO soils the trend starts at the grassland, moving to forest, ending at the agricultural land. At Koiliaris CZO and Clear Creek the trend starts at natural ecosystem ending at cultivation. At Milia the trend of soil restoration starts at low C amendment addition ending at higher application. It appears that the first two principal components can account for the variability that describes the evolutionary nature of soil from formation to cultivation as well as a reverse in this trajectory when moving from intensive utilization with soil degradation to restoration.

Fig. 8 presents the variables with the higher values, either positive or negative, of the factors with the higher impact on soil structure dynamics and C sequestration. These variables include soil bulk density, clay content, temperature, ET, initial SOC mass and AC3 fraction, decomposition rates of the BIO and HUM pools, the contribution of plant litter and silt–clay-sized aggregates on AC2 and AC3 formation, the formation rate of AC3 aggregates, and the fragmentation of RPM through fauna. Briefly the factors controlling soil structure and C dynamics are related to soil properties, climatic conditions, decomposition rates of OM, and the formation rates and the predominant buildersers of macroaggregates.

4. CONCLUSIONS

In this study, the CAST model was modified to incorporate the impact of tilling and frozen soil conditions on aggregate formation and SOC sequestration. The modified CAST model was used to simulate 20 cases of different soil management practices located at 7 sites across a wide range of climatic conditions, land uses, and parent material. The CAST model can successfully simulate and predict aggregate formation and C sequestration on soils across the world, under a variety of climatic, lithological, and agricultural conditions. The PCA indicated the predominant factors controlling soils structure and C sequestration to be the plant litter contribution to aggregate formation, the decomposition rates of the humified and bacterial C pools, the initial state of soil in terms of SOC, silt and clay, and macroaggregate contents, the fragmentation rate of plant material through earthworms and other fauna, the macroaggregate formation rates, the temperature and ET, and the soil bulk density. The model reliably simulated soil C and soil structure dynamics for a wide variety of land uses, climatic and lithological conditions, and soil properties providing support for the underlying conceptual and mathematical model, and evidence that the CAST model provides a valuable quantitative analysis tool to interpret soil structure formation processes and to aid the design of sustainable soil management practices which support WSA formation.

ACKNOWLEDGMENT

Funding for this work was provided by the EU FP7-ENV-2009 Project SoilTrEC "Soil Transformations in European Catchments" (Grant #244118).

REFERENCES

Abiven, S., Menasseri, S., Angers, D.A., Leterme, P., 2008. A model to predict soil aggregate stability dynamics following organic residue incorporation under field conditions. Soil Sci. Soc. Am. J. 72, 119–125. http://dx.doi.org/10.2136/sssaj2006.0018.

Abiven, S., Menasseri, S., Chenu, C., 2009. The effects of organic inputs over time on soil aggregate stability—a literature analysis. Soil Biol. Biochem. 41, 1–12. http://dx.doi.org/10.1016/j.soilbio.2008.09.015.

Álvaro-Fuentes, J., Paustian, K., 2010. Potential soil carbon sequestration in a semiarid Mediterranean agroecosystem under climate change: quantifying management and climate effects. Plant Soil 338, 261–272. http://dx.doi.org/10.1007/s11104-010-0304-7.

Andrianaki, M., Bernasconi, S.M., Nikolaidis, N.P., 2017. Quantifying the incipient development of soil structure and functions within a glacial forefield chronosequence. Adv. Agron. 142, 215–239.

Banwart, S., Bernasconi, S.M., Bloem, J., Blum, W., Brandao, M., Brantley, S., Chabaux, F., Duffy, C., Kram, P., Lair, G., Lundin, L., Nikolaidis, N., Novak, M., Panagos, P., Ragnarsdottir, K.V., Reynolds, B., Rousseva, S., de Ruiter, P., van Gaans, P., van Riemsdijk, W., White, T., Zhang, B., 2011. Soil processes and functions in critical zone observatories: hypotheses and experimental design. Vadose Zo. J. 10, 974–987. http://dx.doi.org/10.2136/vzj2010.0136\r10.1002/hyp.7380.

Bernasconi, S.M., Bauder, A., Bourdon, B., Brunner, I., Bünemann, E., Chris, I., Derungs, N., Edwards, P., Farinotti, D., Frey, B., Frossard, E., Furrer, G., Gierga, M., Göransson, H., Gülland, K., Hagedorn, F., Hajdas, I., Hindshaw, R., Ivy-Ochs, S., Jansa, J., Jonas, T., Kiczka, M., Kretzschmar, R., Lemarchand, E., Luster, J., Magnusson, J., Mitchell, E.A.D., Venterink, H.O., Plötze, M., Reynolds, B., Smittenberg, R.H., Stähli, M., Tamburini, F., Tipper, E.T., Wacker, L., Welc, M., Wiederhold, J.G., Zeyer, J., Zimmermann, S., Zumsteg, A., 2011. Chemical and biological gradients along the Damma Glacier soil chronosequence, Switzerland. Vadose Zo. J. 10, 867–883. http://dx.doi.org/10.2136/vzj2010.0129.

Brantley, S.L., Goldhaber, M.B., Vala Ragnarsdottir, K., 2007. Crossing disciplines and scales to understand the critical zone. Elements 3, 307–314. http://dx.doi.org/10.2113/gselements.3.5.307.

Carvalho Leite, L.F., De Sá Mendonça, E., Oliveirade De Almeida MacHado, P.L., Inácio Fernandes Filho, E., Lima Neves, J.C., 2004. Simulating trends in soil organic carbon of an acrisol under no-tillage and disc-plow systems using the century model. Geoderma 120, 283–295. http://dx.doi.org/10.1016/j.geoderma.2003.09.010.

CEC, 2006. Communication from the Commission to the Council (CEC), the European Parliament, the European Economic and Social Committee and the Committee of the Regions: Thematic Strategy for Soil Protection. Commission of the European Communities, COM, Brussels.

Coleman, K., Jenkinson, D.S., 1999. RothC-26.3—a model for the turnover of carbon in soil: model description and windows users guide. Lawes Agricultural Trust, Harpenden. ISBN: 0-951-4456-8-5. November 1999 issue.

Dou, F., Wight, J.P., Wilson, L.T., Storlien, J.O., Hons, F.M., Sainju, U.M., 2014. Simulation of biomass yield and soil organic carbon under bioenergy sorghum production. PLoS One 9. http://dx.doi.org/10.1371/journal.pone.0115598.

Elliott, E.T., 1986. Aggregate structure and carbon, nitrogen, and phosphorus in native and cultivated soils. Soil Sci. Soc. Am. J. 50, 627–633.

Finke, P.A., 2012. Modeling the genesis of luvisols as a function of topographic position in loess parent material. Quat. Int. 265, 3–17. http://dx.doi.org/10.1016/j.quaint.2011.10.016.

Finke, P.A., Hutson, J.L., 2008. Modelling soil genesis in calcareous loess. Geoderma 145, 462–479. http://dx.doi.org/10.1016/j.geoderma.2008.01.017.

Galdos, M.V., Cerri, C.C., Cerri, C.E.P., Paustian, K., Van Antwerpen, R., 2009. Simulation of soil carbon dynamics under sugarcane with the CENTURY model. Soil Sci. Soc. Am. J. 73, 802–811. http://dx.doi.org/10.2136/sssaj2007.0285.

Gao, C.S., Wang, J.G., Zhang, X.Y., Sui, Y.Y., 2008. The evolution of organic carbon in Chinese mollisol under different farming systems: validation and prediction by using century model. Agric. Sci. China 7, 1490–1496. http://dx.doi.org/10.1016/S1671-2927(08)60407-1.

General Assembly, United Nations, 2015. Transforming our world: the 2030 agenda for sustainable development. http://dx.doi.org/10.1007/s13398-014-0173-7.2 https://sustainabledevelopment.un.org/content/documents/7891Transforming%20Our%20World.pdf.

Gilhespy, S.L., Anthony, S., Cardenas, L., Chadwick, D., del Prado, A., Li, C., Misselbrook, T., Rees, R.M., Salas, W., Sanz-Cobena, A., Smith, P., Tilston, E.L.,

Topp, C.F.E., Vetter, S., Yeluripati, J.B., 2014. First 20 years of DNDC (DeNitrification DeComposition): model evolution. Ecol. Modell. 292, 51–62. http://dx.doi.org/10.1016/j.ecolmodel.2014.09.004.

Goglio, P., Grant, B.B., Smith, W.N., Desjardins, R.L., Worth, D.E., Zentner, R., Malhi, S.S., 2014. Impact of management strategies on the global warming potential at the cropping system level. Sci. Total Environ. 490, 921–933. http://dx.doi.org/10.1016/j.scitotenv.2014.05.070.

Jastrow, J.D., Amonette, J.E., Bailey, V.L., 2007. Mechanisms controlling soil carbon turnover and their potential application for enhancing carbon sequestration. Clim. Change 80, 5–23. http://dx.doi.org/10.1007/s10584-006-9178-3.

Jenkinson, D.S., 1990. The turnover of organic carbon and nitrogen in soil. Philos. Trans. R. Soc. Lond. B Biol. Sci. 329, 361–368. http://dx.doi.org/10.1098/rstb.1990.0177.

Jenkinson, D.S., Coleman, K., 2008. The turnover of organic carbon in subsoils. Part 2. Modelling carbon turnover. Eur. J. Soil Sci 59, 400–413. http://dx.doi.org/10.1111/j.1365-2389.2008.01026.x.

Keating, B.A., Carberry, P.S., Hammer, G.L., Probert, M.E., Robertson, M.J., Holzworth, D., Huth, N.I., Hargreaves, J.N.G., Meinke, H., Hochman, Z., McLean, G., Verburg, K., Snow, V., Dimes, J.P., Silburn, M., Wang, E., Brown, S., Bristow, K.L., Asseng, S., Chapman, S., McCown, R.L., Freebairn, D.M., Smith, C.J., 2003. An overview of APSIM, a model designed for farming systems simulation. Eur. J. Agron. 18, 267–288. http://dx.doi.org/10.1016/S1161-0301(02)00108-9.

Kotronakis, M., Giannakis, G.V., Nikolaidis, N.P., Rowe, E.C., Valstar, J., Paranychianakis, N.V., Banwart, S.A., 2017. Modeling the impact of carbon amendments on soil ecosystem functions using the 1D-ICZ model. Adv. Agron. 142, 315–352.

Lal, R., 2004. Soil carbon sequestration impacts on global climate change and food security. Science 304, 1623–1627. http://dx.doi.org/10.1126/science.1097396.

Lal, R., 2013. Food security in a changing climate. Ecohydrol. Hydrobiol. 13, 8–21. http://dx.doi.org/10.1016/j.ecohyd.2013.03.006.

Lal, R., 2015. Restoring soil quality to mitigate soil degradation. Sustainability 7, 5875–5895. http://dx.doi.org/10.3390/su7055875.

Li, C., Trettin, C., Sun, G., McNulty, S., Butterbach-Bahl, K., 2005. Modeling carbon and nitrogen biogeochemistry in forest ecosystem. In: International Nitrogen Conference, pp. 893–898.

Li, S., Li, Y., Li, X., Tian, X., Zhao, A., Wang, S., Wang, S., Shi, J., 2016. Effect of straw management on carbon sequestration and grain production in a maize-wheat cropping system in anthrosol of the Guanzhong plain. Soil Tillage Res. 157, 43–51. http://dx.doi.org/10.1016/j.still.2015.11.002.

Li, N., You, M.-Y., Zhang, B., Han, X.-Z., Panakoulia, S.K., Yuan, Y.-R., Liu, K., Qiao, Y.-F., Zou, W.-X., Nikolaidis, N.P., Banwart, S.A., 2017. Modeling soil aggregation at the early pedogenesis stage from the parent material of a Mollisol under different agricultural practices. Adv.Agron. 142, 181–214.

Liao, Y., Wu, W.L., Meng, F.Q., Smith, P., Lal, R., 2015. Increase in soil organic carbon by agricultural intensification in northern China. Biogeosciences 12, 1403–1413. http://dx.doi.org/10.5194/bg-12-1403-2015.

Lichter, K., Govaerts, B., Six, J., Sayre, K.D., Deckers, J., Dendooven, L., 2008. Aggregation and C and N contents of soil organic matter fractions in a permanent raised-bed planting system in the highlands of central Mexico. Plant Soil 305, 237–252. http://dx.doi.org/10.1007/s11104-008-9557-9.

Luo, Z., Wang, E., Bryan, B.A., King, D., Zhao, G., Pan, X., Bende-Michl, U., 2013. Metamodeling soil organic carbon sequestration potential and its application at regional scale. Ecol. Appl. 23, 408–420. http://dx.doi.org/10.1890/12-0672.1.

Malamoud, K., McBratney, A.B., Minasny, B., Field, D.J., 2009. Modelling how carbon affects soil structure. Geoderma 149, 19–26. http://dx.doi.org/10.1016/j.geoderma.2008.10.018.

Milne, E., Banwart, S.A., Noellemeyer, E., Abson, D.J., Ballabio, C., Bampa, F., Bationo, A., Batjes, N.H., Bernoux, M., Bhattacharyya, T., Black, H., Buschiazzo, D.E., Cai, Z., Cerri, C.E., Cheng, K., Compagnone, C., Conant, R., Coutinho, H.L.C., de Brogniez, D., Balieiro, F., de, C., Duffy, C., Feller, C., Fidalgo, E.C.C., da Silva, C.F., Funk, R., Gaudig, G., Gicheru, P.T., Goldhaber, M., Gottschalk, P., Goulet, F., Goverse, T., Grathwohl, P., Joosten, H., Kamoni, P.T., Kihara, J., Krawczynski, R., La Scala, N., Lemanceau, P., Li, L., Li, Z., Lugato, E., Maron, P.A., Martius, C., Melillo, J., Montanarella, L., Nikolaidis, N., Nziguheba, G., Pan, G., Pascual, U., Paustian, K., Pineiro, G., Powlson, D., Quiroga, A., Richter, D., Sigwalt, A., Six, J., Smith, J., Smith, P., Stocking, M., Tanneberger, F., Termansen, M., van Noordwijk, M., van Wesemael, B., Vargas, R., Victoria, R.L., Waswa, B., Werner, D., Wichmann, S., Wichtmann, W., Zhang, X., Zhao, Y., Zheng, J., Zheng, J., 2015. Soil carbon, multiple benefits. Environ. Dev. 13, 33–38. http://dx.doi.org/10.1016/j.envdev.2014.11.005.

Nadeu, E., Gobin, A., Fiener, P., van Wesemael, B., van Oost, K., 2015. Modelling the impact of agricultural management on soil carbon stocks at the regional scale: the role of lateral fluxes. Glob. Chang. Biol. 21, 3181–3192. http://dx.doi.org/10.1111/gcb.12889.

Nelson, E., Mendoza, G., Regetz, J., Polasky, S., Tallis, H., Cameron, D.R., Chan, K.M.A., Daily, G.C., Goldstein, J., Kareiva, P.M., Lonsdorf, E., Naidoo, R., Ricketts, T.H., Shaw, M.R., 2009. Modeling multiple ecosystem services, biodiversity conservation, commodity production, and tradeoffs at landscape scales. Front. Ecol. Environ. 7, 4–11. http://dx.doi.org/10.1890/080023.

Nikolaidis, N.P., Bidoglio, G., 2013. Soil organic matter dynamics and structure. Sustain. Agric. Rev 12, 175–199. http://dx.doi.org/10.1007/978-94-007-5961-9_6.

Parton, W.J., Schimel, D.S., Cole, C.V., Ojima, D.S., 1987. Analysis of factors controlling soil organic matter levels in great plains grasslands. Soil Sci. Soc. Am. J. 51, 1173–1179. http://dx.doi.org/10.2136/sssaj1987.03615995005100050015x.

Poeplau, C., Don, A., 2015. Carbon sequestration in agricultural soils via cultivation of cover crops—a meta-analysis. Agric. Ecosyst. Environ. 200, 33–41. http://dx.doi.org/10.1016/j.agee.2014.10.024.

Rampazzo Todorovic, G., Lair, G.J., Blum, W.E.H., 2014. Modeling and prediction of C dynamics in soil chronosequences of the critical zone observatory (CZO) Marchfeld/Austria. Catena 121, 53–67. http://dx.doi.org/10.1016/j.catena.2014.05.002.

Ross, C.W., Grunwald, S., Myers, D.B., Xiong, X., 2015. Land use, land use change and soil carbon sequestration in the St. Johns River Basin, Florida, USA. Geophys. J. Roy. Astron. Soc. 7, 19–28. http://dx.doi.org/10.1016/j.geodrs.2015.12.001.

Segoli, M., De Gryze, S., Dou, F., Lee, J., Post, W.M., Denef, K., Six, J., 2013. AggModel: a soil organic matter model with measurable pools for use in incubation studies. Ecol. Modell. 263, 1–9. http://dx.doi.org/10.1016/j.ecolmodel.2013.04.010.

Six, J., Paustian, K., Elliott, E.T., Combrink, C., 2000. Soil structure and organic matter: I. Distribution of aggregate-size classes and aggregate-associated carbon. Soil Sci. Soc. Am. J. 64, 681–689. http://dx.doi.org/10.2136/sssaj2000.642681x.

Smith, P., Smith, J.U., Powlson, D.S., McGill, W.B., Arah, J.R.M., Chertov, O.G., Coleman, K., Franko, U., Frolking, S., Jenkinson, D.S., Jensen, L.S., Kelly, R.H., Klein-Gunnewiek, H., Komarov, A.S., Li, C., Molina, J.A.E., Mueller, T., Parton, W.J., Thornley, J.H.M., Whitmore, A.P., 1997. A comparison of the performance of nine soil organic matter models using datasets from seven long-term experiments. Geoderma 81, 153–225. http://dx.doi.org/10.1016/S0016-7061(97)00087-6.

Smittenberg, R.H., Gierga, M., Göransson, H., Christl, I., Farinotti, D., Bernasconi, S.M., 2012. Climate-sensitive ecosystem carbon dynamics along the soil chronosequence of the Damma glacier forefield, Switzerland. Glob. Chang. Biol. 18, 1941–1955. http://dx.doi.org/10.1111/j.1365-2486.2012.02654.x.

Stamati, F.E., Nikolaidis, N.P., Banwart, S., Blum, W.E.H., 2013a. A coupled carbon, aggregation, and structure turnover (CAST) model for topsoils. Geoderma 211–212, 51–64. http://dx.doi.org/10.1016/j.geoderma.2013.06.014.

Stamati, F.E., Nikolaidis, N.P., Schnoor, J.L., 2013b. Modeling topsoil carbon sequestration in two contrasting crop production to set-aside conversions with RothC—calibration issues and uncertainty analysis. Agric. Ecosyst. Environ. 165, 190–200. http://dx.doi.org/10.1016/j.agee.2012.11.010.

Tiessen, H., Cuevas, E., Chacon, P., 1994. The role of soil organic-matter in sustaining soil fertility. Nature 371, 783–785. http://dx.doi.org/10.1038/371783a0.

Tisdall, J., Oades, J., 1982. Organic matter and water stable aggregates in soils. J. Soil Sci. 33, 141–163.

Udom, B.E., Nuga, B.O., Adesodun, J.K., 2016. Water-stable aggregates and aggregate-associated organic carbon and nitrogen after three annual applications of poultry manure and spent mushroom wastes. Appl. Soil Ecol. 101, 5–10. http://dx.doi.org/10.1016/j.apsoil.2016.01.007.

Vavlas, N.C., Nikolaidis, N.P., Paranychianakis, N., Kotronakis, M., 2014. Modeling of soil structure and soil carbon sequestration in intensively cultivated fields using organic amendments. In: Proceedings of the 12th International Conference on Protection and Restoration of the Environment, pp. 377–384.

Zhang, L., Zhuang, Q., Li, X., Zhao, Q., Yu, D., Liu, Y., Shi, X., Xing, S., Wang, G., 2016. Carbon sequestration in the uplands of Eastern China: an analysis with high-resolution model simulations. Soil Tillage Res. 158, 165–176. http://dx.doi.org/10.1016/j.still.2016.01.001.

CHAPTER TEN

Integrated Critical Zone Model (1D-ICZ): A Tool for Dynamic Simulation of Soil Functions and Soil Structure

G.V. Giannakis*[,1], N.P. Nikolaidis[†], J. Valstar[‡], E.C. Rowe[§], K. Moirogiorgou[¶], M. Kotronakis*, N.V. Paranychianakis[†], S. Rousseva[||], F.E. Stamati*, S.A. Banwart[#]

*Technical University of Crete, School of Environmental Engineering, Chania, Greece
[†]School of Environmental Engineering, Technical University of Crete, University Campus, Chania, Greece
[‡]Deltares Subsurface and Groundwater Systems, Utrecht, The Netherlands
[§]Centre for Ecology & Hydrology, ECW, Bangor, United Kingdom
[¶]Technical University of Crete, School of Electronic & Computer Engineering, Chania, Greece
[||]Institute of Soil Science, Agrotechnology and Plant Protection "N. Poushkarov", Sofia, Bulgaria
[#]School of Earth and Environment, University of Leeds, Leeds, United Kingdom
[1]Corresponding author: e-mail address: gegiannakis@isc.tuc.gr

Contents

Abstract

Food security should be addressed in relation to soil sustainability and sustainable land care, and examined within the science framework of Earth's critical zone as an integrated system that includes Earth surface interactions, connected to soil functions, and ecosystem services. There is a great need to develop critical zone mathematical

Advances in Agronomy, Volume 142
ISSN 0065-2113
http://dx.doi.org/10.1016/bs.agron.2016.10.009

models that will simulate and quantify soil functions and that can be used as management tools to address soil sustainability and land care practices. The integrated critical zone model, 1D-ICZ, couples computational modules for soil organic matter dynamics, soil aggregation and structure dynamics, bioturbation, plant productivity and nutrient uptake, water flow, solute speciation and transport, and mineral weathering kinetics. The 1D-ICZ model, coupled with new pedotransfer functions to predict bulk soil properties, introduces for the first time a model that dynamically links soil structure characteristics and hydraulic soil properties by simulating their changes under varying meteorological conditions and plant growth. Field data from a Mediterranean olive grove at the Koiliaris Critical Zone Observatory (CZO) were used to simulate carbon addition to soil and agricultural management scenarios, in order to illustrate the model's ability to quantify soil management impact on soil functions and biogeochemical transformations and fluxes. The 1D-ICZ model can be used to assess, understand, and quantify the complex interactions between the different processes in the soil-plant-water system and can be applied as a tool to design sustainable agricultural management practices, taking into consideration synergy and trade-offs among soil functions.

1. INTRODUCTION

1.1 Food Security in Relation to Soil Sustainability and Sustainable Land Care

The human population is estimated to approach 10 billion by the middle of this century (Godfray et al., 2010), with consequent increases of 50–100% in food demand (Royal Society of London, 2009) and increases of more than 50% until 2030 in water demand (Beddington, 2009). While the demand for food is expected to increase significantly, the land available for food production is decreasing due to urban sprawl and growth in biofuel production. Food security is perceived as one of the grand challenges for the decades to come for humanity (UN FAO, 2011). Meeting food demand through continued expansion of agriculture into marginal lands would cause drastic increases in greenhouse gas emissions (Tilman et al., 2011), unsustainable biodiversity loss (Baudron and Giller, 2014), and disruption to soil processes and therefore biogeochemical cycles (Amundson et al., 2015). However, to sustainably intensify production from existing agricultural lands requires improved understanding of how to maintain soil function while increasing resource use efficiency (Garnett et al., 2013). Soil structure, i.e., the physical distribution of primary particles, their aggregation into large particles, and the resulting distribution and connectedness of pore space, is both a key control and indicator of the effectiveness of many soil functions (Tisdall and Oades, 1982).

Soil is one of the most important components of the natural environment. Soil functions are derived from the complex interactions between the heterogeneous near-surface system, including plants, animals, and the microbial community, and abiotic factors such as rock, atmosphere, and water. Soil functions are inherent capabilities of the soil that include biomass and food production, maintaining soil biodiversity, carbon and nutrient sequestration, water filtration and transformation, landscape and heritage, and source of raw materials. This environmental system and its interactions are defined as the Earth's Critical Zone (CZ) (National Research Council, 2001). Through this definition, it is possible to conceptualize a holistic research approach, which places the soil within the context of integrated geologic, ecologic, and freshwater systems. The interfaces between soil and the other parts of the CZ are where soil functions operate and ecosystem services are derived, delivering the benefits of soil formation, plant nutrition, atmospheric greenhouse gas removal, storage of carbon (C), nitrogen (N), and phosphorus (P) and their transformation to plant-available forms, water yield and quality, and providing habitat for maintaining biodiversity (Banwart et al., 2011). It is generally accepted that human and ecosystem sustenance is strongly dependent on soil functions, while human activities can adversely affect soil functions. In many cases, intensification of agriculture has significantly impacted functions and resulted in degraded soils, even though different practices may have different effects: maintaining productivity through irrigation and fertilization can result in increased soil organic matter (SOM) contents (Vanlauwe and Giller, 2006), but tillage and the exposure of soil to wind and rain can cause severe losses of soil, SOM, and soil structure (Hobbs et al., 2008).

Food security can only be addressed in the context of soil sustainability and sustainable land care. Soil sustainability refers to maintaining and enhancing soil functions, while sustainable land care refers to the management of carbon and nutrient fluxes in order to provide appropriate habitats for agricultural and natural plant communities while minimizing the impacts to the environment. It clearly requires the optimization of multiple soil functions. There is a great need to develop critical zone mathematical models that will simulate and quantify soil functions and that can be used as management tools to assess and implement beneficial changes to enhance soil sustainability and land care.

1.2 Soil Ecosystem and Critical Zone Modeling

Although the development of soil and ecosystem models has contributed greatly to improved understanding and management of agroecosystems, many models are focused on specific aspects of the critical zone and there is a need to better integrate knowledge of physical, biological, and chemical systems (Nikolaidis and Bidoglio, 2013). In particular, there is a major need to account for soil structure addressed through particle aggregation dynamics, and its influence on soil water flow and retention, C and nutrient storage, fertility, and biomass production (Banwart et al., 2011). Recent models that simulate the dynamic interaction of SOM and soil aggregates (Segoli et al., 2013; Stamati et al., 2013) provide a significant step forward in the translation of conceptual models of soil aggregation and disaggregation into mathematical equations and parameter sets that are able to calculate the dynamics of soil aggregation and related carbon and nutrient sequestration and turnover. However, these dynamic models of soil structure do not simulate other important soil functions that drive soil structure development such as biomass production, water filtration, nutrient, and pollutant transformations, and biodiversity.

Process models can be used to quantify soil state and functions, and thereby, coupling changes in soil composition to changes in soil structure, carbon dynamics, and biodiversity. Dynamic modeling of soil aggregation and disaggregation is necessary, since the abundance and an adequate distribution of soil macroaggregates (>250 μm), define soil as "agronomically favorable" and fertile (Banwart et al., 2011; Shein, 2005). In addition, aggregate size distribution is linked directly to the inflection point of the soil water retention curve and the water holding capacity of the soil (Dexter, 2004a,b,c).

1.3 Integrated Modeling of Critical Zone

The aim of the current study is to develop a tractable mathematical model that links soil aggregate formation and soil structure development to nutrient dynamics, plant nutrition, and water flow and mass transport. This study describes the development of the one-dimensional Integrated Critical Zone model (1D-ICZ). The model simulates and quantifies four of the main soil ecosystem functions (biomass production, carbon/nutrient sequestration in relation to soil structure, water filtration, and biodiversity) by accounting for interactions between water and heat flow, solute transport, soil structure dynamics, carbon and nutrient dynamics, and plant biomass production. New empirical pedotransfer functions (PTFs) have been also developed using field data collected in the SoilTrEC project. These PTFs use dynamic state variables (soil bulk density, organic matter content SOM, and mass distribution of water stable aggregates—WSA) calculated by the 1D-ICZ model to estimate the van Genuchten parameters for equations that define soil water retention (van Genuchten, 1980) and in turn, calculate dynamic changes to saturated hydraulic conductivity and volumetric water content.

The study also aimed to address the hypothesis that it is necessary to take aggregate dynamics into account for accurate prediction of the effects of carbon addition and different management practices on soil functions, including soil organic carbon (SOC) storage and turnover, and plant productivity. Field data from an olive grove in a Mediterranean site in the Koiliaris CZO were used to simulate scenarios of carbon addition and agricultural management practices in order to illustrate the model's ability to quantify their impact on soil functions.

2. METHODOLOGY

2.1 Model Description

The 1D-ICZ model is comprised of submodules linked together to simulate the interactions of biotic and abiotic processes above and below ground in order to simulate predominant soil functions as well as the dynamics of soil hydraulic conductivity and water holding capacity. The core conceptual model is schematically presented in Fig. 1. SOM and aggregation dynamics were simulated using a modified version of the CAST model (Stamati et al., 2013). The model considers that SOM exists in four carbon pools: the plant litter pools composed of decomposable (DPM) and resistant (RPM) plant material fractions, the microorganism (including the saprophytic fungi) pools (BIO) and the humus (HUM). The DPM and RPM fractions are incorporated into the soil and colonized by microbial decomposers. Microbial metabolites, fungal hyphae, and root exudates act as binding agents for soil particles, forming macroaggregates around particulate organic matter (POM). The aggregated POM is then decomposed and fragmented, enhancing the formation of microaggregates within the macroaggregates. When the DPM within macroaggregates is further decomposed and decreases in importance as a source of carbon and energy for the heterotrophic microbial community, microbial activity decreases and macroaggregate stability declines. At this stage, when slaking with water occurs, macroaggregates release microaggregates, silt-clay-sized aggregates, and degraded POM. When new plant litter enters the system, these materials can again form macroaggregates. Humus depolymerization forms low-molecular-weight organic N (LMWN) and P (LMWP) compounds which are subsequently

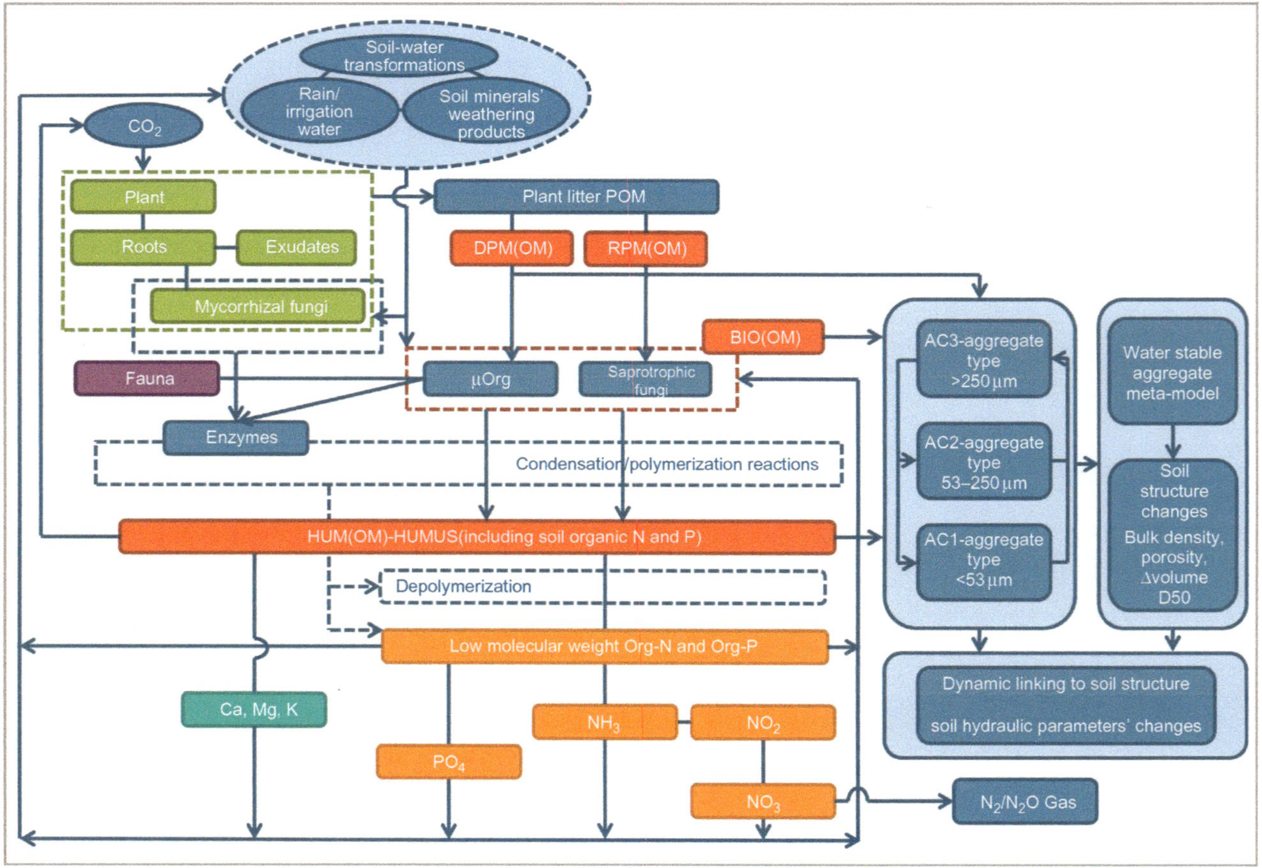

Fig. 1 Schematic representation of the modified CAST model used in the 1D-ICZ model.

mineralized to their inorganic species. During the above biological processes, N, P, potassium (K), calcium (Ca), and magnesium (Mg) are released and can contribute to plant nutrition. Nutrients and other inorganic species also enter the system through mineral weathering, atmospheric deposition, and irrigation.

Soil structure (e.g., macroaggregate content) depends strongly on SOC content which in turn influences soil hydraulic properties. Empirical rules for these interrelationships are defined through empirical pedotransfer functions that calculate dynamically, the soil hydraulic properties throughout the soil profile.

A full description of the submodels of 1D-ICZ model can be found in the model's manual (Nikolaidis et al., 2014). The submodels are run sequentially using a mixed time-step algorithm: CAST, PROSUM, chemical weathering, and bioturbation models use a fixed time step of 1 month, whereas the model of water and solute flow and transport uses a variable time step. The procedure is the following. All variables that are updated in the modules with a monthly time step are translated to production/consumption rates and after each time step in Hydrus, the impact of these rates is applied to the variables. A brief summary of each submodel and of the development of the pedotransfer functions follows.

2.1.1 Flow, Transport, and Bioturbation Models

The flow and transport model based on Hydrus-1D (Šimůnek et al., 2009) tracks the flow of water, heat, and solute transport. The water flow is described by Richard's equation:

$$\frac{\partial \theta}{\partial t} = \frac{\partial}{\partial z}\left(k\frac{\partial h}{\partial z}\right) + S \quad (1)$$

where θ is the volumetric water content ($L^3\,L^{-3}$), t is time [T], z is the vertical coordinate [L], k is the unsaturated hydraulic conductivity [$L\,T^{-1}$], h is the pressure head [L], and S is a water mass source (positive) or sink (negative) ($L^3\,L^{-3}\,T^{-1}$). The parameter k and variable h are dependent on the volumetric water content θ.

The source/sink term S also depends on the water content as the water uptake of plants can be limited due to water availability. PROSUM calculates the actual uptake by plants as a function of the vertical coordinate (S_{PROSUM}). These fluxes then are incorporated as prescribed fluxes within the water flow model (Eq. 2):

$$S(z,t) = S_{PROSUM}(z,t) \qquad \forall\, 0 < z < L \quad (2)$$

At the top cell of the model, a source (precipitation)/sink (evaporation) $S(0)$ is used as boundary condition, in which the precipitation P is prescribed and the actual evaporation E_a also depends on the water availability in the top cell and the prescribed potential evaporation (Eq. 3):

$$S(0,t) = P(t) - E_a(t) + S_{PROSUM}(0,t) \tag{3}$$

At the lower cell, a free drainage boundary condition is used (Eq. 4).

$$\frac{\partial h}{\partial z} = 0 \quad z = L \tag{4}$$

The heat transport is described by the convection–diffusion equation (Eq. 5):

$$\frac{\partial C_p(\theta)T}{\partial t} = \frac{\partial}{\partial z}\left(\lambda(\theta)\frac{\partial T}{\partial z}\right) - C_w\frac{\partial qT}{\partial z} - C_w ST \tag{5}$$

where $C_p(\theta)$ is the volumetric heat capacity of the porous medium $[\mathrm{M\,L^{-1}\,T^{-2}\,K^{-1}}]$, T is the temperature [K], $\lambda(\theta)$ is the coefficient of the apparent thermal conductivity of the soil $[\mathrm{M\,L\,T^{-3}\,K^{-1}}]$, C_w is the volumetric heat capacity of water $[\mathrm{M\,L^{-1}\,T^{-2}\,K^{-1}}]$, and q is the water flux obtained by Darcy's Law ($q = k\,dh/dz$) $[\mathrm{L\,T^{-1}}]$.

The volumetric heat capacity $C_p(\theta)$ is obtained by

$$C_p(\theta) = C_n\theta_n + C_o\theta_o + C_w\theta \tag{6}$$

where C_n and C_o are the volumetric heat capacity of the solid phase and organic fraction $[\mathrm{M\,L^{-1}\,T^{-2}\,K^{-1}}]$ and θ_n and θ_o are the volumetric fractions of the solid phase and organic matter, respectively [unitless]. The latter values are provided by the C/N/P dynamics and structure model, CAST.

At the upper boundary, the temperature is described as a fixed location, time varying, boundary condition depending on the atmospheric temperature. The lower boundary flux is described by free drainage without a convective thermal flux.

The solute transport for each chemical species k is described by the advection-dispersion equation. For non-volatile components the equation is:

$$\frac{\partial \theta c_k}{\partial t} = \frac{\partial}{\partial z}\left(\theta D_k^w \frac{\partial c_k}{\partial z}\right) - \frac{\partial q c_k}{\partial z} + S c_k^S - r_k + p_k \tag{7}$$

where c_k is the concentration $[\mathrm{mol\,L^{-3}}]$ of component k, D^w is the dispersion/diffusion coefficient for the liquid phase $[\mathrm{L^2\,T^{-1}}]$, $c_k{}^S$ is the concentration at the source (prescribed for atmospheric deposition) or the lower boundary sink (value calculated within the model), r is the rate of nutrient uptake by plants [mol L^{-3} T^{-1}], and p is the production rate [mol L^{-3} T^{-1}]. The latter two rates are provided by the PROSUM, CAST, and weathering submodels, respectively. Mass exchange of the components with the solid phase is calculated in the chemical equilibrium model.

For volatile components, equilibrium between the aqueous and gas phase is assumed. For these components, the advection-dispersion equation is applied (Eq. 8):

$$\frac{\partial(\theta+\theta_g H_k)c_k}{\partial t}=\frac{\partial}{\partial z}\left(\left(\theta D_k^w+\theta_g H D_k^g\right)\frac{\partial c_k}{\partial z}\right)-\frac{\partial q c_k}{\partial z}+Sc_k^S-r_k+p_k \quad (8)$$

In this equation, H_k is the Henry constant defined as the concentration in the gas phase divided by the concentration in the liquid phase, θ_g is the volumetric volume of the gas phase, and D^g is the diffusion coefficient in the gas phase.

At the upper boundary, the aqueous concentration is defined as a constant concentration boundary condition, depending on the atmospheric gas concentration, C^g:

$$C_k(0,t)=\frac{C_k^g(t)}{H} \quad (9)$$

The concentration is not directly prescribed, but it is assumed constant at a certain distance (1 cm is recommended) and a diffusive flux is prescribed as the boundary condition. The Hydrus-1D model code was adapted, so that CO_2 and NH_3 gas concentrations can be set at the surface boundary.

The bioturbation model was coded based on the procedure described in the SoilGen model (Finke and Hutson, 2008). Within each monthly time step, it is assumed that a fraction of the solid and aqueous components are perfectly mixed by bioturbation, within a depth that is influenced by the soil fauna.

2.1.2 Chemical Equilibrium and Weathering Model

The chemical equilibrium model was developed using BRNS (Aguilera et al., 2005) and was adapted to account for the effects of time-varying water saturation and exchange with the gas phase. The weathering submodel makes use of the SAFE chemical weathering module from ForSAFE (Wallman et al., 2005) taking into account the relevant source minerals (Table 1) in the soil parent material and the mass transfer to soil solution

Table 1 The Chemical Components Used in the Chemical Equilibrium Model, and Mineral Phases Used in Weathering Module

In the aqueous phase	Ca^{2+}, Mg^{2+}, Na^{+}, K^{+}, H^{+}, Al^{3+}, Cl^{-}, $SO_4{}^{2-}$, HCO_3^{-}, OH^{-}, F^{-}, A^{3-} (organic anion), H_2CO_3, $CaCO_3$(l), $CaHCO_3{}^{+}$, $CaOH^{+}$, $CaSO_4$(l), $MgCO_3$, $MgHCO_3{}^{+}$, $MgOH^{+}$, $MgSO_4$, $NaSO_4{}^{-}$, $NaCO_3{}^{-}$, $KSO_4{}^{-}$, $Al(OH)^{2+}$, $Al(OH)_2{}^{+}$, $Al(OH)_3$, $Al(OH)_4{}^{+}$, AlF^{2+}, $AlF_2{}^{+}$, AlF_3, $AlF_4{}^{+}$, $AlF_5{}^{2-}$, $AlSO_4{}^{+}$, $Al(SO_4)_2{}^{-}$, AlA, $Al(HA)^{+}$, HA^{2-}, H_2A^{-}, H_3A, NH_3, and $NH_4{}^{+}$
In the cation exchange phase	X-Ca^{2+}, X-Mg^{2+}, X-Na^{+}, X-K^{+}, X-Al^{3+}, Surface-O^{-}, Surface-OH, Surface-OH_2^{+}, and Surface-$OH_2SO_4^{-}$
In the solid phase	$CaCO_3$ (calcite), $CaSO_4$ (gipsum), and $Al(OH)_3$ (Gibbsite)
In the gaseous phase	CO_2 and NH_3
Mineral phases used in weathering kinetics	$KAlSi_3O_8$ (K-Feldspar), $NaAlSi_3O_8$ (Na-Feldspar), $CaMg(CO_3)_2$ (dolomite), $Al_2Si_2O_5(OH)_4$ (kaolinite), $KAl_3Si_3O_{10}(OH)_2$ (muscovite), and $NaAl_2(Si_3Al)O_{10}(OH)_2$ (paragonite)

of Ca^{2+}, Na^{+}, K^{+}, Mg^{2+}, Al^{3+}, H_4SiO_4, and $PO_4{}^{3-}$ components and also accounts for the production and consumption of H^{+}. The chemical equilibrium constants used, were obtained from the PhreeqC thermodynamic database (Parkhurst and Appelo, 1999) and from Krám et al. (1999). The model uses the components in the aqueous, cation exchange, solid, and gaseous phases shown in Table 1.

2.1.3 Carbon and Nutrient Dynamics in Vegetation, PROSUM

A simple submodel was developed to simulate plant growth, water and nutrient acquisition, and litter production above- and below ground. The PROSUM model predicts responses of these key vegetation processes to changes in temperature and availability of light, water, CO_2, and the nutrient elements N, P, Ca, Mg, and K for a range of herbaceous and woody vegetation types. Grazing and management events such as tillage and harvest are also simulated. Nutrient limitation essentially follows a Liebig minimum-element approach, but the model incorporates current understanding that plants and their mycorrhizal fungi symbionts can manipulate nutrient availability to avoid single element limitation (Elser et al., 2007; Rowe et al., 2008) by calculating root and fungal exudate fluxes which can increase weathering. The effects of soil exploration efficiency by plants and fungi on nutrient and water uptake are also accounted for, by calculating root and mycorrhizal fungal hyphae length and thereby the proportion of the soil volume from which nutrients can be acquired. Nutrients and water are assumed to be accessible within a volume of soil equal to the total length of roots and mycorrhizae multiplied by an area set by a resource-specific diffusion radius: 10 mm for water, 8 mm for N, 0.02 mm for P, 0.1 mm for Ca and Mg, and 0.15 mm for K. Changes in element stocks are calculated monthly, for different parts of the plant: "Leaf," which includes all nonwoody shoot material; "Wood"; and "Root." Mycorrhizal fungi biomass is conceptualized and simulated as if it were part of the plant. Photosynthesized C is partitioned between plant parts, mycorrhizae, and root exudates, on the basis of typical allocations derived from Chapin et al (2012) for different plant functional types, i.e. herbaceous, woody deciduous, or woody evergreen; and whether or not the plants are mycorrhizal (Table 2).

A potential production (net C fixation) rate is assumed, on the basis of the maximum value of 53 mol C m^{-2} mo^{-1} observed in a review of productivity in global natural ecosystems by Kerkhoff et al. (2005). The observed value is used at a CO_2 concentration of 393 μL L^{-1} air and is increased or decreased according to the current CO_2 concentration. Simple response

Table 2 Percent Allocation of net Primary Production to Herbaceous Growth, Woody Growth, Root Growth, Root Exudates, and Root Transfers to Mycorrhizae

	Woody Mycorrhizal	Herbaceous Mycorrhizal	Woody Nonmycorrhizal	Herbaceous Nonmycorrhizal
Carbon allocation (%)				
Herbaceous	33.0	37.0	38.6	44.2
Woody	10.9	0.0	12.7	0.0
Root	25.4	28.5	29.7	34.0
Exudates	14.5	16.3	16.9	19.4
Mycorrhizae	14.5	16.3	0.0	0.0
Volatile emissions	1.8	2.0	2.1	2.4

Recalculated From Chapin, F.S., Matson, P.A., Vitousek, P.M., 2012. Principles of Terrestrial Ecosystem Ecology, second ed. Springer, New York. doi:10.1007/978-1-4419-9504-9.

functions are then applied to determine the reduction in this potential production due to the primary limitations of temperature, light, and water. Available water is calculated in the 1D-ICZ as the amount (mm mo^{-1}) of water that is held in the soil between saturated water content and wilting point, and accessible water is determined by the soil volume explored by roots and mycorrhizae, as explained earlier. The maximum water-limited production is determined using a simple water use efficiency approach. A maximum water use efficiency of 522 moles total C fixation m^{-3} water was assumed, based on the converting the maximum yield observed in a review of Mediterranean wheat production (Katerji et al., 2008) to total C production by assuming 0.41 g grain g^{-1} total biomass (Van Duivenbooden et al., 1996).

Nutrient-limited productivity is then calculated by determining nutrient availability within the time step (accessible nutrient stock plus existing nutrient stocks within the plant), and calculating maximum potential productivity on the basis of a weighted mean of the maximum carbon/nutrient ratios for each plant part. For example, for N (Eq. 10),

$$WeightedMaxCN = \frac{\sum_{1}^{b} Palloc_b}{\sum_{1}^{b}\left(\frac{Palloc_b}{MaxCN_b}\right)} \tag{10}$$

where b is the number of plant parts, $Palloc_b$ is the proportional allocation of new C to that plant part, and $MaxCN_b$ is the maximum C/N ratio for that plant part.

Actual productivity is set as the minimum of these element-limited values, unless it is more strongly constrained by light or water availability. Actual nutrient uptake is then calculated, assuming "luxury" uptake of nutrient elements that are in excess, up to a maximum set by the minimum carbon/nutrient ratios for each plant part. For example for N (Eq. 11),

$$MaxUpt = \sum_{1}^{b} \frac{C_b}{MinCN_b} \tag{11}$$

where $MaxUpt$ is the maximum uptake of N in a month, C_b is the current stock of C in plant part b (including the C fixation that was just calculated) and $MinCN_b$ is the minimum C/N ratio for the specific plant part. The actual uptake is then calculated and distributed over the plant parts.

Harvest and tilling events can occur within any month. Harvest involves complete removal of user-defined proportions of plant parts, and also transfers some biomass into litter pools. Tilling transfers all above-ground biomass and most root and mycorrhizal biomass into litter pools.

2.1.4 C/N/P Dynamics and Structure Model, CAST

The CAST model (Stamati et al., 2013) extends the representation of C pools with different turnover rates that is used in ROTH-C (Coleman and Jekinson, 1999) to account for dynamic soil aggregation/disaggregation, influences of soil biota including mycorrhizal fungi and fauna (Nikolaidis and Bidoglio, 2013; Pohlert et al., 2005, 2007). The CAST model simulates the formation of macroaggregates around POM followed by the release of microaggregates. The HUM (humified OM) pool is enzymatically depolymerized to LMWN and LMWP compounds which then are mineralized to inorganic N and P. This model is driven by plant litter input at the surface and is applied at the top layer of the soil. Below the top layer, input of OM is attributed to the dissolved phase from root litter or from redistributed OM. A schematic of the CAST model and its linking with the N and P cycles is presented in Fig. 1.

Organic N and P stocks for each pool are calculated using C/N and C/P ratios. In other words, it is assumed that the OM sequestered in the different size aggregates has the same composition.

LMWN and LMWP are calculated from N and P pools in the silt-clay-related fraction and the BIO and HUM pools of the three aggregate classes (AC1—silt-clay-sized aggregates, AC2—microaggregates, AC1 in AC3 and AC2 in AC3—macroaggregates). Each of the N and P pools results in the production of LMWN and LMWP, respectively, using first-order kinetics. As an example, the LMWN production from BIO pool of AC1 aggregate class follows the equation (12):

$$BIO_{AC1}(N)\_to\_LMWN = BIO_{AC1}(N)*\left(1 - e^{-k_{LMWN}*\frac{abc}{12}}\right) \tag{12}$$

where $BIO_{AC1}(N)$ is the N content of the BIO carbon pool of the AC1 aggregate class, k_{LMWN} is the LMWN production rate constant which is corrected by the product (*abc*) of three major factors determining microbial activity ("*a*" for temperature, "*b*" for water deficit, and "*c*" for soil cover) for each monthly time step.

LMWN and LMWP are linearly partitioned between the dissolved and particulate phase as shown in the following example (Eq. 13):

$$BIO_{AC1\_dissolved} = \frac{BIO_{AC1}}{\left(1 + k_{partition\_AC1}\right)^{*} D_m} \tag{13}$$

where D_m is the particle density of the mineral phase and $k_{partition\_AC1}$ is the partitioning constant between the BIO dissolved and particulate phase of the AC1 fraction.

Once organic N is in the dissolved phase, mineralization (Eqs. 14 and 15), nitrification (Eqs. 22–25), denitrification (Eqs. 26–30), ammonium adsorption (Eqs. 19–21), and ammonia volatilization follow. Phosphorus mineralization and adsorption reactions are treated in the same way (Nikolaidis et al., 2014). Plants can uptake both LMWN and inorganic N as well as LMWP and inorganic P available in the dissolved phase.

2.1.4.1 N Mineralization

The mass balance equation of LMWN for each monthly time step (i) and the kinetics of mineralization are presented below:

$$LMWN_i = LMWN_{i-1} - LMWN_{plant} + LMWN_{prec} + LMWN_{fert} + LMWN_{prod} \tag{14}$$

$$LMWN\_to\_NH_4 - N = LMWN_i{}^{*}\left(1 - e^{-\frac{k_{mineralization}}{12}{*}\frac{i}{\left(1/f_T + 1/f_\theta\right)}}\right) \tag{15}$$

where the subscripts *plant*, *prec*, *fert*, and *prod* stand for plant uptake, input from precipitation and fertilization, and produced LMWN in the specific month, respectively. The temperature (f_T) and moisture (f_θ) factors are given by the following equations. The temperature reduction function is calculated by the O'Neill-function (Eq. 16) as proposed by Stange (2001) and used by Pohlert et al. (2007),

$$f_T = \left(\frac{T_{MX} - T_{Soil}}{T_{MX} - T_{Opt}}\right)^{T_A} {*} e^{\left(T_A{*}\frac{T_{Soil} - T_{Opt}}{T_{MX} - T_{Opt}}\right)} \tag{16}$$

where T_{Soil} is the actual temperature, T_{MX} is the maximum temperature for decomposition (58°C), T_{Opt} is the optimum temperature for decomposition (34.15°C), and T_A is the scale parameter of the O'Neill-function (5.39).

For the calculation of soil moisture factor, the volumetric soil moisture content is converted to water filled pore space (WFPS). Afterward, the soil moisture factor is calculated using the Weibull-function (Eqs. (17) and (18)), see Pohlert et al. (2007) and Stange (2001),

$$A = \frac{\theta_\nu}{\varphi} \tag{17}$$

and

$$f_\theta = 1 - \left(1 + e^{\frac{A - M_{Crit}}{\gamma}}\right)^{-1} \tag{18}$$

where A is the water filled pore space, θ_ν is the volumetric water content ($m^3\,m^{-3}$), φ is the porosity, M_{Crit} is the half optimum soil moisture (WFPS), and γ is the scale parameter of the Weibull-function. According to Stange (2001) and Pohlert et al. (2007), the following estimations can be made: $M_{Crit} = 0.425A$ and $\gamma = 0.15A$.

2.1.4.2 Ammonium Adsorption

A part of the released ammonium is absorbed on clay minerals following second-order adsorption kinetics. Ammonium adsorption is a function of dissolved ammonium and remaining unoccupied clay sites. The total dissolved ammonium concentration $NH_4(i)$ in (t ha^{-1}) is updated during the time step by including the ammonium loads due to precipitation (NH_{4_prec}), fertilization (NH_{4_fert}), mineralization ($LMWN\_to\_NH_4$), and fauna denudation ($fauna\_to\_NH_4$) (Eq. 19):

$$\begin{aligned} NH_{4(i)} = {} & NH_{4(i-1)} + NH_{4\_prec(i,j)} + NH_{4\_fert(i,j)} \\ & + LMWN\_to\_NH_{4(i,j)} + fauna\_to\_NH_{4(i,j)} \end{aligned} \tag{19}$$

Then the amount of ammonium adsorbed to clay, $NH_4\_to\_clay_{(i,j)}$ is calculated in t ha^{-1} mo^{-1} (Eq. 20),

$$NH_4\_to\_clay_{(i,j)} = k_{ads} * \left(NH_4\_clay_{\max(i)} - NH_4\_to\_clay_{(i,j-1)}\right) \tag{20}$$

where $NH_4\_to\_clay_{(i,j)}$ is the ammonium adsorbed onto clay, t ha^{-1} mo^{-1}, k_{ads} is the second-order rate constant, ha mo t^{-1}, $NH_4\_clay\__{max(i)}$ is the adsorption capacity of clay for ammonium with units of t ha^{-1} calculated as follows (Eq. 21):

$$NH_4\_clay_{\max(i)} = ads\_site\_density_{(i)} * clay_{(i)} \tag{21}$$

where *ads_site_density* in tN ha^{-1} clay (%)$^{-1}$ is the adsorption site density and *clay*(i) is the % clay content of the soil layer.

2.1.4.3 Nitrification

Nitrification is described by the following first-order kinetics (estimation of gross nitrification) (Pohlert et al., 2007; Zhang et al., 2002; Eq. 22),

$$NH_4{-}N\_to\_NO_3{-}N = \left(NH_4 - N*\left(1 - e^{-\frac{k_{nitrification}}{12}*f_{TNI}}\right)\right)*f_{\theta NI}*f_{pHNI} \tag{22}$$

where the temperature (f_{TNI}), moisture ($f_{\theta NI}$), and pH (f_{pHNI}) factors are given by the following described equations. The temperature reduction function is calculated by the O'Neill-function (Eq. 23) as proposed by Stange (2001) and used by Pohlert et al (2007),

$$f_T = \left(\frac{T_{MXNI} - T_{Soil}}{T_{MXNI} - T_{OptNI}}\right)^{T_{ANI}} * e^{\left(T_A * \frac{T_{Soil} - T_{OptNI}}{T_{MXNI} - T_{OptNI}}\right)} \tag{23}$$

where T_{Soil} is the actual temperature, T_{MX} is the maximum temperature of the O'Neill-function for nitrification (45°C), T_{OPT} is the optimum temperature for nitrification (33.33°C), and T_{ANI} is the scale parameter (1.90). The moisture reduction function ($f_{\theta NI}$, Eq. 24), and pH-specific reduction value (f_{pHNI}, Eq. 25), for nitrification, are calculated as a fourth- and second-order polynomial according to Zhang et al. (2002) after Pohlert et al. (2007):

$$f_{\theta NI} = -12.904*A^4 + 17.651*A^3 - 5.5368*A^2 + 0.9975*A - 0.0243 \tag{24}$$

$$f_{pHNI} = -0.0604*pH^2 + 007347*pH - 1.2314 \tag{25}$$

2.1.4.4 Denitrification

Denitrification is calculated as described by Pohlert et al. (2007), by the following equations. During each time step, the nitrates that can be lost as nitrogen are updated (Eq. 26):

$$NO_3 \rightarrow N_{2_{(i)}} = NO_3\_to\_N_{2_{(i-1)}} + NO_{3\_prec_{(i,j)}} + NO_{3\_fert_{(i,j)}} + NH_4 - N\_to\_NO_3 - N_{(i,j)} \tag{26}$$

and the gaseous loss due to denitrification is calculated (Eq. 27):

$$NO_3\_to\_N_2 = NO_3 \rightarrow N_2 * \left(1 - e^{-\frac{k_{denitrification} * f_{TDNI} * f_{\theta DNI}}{12}}\right) \tag{27}$$

where the temperature (f_{TDNI}) and moisture ($f_{\theta DNI}$) factors are given by the following described equations. The temperature factor is calculated for temperatures below and above 10°C using Eqs. (28) and (29).

$$if\ T_{SOIL} \leq 10^oC,\ \ f_{TDNI} = 0.67 * e^{(0.43*(T_{SOIL}-10)} \tag{28}$$

$$if\ T_{SOIL} > 10^oC,\ \ f_{TDNI} = e^{(0.08*(T_{SOIL}-15)} \tag{29}$$

The moisture factor is given by Eq. (30):

$$f_{\theta DNI} = e^{0.304 + 2.94*(\theta_{vsat} - \theta_v) - 47*(\theta_{vsat} - \theta_v)^2} \tag{30}$$

where θ_v is the volumetric water content (m^3 m^{-3}) and θ_{vsat} is the volumetric water content at saturation (m^3 m^{-3}).

2.1.4.5 Impact on Soil pH

Soil pH is affected by the production or consumption of hydrogen ions in the C and N cycles due to production, decomposition, mineralization, nitrification, and denitrification. During decomposition it is assumed that 2 mol H^+ are produced for every CO_2 molecule produced. Two H^+ moles are also assumed to be produced for every LMWN mole mineralized and 1:1 stoichiometry is assumed between the NH_4 to NO_3 transformation during nitrification. Lastly, 6 mol H^+ are consumed for every NO_3 molecule converted to N_2 by denitrification. These changes in H^+ production/consumption due to organic matter transformations (CAST module) are then being transferred in the chemical equilibrium model for a comprehensive evaluation of soil pore water pH changes.

2.1.4.6 Base Cations Ca^{2+}, Mg^{2+}, K^+

Base cations, Ca^{2+}, Mg^{2+}, and K^+ are released during the organic matter decomposition using the respective C/Cation ratio:

$$Cation(mg/L) = \frac{SOC_{dissolved}}{C/Cation}$$

Finally, tillage is incorporated into the model by making the WSA destruction parameters time variable.

2.1.5 New Pedotransfer Functions Development for Dynamic Simulation of Soil Structure

Pedotransfer functions can be used to predict soil hydraulic parameters from easily measured soil properties (Wösten et al., 2001). Properties that influence hydraulic responses are: soil bulk density (BD), particle size distribution, organic matter (OM) content, calcium carbonate content, porosity, aggregates content, mean particle diameter, pH, and cation exchange capacity (Khodaverdiloo et al., 2011; Wösten et al., 2001).

In this study, the SoilTrEC project database of soil properties, including 54 soil profile data, across 4 CZOs exhibiting a wide range of values for state factors of soil formation, was used to develop equations predicting the van Genuchten's (1980) equation parameters (Table 3; Eq. 31). These parameters are described as a function of at least three soil properties: (a) the percentage of water stable macroaggregates (AC3—diameter >250 μm), (b) BD, and (c) OM content. The CAST module within 1D-ICZ model predicts the dynamics of these soil properties, and thus, the new PTFs can link soil structural changes to changes in soil hydraulic properties.

The primary database included only soil water content data at different pressure heads but no hydraulic conductivity data. To overcome this, the PTFs of Wösten et al. (1999) are used for predicting saturated hydraulic

Table 3 van Genuchten (1980) Equations for Water Content and Hydraulic Conductivity Calculation

$$\theta(h) = \theta_r + \frac{(\theta_s - \theta_r)}{[1 + |\alpha h|^n]^{1-(1/n)}} \tag{31}$$

$$\theta(h) = \frac{\theta_s}{[1 + |\alpha h|^n]^{1-(1/n)}} \tag{32}$$

$$K(h) = Ks\frac{\left[(1 + |\alpha h|^n)^{(1-(1/n))} - |ah|^{(n-1)}\right]^2}{(1 + |\alpha h|^n)^{(1-(1/n))(l+2)}} \tag{33}$$

θ is the volumetric water content ($m^3\ m^{-3}$), h is the soil water pressure head (m), θ_r and θ_s are the residual and saturated water contents ($m^3\ m^{-3}$), respectively, and α (m^{-1}) and n (−) are shape factors.

conductivity (K_s) and the shape parameter l for the soil water retention curves. The entire data base (203 soil samples) was divided into two subsets as a function of soil depth. The first subset (88 soil samples) considered the topsoils (0–20 cm) and the second (76 soil samples), the subsoils (max soil profile depth 130 cm).

The nonlinear least-squares optimization program RETC (van Genuchten et al., 1991) was used to fit the hydraulic parameters of Eq. (31) (Table 3), to the measured water retention data and to optimize the values of the θ_r, θ_s, α, and n parameters. The water retention data that were used included water contents at pressure heads of 0.25, 1, 5, 10, 33 kPa. The fitting procedure showed very low or zero values of soil residual water content (θ_r) so it was set to zero, and the best fitting model was Eq. (32) (Table 3). The R^2 values of the nonlinear regression varied from 0.84 to 0.999 with an average value of 0.977 indicating close agreement between the calculations with the fitted equation and the observed variation in the data.

2.2 Coupling of Submodels

A main program drives the HYDRUS, chemical weathering, CAST, PROSUM, and bioturbation submodels. A time step algorithm is used to sequentially run the submodels, following the initialization and required input data. A fixed time step of one month (365.25 days divided by 12 months) is used for the CAST, PROSUM, chemical weathering, and bioturbation models whereas a smaller variable time step is used for the flow and transport models. Fig. 2 presents a flow chart describing the time step algorithm.

At the start of each month, water and solutes quantities available for plant growth, in each of the soil layers, are calculated within the flow and transport model. The model then calculates the production and consumption rates for chemical weathering, CAST and PROSUM submodels. The nutrients produced by the chemical weathering and CAST models, during the month, become now available for uptake by plant. After PROSUM calculations, production/consumption rates of water and all the solutes during the month are known. In the following step, a number of iterations for the flow and transport model are performed until the simulated time steps sum up to 1 month. During these time steps, the production and consumption rates are accounted for by multiplying them with the appropriate time step size and update the soil water content and solute concentrations. Each time step

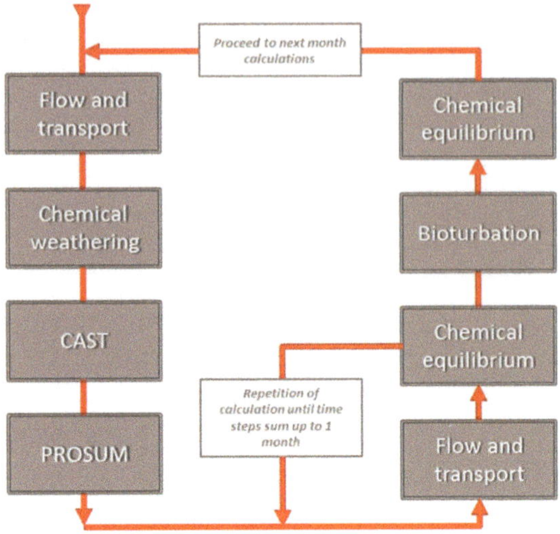

Fig. 2 Submodels coupling and time step handling in the model. The PROSUM submodel simulates carbon and nutrient dynamics in vegetation and the CAST submodel simulates C/N/P dynamics and soil structure.

of the flow and transport model is followed by the chemical equilibrium model in order to determine the division of the chemical components between mobile (dissolved) and immobile (precipitated or adsorbed) phases. At the end of the month, the bioturbation model is used to redistribute water and solute concentrations. Finally, the chemical concentrations are equilibrated again and the calculation continues. If during the transport of solutes less nutrients are available, during the month than it was foreseen at the beginning of the month, then the deficit is recorded and it is corrected during the next month.

2.3 1D-ICZ Model Simulations—Case Study

2.3.1 Site Description and Field Data Used for Model Evaluation

The data used for model evaluation were derived from of an olive grove site in Koiliaris CZO (Chania, Crete, Greece). The area is representative of degraded soils due to intensive and long-term agricultural processes, primarily horticulture and grazing. The area's land uses, include herbaceous and shrub vegetation associations (58%) (particularly at higher elevation), cultivated agricultural areas (29.4%), forests (8.5%), urban areas (2.8%), and aquatic environments (0.6%). The land surface slope ranges from 1% (at low elevations) to 43% (at high elevations up to 2014 m a.s.l.). The climate in the region is characterized as typical semiarid Mediterranean with mean annual temperature of 18.6°C and mean annual precipitation of 713 mm (94% October to May) (Banwart et al., 2011).

The olive grove site (35°26.66 N, 24°08.209E, 9 m a.s.l.) used for model evaluation, occupies an area of about 3400 m^2 which has 71 olive trees with approximate age of 20 years and irrigation during the summer period. The olive harvest takes place every second year from November to February, followed by pruning. Sheep and goats graze the field adding approximately 0.26 tC ha^{-1} $year^{-1}$ as manure.

Soil sampling at this site took place during the spring of 2010 and the summer of 2014. The first sampling included the characterization of the whole soil profile as described (K2 site) by Moraetis et al. (2015), and these data were used for model initialization. Data from the second sampling were used for calibration; only the topsoil (0-20 cm) was sampled, assuming that subsoil conditions were unlikely to change significantly during the relatively short period (56 months) between the two samplings. The second sampling analyses included WSA distribution (Elliot, 1986) and SOC content determination within each WSA fraction (multi N/C® 2100S Analyzer, AnalyticJena). Soil solution was sampled on five occasions in 2012–2013 using ceramic suction lysimeter cups connected with a manual vacuum pump. The sampling depth was 30–40 cm of soil profile and the samples were analyzed for physicochemical parameters.

2.3.2 Initialization and Calibration of the Model

The soil profile simulated in this study had a depth of 60 cm and was divided into six layers for simulation of water transport and solute transport. Soil carbon and structure were simulated using three soil layers as shown in Table 4. Meteorological data for precipitation, temperature, and also potential evapotranspiration were derived from a local meteorological station. Water inputs from irrigation and rain were measured. Regarding the tree biomass, initial values for wood, leaf, root, and mycorrhizal fungi were estimated for a 20-year-old olive tree (Ilarioni et al., 2013).

The 1D-ICZ model was calibrated using the 2014 data for SOC content and organic carbon (OC) distribution in each aggregate fraction (AC3 > 250 μm, 53 μm < AC2 < 250 μm, AC1 < 53 μm) as well as WSA size classes distribution by mass. Soil solution sampling data from a soil profile depth of 40 cm, at five dates during the 2012–2013 period were used for solutes calibration (Table 4).

Table 4 Values for Model Initialization and Calibration

Profile Discretization (cm)								Macroaggregates AC3 (>250 μm)						Microaggregates AC2 (53–250 μm)						Silt-Clay-Sized AC1 (<53 μm)					
								Initial Value		Calibration Value		Simulated Value		Initial Value		Calibration Value		Simulated Value		Initial Value		Calibration Value		Simulated Value	
Water and Solutes Transport	SOC and Structure Dynamics	Clay (%)	Silt (%)	Sand (%)	SOC (g kg^{-1})	BD (kg m^{-3})	Porosity (%)	WSA (%)	OC (tC ha^{-1})	WSA (%)	OC (tC ha^{-1})	WSA (%)	OC (tC ha^{-1})	WSA (%)	OC (tC ha^{-1})	WSA (%)	OC (tC ha^{-1})	WSA (%)	OC (tC ha^{-1})	WSA (%)	OC (tC ha^{-1})	WSA (%)	OC (tC ha^{-1})	WSA (%)	OC (tC ha^{-1})
0–20	0–20	5.1	30.02	64.88	17.81	1220	55	58.2 (±8.1)	30.5	59.9 (±3.69)	31.4 (±2.21)	60.97	26.84	29.6 (±5.2)	8.12	29.9 (±2.79)	7.6 (±0.23)	28.44	8.51	12.2	4.88	10.2 (±1.81)	4.4 (±0.72)	10.6	5.78
20–40	20–40	4.95	28.42	66.63	10.73	1296	52	41.06	14.61	ns		—	—	43.43	8.65	ns		—	—	15.5	4.56	ns		—	—
40–50																									
50–55	40–60	5.23	27.03	67.73	4.82	1359	50	—	—	ns		—		—	—	ns		—		—	—	ns		—	
55–60																									

2.3.3 Simulation Scenarios

The model's ability to simulate changes in soil structure, hydraulic parameters, and other soil properties was assessed through simulation of four different soil management scenarios. The No Carbon addition-No Tilling, "NC-NT," scenario, describes current practices and corresponds to a 10-year simulation of the calibration run. The second scenario, No Carbon addition and Tilling (NC-T), includes the same practices but with a light tillage procedure in May of each year. Two further scenarios were developed, based on the NC-T scenario but with organic matter incorporation into the 0–20 cm soil layer by tillage, at rates of 5.12 tC ha^{-1} year^{-1} (25-T) or 10.2 tC ha^{-1} year^{-1} (50-T). The organic matter was applied as a mixture of municipal solid waste compost and sheep/goat manure at a ratio of 70:30.

3. RESULTS AND DISCUSSION

3.1 Pedotransfer Functions Development

Multiple linear regression analysis was performed on the two subsets of data to set relationships between the θ_s, α, and n parameters and the measured soil physicochemical parameters for the topsoil and subsoil. The physicochemical parameters used were sand, silt, clay, and OM contents; BD; and macroaggregates mass distribution. The relationships obtained and the associated R^2 values for topsoil and subsoil are summarized in Table 5.

Since no data were available for hydraulic conductivity, Eq. (33) (Table 3), was used to quantify hydraulic conductivity using the α and n parameters obtained from the developed PTFs of this study. The required additional values such as the saturated hydraulic conductivity (K_s) and the shape parameter l were derived using the PTFs of Wösten et al. (1999). In Fig. 3, a comparison between the two groups of PTFs for predicting θ_s, α, and n shape parameters is presented. To evaluate the reliability of developed PTFs, the volumetric water content at different pressure heads was calculated for the whole dataset and compared with the measured values.

In Fig. 4, the volumetric water content at −33 kPa is shown with the R^2 and RMSE. For topsoil the R^2 was 0.72 and RMSE 0.06 and for subsoil these values were 0.69 and 0.04, respectively, implying that the PTFs are reliable and can be incorporated in the 1D-ICZ model. The 1D-ICZ model uses as state variables the OM and macroaggregate contents, and estimate changes in BD. By incorporating the PTFs in the model, a dynamic linkage between changes of soil structure and soil hydraulic dynamics is achieved.

Table 5 New Pedotransfer Functions for Topsoils (<20 cm Depth) and Subsoils (>20 cm Depth)

Topsoils	$\boldsymbol{\theta}_s = 0.534446 + 0.0196039*\ln(1 - \mathbf{WSA}/100) + 0.287781*\mathbf{BD} + 0.00108097*\mathbf{OM} - 0.250882*\mathbf{BD}^2 - 0.00460188*\mathbf{BD}*\ln(\mathbf{OM}) + 0.00533532*\ln(1 - \mathbf{WSA}/100)*\mathbf{BD}*\ln(\mathbf{OM})$ ($R^2 = 0.60$)
	$\ln(\boldsymbol{n}) = 1.63695 + 0.0174616*\ln(1 - \mathbf{WSA}/100) - 1.48231*\mathbf{BD} - 0.201182*\ln(\mathbf{OM}) - 0.0910037*\ln(\mathbf{CLAY}) + 0.465118*\mathbf{BD}^2 + 0.117899*\mathbf{BD}*\ln(\mathbf{OM}) + 0.0176379*\ln(\mathbf{CLAY})*\mathbf{BD}*\ln(\mathbf{OM}) + 0.00315785*\ln(1 - \mathbf{WSA}/100)*\ln(\mathbf{CLAY})*\mathbf{BD}*\ln(\mathbf{OM})$ ($R^2 = 0.48$)
	$\ln(\boldsymbol{\alpha}) = 6.45592 - 0.111951*\ln(1 - \mathbf{WSA}/100) - 1.08597*\ln(\mathbf{BD}) - 0.0196937*\mathbf{OM} - 0.0820441*\mathbf{SILT} - 0.0460959*\mathbf{SAND}$ ($R^2 = 0.37$)
Subsoils	$\ln(\boldsymbol{\theta}_s) = -0.861982 + 1.86647*(\mathbf{WSA}/100) - 0.448491*(\mathbf{WSA}/100)*(\mathbf{WSA}/100) - 1.09469*(\mathbf{WSA}/100)*\mathbf{BD} - 0.0297148\ \mathbf{OM}*(\mathbf{WSA}/100)*\mathbf{BD}$ ($R^2 = 0.51$)
	$\ln(\boldsymbol{\alpha}) = 7.80073 - 2.38738*(\mathbf{WSA}/100) - 5.56583*\mathbf{BD} - 0.146628*\mathbf{OM}^2 + 1.5992*\mathbf{BD}^2 + 1.75745*\mathbf{OM} - 0.0239936*\mathbf{SAND} - 0.060879*\mathbf{SILT} - 0.327761*\mathbf{BD}*\mathbf{OM} + 0.00219901*(\mathbf{WSA}/100)*\mathbf{SAND}*\mathbf{SILT} - 0.000291188*\mathbf{OM}*\mathbf{BD}*\mathbf{SAND}*\mathbf{SILT}$ ($R^2 = 0.6$)
	$\ln(\boldsymbol{n}) = 0.248268 + 0.00171927*\mathbf{SILT} - 0.0039649*\mathbf{CLAY} - 0.00415943*\mathbf{WSA} - 0.049055*\mathbf{BD}^2 - 0.0758701*\ln\mathbf{OM} + 7.40763*\mathbf{SILT}^{-1} + 0.966634*\mathbf{CLAY}^{-1} - 6.4206e{-05}*\mathbf{SILT}*\mathbf{CLAY} + 0.000287875*\mathbf{SILT}*\mathbf{CLAY}*(\mathbf{WSA}/100) + 0.0592793\ \mathbf{BD}^2*\ln\mathbf{OM}*(\mathbf{WSA}/100)$ ($R^2 = 0.83$)

BD: bulk density (g cm^{-3}), **OM**: organic matter content (%, w/w of dry soil), **WSA**: water stable aggregates >250 μm (%, w/w of dry soil), **SAND**, **SILT**, and **CLAY** contents (%, w/w of dry soil).

3.2 Model Calibration

The model calibration procedure included the calibration of OC content in each of the WSA fractions as well as the WSA distribution using the data obtained from topsoil (20 cm). An iterative methodology was adapted for model calibration. The procedure included the calibration of CAST module parameters to simulate the SOC and WSA fractionation, followed by calibration of the plant module (PROSUM) parameters, which set the litterfall and nutrient uptake. Nutrient availability for plant growth depends on input through rain and irrigation as well as solute production by CAST and

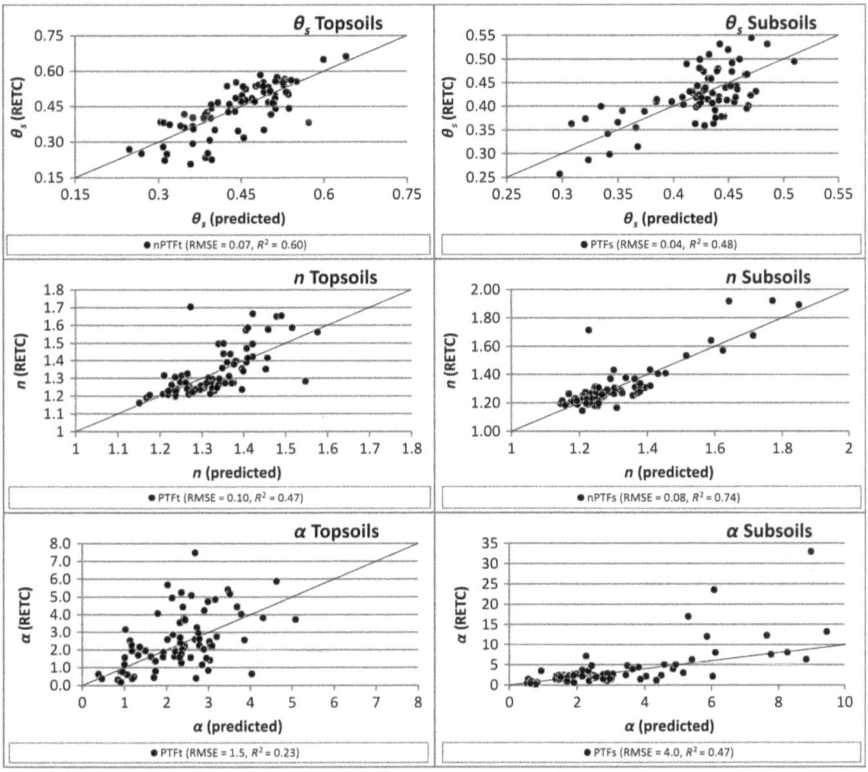

Fig. 3 The θ_s, α, and n shape parameters calculated with the developed PTFs for topsoils and subsoils, plotted against the values obtained by measured values fitted with RETC.

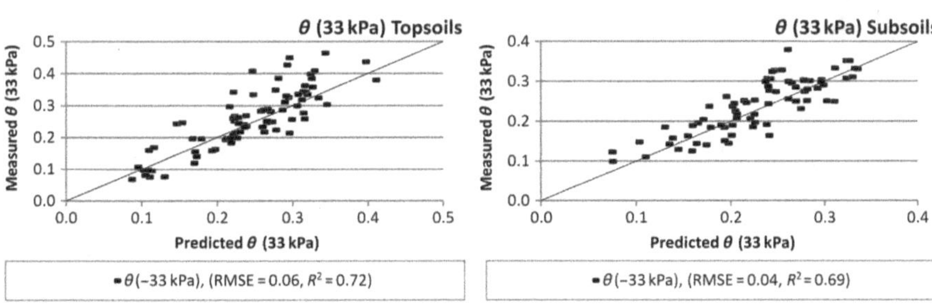

Fig. 4 Predicted compared with measured volumetric water content at pressure head of 33 kPa. Predicted values derived using PTF calculated parameters.

weathering. As a result, weathering rate constants were calibrated next. This procedure was repeated until an acceptable comparison between modeled and field data for carbon and WSA fractionation, biomass production, and solute concentrations was achieved.

The calibration graphs which are shown in Fig. 5 present the comparison of model calibration results and field data including (a) SOC; (b) WSA mass distribution; (c) plant carbon; and (d) N, P, K, Water, and overall (All) limitation of plant C fixation rate.

The model was able to effectively simulate the OC content in the WSA fractions (Fig. 5A) as well as the total SOC (Table 4). The simulated values of OC showed a slight deviation from the observed values in the AC2 and AC3 fractions, which were not considered significant due to low absolute differences and the seasonal variability which is attributed to the high sensitivity of the model to carbon turnover calculations. Regarding the WSA fractions distribution (Fig. 5B), the calibration was successful with none of the simulated values exceeding the deviation of the measured data. Table 6 presents the mean and standard deviation of the measured solute concentrations and the model simulations at the 30–40 cm depth for the main plant nutrition elements and pH. Nitrogen in terms of NO_3-N and NH_4-N, phosphorus (PO_4-P) and potassium measured concentrations in the lysimeters, were used. The results show that the measured and calibrated means were statistically similar at the 95% confidence level. The pH was also calibrated to a value within $\pm$ one standard deviation of the mean lysimeter measurement. Seasonal changes of biomass (leaf, wood, root, and mycorrhizae) stocks, due to litterfall, harvest, and pruning, were well represented (Fig. 5C). Restrictions on plant growth are illustrated in Fig. 5D; the main limitations were due to water availability in summer months and restricted availability during winter months due to N, P, and K, nutrients.

3.3 Soil Management Scenarios

The simulations of the four theoretical management scenarios are presented in Figs. 5–6. The first scenario (Fig. 6) included yearly tilling as a management practice for pest destruction and plant litter incorporation to the first 20 cm of the soil profile. An exhaustive review of the impacts of plant residue incorporation to soil, on all aspects of soil properties is given by Kumar and Goh (1999). Plant litter incorporation to soil can lead to increase in SOM and WSA distribution depending on the residue characteristics (quantity, quality), regional climate conditions, and incorporation method. Simulated OC and WSA dynamics suggest that plant litter incorporation into soil through tilling was not enough to maintain SOC (Figs. 6A and 5B), since the OC incorporated to soil as plant litter decomposed quickly (enhanced by tilling) leading to an 11% decrease of SOC after 10 years.

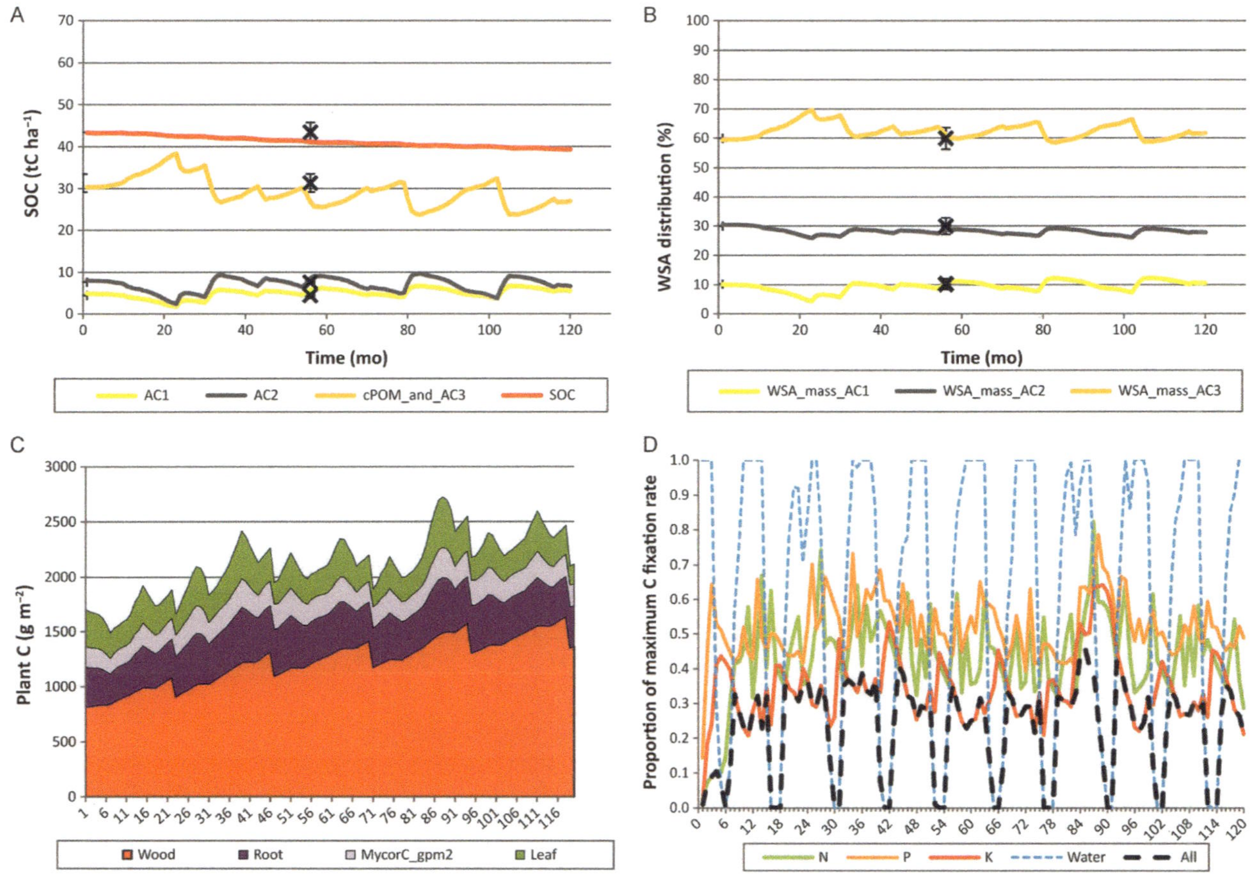

Fig. 5 Model calibration: (A) SOC and OC content in each WSA fraction (*crosses* show the observed values), (B) WSA distribution (*crosses* show the observed values), (C) plant carbon (stock) during the simulation, and (D) N, P, K, Water, and overall (All) limitation of plant C fixation rate which represents the lowest value used to set the physiological limits for plant production.

Table 6 Solute Samples Characteristics and Simulated Solutes

	Soil Solution Samplings—Solutes						Simulation—Solutes
Parameter	**05-Apr-12**	**26-Oct-12**	**13-Dec-12**	**24-Jan-13**	**21-Feb-13**	**Mean (±stdDev.)**	**Mean (±stdDev.) Values at the 30–40 cm Soil Depth**
pH	8.16	Not measured	Not measured	7.82	8.30	8.09 (±0.25)	8.16 (±0.16)
N-NO_3 (mg L^{-1})	0.13	<0.1	3.64	0.78	<0.1	0.95 (±1.53)	0.53 (±2.58)
N-NH_4 (mg L^{-1})	0.44	<0.02	<0.02	10.98	<0.02	2.29 (±4.86)	3.79 (±17.43)
P-PO_4 (mg L^{-1})	0.59	<0.07	<0.07	1.50	0.48	0.54 (±0.59)	0.22 (±0.2)
K (mg L^{-1})	Not measured	Not measured	6.92	25.6	4.05	12.2 (±11.7)	6.71 (±9.05)

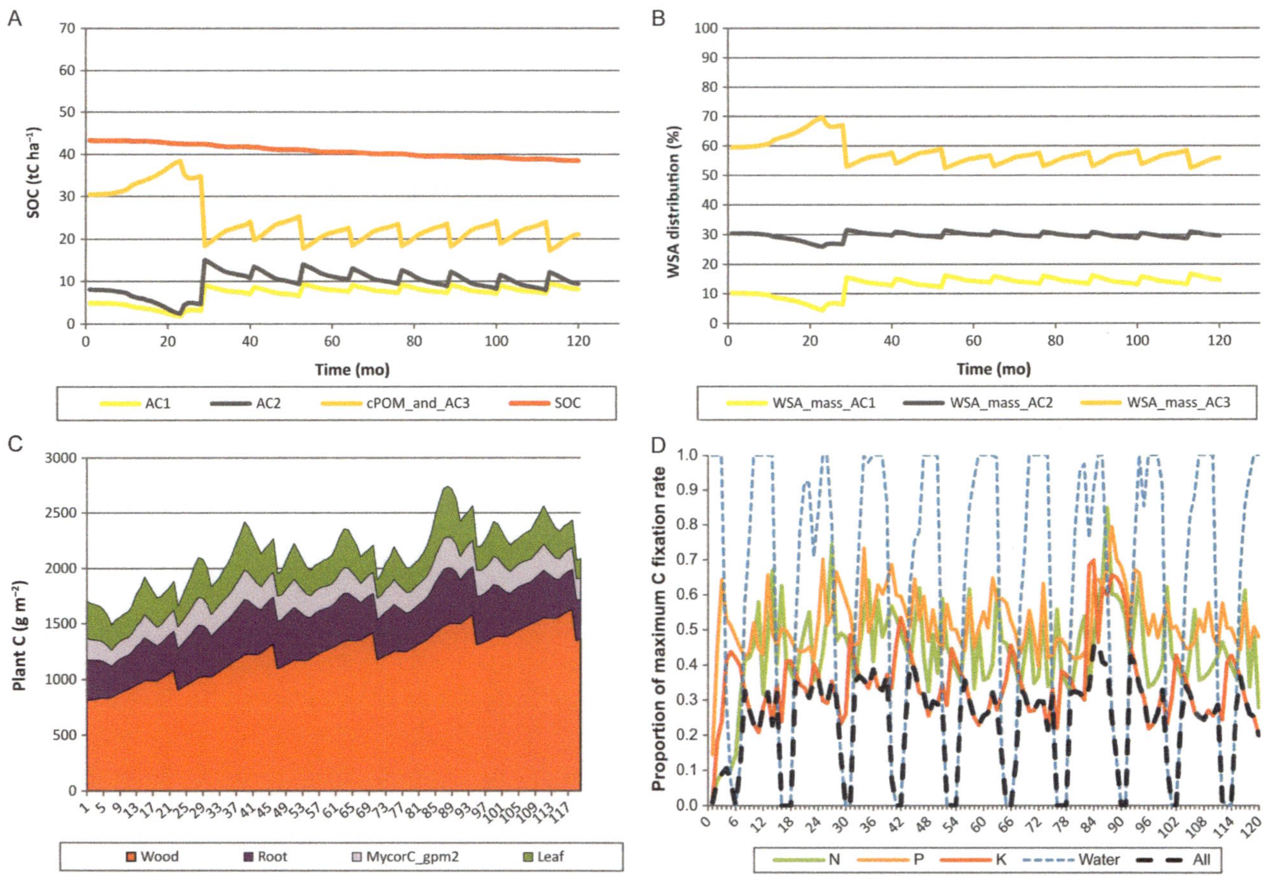

Fig. 6 No carbon addition with tilling (NC-T) scenario: (A) SOC and OC content at each WSA fraction with the observed values, (B) WSA distribution with the observed values, (C) plant carbon (stock) during the simulation, and (D) N, P, K, Water, and overall (All) limitation of plant C fixation rate which represents the lowest value used to set the physiological limits for plant production.

By contrast, WSA were approximately unchanged during the same period although there was a slight increase in the silt-clay sized aggregate class (AC1).

In Figs. 7A and 8A, the SOC simulations of the two scenarios which include carbon addition are presented. In both simulations, SOC content showed an increase after 10 years as a result of increased OC incorporated to soil. Changes in the inter-annual distribution of WSA were observed due to macroaggregate destruction resulting from tilling (Figs. 7B and 8B). In general, there was approximately a 10% increase over the duration of the simulation in the macroaggregate content (WSA mass AC3).

The two C addition scenarios showed annual average plant biomass production of 246.7 g C m^{-2} and 247.5 g C m^{-2} for "25-T" and "50-T," respectively, which exceeded the respective production rates of the two runs without C addition (241.6 gC m^{-2} and 239.9 gC m^{-2} for "NC-NT" and "NC-T," respectively). Limitations of nutrients and water availability on plant growth are shown in Figs. 5D, 6D, 7D, and 8D. Water was the main limiting factor of growth during spring and early summer before irrigation began, and K was the most limiting out of all nutrients. The biennial harvest of olives was simulated using a constant ratio of the leaves harvested on November of every second year. During harvest events, a 35.5% of the total "leaf" biomass was assumed to be harvested representing olives and pruning. The average harvested olives in wet mass was simulated to be 17.44 kg $tree^{-1}$ for the "50-T" scenario, 16.85 kg $tree^{-1}$ for the "25-T," and 17.25 and 17.35 kg $tree^{-1}$ for the "NC-T" and "NC-NT" scenarios, respectively.

The results of the four management scenarios are presented in Fig. 9 and illustrate the simulated impacts of soil management in terms of carbon addition and tilling. Carbon input increased SOC by 43% and 85% for the "25-T" and "50-T" scenarios, respectively, whereas in the "NC-T" scenario there was an 11% decline in the initial carbon storage. Carbon losses due to tillage (NC-T) were slightly greater than in the "NC-NT" (−9%) scenario. The impact of tillage is shown in Fig. 9B, where the macroaggregates (AC3) decreased by 6% in the "NC-T" scenario after 10 years of simulation. The tilling impact on soil structure was also apparent in the "25-T" and "50-T" scenarios where the tilling events were followed by a sharp decrease in the macroaggregate content, but then followed by a quick recovery due to carbon addition, which favors the formation of macroaggregates.

The impact of SOC and macroaggregates on saturated water content and hydraulic conductivity is shown in Fig. 9C and D. The SOC storage and

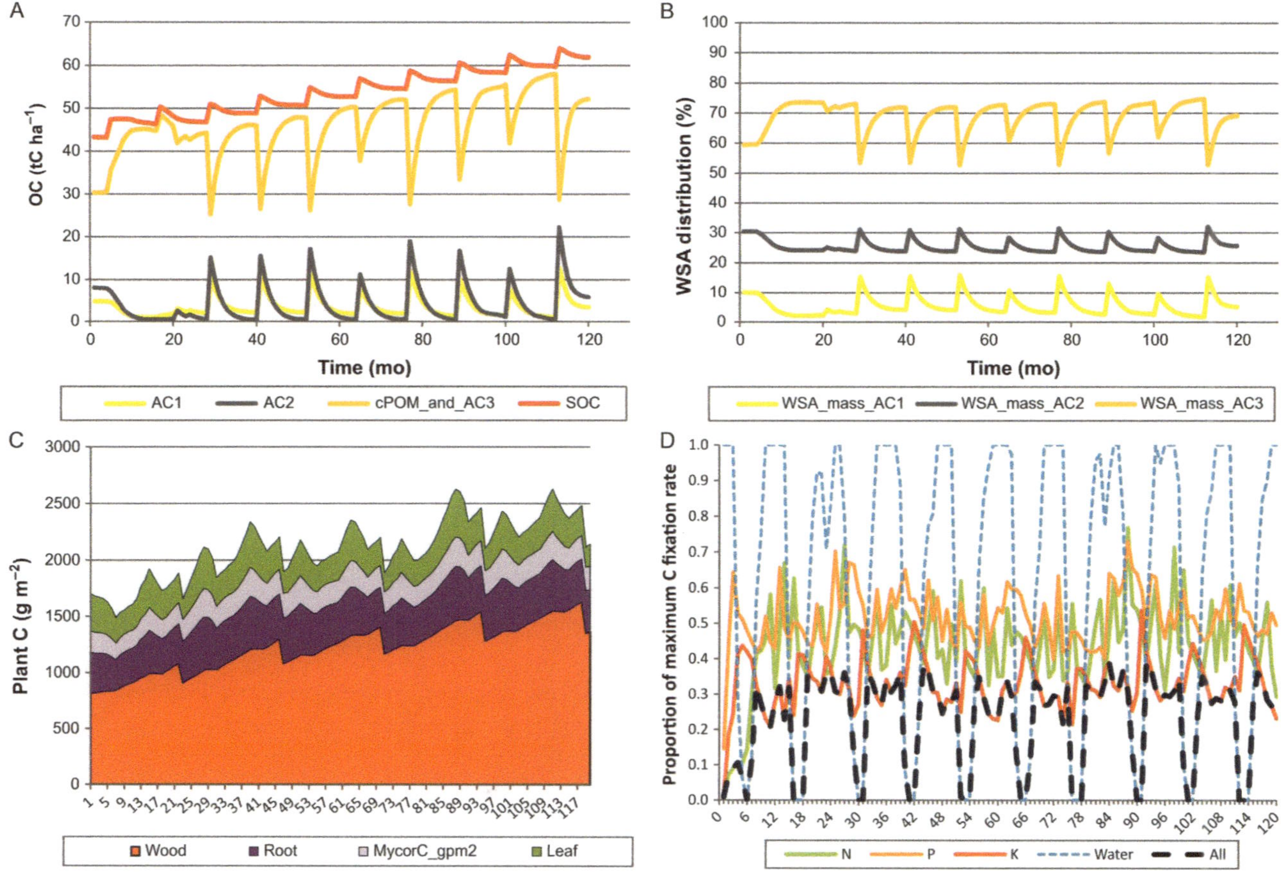

Fig. 7 5.12 tC ha^{-1} Addition with tilling (25-T) scenario: (A) SOC and OC content at each WSA fraction with the observed values, (B) WSA distribution with the observed values, (C) plant carbon (stock) during the simulation, and (D) N, P, K, Water, and overall (All) limitation of plant C fixation rate which represents the lowest value used to set the physiological limits for plant production.

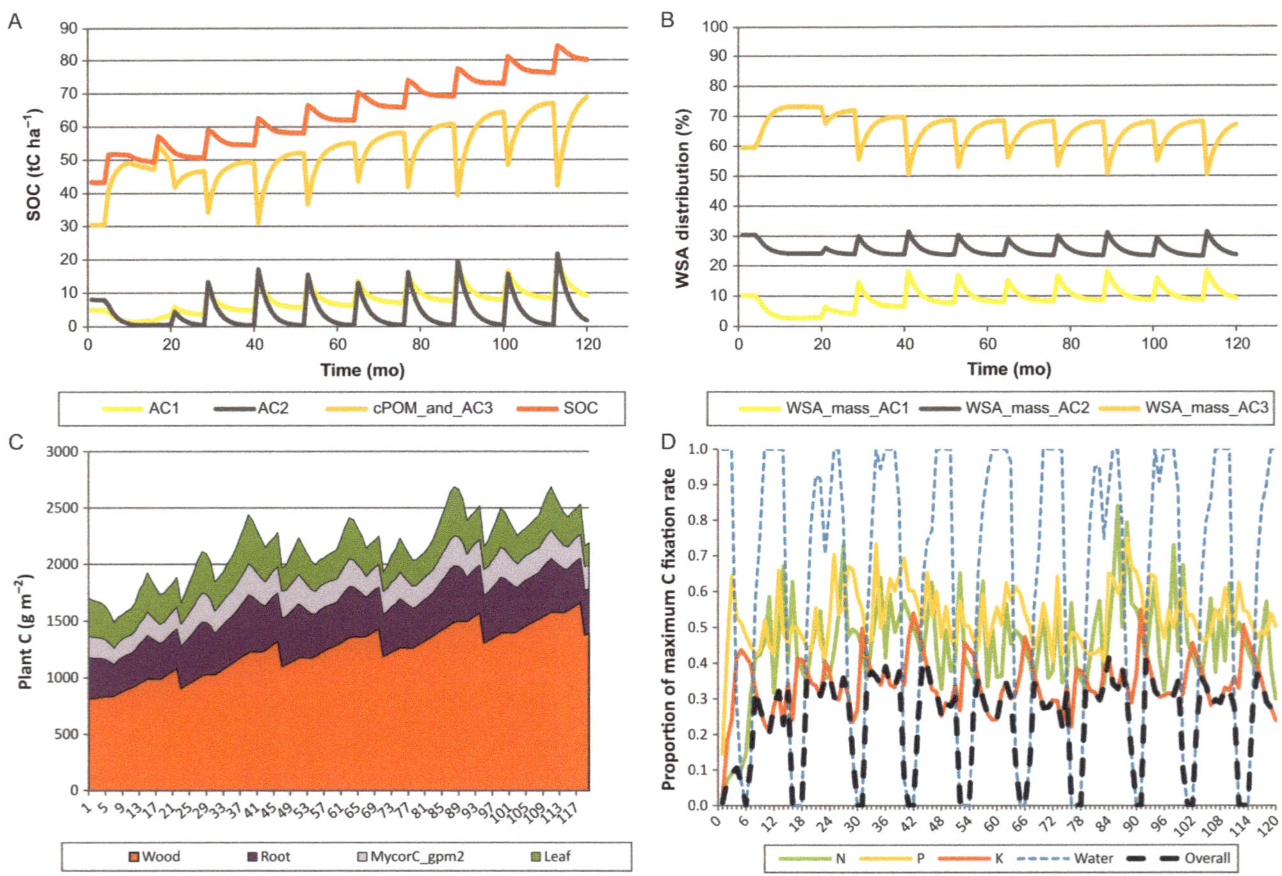

Fig. 8 10.2 tC ha^{-1} Addition with tilling (50-T) scenario: (A) SOC and OC content at each WSA fraction with the observed values, (B) WSA distribution with the observed values, (C) plant carbon (stock) during the simulation, and (D) N, P, K, Water, and overall (All) limitation of plant C fixation rate which represents the lowest value used to set the physiological limits for plant production.

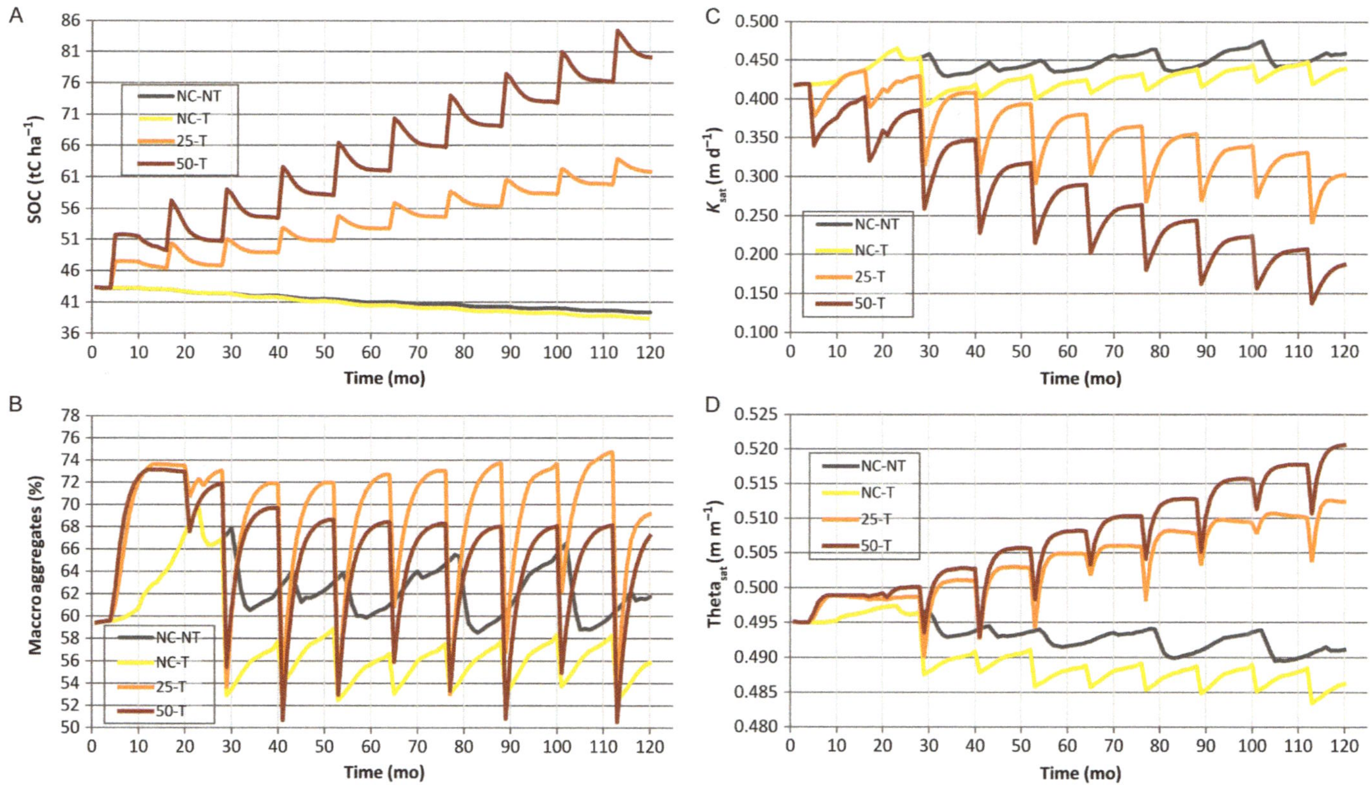

Fig. 9 Soil structure parameters simulated for each scenario: (A) SOC content, (B) macroaggregates (>250 μm) percentage, (C) saturated hydraulic conductivity—K_{sat}, and (D) saturated volumetric water content-Theta$_{sat}$ ("NC-NT": No extra Carbon addition-No tilling; "NC-T": No extra Carbon addition-With tilling; "25-T": 5.12 tC ha^{-1} year^{-1} With tilling, "50-T": 10.2 tC ha^{-1} year^{-1} With tilling).

macroaggregate formation in the "25-T" and "50-T" scenarios affected the hydraulic properties of the soil by increasing water retention. On the other hand, tilling disturbed soil structure (macroaggregate collapse) and soil saturated hydraulic conductivity decreased significantly. During macroaggregate formation, an increase in saturated hydraulic conductivity is observed as macroaggregate presence in soil increase its permeability and drainage ability. As it is shown in Fig. 9B, the rate of macroaggregate destruction is higher than the rate of formation and that is also reflected in the rate of decrease and increase of K_{sat}.

The 1D-ICZ model, coupled with the PTF, introduces for the first time, a model that dynamically links soil structure characteristics and hydraulic soil properties by simulating their changes.

Fig. 10 presents the quantification of soil functions for the four simulated scenarios. The SOC stock was enhanced through carbon input in the "25-T" and "50-T" scenarios, as was the soil structure in terms of macroaggregate formation. Carbon addition enhanced plant biomass production rate, especially at the high carbon addition rate (50-T). The same effect was observed for the soil microorganisms and fauna (BIO and Fauna) C content. The dynamic changes in microorganisms and fauna stocks in the soil can be used as a first proxy indicator for biodiversity. In this way, the 1D-ICZ model can quantify multiple soil functions such as biomass production, carbon sequestration, water filtration and transformation, and a proxy indicator for soil biodiversity, and how these vary due to the dynamic changes in soil structure.

Soil respiration rate showed a twofold increase for the "25-T" scenario compared to "NC-NT," while for the "50-T" scenario the increase was more than five times. Tillage without carbon (NC-T) slightly increased the CO_2 efflux rate and decreased the carbon storage compared to no tillage (NC-NT). Tillage decreased carbon stocks, macroaggregate content, and plant biomass production while it increased the BIO and Fauna carbon content. Moreover, compared to the "NC-NT," tillage enhanced the leaching of NO_3 and PO_4 while K leaching rate increased due to carbon addition. NO_3 and PO_4 leaching rates were slightly higher in the "25-T" and "50-T" scenarios as a result of the increased plant uptake.

Overall, tillage increased the leaching of solutes and CO_2 efflux, reduced soil aggregation, and plant production. Carbon addition enhanced plant production, soil structure, and microorganism growth, it also stimulated CO_2 emissions and the leaching rates of solutes.

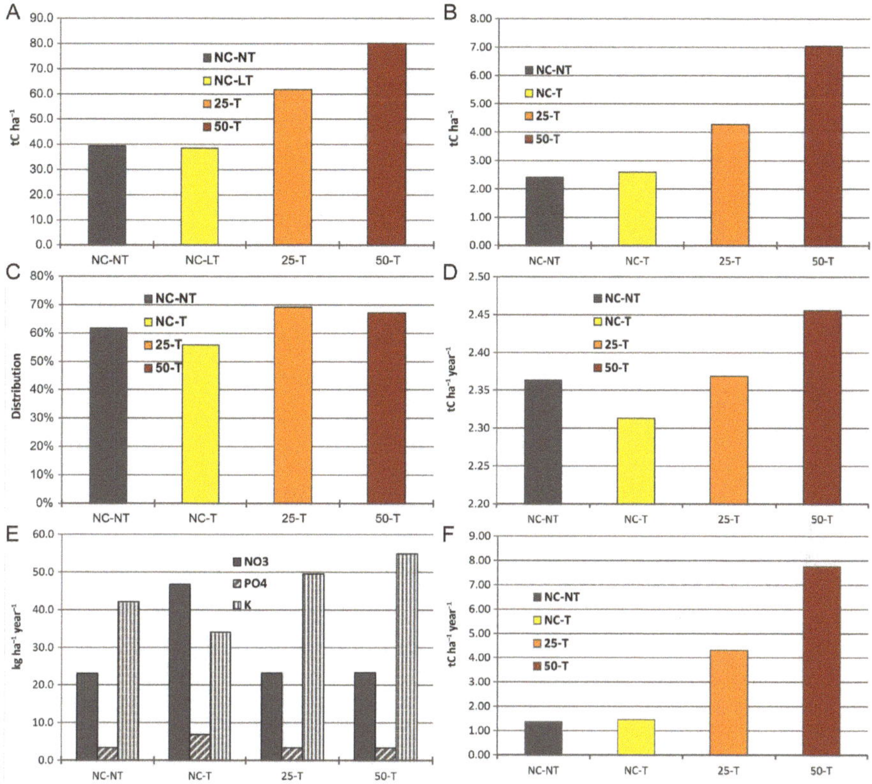

Fig. 10 Simulation of the material stocks and fluxes for the four scenarios: (A) carbon stocks, (B) bio and fauna stock (a proxy for biodiversity), (C) macroaggregate fraction, (D) plant biomass production rate, and (E) leaching rates of N, P, K, and (F) CO_2 flux rate ("NC-NT": No extra Carbon addition-No tilling; "NC-T": No extra Carbon addition-With tilling; "25-T": 5.12 tC ha^{-1} year^{-1} With tilling, "50-T": 10.2 tC ha^{-1} year^{-1} With tilling).

4. CONCLUSIONS

The 1D-ICZ model is the first attempt to simulate key processes necessary for integrated understanding of soil functioning. The scientific advancement made through the development of the 1D-ICZ model is the rigorous simulation and quantification of important soil functions such as the cycling of nutrients (C, N, and P) accounting for the effect of soil structure; biomass production including effects of mycorrhizae and exudates on nutrient uptake and assimilation; and water transformations and filtration; and weathering of cations and other elements. Thus the

1D-ICZ model can be used to assess, understand, and quantify the complex interactions between the different process in the soil-plant-water system as well as providing a tool to design sustainable agricultural management practices by taking under consideration the synergies and trade-offs of the different soil functions.

ACKNOWLEDGMENTS

The authors acknowledge funding support from the European Commission FP 7 Collaborative Project "Soil Transformations in European Catchments" (SoilTrEC) (Grant Agreement No. 244118).

REFERENCES

Aguilera, D.R., Jourabchi, P., Spiteri, C., Regnier, P., 2005. A knowledge-based reactive transport approach for the simulation of biogeochemical dynamics in Earth systems. Geochem. Geophys. Geosyst. 6, Q07012. http://dx.doi.org/10.1029/2004GC000899.

Amundson, R., Berhe, A.A., Hopmans, J.W., Olson, C., Sztein, A.E., Sparks, D.L., 2015. Soil and human security in the 21st century. Science 348, 1261071. http://dx.doi.org/10.1126/science.1261071.

Banwart, S., Bernasconi, S.M., Bloem, J., Blum, W., Brandao, M., Brantley, S., Chabaux, F., Duffy, C., Kram, P., Lair, G., Lundin, L., Nikolaidis, N., Novak, M., Panagos, P., Ragnarsdottir, K.V., Reynolds, B., Rousseva, S., de Ruiter, P., van Gaans, P., van Riemsdijk, W., White, T., Zhang, B., 2011. Soil processes and functions in critical zone observatories: hypotheses and experimental design. Vadose Zone J. 10, 974–987. http://dx.doi.org/10.2136/vzj2010.0136.

Baudron, F., Giller, K.E., 2014. Agriculture and nature: trouble and strife? Biol. Conserv. 170, 232–245. http://dx.doi.org/10.1016/j.biocon.2013.12.009.

Beddington, J., 2009. Food, Energy, Water and the Climate: A Perfect Storm of Global Events? Government Office for Science, London, pp. 1–9.

Chapin, F.S., Matson, P.A., Vitousek, P.M., 2012. Principles of Terrestrial Ecosystem Ecology, second ed. Springer, New York. http://dx.doi.org/10.1007/978-1-4419-9504-9.

Coleman, K., Jekinson, D.S., 1999. RothC-26.3—a model for the turnover of carbon in soil: model description and windows users guide. November 1999 issue.

Dexter, A.R., 2004a. Soil physical quality: part I. Theory, effects of soil texture, density, and organic matter, and effects on root growth. Geoderma 120, 201–214. http://dx.doi.org/10.1016/j.geodermaa.2003.09.005.

Dexter, A.R., 2004b. Soil physical quality: part II. Friability, tillage, tilth and hard-setting. Geoderma 120, 215–225. http://dx.doi.org/10.1016/j.geoderma.2003.09.005.

Dexter, A.R., 2004c. Soil physical quality: part III. Unsaturated hydraulic conductivity and general conclusions about S-theory. Geoderma 120, 227–239. http://dx.doi.org/10.1016/j.geoderma.2003.09.004.

Elliot, E.T., 1986. Aggregate structure and carbon, nitrogen, and phosphorus in native and cultivated soils. Soil Sci. Soc. Am. J. 50, 627–633.

Elser, J.J., Bracken, M.E.S., Cleland, E.E., Gruner, D.S., Harpole, W.S., Hillebrand, H., Ngai, J.T., Seabloom, E.W., Shurin, J.B., Smith, S.E., 2007. Global analysis of nitrogen and phosphorus limitation of primary producers in freshwater, marine and terrestrial ecosystems. Ecol. Lett. 10, 1135–1142. http://dx.doi.org/10.1111/j.1461-0248.2007.01113.x.

Finke, P.A., Hutson, J.L., 2008. Modelling soil genesis in calcareous loess. Geoderma 145, 462–479. http://dx.doi.org/10.1016/j.geoderma.2008.01.017.

Garnett, T., Appleby, M.C., Balmford, A., Bateman, I.J., Benton, T.G., Bloomer, P., Burlingame, B., Dawkins, M., Dolan, L., Fraser, D., Herrero, M., Hoffmann, I., Smith, P., Thornton, P.K., Toulmin, C., Vermeulen, S.J., Godfray, H.C.J., 2013. Sustainable intensification in agriculture: premises and policies. Science 341, 33–34. http://dx.doi.org/10.1126/science.1234485.

Godfray, H.C.J., Beddington, J.R., Crute, I.R., Haddad, L., Lawrence, D., Muir, J.F., Pretty, J., Robinson, S., Thomas, S.M., Toulmin, C., 2010. Food security: the challenge of feeding 9 billion people. Science 327, 812–818. http://dx.doi.org/10.1126/science.1185383.

Hobbs, P.R., Sayre, K., Gupta, R., 2008. The role of conservation agriculture in sustainable agriculture. Philos. Trans. R. Soc. B Biol. Sci. 363, 543–555. http://dx.doi.org/10.1098/rstb.2007.2169.

Ilarioni, L., Nasini, L., Brunori, A., Proietti, P., 2013. Experimental measurement of the biomass of Olea europaea L. Afr. J. Biotechnol. 12, 1216–1222. http://dx.doi.org/10.5897/AJB12.2682.

Katerji, N., Mastrorilli, M., Rana, G., 2008. Water use efficiency of crops cultivated in the Mediterranean region: review and analysis. Eur. J. Agron. 28, 493–507. http://dx.doi.org/10.1016/j.eja.2007.12.003.

Kerkhoff, A.J., Enquist, B.J., Elser, J.J., Fagan, W.F., 2005. Plant allometry, stoichiometry and the temperature-dependence of primary productivity. Glob. Ecol. Biogeogr. 14, 585–598. http://dx.doi.org/10.1111/j.1466-822X.2005.00187.x.

Khodaverdiloo, H., Homaee, M., van Genuchten, M.T., Dashtaki, S.G., 2011. Deriving and validating pedotransfer functions for some calcareous soils. J. Hydrol. 399, 93–99. http://dx.doi.org/10.1016/j.jhydrol.2010.12.040.

Krám, P., Santore, R.C., Driscoll, C.T., Aber, J.D., Hruška, J., 1999. Application of the forest-soil-water model (PnET-BGC/CHESS) to the Lysina catchment, Czech Republic. Ecol. Model. 120, 9–30. http://dx.doi.org/10.1016/s0304-3800(99)00064-2.

Kumar, K., Goh, K.M., 1999. Crop residues and management practices: effects on soil quality, soil nitrogen dynamics, crop yield, and nitrogen recovery. Adv. Agron. 68, 197–319. http://dx.doi.org/10.1016/S0065-2113(08)60846-9.

Moraetis, D., Paranychianakis, N.V., Nikolaidis, N.P., Banwart, S.A., Rousseva, S., Kercheva, M., Nenov, M., Shishkov, T., de Ruiter, P., Bloem, J., Blum, W.E.H., Lair, G.J., van Gaans, P., Verheul, M., 2015. Sediment provenance, soil development, and carbon content in fluvial and manmade terraces at Koiliaris River Critical Zone Observatory. J. Soil. Sediment. 15, 347–364. http://dx.doi.org/10.1007/s11368-014-1030-1.

National Research Council, 2001. Basic Research Opportunities in the Earth Sciences. National Academic Press, Washington, DC.

Nikolaidis, N.P., Bidoglio, G., 2013. Soil organic matter dynamics and structure. Sustain. Agric. Rev. 12, 175–199. http://dx.doi.org/10.1007/978-94-007-5961-9.

Nikolaidis, N.P., Valstar, J., Rowe, E.C., Moirogiorgou, K., Kotronakis, E., Giannakis, G.V., Stamati, F.E., Banwart, S.A., 2014. Integrated Critical Zone Model (1D-ICZ Model)—a tool for dynamic simulation of soil functions and soil structure: technical manual. School of Environmental Engineering Technical Universty of Crete, Chania, Greece. http://www.herslab.tuc.gr/downloads/1D-ICZ-model.

Parkhurst, D.L., Appelo, C.A.J., 1999. User's guide to PhreeqC (version 2)—a computer program for speciation, batch-reaction, one-dimensional transport, and inverse geochemical calculations: USGS, Water-Resources Investig. Rep. 99–4259.

Pohlert, T., Huisman, J.A., Breuer, L., Frede, H.-G.T., 2005. Evaluation of the soil nitrogen balance model in SWAT with lysimeter data. In: Srinivasan Jacobs, J., Day, D., Abbaspour, K.R. (Eds.), 3rd International SWAT Conference, Zurich, Switzerland, pp. 496–508.

Pohlert, T., Huisman, J.A., Breuer, L., Frede, H.G., 2007. Integration of a detailed biogeochemical model into SWAT for improved nitrogen predictions—model development,

sensitivity, and GLUE analysis. Ecol. Model. 203, 215–228. http://dx.doi.org/10.1016/j.ecolmodel.2006.11.019.
Rowe, E.C., Smart, S.M., Kennedy, V.H., Emmett, B.A., Evans, C.D., 2008. Nitrogen deposition increases the acquisition of phosphorus and potassium by heather Calluna vulgaris. Environ. Pollut. 155, 201–207. http://dx.doi.org/10.1016/j.envpol.2007.12.008.
Royal Society of London, 2009. Reaping the Benefits: Science and the Sustainable Intensification of Global Agriculture. Royal Society, London.
Segoli, M., De Gryze, S., Dou, F., Lee, J., Post, W.M., Denef, K., Six, J., 2013. AggModel: a soil organic matter model with measurable pools for use in incubation studies. Ecol. Model. 263, 1–9. http://dx.doi.org/10.1016/j.ecolmodel.2013.04.010.
Shein, E.V., 2005. Course of Physics (in Russian). Moscow State University Publications, Moscow.
Šimůnek, J., Šejna, M., Saito, H., Sakai, M., van Genuchten, M.T., 2009. The HYDRUS-1D Software Package for Simulating the One-Dimensional Movement of Water, Heat, and Multiple Solutes in Variably-Saturated Media Version 4.08. Department of Environmental Sciences, University of California Riverside, Riverside.
Stamati, F.E., Nikolaidis, N.P., Banwart, S., Blum, W.E.H., 2013. A coupled carbon, aggregation, and structure turnover (CAST) model for topsoils. Geoderma 211–212, 51–64. http://dx.doi.org/10.1016/j.geoderma.2013.06.014.
Stange, F., 2001. Entwicklung und Anwendung eines prozeíorientierten Modells zur Beschreibung der N2O und NO-Emissionen aus Böden temperater Wälder. Univ. of Freiburg, Garmisch-Partenkirchen.
Tilman, D., Balzer, C., Hill, J., Befort, B.L., 2011. Global food demand and the sustainable intensification of agriculture. Proc. Natl. Acad. Sci. U.S.A. 108, 20260–20264. http://dx.doi.org/10.1073/pnas.1116437108.
Tisdall, J.M., Oades, J.M., 1982. Organic matter and water-stable aggregates in soils. J. Soil Sci. 33, 141–163.
UN FAO, 2011. Food agriculture and cities: challenges of food and nutrition security, agriculture and ecosystem management in an urbanizing world—food for the cities multidisciplinary initiative.
Van Duivenbooden, N., De Wit, C.T., Van Keulen, H., 1996. Nitrogen, phosphorus and potassium relations in five major cereals reviewed in respect to fertilizer recommendations using simulation modelling. Fertil. Res. 44, 37–49.
van Genuchten, M.T., 1980. Closed-form equation for predicting the hydraulic conductivity of unsaturated soils. Soil Sci. Soc. Am. J. 44, 892–898.
van Genuchten, M.T., Leij, F.J., Yates, S.R., 1991. The RETC code for quantifying the hydraulic functions of unsaturated soils: Rep. EPA/600/2–91/065. R.S. Kerr Environ. Res. Lab. USEPA, Ada, OK.
Vanlauwe, B., Giller, K.E., 2006. Popular myths around soil fertility management in sub-Saharan Africa. Agric. Ecosyst. Environ. 116, 34–46. http://dx.doi.org/10.1016/j.agee.2006.03.016.
Wallman, P., Svensson, M.G.E., Sverdrup, H., Belyazid, S., 2005. ForSAFE—an integrated process-oriented forest model for long-term sustainability assessments. For. Ecol. Manage. 207, 19–36. http://dx.doi.org/10.1016/j.foreco.2004.10.016.
Wösten, J.H.M., Lilly, A., Nemes, A., Le Bas, C., 1999. Development and use of a database of hydraulic properties of European soils. Geoderma 90, 169–185. http://dx.doi.org/10.1016/s0016-7061(98)00132-3.
Wösten, J.H.M., Pachepsky, Y.A., Rawls, W.J., 2001. Pedotransfer functions: bridging the gap between available basic soil data and missing soil hydraulic characteristics. J. Hydrol. 251, 123–150. http://dx.doi.org/10.1016/s0022-1694(01)00464-4.
Zhang, Y., Li, C., Zhou, X., Moore Iii, B., 2002. A simulation model linking crop growth and soil biogeochemistry for sustainable agriculture. Ecol. Model. 151, 75–108. http://dx.doi.org/10.1016/s0304-3800(01)00527-0.

CHAPTER ELEVEN

Modeling the Impact of Carbon Amendments on Soil Ecosystem Functions Using the 1D-ICZ Model

M. Kotronakis*, G.V. Giannakis*, N.P. Nikolaidis[†,1], E.C. Rowe[‡], J. Valstar[§], N.V. Paranychianakis[†], S.A. Banwart[¶]

*Technical University of Crete, School of Environmental Engineering, Chania, Greece
[†]School of Environmental Engineering, Technical University of Crete, University Campus, Chania, Greece
[‡]Centre for Ecology & Hydrology, ECW, Bangor, United Kingdom
[§]Deltares Subsurface and Groundwater Systems, Utrecht, The Netherlands
[¶]School of Earth and Environment, University of Leeds, Leeds, United Kingdom
[1]Corresponding author: e-mail address: nikolaos.nikolaidis@enveng.tuc.gr

Contents

Abstract

In the next four decades, humanity needs to double food and energy production and increase the supply of clean water by over 50% while mitigating and adapting to climate change. A central element in the strategy of addressing these major environmental challenges is to maintain the central role of Earth's essential soil functions and related ecosystem services. Many soil functions are affected by soil structure in terms of particle aggregation and porosity. The objective of this work is to model soil structure and biomass dynamics, nutrients uptake, and yields using the 1D Integrated Critical Zone (1D-ICZ) model which is a mechanistic mathematical description of soil processes and functions. The 1D-ICZ model simulates the coupled processes that underpin major soil functions including water flow and storage, biomass production, carbon and

Advances in Agronomy, Volume 142
ISSN 0065-2113
http://dx.doi.org/10.1016/bs.agron.2016.10.010

nutrient sequestration, pollutant transformation, and supporting biological processes, and thus is capable of quantifying essential soil ecosystem services. The model was validated using data derived from a field experiment where tomato plants were grown using different treatments of commercial mineral fertilizers, compost, manure, and a 30% manure–70% compost amendment. Detailed data have been collected over four growing seasons on soil and soil solution chemistry, aggregate formation, and plant production. The model has been able to capture the biomass production, the temporal dynamics of the water-stable aggregate formation and the dynamics of carbon and nutrient sequestration in the different sizes aggregates as well as the variability of water filtration and transformation efficiency in the different amendment treatments. The model results demonstrate the value of applying computational simulation tools such as the 1D-ICZ model to test options for improved land management measures and to support sustainable land care practices.

1. INTRODUCTION

Food security has been recognized as one of the major socioenvironmental challenges of this century. The combination of increasing global population and wealth with the increasing need for food supplies originating from land cultivation exerts extreme pressure on the agricultural sector to meet this demand. This pressure is enhanced by the increasing energy demand for biofuels and energy crops. In addition, current predictions of climate change impacts impose further, significant pressure on agriculture in the coming decades including increasing greenhouse gas (GHG) emissions, biodiversity loss, and land and freshwater degradation (Foley et al., 2011).

The expansion of agricultural land by converting Earth's grasslands, savanna, and deciduous and tropical forests during the last century is responsible for adverse impacts on soil carbon storage, fertility, and biodiversity. This expansion is mainly taking place in the tropics and is strongly contributing to GHG emissions and loss of ecosystem services (Foley et al., 2011). However, even with the expansion of productive land, there is strong market pressure for intensification of agriculture to meet the ever-growing demand for crops, by boosting the agricultural output through intensified inputs of nutrients, water, and new plant traits.

Agricultural intensification has coincided with the development of agricultural machinery and inorganic fertilization (IF) and has contributed significantly to increased food production (Amundson et al., 2015). Foley et al. (2011) reported that in the past five decades, the increase in fertilizer used worldwide was 500% which in addition to the nearly doubling of the irrigated cropland areas has contributed to the degradation of water quality and aquatic ecosystems as well as the soil's ability to retain carbon and to compensate for soil erosion (Amundson et al., 2003, 2015). In addition, conventional agriculture has degraded soil fertility through reduced carbon inputs (and thus nutrient stocks) and mechanical tillage that breaks down soil aggregates and soil structure (Luo et al., 2010).

It is important to reexamine agricultural practices in order to improve soil fertility and increase agricultural yields while minimizing the environmental impacts over intergenerational time scales, during which climate change impacts are expected to create additional challenges for soil management and agricultural production. Agroecological practices and active carbon amendments can be used to restore soil functions and improve soil fertility and reverse some of the harmful impacts of intensification (Lal, 2006; Martin et al., 2010). Management practices such as organic farming, precision agriculture, and conservation agriculture have been suggested as promising alternatives for sustainable agricultural intensification (Foley et al., 2011; Garnett et al., 2013; Hobbs et al., 2008).

Organic amendments were first used in agriculture thousands of years ago in order to improve soil quality and fertility. The heterotrophic soil microbial community decomposes soil organic carbon (SOC) slowly releasing nutrients which benefit plant growth and enhance soil structure by priming aggregate formation and thus controlling water filtration (Milne et al., 2015). Organic amendments such as animal manure, green manure and cover crops, municipal biosolids, manufacturing processes' wastes, and composts are being used extensively in agriculture (Scotti et al., 2015). The benefits of organic amendments on soil fertility vary among different types of amendments, and they are characterized by the high organic carbon content which improves the physical, chemical, and biological properties of the soil. The decomposition of organic amendments provides a source of slowly released nutrients for plant growth and stimulates microbial biomass (BIO) and enzyme activity which catalyze the biochemical alteration of particulated organic matter and the creation of humic substances (Hargreaves et al., 2008).

The identification of the effects of organic amendments on soil through experimentation and the quantitative, mechanistic interpretation of the resulting observed changes in soil processes and functions are crucial for planning best soil management practices. The potential ability of organic amendments to increase crop yield could contribute toward the mitigation of food security concerns by improving soil quality and functioning. On the other hand, this opportunity has to be examined under the prism of its impacts on the environment and the practicality to sustain these benefits over time—i.e., to make soil functions last. These trade-offs are dependent on numerous factors and experiments cannot capture the whole range of responses that each applied practice can have on the soil. Mathematical models that simulate the interactions that occur within the plant–soil–water–atmosphere continuum can contribute to quantifying the soil functions, and to carry out hypothesized changes in practice that can balance the benefits and potential trade-offs, and redesign the agricultural system.

Results are presented from a 4-year field-scale horticulture experiment which included four fertilization treatments of tomato plants: IF, municipal solid waste compost (MSWC), manure, and a 70:30 ratio of MSWC: manure. The experimental design was based on the hypothesis that SOM replenishment as an agricultural practice can improve soil fertility as demonstrated by increased crop yields from an initial baseline following carbon amendment. Data from the experiment were used to set up the 1D Integrated Critical Zone (1D-ICZ) model (see Chapter "Integrated Critical Zone Model (1D-ICZ): A Tool for Dynamic Simulation of Soil Functions and Soil Structure" by Giannakis et al.). Vegetation dynamics, carbon and nutrient transformations and sequestration in relation to soil structure dynamics, water filtration, and simplified descriptions of functional biodiversity can be simulated, accounting for interactions between water and heat flow, solute transport and geochemical reactions, soil structure dynamics, carbon and nutrient dynamics, and plant and soil microbe biomass production and decomposition. Lastly, trade-offs between treatments are discussed, illustrating the use of mathematical tools to contribute to the design of sustainable agricultural management practices.

2. MATERIALS AND METHODS

2.1 Field Experiment

2.1.1 Description of the Study Area

The study was conducted over 4 years (2011–2014) at the plain of Koiliaris River Basin (KRB) within the study area of the Koiliaris Critical Zone Observatory (CZO) to assess the effects of the application of MSWC and manure on soil properties, plant nutrient status, and crop yield. The experimental field is located 14 km south-east from Chania, Greece (35° 26′ 14″ N, 24° 08′ 34″ E) at an altitude of about 14 m above sea level.

2.1.2 Climate and Meteorological Condition

The dry summer season at the site lasts for about 6 months with no significant rainfall throughout this period. The meteorological data used were derived from the weather station in Chania (Elev: 10 m, 35° 30′ 28″ N, 24° 00′ 22″ E) operated by the National Observatory of Athens. The daily mean temperature during the 4-year period ranged between 11.7 and 26.9°C, and the mean annual rainfall in the corresponding period was 648 mm.

2.1.3 Geological and Soil Characterization

The study area lies on recent alluvial deposits (Quaternary formations) and soils are characterized as Calcaric Regosols (pararendzinas, Entisols). Soil texture was classified as silty–loam in the upper (0–60 cm) ploughed layer (82 g kg^{-1} clay-sized content, 619 g kg^{-1} silt-sized content, and 299 g kg^{-1} sand-sized content), with soil bulk density (BD) of 1380 kg m^{-3} (measured by the excavation method; Maynard and Curran, 2006). The soil pH was 7.9 measured in soil and distilled water solution at a ratio of 1:2.5. The SOC content at the beginning of the experiment was 28.2 g kg^{-1}, while the initial total nitrogen (TN) content was 2.5 g kg^{-1}. Bulk elemental analyses of soil reported as mass fraction of oxide compounds (Na_2O, MgO, K_2O, CaO, TiO_2, MnO, Fe_2O_3, Al_2O_3, SiO_2, and P_2O_5) were conducted by energy dispersive X-ray fluorescence spectroscopy (EDS-XRF), using a Bruker S2 Ranger XRF spectrometer. The mineralogical composition was determined by quantitative X-ray diffraction techniques using a Bruker D8 Advance powder diffractometer. The mineralogical composition of bulk soil is dominated by quartz along with the presence of calcite, kaolinite, paragonite, dolomite, and muscovite (Table 1).

Table 1 Mineralogical Composition of Bulk Soil

Minerals	Composition (%)
Quartz	59.4
Orthoclase	0.5
Albite	0.6
Calcite	10.5
Dolomite	7.1
Kaolinite (BISH)	7.9
Muscovite 2M1	6.9
Paragonite 1M	7.0

2.1.4 Experimental Field Preparation, Experimental Design, and Agricultural Treatments Applied

The experimental field was a former orange tree plantation which had been abandoned for approximately 25 years prior to the start of the experiment. One month before the onset of the experiment the vegetation was totally removed and the field was ploughed to 60 cm. Prior to ploughing, three soil profile pits were dug, and bulk soil samples were taken from six distinct soil horizons (0–3, 3–15, 15–30, 30–50, 50–70, and 70–120 cm). The soil samples were prepared according to Regelink et al. (2015) for further analysis of soil texture, water-stable aggregates (WSAs), and soil chemical properties.

The experimental design is shown in Fig. 1. This included three treatments with incorporation in the soil of organic amendments and one control with only the addition of IF at a rate of 200 kg N ha^{-1}, which is the typical rate applied by the local farmers. The mass loading of organic amendments was based on bulk dry weight of each material. The organic amendment treatments included (i) MSWC at a rate of 50 t dry weight ha^{-1}, (ii) sheep/goat manure (M) at the same rate of 50 t dry weight ha^{-1}, and (iii) the last, amended with a 70%/30% mixture of manure and compost (MSWC + M) at the same rate (50 t dry weight ha^{-1}).

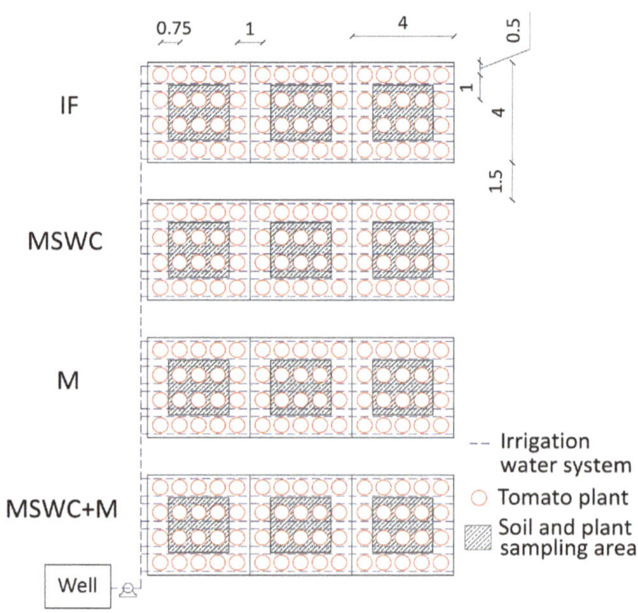

Fig. 1 Experimental plan view design.

Due to management constraints, only a single plot was used per treatment, each of $4 \times 12\ m^2$ (Fig. 1). However, the existing variation in soil characteristics was minimal, as shown by the values for SOC and TN at the start of the experiment (see Figs. 8 and 9), and treatments were randomly allocated to the four plots in the field (Fig. 1). Each of the four plots was further divided into three subareas (subplots) to obtain an estimate of observational error. The amendments were applied evenly manually at the surface of each plot in order to ensure a uniform distribution and incorporated to the top 7 cm of the soil by tilling before planting. Tilling was also applied to the IF treatment.

Productivity was assessed using tomato plants, which are grown extensively in the region and have a short growing cycle, enabling assessment of effects of the organic and inorganic fertilization on the plant development on an annual basis. Tomato seedlings (*Solanum lycopersicum* "Bobcat"; 60 per treatment) were transplanted in early August and harvested in November in the first year of the experiment, and in subsequent years the growing period was from early May until October. Groundwater, as characterized in Table 2, was used for irrigation through the installed irrigation drip pipe network (2 pipes/row of plants, see Fig. 1). The data on irrigation management (dose of irrigation applied in all treatments on monthly basis) are provided in Table 3. One soil profile moisture probe (PR2 Profile Probe

Table 2 Chemical Characterization of Rain and Irrigation Water Used for Simulation

Species	Irrigation Water (mol L^{-1})	Rain Water (mol L^{-1})
Na^+	2.80E−04	8.83E−06
K^+	7.26E−05	1.26E−05
Mg^{2+}	7.94E−04	1.78E−05
Al^{3+}	4.58E−09	2.33E−08
Ca^{2+}	3.87E−03	4.12E−05
SO_4^{-2}	4.83E−04	1.80E−05
HCO_3^-	3.89E−03	1.87E−04
H^+	1.00E−07	1.00E−06
NO_3^-	1.15E−03	3.61E−04
NH_3	2.77E−05	2.79E−05
PO_4^{-3}	6.22E−07	2.11E−07

Table 3 Irrigation Schedule During the 4-Year Experiment

2011		2012		2013		2014	
Month	**Irrig. (mm)**	**Month**	**Irrig. (mm)**	**Month**	**Irrig. (mm)**	**Month**	**Irrig. (mm)**
		May-12	97.5	May-13	166.7	May-14	146.4
		Jun.-12	136.0	Jun.-13	429.7	Jun.-14	403.6
		Jul.-12	230.8	Jul.-13	484.4	Jul.-14	622.4
Aug.-11	578.1	Aug.-12	299.5	Aug.-13	494.8	Aug.-14	588.5
Sep.-11	395.8	Sep.-12	218.8	Sep.-13	382.8	Sep.-14	57.3
Oct.-11	15.6	Oct.-12	46.9				
Sum	989.6		1029.4		1958.3		1818.2

Irrig., irrigation water.

Systems 100 cm by Delta-T) was installed in each treatment at the beginning of the experiment and 12 (3 per treatment) ceramic suction lysimeter cups (by Eijkelkamp) were installed between tomato plants at three depths (10, 30, and 60 cm) in June 2012 for soil solution extraction. Weeds were removed manually.

2.1.5 Inorganic Fertilizer and Organic Amendments Application and Characteristics

The fertilizer dose applied in the IF treatment was 200 kg total available N ha^{-1} in each growing cycle except for the first year, which was 30 kg N ha^{-1}. Two synthetic soluble inorganic fertilizers were applied through irrigation. The first fertilizer was a (30–10–10) consisting of 30% total available N (in the form of 2.8% NO_3, 4.8% NH_4, and 23% urea), 10% available P as P_2O_5, and 10% available K as K_2O plus micronutrients, and was applied at 10-day intervals in the first two stages of plant development (initial and development season). The second fertilizer was a (20–20–20), and it was applied at 10-day intervals in the last two stages of plant development (mid- and late season) and contained 20% N (in the form of 3.9% NO_3, 3.8% NH_4, and 12.3% urea), 20% P, 20% K, and micronutrients. In total, the ratio of the (30–10–10) and (20–20–20) applied in each growing cycle was 1:3.

The MSWC used was provided by the Municipal Company of Solid Waste Disposal (DEDISA) of Chania. The compost was made of residual organic municipal waste together with green waste. The farmyard manure was collected from sheepfolds and included fresh sheep and goat manure. Four oven-dried samples (45°C) of compost and manure were ground to pass through a 2-mm sieve and stored for physicochemical analyses. Total organic C and TN contents were determined after further grinding to pass through a 0.2-mm sieve followed by combustion on an N/C analyzer (Multi N/C 2100 by Analyticjena). Samples were digested in a microwave digester (Multiwave 3000 by Anton Paar), and TN and total phosphorus (TP) in the digests were determined by the micro-Kjeldahl (Bremner and Mulvaney, 1982) and Olsen method (Olsen et al., 1954), respectively. The physical and chemical properties of the soil and the organic amendments used in the experiment are presented in Table 4.

2.1.6 Soil Sampling Schedule, Method, and Analyses

Three core samples were taken (using a soil auger) from each subplot and two soil depths (0–15 and 15–30 cm) on a monthly basis for the first 2 years. In the next 2 years, soil sampling took place three times per year: before and after the incorporation of the organic amendment and at the end of the growing season. The core samples were combined to make a sample of 200 g for each subplot and depth to be analyzed. A subsample was used for moisture content determination by oven drying at 105°C. The

Table 4 Chemical Parameters of the Soil and the Organic Amendments (Compost and Manure)

	Soil	MSWC	M
	Mean (s.e.)	**Mean (s.e.)**	**Mean (s.e.)**
pH (in H_2O)	7.87 (0.1)	7.95 (0.03)	7.60 (0.02)
Electrical conductivity (mS cm^{-1})	0.2 (0.01)	14.9 (0.4)	15.2 (0.3)
TOC (g kg^{-1})	28.2 (2.1)	253 (16.9)	279 (15.6)
TN (g kg^{-1})	2.5 (0.1)	17.4 (3.5)	17.9 (2.6)
TP (g kg^{-1})	0.03 (0.004)	2.6 (0.2)	20.9 (1.1)
TK (g kg^{-1})	0.45 (0.04)	15.5 (0.4)	14.2 (1.3)
C/N (g g^{-1})	11.3	14.6	15.7
C/P (g g^{-1})	997.5	99.3	13.4
C/K (g g^{-1})	62.6	16.3	19.7

M, manure; *MSWC*, municipal solid waste compost; *TK*, total potassium; *TN*, total nitrogen; *TOC*, total organic carbon; *TP*, total phosphorus.

remaining sample was air dried, passed through a 2-mm sieve, and then stored at room temperature in air sealed bags for chemical analyses. In addition, three soil samples of 300 g were taken from the plots of each treatment for aggregate fractionation at the beginning and at the end of each growing season. An extra sampling was carried out in the third year, before tilling, to assess its effect on soil aggregate stability. The field-moist samples were gently broken up by hand in order to remove organic residues, larger roots, and any rock fragments. Then the samples were air dried and sieved through an 8 mm sieve, and the samples were stored at room temperature. Undisturbed soil cores (100 cm^3) were taken for BD analysis.

Soil pH was determined in a 1:2.5 (v/v) soil and distilled water suspension using a combined glass electrode and pH meter calibrated to standard pH values on the hydrogen ion activity scale. A soil to water suspension of 1:5 was used for electrical conductivity (EC) determination. BD was calculated using 100 cm^3 cores. SOC and TN content were measured using an N/C analyzer (Multi N/C 2100 by Analyticjena). Nitrates and ammonium concentrations were determined colorimetrically in soil extracts (in 2 M KCl) using a Lambda 25 spectrometer (Perkin-Elmer) following the cadmium reduction-naphthyl-ethylenediamine and Nessler methods, respectively. This is a wet extraction method that estimates the availability of nutrients to plants, and the results were used for mathematical model calibration. The available P was determined by the Olsen method (Olsen et al., 1954) after extraction with 0.5 M sodium bicarbonate (pH 8.5).

WSA determination was based on a physical fractionation scheme described by Stamati et al. (2013b). Briefly, the samples were separated into five WSA fractions according to the procedure presented by Elliott (1986): (i) large macroaggregates (>2000 μm), (ii) medium macroaggregates (1000–2000 μm), (iii) small macroaggregates (250–1000 μm), (iv) microaggregates (53–250 μm), and (v) silt–clay-sized microaggregates and mineral particles (<53 μm). The sand content of each WSA fraction was determined after dispersing a subsample or each fraction in sodium hexametaphosphate (0.5%) for 18 h on a rotary shaker at 190 rpm (Stamati et al., 2013b). The percentage ratio of aggregates was measured after correction for sand content. C and N content was measured in all WSA fractions.

2.1.7 Soil Solution Sampling and Analyses

Soil pore water sampling from suction lysimeters was carried out at the same time as soil sampling, on a monthly basis. Soil pore water pH and EC were measured in the field using portable pH/EC probes. The pore water dissolved organic carbon (DOC) and dissolved nitrogen content were measured using the N/C analyzer (Multi N/C 2100 by Analyticjena). Nitrate (NO_3^-), ammonium (NH_4^+), phosphate (PO_4^{-3}), and sulfate (SO_4^{-2}) concentrations were determined using a spectrophotometer HACH 2800DR following the cadmium reduction, Nessler, ascorbic acid, and SulfaVer 4 methods, respectively. Concentrations of Na^+, K^+, Ca^{2+}, Mg^{2+}, and total Al were determined by ICP-MS (Agilent 7500-CX).

2.1.8 Plant Tissue Sampling and Analyses

Tomato leaves were sampled three times within the growing period, while mature tomato fruits were regularly collected throughout the growing period to assess yield. At the end of each growing period, 12 tomato plants from each treatment (four plants per subplot) were removed and weighed for fresh weight. Subsamples from fruit, shoot, and leaves were collected, oven-dried at 40°C, ground to a fine powder, and stored for further analyses.

Plant C was analyzed in plant tissue using the N/C analyzer (Multi N/C 2100 by Analyticjena). Total N contents within each plant part were assessed with the micro-Kjeldahl method (Bremner and Mulvaney, 1982) and total P contents using the ammonium molybdate–ascorbic acid method. Plant tissue analysis for other nutrient elements (Ca, Mg, Fe, K) was conducted after microwave-assisted (Anton Paar Multiwave 3000) acid digestion, using ICP-MS (Agilent Technologies 7500CX ICP-MS) analysis of the digestate solution.

2.2 1D-ICZ Model Description and Simulations

2.2.1 Model Description

The 1D-ICZ model (see Chapter "Integrated Critical Zone Model (1D-ICZ): A Tool for Dynamic Simulation of Soil Functions and Soil Structure" by Giannakis et al.). links the development and loss of soil structure to nutrient dynamics, plant growth, and soil hydrology. It comprises four main submodels: (i) flow and transport (HYDRUS-1D) (Šimůnek et al., 2009); (ii) bioturbation, chemical speciation equilibrium, and mineral dissolution kinetics (SAFE) (Wallman et al., 2005); (iii) C/N/P/Ca/Mg dynamics and aggregate formation, CAST (Stamati et al., 2013a); and (iv) vegetation dynamics (PROSUM). In the CAST model, the C pools of RothC model (decomposable plant material (*DPM*), resistant plant material (*RPM*), *BIO*, and humified organic matter (*HUM*)) have been adapted and have extended by additional pools of functional organism (FAUNA) to reflect the biomass of the soil fauna that subsists on BIO. Three types of WSA are simulated: AC1 (small WSA, <53 μm) consisting of BIO and HUM; AC2 (water-stable microaggregates, 53–250 μm) consisting of BIO, HUM, and fine DPM (DPM_f) and RPM (RPM_f) pools; and AC3 aggregate type (water-stable macroaggregates, >250 μm) consisting of BIO, HUM, DPM_f, RPM_f, and the coarser fractions in the DPM (DPM_c) and RPM (RPM_c) pools. The C within each aggregate type decomposes by a first-order kinetic mass action law with each aggregate type having its own characteristic rate constant. The C is mineralized with mass fluxes produced CO_2, BIO (living biomass), and HUM (humified material). The value of the decomposition rate constant (k) for each pool is modified by the product of three correction factors (a, b, c with values in the range of 0–1.0) for the major factors determining microbial activity: temperature, water deficit, and cover crops, as in the RothC model. The 1D-ICZ model implements a version of the CAST model which uses the soil temperature of each layer, while water deficit is calculated as the available soil moisture minus the field capacity of the layer and soil cover is provided by PROSUM as a function of plant biomass. The plant dynamics subroutine (PROSUM) calculates net primary production (NPP) as a function of potential production, primary limitations, nutrient limitations, and physiological growth reduction factors. During litterfall and harvest events, PROSUM transfers C and nutrients (i.e., in leaf, wood, and root) into the CAST litter pools.

The model quantifies the dynamics of C, N, and P sequestration in soils in relation to soil structure and protection of organic matter within aggregates; above- and belowground C stocks including highly simplified functional biodiversity in addition to vegetation productivity, mycorrhizal fungi, the combined activity of decomposer bacteria and fungi, and BIO consumers; and water vertical flow and reactive transport in the soil profile.

2.2.2 Simulation Strategy and Model Setup

The operations that took place during the 4-year experimental period (Table 5) were used to set up the 1D-ICZ model. These included applications of fertilizer and manures, tillage, tomato transplant, incorporation of weeds, and tomato fruit harvesting. Moreover, the tomato and fallow seasons are also depicted. Both tomato plants and the weeds that grew during the winter and spring fallow period were classified for simulation purposes as mycorrhizal herbaceous plants. Tillage was applied before the transplanting of the tomato seedlings (months 8, 17, 29, 41) to incorporate the organic amendments to the soil and prepare it for planting.

Table 5 Operation Implemented and Simulation Structure

Year	2011 (1st Season)											
Month	J	F	M	A	M	Jn	Jl	A	S	O	N	D
Month No.	1	2	3	4	5	6	7	8	9	10	11	12
Plant type	W	W	W	W	W	W	W	T	T	T	T	W
Water input	R	R	R	R	R	R	R	I	R/I	R/I	R	R
Tillage								×				
Fertilization[a]								×	×	×		
Carbon addition[b]								×				
Weeds incorporation							×					
Harvest											×	
Tomato plant removal											×	

Year	2012 (2nd Season)											
Month	J	F	M	A	M	Jn	Jl	A	S	O	N	D
Month No.	13	14	15	16	17	18	19	20	21	22	23	24
Plant type	W	W	W	W	T	T	T	T	T	W	W	W
Water input	R	R	R	R	I	I	I	I	R/I	R/I	R	R
Tillage					×							
Fertilization[a]					×	×	×	×	×			
Carbon addition[b]					×							
Weeds incorporation				×								
Harvest								×	×	×		
Tomato plant removal										×		

Year	2013 (3rd Season)											
Month	J	F	M	A	M	Jn	Jl	A	S	O	N	D
Month No.	25	26	27	28	29	30	31	32	33	34	35	36
Plant type	W	W	W	W	T	T	T	T	T	W	W	W
Water input	R	R	R	R	I	I	I	I	R/I	R	R	R
Tillage					×							
Fertilization[a]					×	×	×	×	×			
Carbon addition[b]					×							
Weeds incorporation				×								
Harvest							×	×	×			
Tomato plant removal									×			

Year	2014 (4th Season)											
Month	J	F	M	A	M	Jn	Jl	A	S	O	N	D
Month No.	37	38	39	40	41	42	43	44	45	47	47	48
Plant type	W	W	W	W	T	T	T	T	T	W	W	W
Water input	R	R	R	R	I	I	I	I	R/I	R	R	R
Tillage					×							
Fertilization[a]					×	×	×	×				
Carbon addition[b]					×							
Weeds incorporation				×								
Harvest							×	×				
Tomato plant removal								×				

[a]Only in IF treatment.
[b]In MSWC, M, and MSWC + M treatments.
I, irrigation water; *R*, rainwater; *R/I*, rainwater with irrigation; *T*, tomatoes; *W*, weeds.

2.2.2.1 Water, Heat, and Solutes Transport (HYDRUS-1D) Setup

The HYDRUS-1D module within 1D-ICZ uses information on the properties of the soil profile to simulate water and heat flow and solutes transport. The soil profile depth was set to 1 m including three different soil horizons characterized by different water and heat transport properties (Table 6). Water flow is described by the van Genuchten–Mualem model (van Genuchten, 1980) equations. The unsaturated soil hydraulic properties for the surface layer were derived from Carsel and Parrish (1988), and these values represent approximate characteristic values for a silty–loam texture soil. Field observations of soil water content were used for fitting the hydraulic parameters of the lower soil layers.

The soil surface boundary condition (upper boundary condition), which involves the soil–air interface, was set to include a surface layer which permits water to build up on the surface. The height of the surface water layer increases due to precipitation and reduces because of infiltration and evaporation. The definition of the upper boundary condition includes the average temperature of the soil surface, which was assumed to be the identical to the air temperature, the potential evaporation rate (m d^{-1}), and the precipitation/irrigation rate (m d^{-1}). Fig. 2 presents the distribution of temperature, evaporation, and water inputs at the upper boundary during the 4 years of simulation. The lower boundary was set to the free drainage boundary condition.

Heat transport in the soil profile is described by the convection–diffusion equation (Chung and Horton, 1987), Eq. (1):

$$\lambda(\theta) = b_1 + b_2\theta + b_3\theta^{0.5} \tag{1}$$

where $\lambda(\theta)$ is the coefficient of soil apparent thermal conductivity, and the b_1, b_2, and b_3 parameters were chosen to characteristic values that represent a loam textured soil (Chung and Horton, 1987). The same soil texture was used to initialize the remaining parameters for the convection–diffusion equation (Table 6). According to the measured data, the volumetric fractions of the solid phase and the organic matter content were set to 0.44 and 0.0425, respectively, for the first soil layer, 0.45 and 0.03 for the second, and 0.50 and 0.01 for the third (Table 6). Field data were used for the initialization of soil solution characteristics. However, suction lysimeters were installed only after the first year, so for the 2011 simulation measured values from December 2012 were used for initialization. This assumed that in December, the soil pore water chemistry had reached equilibrium, and the data could at least be used for the initialization of the approximate

Table 6 Water and Heat Transport Parameterization

Soil Profile Depth (cm)	Soil Material	Water Content	θ_r	θ_s	α	n	K_s	l	Solid Phase	OM	b_1	b_2	b_3	c_n	c_o	c_w
		Water Transport Parameterization									Heat Transport Parameterization					
0–10 10–20 20–30	1	0.2	0.067	0.45	2	1.41	0.108	0.5	0.55	0.042	1.57×10^{14}	2.53×10^{14}	9.89×10^{15}	1.43×10^{16}	1.87×10^{16}	3.12×10^{16}
30–40 40–50 50–60	2	0.3	0.0731	0.4732	1.15	1.43	0.075	0.5	0.5268	0.03	1.57×10^{14}	2.53×10^{14}	9.89×10^{14}	1.43×10^{16}	1.87×10^{16}	3.12×10^{16}
60–70 70–80 80–90 90–100	3	0.4	0.0731	0.4732	1.15	1.43	0.09	0.5	0.5268	0.01	1.57×10^{14}	2.53×10^{14}	9.89×10^{14}	1.43×10^{16}	1.87×10^{16}	3.12×10^{16}

Water content, initial water content (m m^{-1}); θ_r, residual soil water content(m m^{-1}); θ_s, saturated soil water content (m m^{-1}); α, shape parameter of the soil water retention function (m^{-1}); n, shape parameter of the soil water retention function; K_s, saturated hydraulic conductivity (m d^{-1}); l, tortuosity parameter of the conductivity function.

Solid phase, volume fraction of solid phase; *OM*, volume fraction of organic matter; $b_{1,2,3}$, coefficients for the thermal conductivity function (Wm^{-1} K^{-1}); c_n, volumetric heat capacity of the solid phase (Jm^{-3} K^{-1}); c_o, volumetric heat capacity of organic matter (Jm^{-3} K^{-1}); c_w, volumetric heat capacity of the liquid phase (Jm^{-3} K^{-1}).

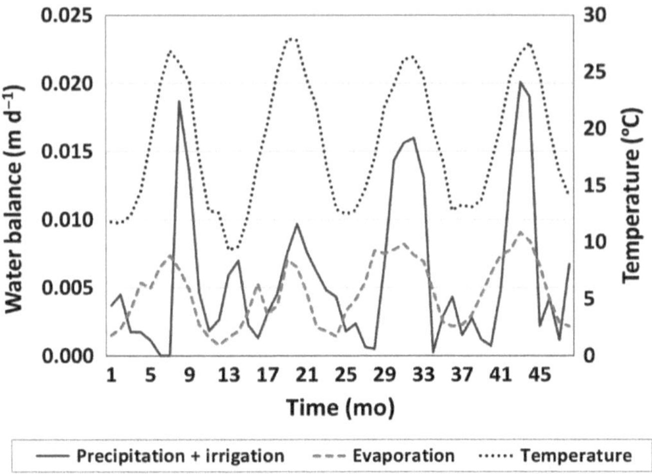

Fig. 2 Soil surface boundary conditions: water input, evaporation, and temperature.

magnitude of solutes concentrations. Table 7 presents the initial values of the chemical species in soil solution.

The ionic composition of precipitation and irrigation water is shown in Table 2. Values represent the averages of measured concentration values. The irrigation water was pumped from the ground water and its chemistry was essentially constant during the year, while rainwater chemical species concentration varied through the year. A value of 80% of water input to the soil was assumed to be available for plant uptake.

2.2.2.2 Plant Dynamics (PROSUM) Setup

The 1D-ICZ model handles vegetation development using the PROSUM subroutine (see Chapter "Integrated Critical Zone Model (1D-ICZ): A Tool for Dynamic Simulation of Soil Functions and Soil Structure" by Giannakis et al.). This subroutine simulates plant growth, nutrient uptake, and litterfall, on the basis of fixed parameters; user-defined time-variable parameters determining mean daytime photosynthetically active radiation (PAR) flux, atmospheric CO_2 concentration, and events such as tillage, harvest, and litterfall; and dynamic inputs from other submodels, e.g., water and nutrient availability. It was assumed that the atmospheric CO_2 concentration was 395 $\mu L\ L^{-1}$. For the estimation of PAR flux, it was assumed that PAR is approximately 47% of solar radiation (Blackburn and Proctor, 1983).

Both tomatoes and fallow/weed species were defined as mycorrhizal herbs. However, the allocation of fixed C to growth of different plant parts

Table 7 Initial Concentrations of the Soil Solution Species

Species	Concentration (mol L^{-1})	Species	Concentration (mol L^{-1})
Ca^{2+}	3.92E−03	$Al(OH)_4^-$	2.53E−10
Mg^{2+}	7.90E−04	$AlSO_4^+$	8.96E−19
Na^+	9.65E−05	$Al(SO_4)_2^-$	6.09E−20
K^+	7.60E−06	H_2A^+	6.29E−14
H^+	1.41E−08	$H_2SiO_4^{2-}$	4.80E−17
Al^{3+}	1.00E−18	H_3A^{2+}	1.77E−19
$CO_2(l)$	5.16E−06	$H_3SiO_4^-$	2.46E−12
SO_4^{-2}	8.75E−04	HA	1.41E−10
HCO_3^-	1.84E−04	**CO_3^{2-}**	7.32E−07
OH^-	7.09E−07	H_4SiO_4	1.00E−10
F^-	1.00E−10	LMWN	1.00E−10
A^-	1.00E−10	**NO_3^-**	1.26E−04
AlA^{2+}	4.88E−22	**NH_3**	6.61E−07
AlF^{2+}	1.05E−21	**NH_4^+**	1.66E−05
AlF_2^+	5.75E−26	X–Ca	1.00E−10
AlF_3	1.07E−31	X–Mg	1.00E−10
AlF_4^-	5.37E−39	X–K	1.00E−10
AlF_5^{2-}	8.32E−48	X–Na	1.00E−10
$AlH_3SiO_4^{2+}$	1.16E−21	X–Al	1.00E−10
$AlHA^{3+}$	8.46E−18	SO–H	1.00E−10
$AlOH^{2+}$	7.26E−16	SO	2.18E−04
$Al(OH)_2^+$	3.73E−13	SO–H_2	1.00E−10
$Al(OH)_3$	2.25E−10	SO–H_2SO_4	3.31E−32

Species in *bold* denote measured values. Concentrations of other species were obtained by chemical equilibration with solid or gas phases.

was adjusted to distinguish the two types of plants (Chapin et al., 2012). Tomato plants are known to have relatively high shoot/root ratios for above- and belowground biomass production, so the proportional allocation to shoot tissue was increased slightly relative to fallow/weed species (Table 8).

Table 8 Proportional Allocation of Net Primary Production to Herbaceous Shoot Growth, Root Growth, Root Exudates, and Mycorrhizae

	Tomato	Weed
Photosynthate allocation to leaf[a]	0.59	0.5
Photosynthate allocation to root growth	0.2	0.26
Photosynthate allocation to mycorrhizae	0.1	0.12
Photosynthate allocation to root exudates	0.1	0.11

[a]"Leaf" is used to denote all nonwoody shoot material, i.e., herbaceous stems, leaves, and fruits.

Harvest events completely remove user-defined proportions of plant parts and also transfer proportions into litter pools. Tillage transfers all aboveground biomass and most root and mycorrhizal biomass into litter pools. In this experiment, fruit harvesting occurred in months November, 2011, October, 2012, October, 2013, and September, 2014 for tomatoes and July, 2011, April, 2012, April, 2013, and April, 2014 for weeds. Field measurements demonstrated out that 40% of the aboveground tomato plant tissue (after harvesting fruits) was returned to the soil through tillage.

The NPP rate in the plant dynamics simulation subroutine is calculated as a function of potential production, primary limitations (water, temperature, light, CO_2), nutrient limitations, and reductions attributed to the pests, diseases, or from grazing or other management practices. The maximum of NPP was set at 26 mol C m^{-2} mo^{-1}. This value represents the maximum aboveground net primary production (ANPP) rate which was observed in the third growing season for the manure treatment (M), after having added the belowground net production rate. Since there were no data available for belowground net production, 50% of the ANPP was assumed. In addition, it was assumed that no herbivores (domestic animals and wild vertebrates and invertebrates) were present. Rabbinge et al. (1989) reported that the maximum NPP estimate for fertilized, irrigated arable crops is between 150 and 350 kg total dry matter ha^{-1} d^{-1} or 18 and 42 mol total C m^{-2} mo^{-1}. Using the belowground biomass assumption of 50%, the tomato max NPP was 216 kg total dry matter ha^{-1} d^{-1} or 26 mol C m^{-2} mo^{-1}. The water use efficiency was set at 625 mol C m^{-3} water based on typical literature values for tomato plants. The values for parameters that were constant for the simulations were set at the beginning of the simulation and are summarized in Table 9.

2.2.2.3 Organic Carbon, Aggregation, and Soil Structure Setup

The data requirements for initialization of this subroutine included the WSA mass fraction for each of the three- size classifications (silt–clay—AC1,

Table 9 Fixed Parameters for PROSUM

Model Parameter	Units	Parameter Description	Value
SpecificRootLength_mpMC	m mol^{-1} root C	Root length per root mass	50
SpecificMycorLength_mpMC	m mol^{-1} mycor C	Length of hyphae per mass of mycorrhizal C	1,400,000
MaxCfix	mol m^{-2} mo^{-1}	Maximum NPP	26
TempRisePos	°C	Low temperature at which productivity is limited by 50%	5.5
TempRiseWidth	°C	Gradualness of the increase in productivity with warmer temperature	4.13
TempFallPos	°C	High temperature at which productivity is limited by 50%	35.25
TempFallWidth	°C	Gradualness of the decrease in productivity with high temperature	4.45
PARexponent	—	Steepness of saturation curve describing light limitation of productivity	−0.00180303
WaterUE_MCpm3	mol C m^{-3} water	Water use efficiency	625
CO2Factor	—	Magnitude of CO_2 effect on productivity	0.424
CO2Exponent	—	Steepness of CO_2 effect on productivity	0.0269
ThresholdPlantCForCover_Mpm2	mol C m^{-2}	Plant total biomass at which the soil is considered protected against erosion	4.2
RootDepthBeta	—	Gradualness of decline in root length density with soil depth	0.000000298
InitLeafC_Mpm2	mol C m^{-2}	Initial stock of leaf (including all nonwoody shoot tissue) biomass	10
InitWoodC_Mpm2	mol C m^{-2}	Initial stock of woody biomass	0
InitRootC_Mpm2	mol C m^{-2}	Initial stock of root biomass	8
InitMycorC_Mpm2	mol C m^{-2}	Initial stock of mycorrhizal biomass	4

microaggregates—AC2, macroaggregates—AC3), the C content of each fraction as well as the silt–clay content of each fraction. The DPM-to-RPM ratio of litterfall was set to 1.8 implying more labile carbon pools in the OM composition of C input. The soil profile was discretized into three layers: 0–20, 20–40, and >40 cm depth according to typical discretization of the 1D-ICZ model. Table 10 presents the SOC pools within the WSA at each soil layer, and Table 11 presents the initial aggregate fractionation and silt and sand content of each aggregate.

2.2.2.4 Model Calibration Strategy

Model calibration was conducted using an iterative methodology by calibrating first the plant dynamics, then soil C and N mass content as well as soil structure, followed by solute concentrations in soil pore water. In the first step, limitations for plant growth (water and nutrients availability) had to be adjusted. The ratios of C to N, P, K, Ca, and Mg input into the soil from organic matter mineralization were adjusted to provide, complementary with input through irrigation and rainwater, the essential nutrients for plant growth. Moreover, parameters determining N and P cycling were properly adjusted, also contributing to nutrient supply. These parameters included rates of P mineralization and rates of organic N breakdown to soluble organic and inorganic N forms such as the rate of transformation of insoluble organic N to dissolved organic nitrogen (DON), DON transformation rate to low-molecular-weight nitrogen (LMWN), the mineralization rate of LMWN to ammonium, and the nitrification and the denitrification transformation rates. The conversion from insoluble organic nitrogen to DON is the initial and rate-limiting step in mineralization (Chapin et al., 2012) and is calculated by manipulating the linear partitioning coefficients ($k_{partition\_AC1}$, $k_{partition\_AC1withinAC2}$, $k_{partition\_AC1withinAC2}$, $k_{partition\_AC1withinAC3}$, and $k_{partition\_AC1withinAC2inAC3}$) that are used to determine the DOC and the carbon to nitrogen (C:N) ratios in the BIO and HUM pools. The C:N ratios in these fractions were set to 10:1 based on data from the available literature (Chapin et al., 2012) and the field measurements.

Following the calculation of the plant growth, the potential for nutrient uptake can be estimated. The model gives the ability to set the maximum nutrient uptake per plant biomass produced. These values which are defined by the "lower limit of C/element (N, P, K, Mg, Ca) ratio" were set to values that were increased to reflect the amounts of nutrients which are accumulated during plant growth. These values correspond to plants like tomatoes

Table 10 SOC Pools Distribution Within the 3 WSA Types at Each Soil Profile Compartment

	AC3										AC2				AC1	
	AC3				AC1 Within AC3		AC2 Within AC3									
Soil Depth (cm)	RPM_c	DPM_c	RPM_f	DPM_f	BIO	HUM	RPM_f	DPM_f	BIO	HUM	RPM_f	DPM_f	BIO	HUM	BIO	HUM
0–20	10.46	1.85	0.00	0.00	0.62	11.70	1.54	1.54	0.15	2.92	3.00	3.00	0.70	13.32	0.20	5.21
20–40	9.68	1.71	0.00	0.00	0.57	10.82	1.42	1.42	0.14	2.70	2.78	2.78	0.65	12.32	0.19	4.82
>40	6.61	1.17	0.00	0.00	0.39	7.39	0.97	0.97	0.10	1.85	1.90	1.90	0.44	8.42	0.13	3.29

All values are in t C ha^{-1}. *AC1*, silt–clay-sized WSA; *AC2*, micro-WSA; *AC3*, macro-WSA; *BIO*, microbial biomass, fragmentation level of plant material; *c*, coarse; *DPM*, decomposable plant material; *f*, fine; *HUM*, humified organic matter; *RPM*, resistant plant material.

Table 11 Initial Distribution of the Silt–Clay and Sand Mass Into the WSA Types at Each Soil Profile Compartment

	Initial Distribution (%) of the Silt–Clay Mass in the Aggregate Fractions				Fraction of Sand Mass in the Aggregate Fractions	
Soil Depth (cm)	AC1	AC2	AC1 Within AC3	AC2 Within AC3	Fraction of Sand Mass in AC2	Fraction of Sand Mass in AC3
0–20	14.87	49.18	28.77	7.19	0.63	0.37
20–40	14.87	49.18	28.77	7.19	0.63	0.37
>40	14.87	49.18	28.77	7.19	0.63	0.37

where fruit biomass accounts for a high percentage of the total aboveground biomass production. Given the calibrated plant biomass and therefore the respective litterfall, the total SOC and C stocks in the WSA size fractions calibration could follow. To achieve this, an adjustment of the composition of C input had to be made. The calibration of OC distribution within the WSA fractions was achieved by regulating factors of macroaggregate formation and destruction. These factors are related to the components of macroaggregates (DPMc, RPMc, AC1, AC2) and their proportional contribution to total macroaggregate mass; the AC3 formation rate as well as destruction rate; and rate of decomposition of carbon within the AC3 WSA fraction ($DPM_{withinAC3}$).

Following the calibration of C content within WSA classes, the changes in WSA fractions during the 4-year simulation could be completed by fitting the rate constants for the silt–clay mass transfer between the WSA fractions. The soil N content and its distribution within WSA fractions were specified by adjusting the C-to-N ratios in the C pools (DPM, RPM, BIO, HUM, etc.). With the completion of the above calibration steps, the nutrients that are simulated by the CAST subroutine become available for the PROSUM vegetation dynamics subroutine, allowing the estimation of the produced biomass, which in turn affects the incorporation of organic carbon in soil. The calibration process continued until the calculated state variable values simulated the best possible fit across all of the variables, to the experimental data.

The above procedure was iteratively applied in order to achieve an acceptable fit to field observations. The root mean squared error (RMSE) and coefficients of variation (CVs) were used to test the goodness of fit in terms of plant production, N, P, K plant uptake; WSA mass distribution by size classification; and the C and N sequestration rates as well as their mass abundance in each WSA fraction. The CV is calculated as the ratio of the RMSE to the mean of the observed values and is expressed as a percentage.

3. RESULTS

3.1 Plant Growth and ANPP

The simulation of aboveground plant C and biomass production for the 4-year experiment is presented in Figs. 3 and 4. The simulation results

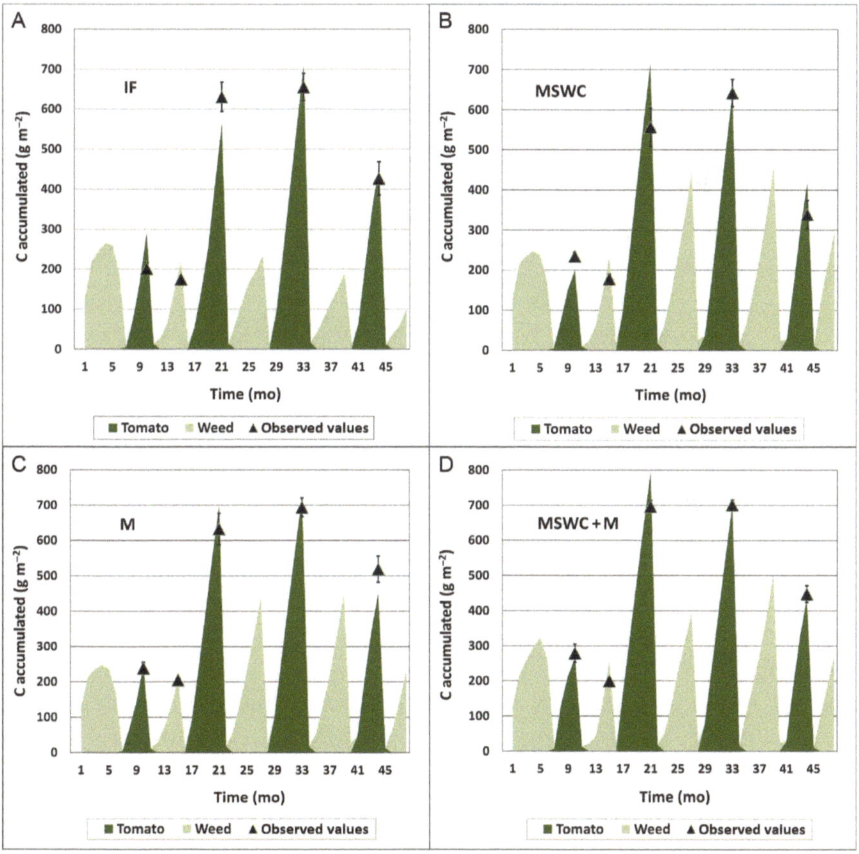

Fig. 3 Simulation results of C accumulated in aboveground plant biomass (tomato and weed) for each treatment over the 4-year period. Point values indicate observed values used for calibration. (A) inorganic fertilization (IF) treatment, (B) municipal solid waste compost addition (MSWC) treatment, (C) manure addition (M) treatment and (D) compost–manure mixture (MSWC+M) treatment. The simulation day 0 refers to January 1, 2011.

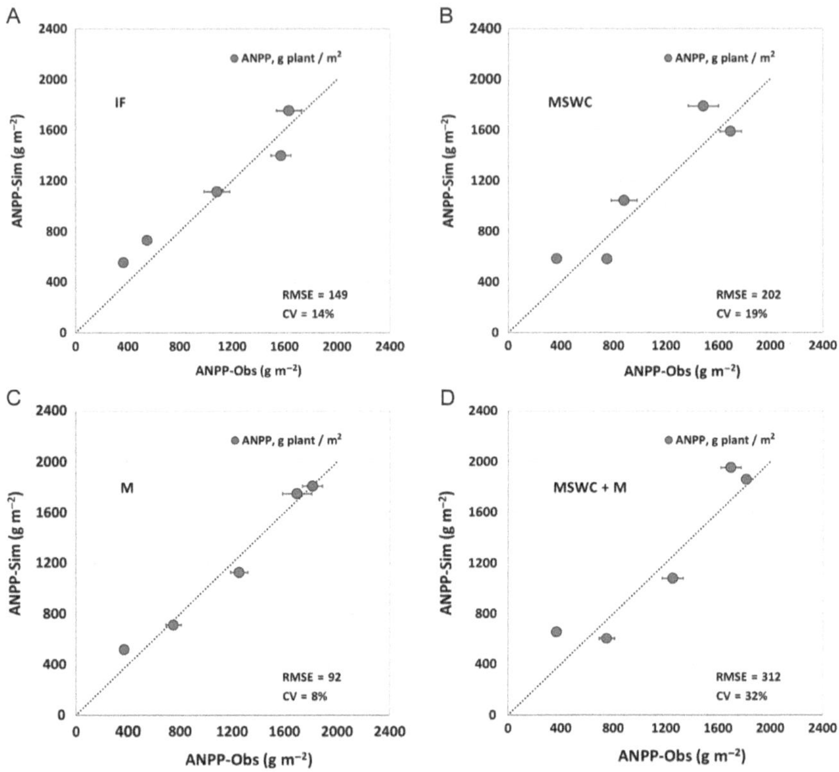

Fig. 4 Comparison between observed and simulated values of aboveground net primary production (g plant m^{-2}). (A) inorganic fertilization (IF) treatment, (B) municipal solid waste compost addition (MSWC) treatment, (C) manure addition (M) treatment and (D) compost–manure mixture (MSWC+M) treatment. Root Mean Squared Error (RMSE) and Variation Coefficient (CV) are presented.

are plotted against observed values of C accumulated in tomato and weeds tissues (Fig. 3) and plant biomass production (Fig. 4). The availability of carbon content data for weeds is limited only to data for the first year. Plant biomass production exhibited variability among different treatments. The tomato shoot biomass production was lower in the first growing season (2011), and this was attributed to the shorter cultivation period (August to November) as well as due to poor condition of this season for optimal tomato cultivation. The next two cultivation cycles (2012 and 2013) showed the highest plant biomass production, while in the last, the production decreased and this was attributed to early and sudden plant collapse due to a microbial pathogen outbreak. As a result, the growing season was shorter by 26 days in average. All these variations were successfully simulated by the 1D-ICZ model.

The most productive year was 2013 for all treatments, whereas the highest biomass production among them was observed in the "M" and the "MSWC + M" treatments. Plants treated with 50 t ha^{-1} MSWC showed lower C accumulated in the aboveground biomass (Fig. 3B) to that produced by plants treated with manure (M) (Fig. 3C) or with the compost–manure mixture (MSWC + M) (Fig. 3D), but similar quantities to the IF treatment (Fig. 4A), thus suggesting that organic amendments can have the potential of providing plants with sufficient quantities of nutrients to sustain plant production.

Overall, the 1D-ICZ model provided very good predictions for the ANPP which is related to plant biomass production (Fig. 4). The goodness of fit estimated using the RMSE and the CV (Fig. 4) ranged between 92 and 312 g m^{-2} and 8% and 32%, respectively.

Fig. 5 presents the simulation of N uptake against the measured N uptake in tomato and weeds tissues. The other nutrients for plant uptake (P, K) were also simulated and are presented in Fig. 6. The highest nutrient accumulation in tomato plants occurred in the second and third growing season following the trend observed for biomass production. The RMSE for the aboveground plant N, P, and K uptake ranged from 5 to 8 g N m^{-2}, 12 to 27 g P m^{-2}, and 12 to 45 g K m^{-2}, respectively. The CV for plant N, P, and K accumulation ranged from 15% to 33%, 44% to 80%, and 23% to 66%, respectively.

When comparing the OA treatments in terms of nutrients uptake by the plant, the "MSWC" treatment exhibited the best performance, where plant nutrients accounted for up to 34%, 93%, and 63% of the applied N, P, and K, respectively, reducing in this way the potential of dissolved N leaching and its relative effects on groundwater. Addition of MSWC to the soil together with manure reduced the plant uptake of P and K to 52% and 59%, respectively, but stimulated the uptake of N to 41%. In accordance, incorporation of manure ("M" treatment) only further reduced P and K uptake (27% and 49%, respectively), while the uptake of N (42%) was maintained at levels similar to that of the "MSWC + M" treatment.

In general, the 1D-ICZ model was able to follow the temporal trends and simulate the accumulation of N in both the tomato plants and weeds to a satisfactory degree compared to the field data. The model underestimated the plant K content of the first year due to limitations of water availability and simulated successfully the plant K content for the following 3 years. Similarly, the model underestimated the plant and weed P contents in the second year but provided good predictions for the following years.

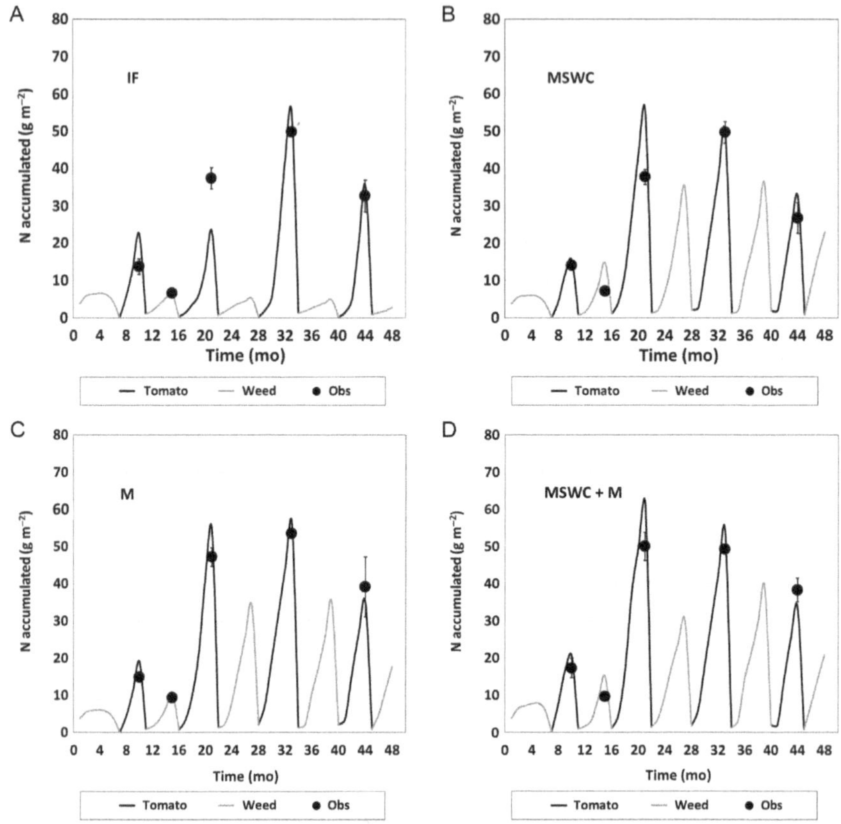

Fig. 5 Simulation results of accumulated N in aboveground plant biomass (tomato and weed) for each treatment over the 4-year period. Point values indicate observed values used for calibration. (A) inorganic fertilization (IF) treatment, (B) municipal solid waste compost addition (MSWC) treatment, (C) manure addition (M) treatment and (D) compost–manure mixture (MSWC+M) treatment.

3.2 SOC and Aggregate-Associated SOC

The soil application of organic amendments enhanced C sequestration. After 4 years of annual addition of soil organic amendments the total carbon introduced in these treatments was 47.8 t C ha^{-1} for "MSWC," 54.4 t C ha^{-1} for "M," and 49.6 t C ha^{-1} for "MSWC+M." Although the total C input was different among the OA treated soils, the results indicated that there was no significant difference in SOC at the end of the experiment. The C sequestered after the 4-year application of OA was 30.4 t C ha^{-1} for "MSWC+M," 28.9 t C ha^{-1} for "M," and 26.0 t C ha^{-1} "MSWC." Regarding the "IF" treatment, there was no evidence of SOC sequestration

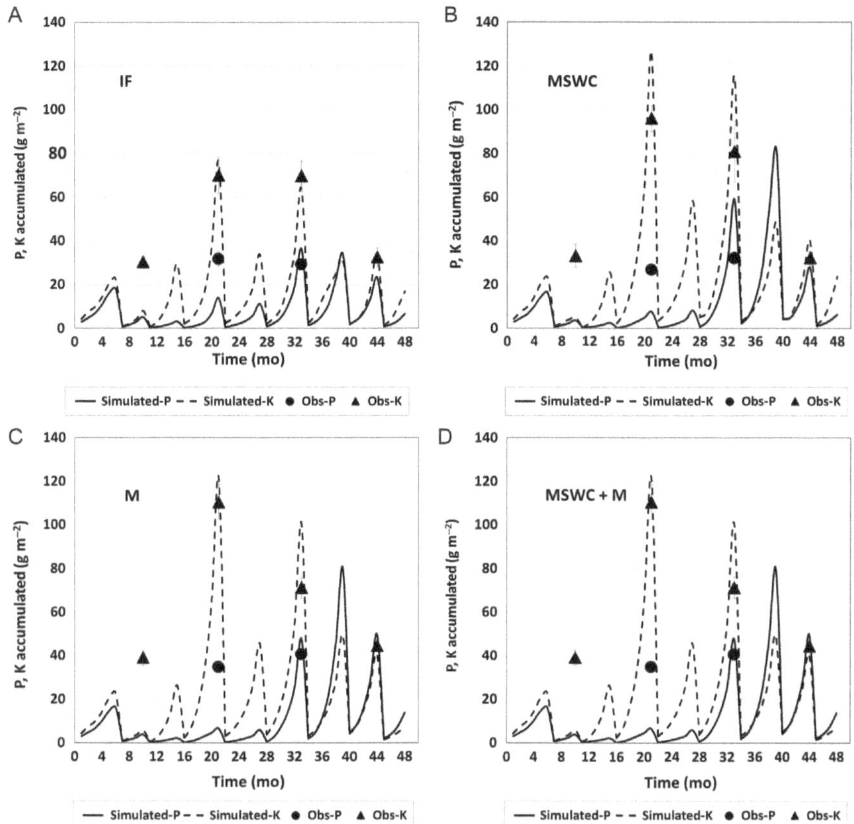

Fig. 6 Simulation results of accumulated P and K in aboveground plant biomass for each treatment over the 4-year period. Point values indicate observed values used for calibration. (A) inorganic fertilization (IF) treatment, (B) municipal solid waste compost addition (MSWC) treatment, (C) manure addition (M) treatment and (D) compost–manure mixture (MSWC+M) treatment.

despite the fact that tomato plant and weed residues were incorporated every year in the soil. The plant biomass incorporated to the soil through residues was calculated to be 40% of the annual biomass production.

Results of the SOC distribution within the WSA fractions revealed that OC was preferentially occluded in the macroaggregate fraction. At the end of the experiment, the OC stock in AC3 was highest for "MSWC" (61.8 t C ha^{-1}), followed by "M" (57.2 t C ha^{-1}) and "MSWC+M" (52.4 t C ha^{-1}), showing a twofold (on average) increase from the initial values. On the other hand, the "IF" treatment did not affect the carbon content of WSA fractions. The SOC increased with increasing aggregate

Table 12 Four-Year Mean Values of SOC Concentration and C/N Ratio in Aggregate Fractions (AC1, AC2, AC3)

	SOC (g kg^{-1})			*C/N* Ratio		
	AC3	**AC2**	**AC1**	**AC3**	**AC2**	**AC1**
IF	15.9 (0.5)	9.0 (0.4)	2.0 (0.2)	11.1 (0.7)	9.4 (0.7)	9.8 (0.4)
MSWC	24.1(1.4)	10.1 (0.7)	2.3 (0.3)	12.0 (0.7)	9.0 (0.6)	9.3 (0.5)
M	22.4 (1.3)	11.5 (1.0)	2.1 (0.2)	13.3 (0.8)	10.3 (0.8)	9.7 (0.4)
MSWC+M	20.8 (1.4)	11.9 (0.8)	2.3 (0.2)	13.0 (1.2)	8.5 (0.5)	9.2 (0.4)

Values in brackets are the standard error of the mean values.

class and the C/N ratio was higher in AC3 compared to AC2 in all treatments (Table 12), suggesting a higher content of fresh and labile C in macroaggregates (AC3).

The model successfully simulated the evolution of SOC stocks in each aggregate fraction over the 4-year period (Fig. 7). The RMSE of SOC stocks in all WSA fractions within treatments ranged between 2.5 and 7.6 t C ha^{-1} which can be considered an acceptable variation. The simulated total SOC sequestration at the end of the experiment was lower by 2 t C ha^{-1} in "IF," 4.6 t C ha^{-1} in "MSWC," 9.9 t C ha^{-1} in "M," and 9.4 t C ha^{-1} in "MSWC+M" in comparison with observed data, but falling within the standard error of the measurements (Fig. 7) denoting acceptable calibration. In the "IF" treatment simulation, in which the only carbon source added to the soil were the plant residues, the sequestration of OC was in accordance with the field observations. The increase in SOC stocks in the other three treatments was also successfully simulated (Fig. 7).

3.3 Soil Nitrogen Stock

Soil total N showed a different pattern to that of the SOC in all treatments (Fig. 8). While SOC remained constant during the study period in the "IF" treatment, N showed a slight increase, reaching its stock of 6.1 t N ha^{-1} at the end of the experiment. In the carbon amended soils, the evolution of N stocks was not linear as was SOC, where a sharp increase was observed in the third year. The highest N sequestration was observed in the "MSWC" treatment (6.4 t N ha^{-1}), followed by "MSWC+M" (5.8 t N ha^{-1}), "M" (4.9 t N ha^{-1}), and "IF" (2.7 t N ha^{-1}). The results point out the beneficial effects of OA on the N status of the soil. Regarding the distribution of N between aggregate fractions, macroaggregates had a greater N content as compared to microaggregates in all the treatments

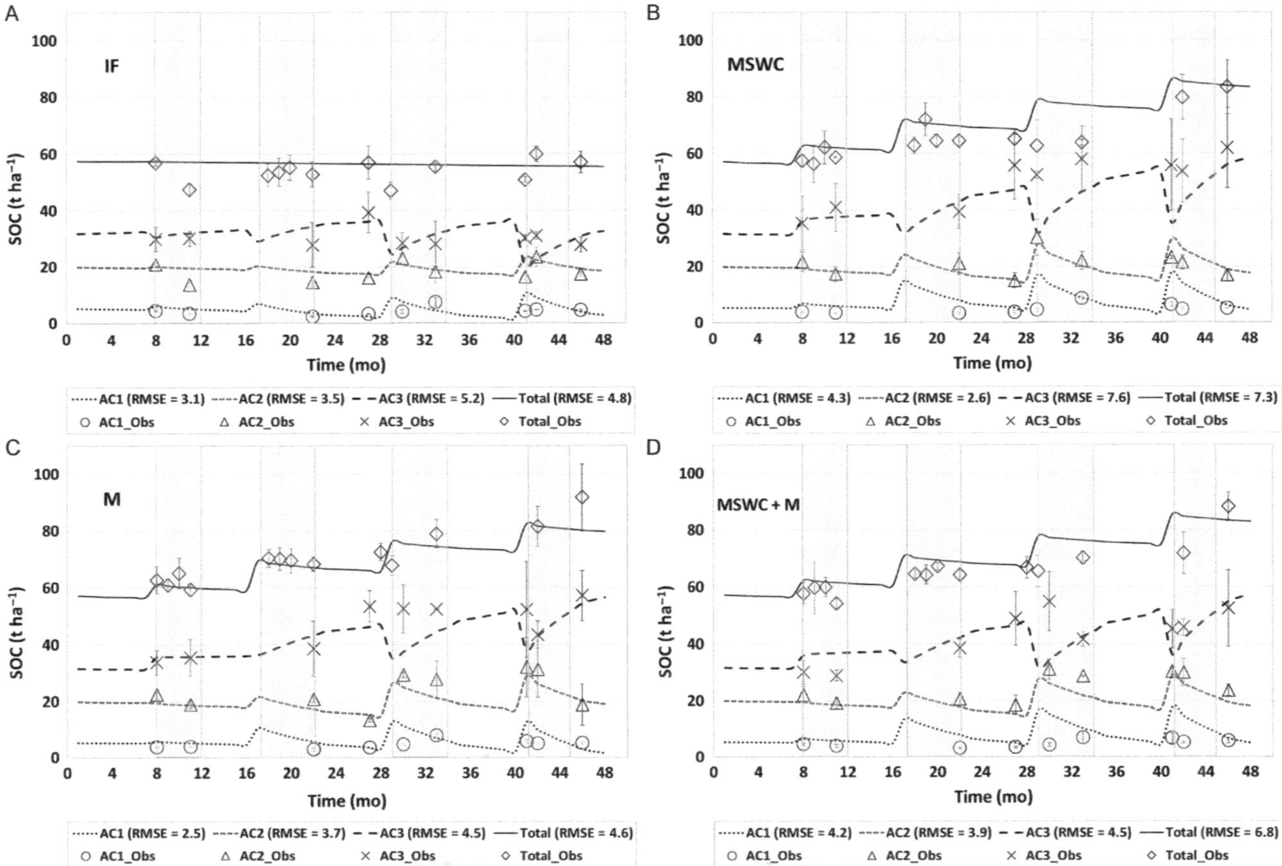

Fig. 7 SOC stock simulation in bulk soil (Total) and aggregate fractions (AC1, AC2, AC3) for each treatment over the 4-year period. The *gray* areas depict the tomato growing cycle periods. Point values present field observations used for calibration, while *lines* the simulation results. (A) inorganic fertilization (IF) treatment, (B) municipal solid waste compost addition (MSWC) treatment, (C) manure addition (M) treatment and (D) compost–manure mixture (MSWC + M) treatment.

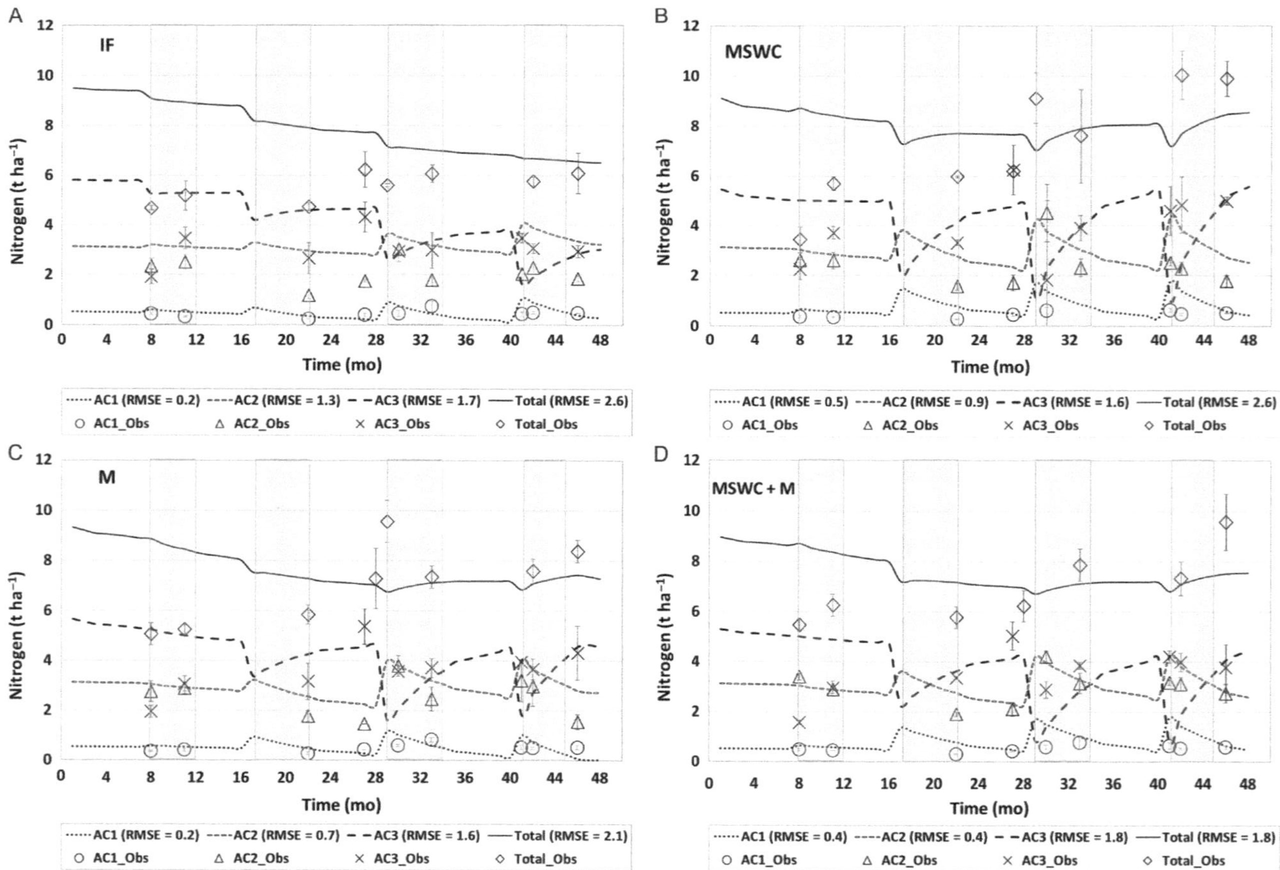

Fig. 8 Soil N stock simulation in bulk soil (Total) and aggregate fractions (AC1, AC2, AC3) for each treatment over the 4-year period. The *gray* areas depict the tomato growing cycle periods. Point values present field observations used for calibration, while *lines* the simulation results. (A) inorganic fertilization (IF) treatment, (B) municipal solid waste compost addition (MSWC) treatment, (C) manure addition (M) treatment and (D) compost–manure mixture (MSWC + M) treatment.

and years, except for the initial conditions, when an opposite trend was observed. The evolution of aggregate-associated N content followed similar trend as the respective SOC content of aggregates.

The model predicted N stocks for all aggregate fractions reasonably well. The RMSE ranged from 0.2 to 1.8 t N ha^{-1} in all treatments denoting acceptable calibration. However, the simulated total N raised significant errors especially in the first 2 years for all treatments. Despite the differences between the simulated and observed N data in the first 2 years, the final 2 years of simulated N values were very close to the field data. When evaluated across treatments, the addition of OA maintained the N content of the soil.

3.4 Soil WSAs

The simulation results as well as the observed data for the WSA distribution is shown in Fig. 9. At the end of the 4-year experiment, AC3 (macroaggregates) content had increased significantly in all treatments. The highest increase was observed in treatments that received the carbon amendments and specifically in the "MSWC + M" treatment, where the increase reached 72.2% relative to the initial AC3 mass fraction. The increase was slightly lower for "M" treatment (67.3%), while an increase of 53.3% was observed for the "MSWC" treatment and 35.7% for the "IF." Overall, the increase of water-stable macroaggregates in the OA amended plots was almost two times higher than in the "IF" treatment, indicating the beneficial effect of OA on soil structure development.

The WSA distribution during the experiment was strongly affected by tillage. AC3 were mechanically destroyed releasing micro- (AC2) and silt–clay-sized (AC1) WSA. The model simulated a significant drop in the AC3 fraction after tillage, which was less pronounced in field data. However, the model was able to simulate the gradual reformation of AC3 after the incorporation of organic amendments (tomato residues, weeds, compost, and manure) through tilling, with acceptable RMSE values (3.9–9.1 %). The ability of the 1D-ICZ model to simulate changes in the distribution of WSA between size classes depends on the modifying factors for the first-order rate constants of macroaggregate formation. Table 13 shows the formation rate constants of the macroaggregates that were used for the parameterization of the 1D-ICZ model. Among all treatments "MSWC" provided higher values of AC3 formation rates.

In general, the 1D-ICZ model was able to capture the WSA dynamics on a subyear basis throughout the simulation period, thus constituting a

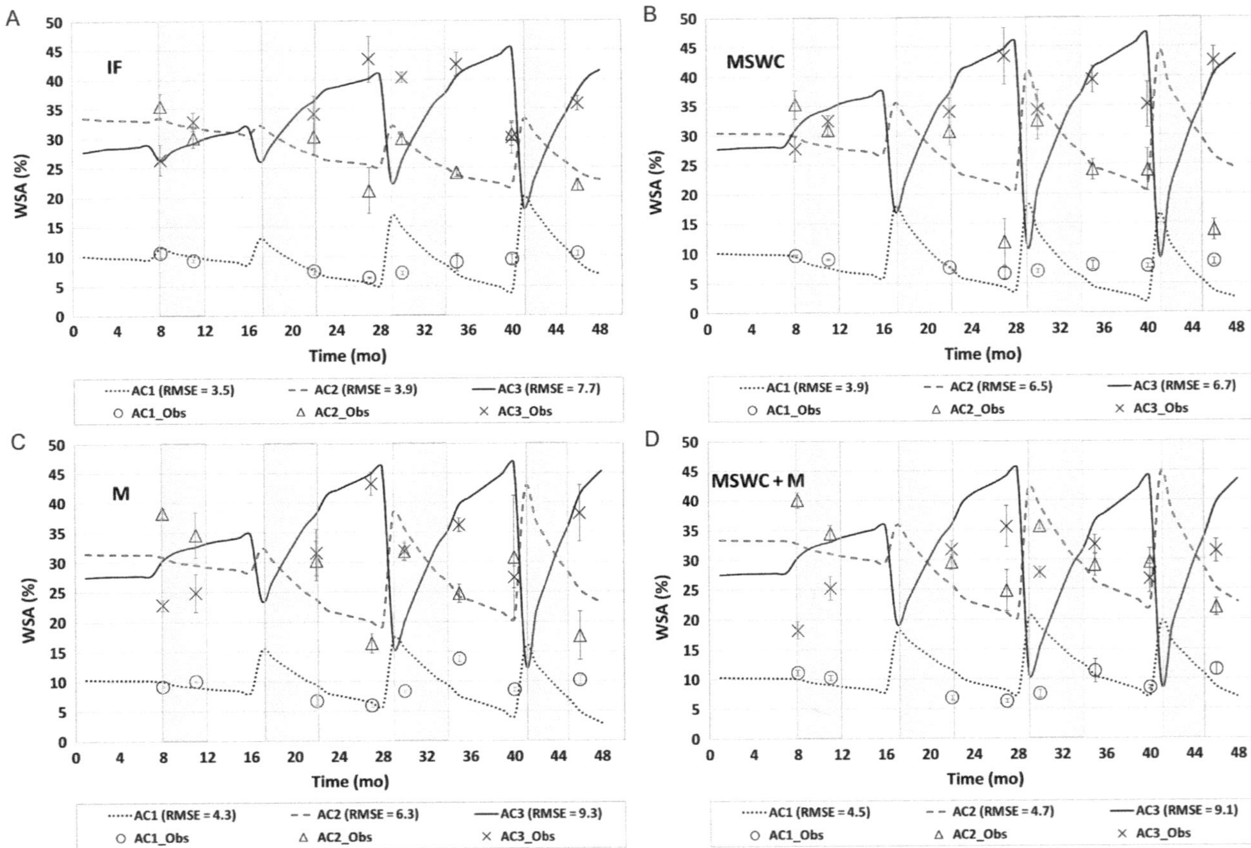

Fig. 9 Simulation of WSA distribution for each treatment over the 4-year period. The *gray* areas depict the tomato growing cycle periods. Point values present field observations used for calibration, while *lines* the simulation results. (A) inorganic fertilization (IF) treatment, (B) municipal solid waste compost addition (MSWC) treatment, (C) manure addition (M) treatment and (D) compost–manure mixture (MSWC+M) treatment.

Table 13 Formation Rate Constants of Macroaggregates (mo^{-1})

	IF	MSWC	M	MSWC+M
First growing period	1.32	1.64	0.51	1.76
Second growing period	1.56	1.58	1.93	0.63
Third growing period	−0.26	1.35	0.86	0.95
Fourth growing period	1.15	1.90	2.15	0.93
Average	0.94	1.61	1.36	1.07

IF, inorganic fertilizers; *M*, manure; *MSWC*, municipal solid waste compost; *MSWC+M*, municipal solid waste compost and manure mixture at a 70:30 ratio.

valuable tool for simulation of agricultural operations and management practices as well as quantification of the impact of the physical processes.

3.5 Toward the Development of a "Soil Sustainability Index"

The concept of sustainable use of soil requires the adoption of management practices that improve/maintain soil functions (Victoria et al., 2012). These practices should target functions like biomass production, carbon sequestration, maintaining soil BIO and functional biodiversity, and water and solute filtration and transformation which can be quantified through a series of model parameters, mass stocks, and fluxes. Table 14 presents the simulated rates of C fluxes (including crop residues, weeds and OA, SOC sequestration, biomass and fauna carbon increase, CO_2 production, OC transport in drainage water), soil structure indices (% WSA increase, % BD decrease), and nutrients (NO_3, PO_4, K) fluxes to the groundwater. A summary of these data for the 4-year simulated period is also depicted in Figs. 10 and 11.

The amount of C input added in the three amended treatments ("MSWC," "M," and "MSWC+M") was approximately the same and included 50 t C ha^{-1} of the respective amendment, and the remaining was the tomato plant residue and grass. The amount of SOC sequestered in the amended treatments varied between 26.3 t C ha^{-1} for the "MSWC" and 21.4 t C ha^{-1} for the "MSWC+M" treatment, corresponding to 42% and 33% sequestration, respectively. On the other hand, the IF treatment received 13.4 t C ha^{-1} as carbon input, and the SOC was depleted by 2 t C ha^{-1}. The CO_2 production varied between 20.6 t C ha^{-1} for the "MSWC" and 25.6 t C ha^{-1} for the "MSWC+M" and 28.3 t C ha^{-1} for the "M" treatment, respectively, as opposed to 1.4 t C ha^{-1} for the "IF" treatment. Active OA of the soil, stimulated CO_2 production that ranged between 33% and 44% of the total C input. The variability of the BIO-FAU carbon increase was

Table 14 Simulated Mass Fluxes (Related to Carbon and N, P, K Leachates) and Soil Structure Indices During the 4-Year Simulation

	Units	IF	MSWC	M	MSWC + M
Carbon fluxes					
C input	t C ha^{-1}	13.4	62.6	63.6	64.7
SOC sequestration	t C ha^{-1}	−2.0	26.3	22.5	21.4
BIO-FAU carbon increase	t C ha^{-1}	0.0	0.9	1.5	1.1
CO_2 production	t C ha^{-1}	1.4	20.6	28.3	25.6
OC leachate	t C ha^{-1}	2.17	1.60	1.60	8.76
Soil structure					
WSA_{AC3} increase	%	49.5	57.3	64.9	58.1
Bulk density decrease	%	2.6	8.0	7.2	6.9
Drainage water nutrients					
NO_3^-	t NO_3 ha^{-1}	0.4	1.7	1.3	2.7
PO_4^{-3}	t PO_4 ha^{-1}	0.8	1.8	1.2	1.2
K^+	t K ha^{-1}	0.02	0.1	0.02	0.2

BIO-FAU carbon increase, biomass and fauna carbon increase; *SOC sequestration*, soil organic carbon sequestration; *WSA_{AC3} increase*, water-stable macroaggregate increase.

small, ranging from 0.9 to 1.5 t C ha^{-1} over the 4 years of the experimental period. Similarly, the rate of OC leachate to the groundwater varied between 1.6 and 8.76 t C ha^{-1} for the various treatments.

The 1D-ICZ model was also able to simulate the dynamic changes in soil structure as depicted in the changes of BD and WSA changes. The "MSWC" showed the highest decrease in BD and the lowest increase in "WSA," while the "M" treatment had the highest increase in WSA and intermediate change of BD, and the mixture ("MSWC + M" treatment) had the lowest decrease in BD and intermediate increase in WSA. The N, P, K concentrations in the drainage water are also presented in the table, showing small but significant variability between the treatments.

The results from this study suggest that 1D-ICZ model could fully quantify the impacts of carbon addition to soil by successfully simulating changes in biomass production and carbon sequestration. The amount of bacterial and fauna biomass carbon also increases with carbon addition, as well as increases in the water-stable macroaggregates, which are considered as an

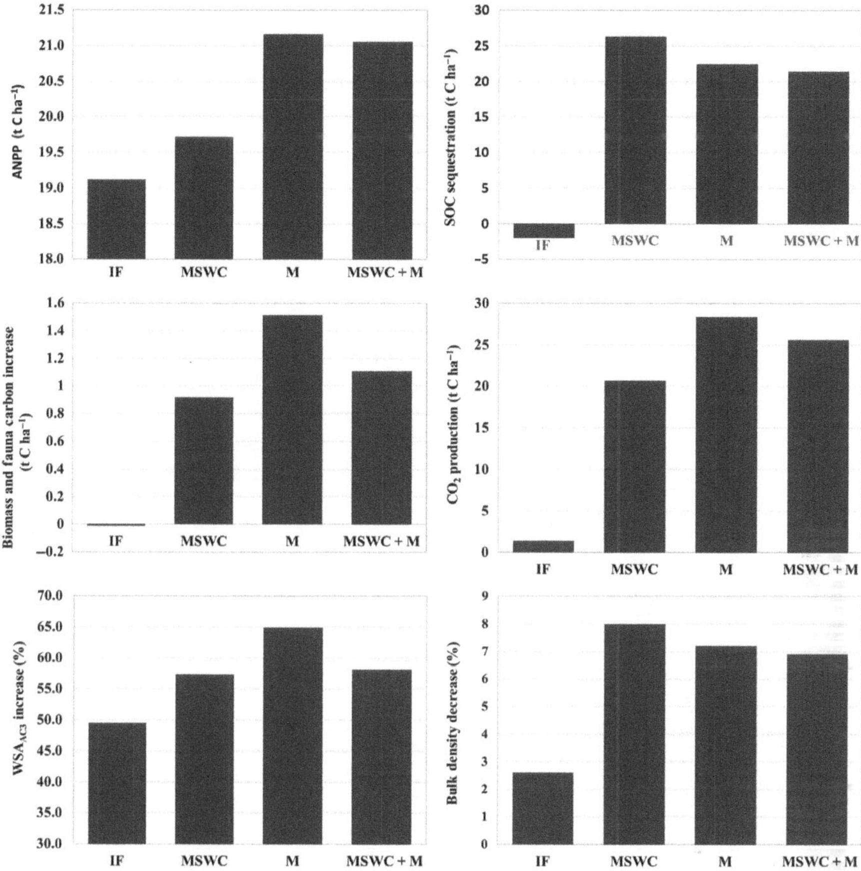

Fig. 10 Sustainability indices (summary for the four tomato growing cycles) in the inorganic fertilization (*IF*), municipal solid waste compost (*MSWC*), manure (*M*), and mixture (*MSWC + M*) treatment.

indicator of improved soil structure. On the other hand, addition of organic carbon increases the nitrate, phosphate, and potassium transport in drainage water contributing to the eutrophication of surface waters. These data can be used to assess the benefits to soil functions due to carbon amendment compared to IF. The results of this study have shown that there are trade-offs that have to be taken under account when managing soil functions. Even though carbon amendments improve significantly most of soil functions, they also adversely contribute to groundwater nutrient pollution. This suggests that the amount of carbon amendment to the soil has to be studied in order to optimize the benefits and the impacts that it has on soil functions and elemental fluxes, respectively. The 1D-ICZ model has been shown to be able to simulate the dynamics of soil functions and predict changes due to land management. The quantification of soil functions can be used to develop soil sustainability indices for the evaluation of trade-offs between different agroecological and land management practices.

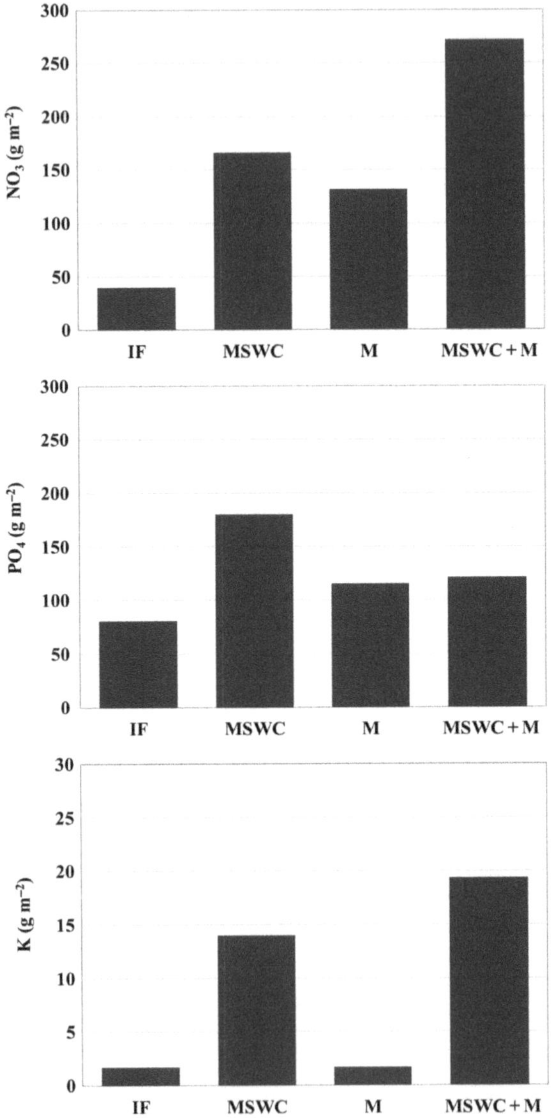

Fig. 11 NO_3, PO_4, and K leaching (summary for the four tomato growing cycles) in the inorganic fertilization (*IF*), municipal solid waste compost (*MSWC*), manure (*M*), and mixture (*MSWC*+*M*) treatment.

4. CONCLUSIONS

This study presented the validation of the 1D-ICZ model using field data from a plot experiment which used carbon amendments in tomato cultivation. The data used were obtained from a field study that was carried out in KRB CZO to assess the effects of different organic amendments on the changes in soil structure and soil (bio)chemical properties related to soil fertility, as well as changes in the yield of tomato plants and the quality of the fruits.

The simulation of the field data for all treatments was very successful. The model has been able to capture the biomass production, the temporal dynamics of the WSA formation, and the dynamics of carbon and nutrient sequestration in the different WSA sizes as well as the variability of water filtration and transformation efficiency in the different treatments. Process rates derived from the calibrated simulations allow summary indices of key soil functions to be quantified. These indices include measures of biomass production, soil carbon storage and GHG mitigation, nutrient sequestration for plant nutrition, and water transmission and filtration of nutrients as contaminants. The relative size of these indices allows a more transparent assessment across the range of soil functions, of the impact of farming practices. This study illustrates the determination of these indices through the comparison between application of inorganic fertilizer and carbon soil amendments as organic fertilizer in order to improve tomato crop yields. The application of the model to other sites will further test the applicability of the model across climatic and lithological conditions and a wider range of agricultural systems and practices.

ACKNOWLEDGMENT

The authors acknowledge funding support from the European Commission FP 7 Collaborative Project "Soil Transformations in European Catchments" (SoilTrEC) (Grant Agreement no. 244118).

REFERENCES

Amundson, R., Guo, Y., Gong, P., 2003. Soil diversity and land use in the United States. Ecosystems 6 (5), 470–482.

Amundson, R., et al., 2015. Soil science. Soil and human security in the 21st century. Science (New York, N.Y.) 348 (6235), 1261071.

Blackburn, J.W., Proctor, T.J., 1983. Estimating photosynthetically active radiation from measured solar irradiance. Sol. Energy 31 (2), 233–234.

Bremner, J.M., Mulvaney, C.S., 1982. Nitrogen—total. In: Page, A.L., Miller, R.H., Keeney, D.R. (Eds.), Methods of Soil Analysis, Part 2. Chemical and Microbiological Properties. American Society of Agronomy, Madison, WI, pp. 595–624.

Carsel, R.F., Parrish, R.S., 1988. Developing joint probability distributions of soil water retention characteristics. Water Resour. Res. 24 (5), 755–769.
Chapin, F.S., Matson, P.A., Vitousek, P.M., 2012. Principles of Terrestrial Ecosystem Ecology. Springer, New York.
Chung, S.O., Horton, R., 1987. Soil heat and water flow with a partial surface mulch. Water Resour. Res. 23 (12), 2175–2186.
Elliott, E., 1986. Aggregate structure and carbon, nitrogen, and phosphorus in native and cultivated soils. Soil Sci. Soc. Am. J. 50 (S3), 627–633.
Foley, J.A., et al., 2011. Solutions for a cultivated planet. Nature 478, 337–342.
Garnett, T., et al., 2013. Sustainable intensification in agriculture: premises and policies. Sci. Mag. 341 (July), 33–34.
Hargreaves, J.C., Adl, M.S., Warman, P.R., 2008. A review of the use of composted municipal solid waste in agriculture. Agr. Ecosyst. Environ. 123 (1), 1–14.
Hobbs, P.R., Sayre, K., Gupta, R., 2008. The role of conservation agriculture in sustainable agriculture. Philos. Trans. R. Soc. B Biol. Sci. 363 (1491), 543–555.
Lal, R., 2006. Enhancing crop yields in the developing countries through restoration of the soil organic carbon pool in agricultural lands. Land Degrad. Dev. 17 (2), 197–209.
Luo, Z., Wang, E., Sun, O.J., 2010. Soil carbon change and its responses to agricultural practices in Australian agro-ecosystems: a review and synthesis. Geoderma 155 (3), 211–223.
Martin, D., et al., 2010. Soil organic carbon storage changes with climate change, landform and land use conditions in Garhwal hills of the Indian Himalayan mountains. Agr. Ecosyst. Environ. 138 (1), 64–73.
Maynard, D.G., Curran, M.P., 2006. Bulk density measurement in forest soils. In: Carter, M.R., Gregorich, E.G. (Eds.), Soil Sampling and Methods of Analysis, second ed. Canadian Society of Soil Science by Taylor & Francis Group, LLC (Chapter 66).
Milne, E., et al., 2015. Soil carbon, multiple benefits. Environ. Dev. 13, 33–38.
Olsen, S.R., Cole, C.V., Watanabe, F.S., Dean, L.A., 1954. Estimation of available phosphorus in soils by extraction with sodium bicarbonate. USDA Cir. No. 939.
Rabbinge, R., Ward, S.A., van Laar, H.H., 1989. Simulation and Systems Management in Crop Protection. Simulation Monographs 32. PUDOC, Wageningen, p. 420.
Regelink, I.C., et al., 2015. Linkages between aggregate formation, porosity and soil chemical properties. Geoderma 247–248, 24–37.
Scotti, R., et al., 2015. Organic amendments as sustainable tool to recovery fertility in intensive agricultural systems. J. Soil Sci. Plant Nutr. 15 (2), 333–352.
Šimůnek, J., Šejna, M., Saito, H., Sakai, M., van Genuchten, M.T., 2009. The HYDRUS-1D Software Package for Simulating the One-Dimensional Movement of Water, Heat, and Multiple Solutes in Variably-Saturated Media Version 4.08. Department of Environmental Sciences, University of California Riverside, Riverside.
Stamati, F.E., Nikolaidis, N.P., et al., 2013a. A coupled carbon, aggregation, and structure turnover (CAST) model for topsoils. Geoderma 211, 51–64.
Stamati, F.E., Nikolaidis, N.P., Schnoor, J.L., 2013b. Modeling topsoil carbon sequestration in two contrasting crop production to set-aside conversions with RothC—calibration issues and uncertainty analysis. Agr. Ecosyst. Environ. 165, 190–200.
van Genuchten, M.T., 1980. A closed-form equation for predicting hydraulic conductivity of unsaturated soils. Soil Sci. Soc. Am. J. 44 (5), 892–898.
Victoria, R., Banwart, S., Black, H., Ingram, J., Joosten, H., Milne, E., Noellemeyer, E., 2012. The benefits of soil carbon: managing soils for multiple economic, societal and environmental benefits. In: UNEP Yearbook 2012: Emerging Issues in Our Global Environment. United Nations Environment Programme, Nairobi, Kenya, p. 80 (Chapter 2).
Wallman, P., et al., 2005. ForSAFE—an integrated process-oriented forest model for long-term sustainability assessments. For. Ecol. Manage. 207 (1), 19–36.

CHAPTER TWELVE

Valuation of Soil Ecosystem Services

J.Ö.G. Jónsson*[,1], B. Davíðsdóttir*, N.P. Nikolaidis[†]

*Environment and Natural Resources, University of Iceland, Gimli, Reykjavík, Iceland
[†]School of Environmental Engineering, Technical University of Crete, University Campus, Chania, Greece
[1]Corresponding author: e-mail address: jog44@hi.is

Contents

Abstract

Soil natural capital and soil ecosystem services (ES) are under increasing pressure because of human activities. Soils provide multiple benefits to humans, and the role of soil in Earth's Critical Zone is fundamental to its functions that provide these benefits. Despite their importance, soils are rarely appreciated for the values they provide. One reason is the absence of their economic value in land-use decision making. We present a framework for categorizing and economically valuing soil ES and illustrating the use of the framework in a case study for three soil ES in the Koiliaris watershed on the Greek island of Crete. The value of the soil ES estimated was crop and livestock biomass 740–7560 id\$ ha^{-1} $year^{-1}$; filtering of nutrients and contaminants 0–278 id\$ ha^{-1} $year^{-1}$; and climate regulation −2200 to −5610 id\$ ha^{-1} $year^{-1}$.

Advances in Agronomy, Volume 142
ISSN 0065-2113
http://dx.doi.org/10.1016/bs.agron.2016.10.011

Highlights

- Soils provide multiple economic benefits that are rarely accounted for.
- A framework for the classification and economic valuation of soil ecosystem services (ES) is presented.
- The soil framework is applied at Koiliaris, Crete, Greece for three soil ES.
- This chapter illustrates soil ES values from −5610 to 7560 id\$ ha^{-1} $year^{-1}$ depending on the specific soil ES.

1. INTRODUCTION

The delivery of ecosystem services (ES) is necessary for maintaining economic systems and is directly and indirectly linked to human well-being. As ES are largely nonmarket goods, they are excluded from formal economic decision making and therefore undervalued and overused. Assessment and economic valuation of ES is an attempt to reverse this trend (Kumar, 2010; MEA, 2005), but comprehensive valuation of soil ES has been somewhat ignored in the literature until now (see Jónsson and Davíðsdóttir, 2016 for an overview). Soils are an important kind of natural capital that has specific functions providing multiple, essential ES (Haygarth and Ritz, 2009). More than 99% of human food (calories) comes from the land and the soil (Pimentel, 2006). Soils filter and clean our drinking water; they deliver plant nutrients needed for vegetation growth (Daily et al., 1997); and they host soil biota that decompose plants when they die. Soils provide habitats for millions of species (Science, 2004) and store twice as much carbon as the biosphere and the atmosphere combined (Bellamy et al., 2005; Scharlemann et al., 2014). Soils also regulate water flows and thereby prevent floods. They provide us with building materials as well as provide the structural foundation for human activities (Frossard et al., 2006). Soils are a source of many current medicines, probiotics and antibiotics (D'Costa et al., 2006; Minton, 2003; Turbé et al., 2010). They store our history, in buried archaeological artifacts and sediments (European Soil Bureau Network, 2005). They are a source of recreation (Decaens et al., 2006) and are a fundamental part of world religions, for example, in Christianity God created man from soil. Derived from this discussion it is clear:

(i) how soils contribute to all ES categories as defined by the Millennium Ecosystem Assessment (MEA, 2005) and

(ii) that soil resources, as other natural resources, are important economic assets, whether or not they are valued via marketplaces.

Given the importance of the multiple services derived from soils, it is evident that they need to be maintained and the only way to do that is to protect soil structure and processes that are the base for their functionality. Unfortunately, soil, as per other types of natural capital, is coming under increased pressure because of human activities. One of the culprits for these degradation processes is the omission of soil ES valuation in land-use decision making. In order to address this deficiency, we propose a valuation framework for soil ES and then illustrate how soil and its various functions that provide ES can be valued economically through a case study in the Koiliaris watershed on the island of Crete, Greece.

1.1 Framework Development

The conceptual framework of a life cycle of soil development describes that soil processes go through a cycle from pedogenenis to degradation of soil functions and eventual physical erosion and loss (Banwart et al., 2012). This life cycle serves as the starting point for the soil ES framework (Fig. 1). This has some parallels with the concepts of Dominati et al. (2010) on formation, maintenance, and degradation of soil natural capital. To expand these concepts, the framework depends heavily on work undertaken by Robinson et al. (2009) and Dominati et al. (2010). There the authors tie together the concept of natural capital of soil and soil ES and how soil threats and soil management influence the soil's natural capital stocks and thus the provision of soil ES (see Jónsson and Davíðsdóttir, 2016). This paper builds on, and adds to, their work by showing how soil ES are linked to beneficiaries and can be valued using standard economic methods. The framework proposed in Fig. 1 consists of three interconnected sections: a biophysical

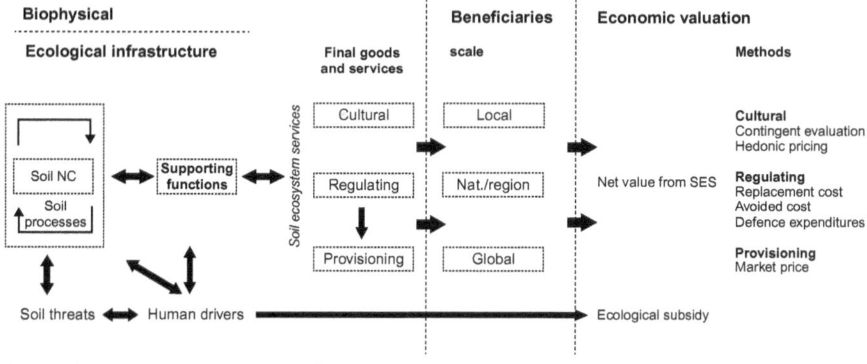

Fig. 1 Soil ecosystem services framework.

model, the identification of beneficiaries, and an economic valuation. The framework is then further subdivided into components within each section. The biophysical section has two main components:

Ecological infrastructure (Ei) of soil (soil natural capital and supporting functions) and *final goods and services* (cultural, regulating, and provisioning) that are the benefits that flow from the Ei (see Table 1).

Table 1 Key Terms

Key Terms	Definitions
Ecological infrastructure of soil	Soil natural capital, its properties, and soil support functions that underlie other ecosystem services and are in a dynamic relationship with soil processes and soil natural capital. Adapted and modified from Bristow et al. (2010)
Ecosystem services	The benefits that people obtain from ecosystems. Ecosystem goods and services are synonymous with ecosystem services (MEA, 2005)
Final goods and services from soil	Benefits that flow from the ecological infrastructure (cultural, regulating, and provisioning services). Adapted and modified from MEA (2005)
Natural capital	Stocks of natural resources found on Earth yielding a flow of valuable ecosystem goods or service into the future (Dominati et al., 2014c)
Soil ecosystem services	The flow from the ecological infrastructure of soil. Soil ecosystem services refer to both ecosystem goods and services from soil. Adapted and modified from MEA (2005) and Bristow et al. (2010)
Soil natural capital	Soil stocks on earth which yield a flow of soil goods and services. Soil natural capital is characterized by soil properties (Dominati et al., 2010)
Soil processes	Any change or reaction which occurs within soils, either physical, chemical, or biological. The complex interactions among the biotic and abiotic elements of the soil. Adapted and modified from de Groot et al. (2002)
Soil properties	The physical, chemical, and biological characteristics of a soil. They can be inherent or manageable (Dominati et al., 2010)
Soil support functions	A subset of interactions between the soil natural capital and soil processes that are required for the production of final soil ecosystem services and goods that satisfy human needs. The support functions are intermediate steps in the stock and flow chain and are therefore neither consumed directly nor valued economically. Adapted and modified from MEA (2005)

Final goods and services are called soil ES in the framework and refer to both ecosystem goods and services from soil. The beneficiaries part shows at what scale (local, regional/national, or global) the recipients of the ES are based. The third section is the economic valuation, where the value for soil ES is established using standard economic methods (e.g., market pricing, avoided cost, and associated methods; see Jónsson and Davíðsdóttir, 2016 for more detail on the different methods for the economic valuation of soil ES).

1.2 Biophysical Section

Soils, along with rivers, aquifers, wetlands, and other landscape elements, are "key components of an 'ecological infrastructure' that supports the continuing delivery of ecosystem services required by natural systems for their survival, and mankind for human well-being" (Bristow et al., 2010, p. 13). This chapter draws on this definition and defines Ei of soil as: soil natural capital, its properties, and soil processes; and soil support functions that underlie other ES and are in a dynamic relationship with soil natural capital. Soil natural capital refers to soil stocks on Earth, which yield a flow of benefits in the form goods and services. Soil natural capital builds on the definition of natural capital (Costanza and Daly, 1992), and it has physical, chemical, and biological properties (Table 2) that can be measured in qualitative or quantitative terms.

The soil support functions (Fig. 1) included in the framework are *soil formation*, *nutrient cycling*, *biodiversity pool*, and *water cycling* (Fig. 1). Soil's Ei creates soil ES (final goods and services).

Soil formation: The chemical, physical, and biological activities lead to the formation of soil over time by weathering of rocks and minerals. This process is affected by relief (terrain), parent material, climate, and geography (Brantley et al., 2007).

Nutrient cycling: The maintenance of soil fertility whereby chemical elements that are essential to the production and functioning of living organisms is moved and chemically reacted through the biotic and abiotic parts of the soil. Microorganisms are key moderators of this service. This cycling is the foundation for many other processes of the soil (Brussaard et al., 2007; Zhang et al., 2007).

Biodiversity pool: Soils are probably the most species-rich habitats of terrestrial ecosystems (Science, 2004; Weber, 2007). This reservoir of biodiversity is important to many soil processes, and many of the species

Table 2 Examples of Soil Properties

Indicators	Metrics
Pedodiversity	Number of soil classes within an area
Aggregate diversity	Mean weight diameter of various aggregates, and aggregate diversity measured with the Shannon–Wiener index
Bulk density	g/cm^3
Changes in topsoil depth	cm
Change in cation exchange capacity (CEC)	Milliequivalents/100 g
Soil contamination	Concentrations in topsoil
Change in topsoil pH	pH
Changes in microbial biomass	C ($mg\ kg^{-1}$)
Change in and absolute level of net N mineralization	$mg\ kg^{-1}$ soil%
Change in total soil organic matter (TSOM)	%

of organisms serve as an essential part of the functional diversity and resilience of the soil.

Water cycling: Water cycling involves the physical process of water moving through the soil. This movement influences the geo-, bio-, and chemical processes in the soil and affects the development of its biodiversity and functions (Dominati et al., 2010).

Soil ES are the flow of benefits from the Ei of soils. The soil ES are split into three categories similar to the MEA (2005) (Table 3): provisioning, regulating, and cultural. For more information about individual soil ES and service category, see Jónsson and Davíðsdóttir (2016).

When assessing soil ES, it is necessary to capture the dynamics of the soil Ei. Thus, the valuation of ES and functions requires a framework that relates land-use management practices to soil functions such as biomass production, carbon and nutrient sequestration, biodiversity, and water transformation as well as to how changes in land management affect these functions. The complexity of soil interactions necessitates the quantification of these functions to be determined through quantification of soil processes, e.g., through empirical measurement and mathematical modeling.

Table 3 MEA Categories (MEA, 2005)

Services	Definitions
Provisioning services	Products people obtain from ecosystems, such as food, fuel, fiber, freshwater
Regulating services	The benefits people obtain from the regulation of ecosystem processes, including air quality maintenance, climate regulation, erosion control, regulation of human diseases, and water purification
Cultural services	The nonmaterial benefits people obtain from ecosystems through spiritual enrichment, cognitive development, reflection, recreation, and esthetic experiences

1.3 Beneficiaries

ES are an anthropocentric concept and are therefore only relevant in the context of human beneficiaries. Beneficiaries are individuals that receive positive change in well-being through final goods and services. Before a value can be placed on ES, the beneficiaries need to be identified and established in their spatial scale. That can be done by looking at the value chain of soil ES and establishing the direct beneficiaries for the final services. This approach follows the distinction between intermediate and final services (Boyd and Banzhaf, 2007) and the establishment of direct beneficiaries (Johnston and Russell, 2011). The beneficiaries can be individuals, commercial entities, or the public sector; they can be distributed at local, national/regional, or global scale, and the benefits can be short or long term (eftec, 2005). The benefits also have ecological, sociocultural, and economic dimensions. An individual beneficiary on a local scale might be a farmer that gains economic benefits from cultivating crops (provisioning services), ecological benefits from the biological control of pests provided in part by a healthy soil biota (regulating services), and sociocultural benefits from living in a pastoral landscape which is considered beautiful (cultural services).

1.4 Economic

The economic valuation of soil ES takes place at the right side of the framework (Fig. 1) where the final goods and services that flow from the soil Ei are valued based on the benefits derived from those services. The economic valuation illustrates the net value of the soil ES. The economic value is derived by, for example, (a) looking at the biophysical data of the particular service (e.g., a regulating service), (b) seeing what would change in the absence of the soil ES (that is, complete loss of the service), and (c) comparing that to the current state of the soil ES. The difference in the biophysical quantity of the service is the foundation for the economic valuation. Support functions within the soil Ei are not valued as this would lead to double counting and this is consistent with standard economic principles (Boyd and Banzhaf, 2007). For a list of potential economic evaluation methods of soil ES, see Jónsson and Davíðsdóttir (2016). Within the economic valuation section, the ecological subsidy is also subtracted for certain services that have a direct human input. This could, as stated before, include, e.g., the use of fertilizer in agricultural fields to increase biomass yields. Here the fertilizer is only partly responsible for the overall output and that needs to be taken into account. If this information is not available from the biophysical proxies, then the price of these inputs is simply subtracted. This is particularly important for valuing trade-off in soil ES as soil ES receiving an ecological subsidy that increases its output and overall value potentially sends the wrong signals to land managers and policy makers about its effectiveness and value.

1.5 Soil Threats and Human Drivers

The status of the soil Ei is affected by external forces, including soil threats and human drivers (Fig. 1). These external forces influence the soil natural capital and the support functions and thus in turn influence the provisioning of final goods and services from soils. As a result, soil threats and human drivers must be included in a framework intended to capture the value of soil ES, and the potential costs associated with their degradation. The soil framework proposed here is able to pick up the influence of these external forces through (i) the ecological subsidy and (ii) the net economic value from soil (see Section 1.4).

1.5.1 Soil Threats

In 2002, the European Commission, in its communication "Towards a Thematic Strategy on Soil Protection," identified eight main threats to soils in the EU (Table 4). Soil degradation is driven or exacerbated by human activities such as inadequate agricultural and forestry practices, industrial activities, tourism, urban and industrial sprawl, road building, soil sealing, and construction work (European Commission, 2006).

These activities negatively affect the status of soil's Ei and its provision of soil ES. The soil threats are made visible in the framework by looking at how they affect the provision of the soil ES and functions and how that results in

Table 4 Soil Threats and Implications for Soil Functions

Soil Threats	Implications for Soil Functions
Erosion	Implications for all functions and services
Decline in organic matter	Climate regulation, biomass production
Soil contamination	Filtering of nutrients and contaminations
Soil sealing	Implications for all functions and services
Soil compaction	Biodiversity pool
Decline in soil biodiversity	Biodiversity pool
Salinization	Filtering of nutrients and contaminations
Floods and landslides	Hydrological control

changes in economic value. Table 4 shows how the soil threats affect the soil functions and services. The soil threats that the framework is able to capture at this moment are organic matter decline and erosion.

1.5.2 Ecological Subsidy

Ecological subsidies are direct human inputs that are intended to influence the supply of certain soil ES (Dale and Polasky, 2007; Tilman et al., 2002; Vitousek et al., 1997). The inputs are *fertilizers*, *pesticides*, and *irrigation*, and they can cause disservices and trade-offs (Power, 2010; Swinton et al., 2007). These inputs need to be accounted for when looking at the outputs of soil systems. Ecological subsidies differ from the human activities that drive or exacerbate soil degradation as those activities overall adversely affect soil Ei and soil ES, while the ecological subsidies enhance the supply of certain soil ES. The ecological subsidy is accounted for in the economic valuation section by looking at the changes in the biophysical output of the soil ES that are attributed to the ecological subsidy. If the information about changes in the biophysical outputs is unavailable, the costs of the inputs can be used as a proxy instead.

1.5.3 Fertilizers

The application of fertilizers, both natural and artificial, influences the provision of soil ES. Their application in an agricultural context makes certain nutrients available for biomass production that might not be so readily available, or even be considerably unavailable, in the natural landscape. This input increases biomass yields, thus potentially increasing the amount of harvestable provisioning services. At the same time, this application can negatively influence many processes of the soil, for example, the supporting functions provided by soil biodiversity (Cao et al., 2011; DEFRA, 2009; Mozumdera and Berrens, 2007; Sarathchandra et al., 2001). The effects of fertilizers are made visible in the framework by looking at the changes in the biophysical outputs of the ES due to the fertilizer inputs. If the information about changes in the biophysical outputs is unavailable, the costs of the inputs can be used as a proxy instead.

1.5.4 Pesticides

Pesticides are any substances or mixture of substances used to prevent, destroy, or control any pest, including vectors of human or animal disease, and unwanted species of plants or animals causing harm or interfering with agricultural food and agricultural commodity production (FAO, 2002). Pesticides can change the natural landscape, flora, and fauna in such a way that it is more suitable to products that humans want, thus increasing yields. This application has its drawbacks as its application is not precise and thus affects biodiversity by killing beneficial organisms (Stockdale et al., 2006) and it can affect the provision of ES (Sandhu et al., 2008). It is difficult to estimate exactly how much pesticides affect the provision of soil ES. Some estimates, based on the energy input in agriculture, attribute 10% of the overall energy inputs to pesticides (Bardi et al., 2013). Because of the lack of quantifiable biophysical data on the effects of pesticides on the provision of soil ES, the costs of the pesticide inputs are used as a proxy for the ecological subsidy.

1.5.5 Irrigation

Irrigation is the artificial application of water to the soil. It is used in agricultural production where there is insufficient water at the right time for crop production and can also be needed where the particular crop is out of its natural range. There are potential risks associated with irrigation. If either the amount of water or the quality is incorrect, the farmer runs the risk of not only wasting water but also damaging the soil, particularly through salinization (Ashman and Puri, 2002). The costs of the irrigation inputs are used as a proxy for the ecological subsidy. When quantifiable biophysical data on the effects of irrigation on the provision of soil ES are available, they are used.

This section has illustrated how the soil framework connects soil natural capital to the provisioning of soil ES. The next section illustrates the use of the framework.

2. METHODOLOGY

In 2013, the soil ES framework was tested in a pilot study in the Koiliaris watershed, a Critical Zone Observatory (CZO) on the island of Crete, Greece. Koiliaris CZO is a partner in the European Union 7th Framework Programme funded research project SoilTrEC (http://www.soiltrec.eu). The rational for selecting Koiliaris CZO as a location for the pilot study was because of data availability and to facilitate synergies with other parts of the SoilTrEC project. SoilTrEC (Soil Transformation in European Catchments) is focused on addressing certain knowledge gaps regarding important soil ES and functions and the importance for mathematical modeling of these as listed by the EU soil thematic strategy (European Commission, 2006). The research in SoilTrEC is driven by the hypothesis that soil processes can be described along a life cycle of soil development (from pedogenenis to erosion) and that the processes and functions of soil can be quantified during this life cycle (Banwart et al., 2012). The biophysical data used for the soil ES economic valuation were obtained from the Koiliaris CZO.

2.1 Area Description

Koiliaris CZO is located in Koiliaris watershed in the northwestern part of Crete, approximately 25 km east of the city of Chania, and the CZO is operated by the Technical University of Crete. The watershed is 130 km^2 and has a very steep rise in elevation ranging from 0 m AMSL on the coast to 2120 m in the White Mountains. Agriculture is the main land use (Fig. 2) and includes cropland and pasture (35%), olive and orange groves (32.1%), shrub land and brush land (32.3%), and mixed forest (0.6%) (Giannakis et al., 2014).

The Koiliaris watershed is of particular interest in the SoilTrEC project because it is an example of long-term intensive land use through farming and intensive grazing. The soils at Koiliaris are under threat both because of these agricultural practices (Nikolaidis et al., 2013) and because of the effects of climate change (IPCC, 2014a). For detailed description of the hydro-, bio-, and geological aspects of Koiliaris CZO, see Moraetis et al. (2015).

2.2 The Integrated Critical Zone Model

The Integrated Critical Zone (ICZ) model developed by the SoilTrEC (Banwart et al., 2012) project is used to quantify soil functions, and it is an

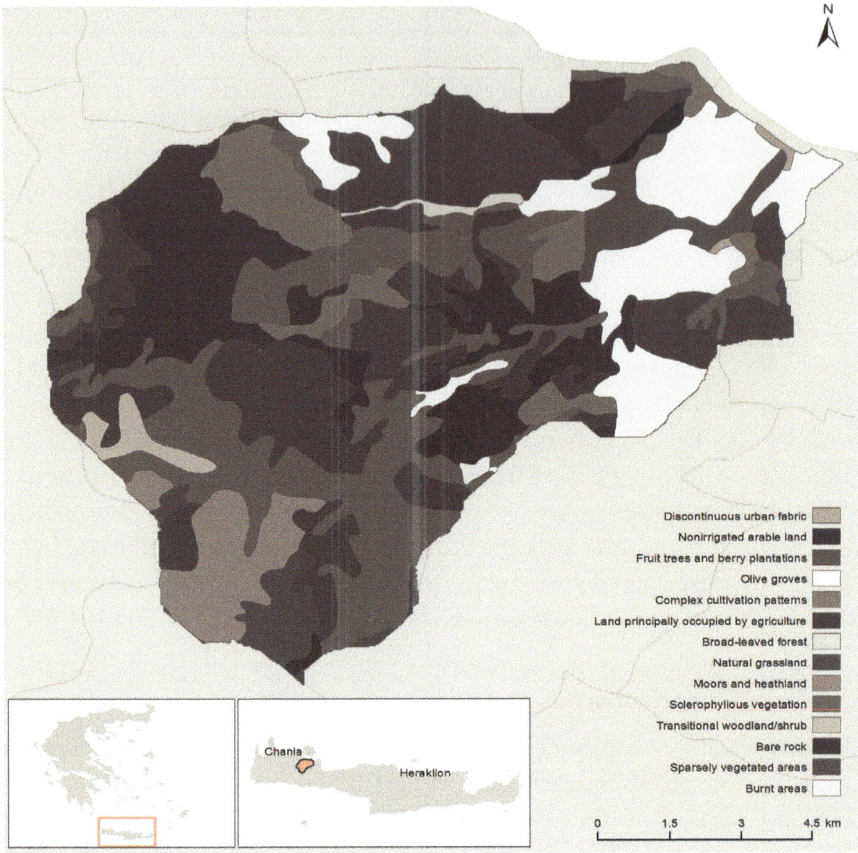

Fig. 2 CORINE land-use map of Koiliaris watershed.

integral part of the proposed framework. The ICZ is a mathematical model that links soil aggregate formation and soil structure to nutrient dynamics, biodiversity, water filtration and transformation, and biomass production. The ICZ model consist of a flow, transport, and bioturbation component; a chemical equilibrium and soil weathering component; a soil aggregation and carbon and nutrient sequestration dynamics component; and a plant growth and nutrient uptake component. The ICZ model was developed in two versions: the 1D-ICZ where it is coupled with HYDRUS-1D and it can simulate soil functions at the soil profile level (Giannakis et al., 2014); and the SWAT-ICZ where it is coupled with the watershed model SWAT to estimate in a semidistributed basis soil functions at the watershed level (Nikolaidis et al., 2013). A brief description of the ICZ follows (Fig. 3) and the details of the model can be found in Giannakis et al. (2017).

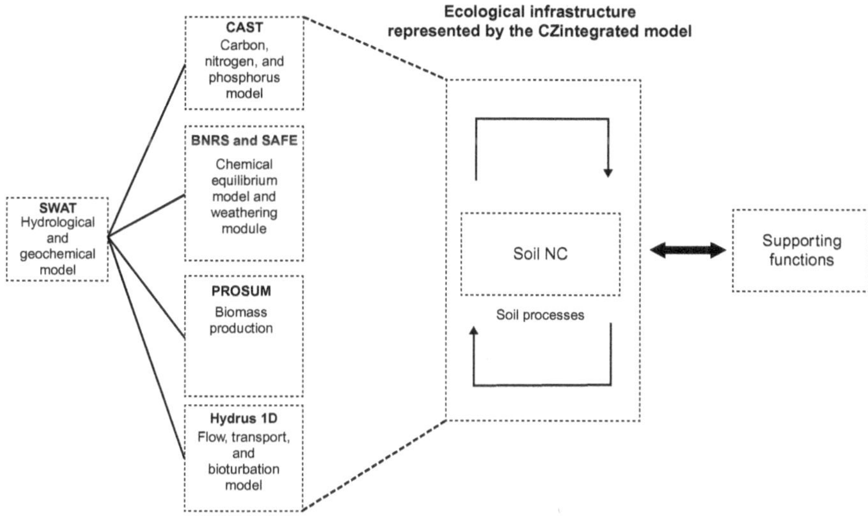

Fig. 3 Submodels of ICZ.

2.2.1 Carbon Sequestration and Nutrient Dynamics

The soil carbon, aggregation, and structure turnover (CAST) model and a simplified mechanistic N and P model (Stamati et al., 2013) were developed for the SoilTrEC project. The CAST model is based on the macroaggregates that are formed around particulate organic matter, followed by the release of microaggregates. In CAST, the transformations of organic matter have been linked with a dynamic model of soil aggregation/disaggregation, a simplified terrestrial ecology model that is comprised of saprophytic fungi, microorganisms (BIO pool), consumers and predators, and a plant/mycorrhizal root/fungi dynamics model. The carbon pools of ROTH-C were adapted to simulate the various organic matter pools which in turn account for the C, N, and P pools using stoichiometric C:N:P ratios. For more information on the CAST model, see Stamati et al. (2013) and Giannakis et al. (2014). This model estimates the dynamics of processes such as vegetation growth and carbon and nitrogen sequestration within the Ei.

2.2.2 Water Filtration and Transformation Dynamics

The chemical equilibrium model BRNS (Regnier et al., 2002) was adapted to account for the effect of water saturation as it varies with time, and exchange equilibria with the gas phase. The adapted code makes use of the SAFE chemical weathering module adapted from the ForSAFE modeling code (Belyazid, 2006). The mineral dissolution kinetics components used within the weathering model are source minerals and the ionic weathering products Ca^{2+}, Na^{+}, K^{+}, Mg^{2+}, Al^{3+}, H_4SiO_4, and PO_4^{3-}. This model relates to soil functions of water filtering and transformation within the Ei.

2.2.3 Plant Dynamics

A simple plant productivity module, PROSUM, was developed and adapted for the SoilTrEC project for inclusion in the 1D-ICZ model. Fixation of carbon (C) by plants is a key process in soil formation. The model is based on the theoretical production ecology principles and predicts the dynamics of key variables (e.g., above- and belowground production of litter C and N; nutrient and water uptake) in response to key physiological drivers and limitation (temperature; availability of light, water, atmospheric CO_2, and the nutrient elements N, P, Ca, Mg, and K; and grazing and management events). Mycorrhizal fungi effects on nutrient acquisition are incorporated into PROSUM on the basis of soil volume explored by the produced microbial biomass, and root exudate fluxes are calculated for use in the weathering component of the 1D transport model.

The ICZ model is coupled in the 1D version to the HYDRUS-1D (for more information on HYDRUS-1D, see Šimůnek et al., 2009) to simulate the flow of water as well as heat and solute transport in the unsaturated zone.

For the upscaled version, the ICZ model is coupled to the watershed model SWAT to estimate soil functions at the watershed level (Nikolaidis et al., 2013). In the SWAT-ICZ model, the SWAT model determines the hydrologic fluxes for every hydrologic response unit (HRU) for three layers (the upper soil, the unsaturated zone, and shallow aquifer) on a monthly basis which are then used by the ICZ model to simulate soil functions. The ICZ model can simulate dynamically soil stocks (such as carbon and nutrient sequestration, biomass production, and bacterial and fauna stocks) and soil fluxes (such as CO_2 emissions and nutrient fluxes to groundwater) and in this way quantify directly the soil functions.

2.3 Soil ES in Koiliaris Watershed

Three soil ES were chosen for evaluation based on the biophysical and modeling data available from our SoilTrEC partners and the soil functions and services emphasized in the SoilTrEC project (Banwart et al., 2012). The services selected were two regulating services (climate regulation and filtering of nutrients and contaminants) and one provisioning (biomass production from crop and livestock).

2.3.1 Filtering of Nutrients and Contaminants

Soils have the ability to control water quality by, to some extent, absorbing, retaining, and chemically transforming solutes (including contaminants), therefore avoiding their release in water bodies such as groundwater, lakes, and rivers. For the Koiliaris watershed, the retention of nitrate (NO_3^-) was assessed as a proxy for the service of filtering nutrients and contaminants. Nitrate retention is of particular interest because nitrates in supplies of drinking water are a well-known health hazard, and their concentration is regulated by the EU and municipalities need to abide by these regulatory limits (Environment, 2000). Another point of interest is that the amount of nitrates in the stream water can be an indication of intensive land use for grazing. Certain parts of the Koiliaris watershed are under intense grazing, especially areas in higher altitudes (Stamati et al., 2011). Well-functioning soils can retain and transform nitrates and prevent their release to waterways. This kind of soil ES therefore provides a benefit to the municipality which would, in the absence of the service, have to remove the nitrates from the drinking water supply. The data on nitrate (NO_3^-) retention were provided on a watershed scale in a tabular and spatial format (ArcGIS) by project partners at the Technical University of Crete, who operate the CZO. The hydrological and chemical aspects of the Koiliaris watershed have been studied in detail (see Kourgialas et al., 2010; Moraetis et al., 2010; Nikolaidis et al., 2013; Sibetheros et al., 2013; Stamati et al., 2011), and the data used here on nitrates (NO_3^-) are derived from that work. The metric representing the service is nitrates flux (NO_3^-) kg ha^{-1} year^{-1}, and it captures the function of soil to retain nitrates (NO_3^-) from the groundwater. The potential maximum amount of leaching in the absence of soil ES vs the current state of nitrates leaching was used. The difference between the potential maximum amount of leaching and current state is the ES that the soil provides. The beneficiaries in this case would include the municipality as there is substantial cost involved cleaning nitrates (NO_3^-) from water sources. Derived from Jónsson and Davíðsdóttir (2016), the method of economic valuation is the avoided cost method, which bases the value of the service on the costs that would be incurred in the absence of the service.

2.3.2 Climate Regulation

The second regulating service assessed is climate regulation, which is defined as the role of soils in regulating global temperatures and precipitation through sequestration of C and N compounds and emissions of the related greenhouse gases. Climate regulation is of concern because of multiple factors. The soils in Crete are under threat because of intensive land use (Nikolaidis, 2011) and climate change (IPCC, 2014b). Soil carbon plays an important role in the stability and fertility of soils (Milne et al., 2015), and soils are an important sink for carbon globally. This can also be a significant source of CO_2 emissions due to mineralization of soil organic matter, if mismanaged (Milne et al., 2015). Soil carbon management is an essential part of climate change mitigation and adaptation, and therefore, soil carbon mineralization was selected as a proxy for climate regulation services. The metric chosen for this service is kg CO_2 ha^{-1} $year^{-1}$, that is, the carbon mineralization rate of the soil. Carbon mineralization can be considered a proxy for CO_2 emissions as it illustrates the amount of CO_2 that is needed to be sequestered at a minimum, in order to keep a carbon neutral balance within the soil system. Data on carbon mineralization (C kg ha^{-1} $year^{-1}$) were provided by project partners at the Technical University of Crete on a watershed scale in a tabular and spatial format (ArcGIS) and were translated into an economic value using the avoided cost method. The avoided cost method applied is based on the cost of GHG mitigation, which refers to the expected cost of mitigating CO_2 emissions. The beneficiaries are regional and national authorities, as they are responsible for carrying out climate change mitigation projects. The use of carbon mineralization rates as a proxy for climate regulations means that costs applied to these rates reflect the maximum value that can be achieved, i.e., from mitigating the entire emissions of CO_2 from soil. Because the data are a proxy for actual CO_2 emissions due to soil carbon mineralization (rather than carbon storage), the value derived from this proxy represents *potential* value of benefits to be achieved by implementing interventions in soil management in order to reduce carbon mineralization rates. It does not reflect the value of climate regulation ES under current practices. Carbon sequestration or net annual C addition to the soil would have been a better proxy for this service, but unfortunately was unavailable for the site at the watershed scale within the current dataset of the SoilTrEC project.

2.3.3 Biomass Production

The third service assessed is biomass production from crops and livestock. Soils provide nutrients, water, and the physical environment for terrestrial biomass production. Humans use biomass in the form of food, wood, fuel, and fiber. Intensive land use in the form of agriculture and livestock production plays an important role in the Crete economy, and this land use impacts the functionality of the soils. Koiliaris watershed produces a plethora of different agricultural products in the form of arable crops, horticulture crops, and livestock. The metric chosen for this service was biomass in kg ha^{-1} year^{-1}, that is, the biomass from crop and livestock produced annually in the watershed. The data on crop production were obtained from the Greek national statistical database on agricultural products (2014), and the data on livestock (sheep and goats) were provided by project partners at the Technical University of Crete. The output for each crop and livestock category was linked to the corresponding CORINE (EEA, 2007) land-use category using expert judgment and values from the literature, resulting in a spatial biomass map for the watershed. Four years (2002–2006) of agricultural statistical data for Chania municipality, obtained from the Greek national statistics database (2014), were used to calculate the mean average production rate in units of kg ha^{-1} year^{-1} for each crop category used in the survey for Chania municipality. The biomass map provided the basis for the economic valuation. The data on biomass production, measured in kg per biomass per production type, were provided on a watershed scale in a tabular and spatial format (ArcGIS) and were then translated into economic value. The economic method chosen for converting biomass production values to economic values utilized producers' prices (see Jónsson and Davíðsdóttir, 2016).

3. RESULTS

3.1 Filtering of Nutrients and Contaminants

The following assumptions are made: (1) nitrate retention is a good proxy for the filtering of nutrients and contaminants, (2) the cost numbers for nitrate pollution removal obtained from the European Nitrogen Assessment (Brink and van Grinsven, 2011) are a good representation of the pollution cost in the watershed, and (3) the biophysical data provide a good approximation of this soil ES service. Nitrate retention as a proxy for filtering of nutrients and contaminants service was previously used by Dominati et al. (2014a,b) (Fig. 4).

The European Nitrogen Assessment was the first continental-scale assessment of reactive nitrogen in the environment, and it included the first cost–benefit analysis for the different forms of reactive nitrogen (Sutton et al., 2011). The biophysical data were derived from a modified SWAT model adapted to the Koiliaris watershed. SWAT models are commonly used when modeling hydrological and geochemical properties of watersheds.

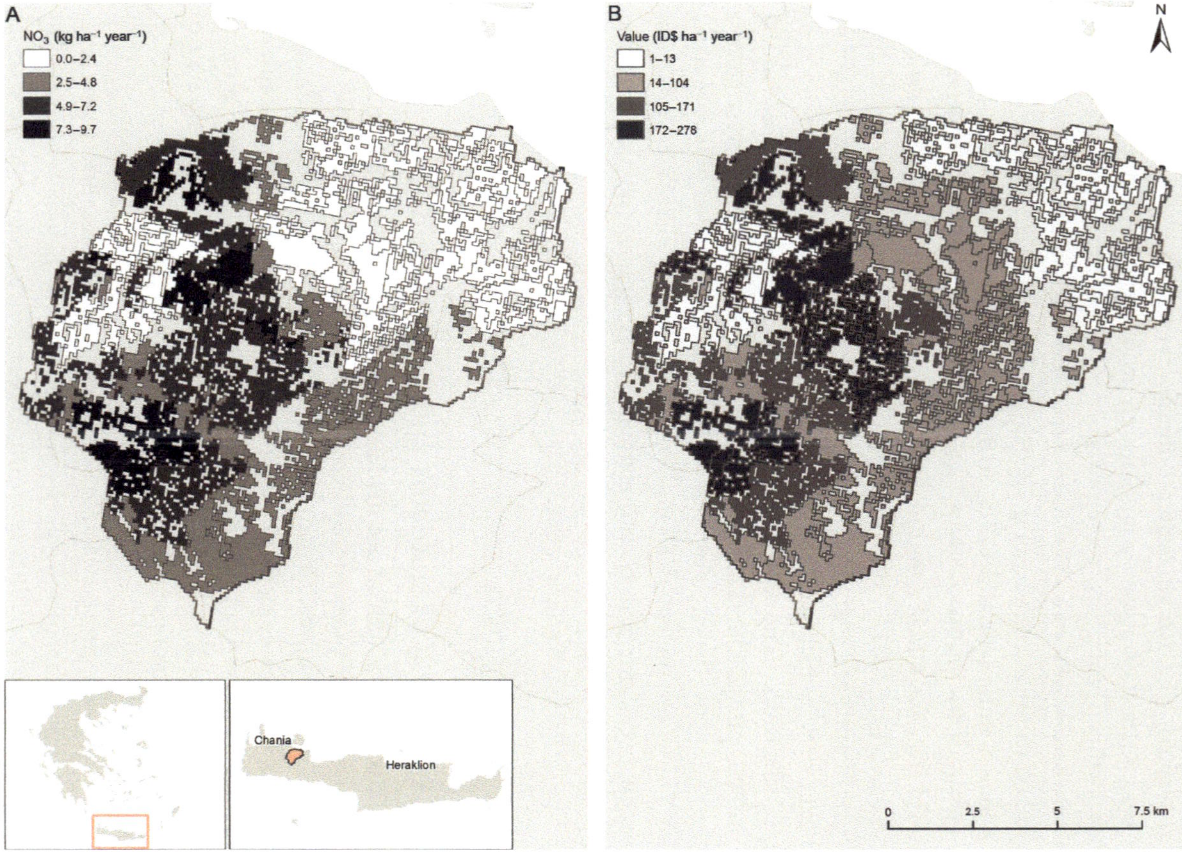

Fig. 4 Filtering of nutrients and contaminants—nitrate retention of soil. (A) Nitrate retention in kg ha^{-1} year^{-1}. (B) Value of nitrate retention in id$ ha^{-1} year^{-1}.

Table 5 Filtering Nutrients and Contaminants in Koiliaris Watershed

Input (kg ha^{-1} $year^{-1}$)	Output (kg ha^{-1} $year^{-1}$)
Nitrogen through fertilizer	Denitrification
NO_3^- through rain	NO_3^- in surface runoff
Organic N in manure from animal grazing	NO_3^- in lateral flow
	NO_3^- leached from the soil

The calculations and results for Koiliaris watershed were reported in Nikolaidis et al. (2013). In the SWAT model, Koiliaris watershed was broken down into 140 HRU. HRU are the smallest units of the SWAT model. The calculations of the HRU as a base unit of the SWAT model involved a number of threshold parameters, including how much each HRU had to contribute to the overall water flow in the watershed. If the HRU contribution was below a certain threshold, the area was ignored (left empty). The input and the output for the calculation of the nitrates flow in the watershed are shown in Table 5.

The soil ES were the difference between the input and the output, thus what the soil was able to retain of the NO_3^- in kg ha^{-1} $year^{-1}$. The potential leaching range maximum (absence of the soil ES) ranged from 0.54 kg to 26.12$(NO_3^-)ha^{-1}year^{-1}$. The current state of leaching ranged from 0.5 to 16.73$(NO_3^-)ha^{-1}year^{-1}$. Thus the soil ES retained from 0.03 to 9.65$(NO_3^-)ha^{-1}year^{-1}$ depending on the HRU. The economic cost numbers are based on estimations on nitrate unit damage cost of the major N pollutants, based on a number of studies reported in the European Nitrogen Assessment (Sutton et al., 2011). This unit damage cost ranged from a low of 6 id$ to a high of 29 id$ kg^{-1} N. Based on the soil retention and the unit damage cost, the estimated avoided cost provided by the soil ES filtering of nutrients and contaminants was in the range 0–280 id$ ha^{-1} $year^{-1}$.

3.2 Climate Regulation

The following assumptions are made: (1) carbon mineralization can be considered a proxy for climate regulation service, (2) the CO_2 mitigation cost numbers obtained from the Greek national bank are an appropriate representation of the value of the service, and (3) the measured range of mineralization in each soil type is an accurate representation of the mineralization of this soil type in the watershed.

Carbon sequestration and carbon accumulation have been used as a proxy for climate regulation service in number of studies (Dominati et al., 2014a; Glenk and Colombo, 2011; Pretty et al., 2000; Rodriguez-Entrena et al., 2012; Sandhu et al., 2008; Xiao et al., 2005). In this study, data on annual soil carbon accumulation were unavailable, and therefore, data on carbon mineralization rate were used to represent CO_2. Agriculture has a significant impact on the state of the carbon in the soil, and this method might serve as way of estimating the carbon cost of different farming methods, seeing whether the agricultural methods are enhancing the capabilities of the soil system to store carbon or if they are causing the system to release more into the atmosphere, thus creating a disservice (for more on ecosystem disservices and agriculture—see, for example, Kragt and Robertson, 2014; Power, 2010; Zhang et al., 2007). The mitigation cost numbers were obtained from the Greek national bank and represent the range of cumulative mitigation cost of a ton of CO_2 over the period of 2010–2050: "The total average cost of reducing greenhouse gas emission under the Mitigation Scenarios is estimated at between €190 and €240 tons of CO_2 (2008 prices), cumulatively for the period 2010–2050" (Bank of Greece, 2011: p. 435). The mitigation cost numbers were converted into 2012 id$, giving a range of 248–314 id$ per tons of CO_2. The mineralization rate of each soil types was based on measurements of the top 10 cm layer of the soil at three sample locations, for three of the four soil types in the watershed. Table 6 shows the soil type and the mineralization rate in t ha^{-1} $year^{-1}$ and the conversion to CO_2 in t ha^{-1} $year^{-1}$ using the carbon to CO_2 mass multiplier of 3.667.

Fig. 5A shows the carbon mineralization rate in C t ha^{-1} $year^{-1}$ on a watershed scale for the different soils (for more information different soil profiles, see Moraetis et al., 2015) at the Koiliaris watershed representing the value of climate regulation services. Fig. 5B maps the estimated value of the carbon mineralization. By using carbon mineralization rates as a proxy for climate regulation, this approach yields negative values (costs) for current practices and reflects the potential for cost avoidance that could be achieved if agricultural practices were to be implemented to reduce to zero this proxy measurement for CO_2 emissions from soil.

The results for climate regulation are in the range of −2200 to −5610 id$ ha^{-1} $year^{-1}$, depending on the mineralization rate of the soil types and the mitigation cost range, high vs low (Table 7). There is uncertainty regarding whether the annual carbon sequestration is enough to keep the soil carbon stock in balance, whether soil carbon is increasing, or if it is

Table 6 Soil Type and Carbon Mineralization Rate

Soil Type	**C Mineralization ($t\ ha^{-1}\ year^{-1}$)**	**CO_2 ($t\ ha^{-1}\ year^{-1}$)**
Calcareous rendzines soil and Mediterranean brown		
Low	2.42	8.89
Middle	2.57	9.42
High	4.71	17.27
Brown and red-brown alkaline Mediterranean		
Low	2.46	9.02
Middle	2.77	10.15
High	2.81	10.30
Podzols mixed with forest acid red		
Low	2.75	10.07
Middle	2.83	10.38
High	4.87	17.87

insufficient to match the carbon loss. The measurements that would substantiate this were unavailable at the time of the study.

3.3 Biomass Production From Crop and Livestock

The following assumptions are made: (1) biomass from crop and livestock is a good proxy for the biomass production service, (2) the biomass numbers from crops and livestock allocated to the different CORINE land-use categories are a sound representation of the biomass production from soils in the Koiliaris watershed, and (3) Greek national prices on agricultural commodities obtained from the FAO database (FAO, 2014) are representative of the producers' prices in the Koiliaris watershed. Agricultural products (from crops and livestock) are often used as proxies for biomass production from ecosystems (Dominati et al., 2014a,b; Porter et al., 2009; Sandhu et al., 2010). The biomass numbers from crops and livestock are formed by an expert judgment, by the authors, and by project partners at the Technical University of Crete, on the biomass output of Koiliaris watershed and are the best information available for the area. The expert judgment is based on national agricultural production statistics for the Chania region

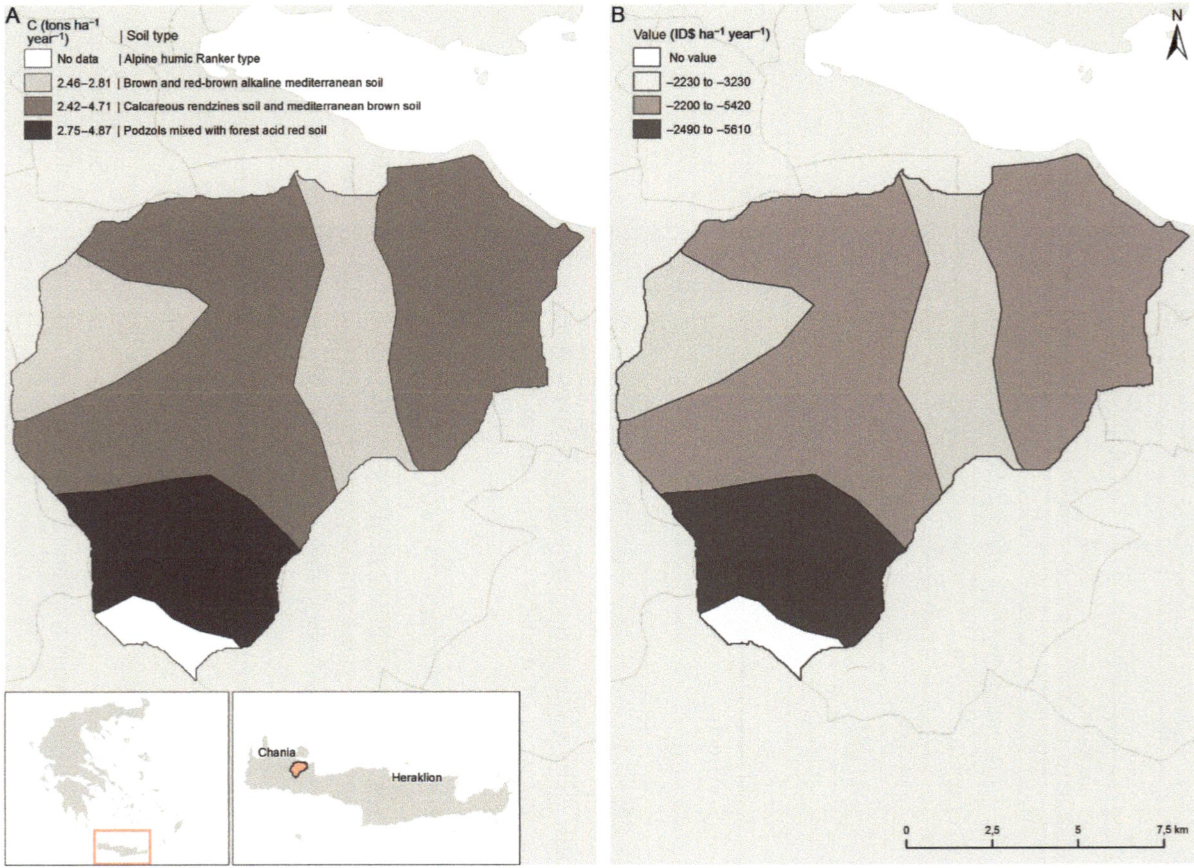

Fig. 5 Climate regulation. (A) Carbon outflow in tons ha^{-1} $year^{-1}$. (B) Value of climate regulation in id\$ ha^{-1} $year^{-1}$.

Table 7 CO_2 and Mitigation Costs

	CO_2 in Soil Types (t ha^{-1} $year^{-1}$)	id$248 Mitigation Cost Low	id$314 Mitigation Cost High
Min	8.89	−2200	−2790
Max	17.87	−4430	−5610

(Hellenic Statistical Authority, 2014), taking into consideration current land use and agricultural practices (see Section 2.3.3 for more detail).

The Greek national prices on agricultural commodities represent the most complete dataset on agricultural commodities for the products produced in Koiliaris. Fig. 6B shows the biomass production and it ranges from 0 to 69,148 in kg ha^{-1} $year^{-1}$.

The estimated average economic values are reported in four main categories:

1. Grassland, shrub land, and pasture 740 id$ ha^{-1} $year^{-1}$ (270–1580).
2. Nonirrigated arable land 2230 id$ ha^{-1} $year^{-1}$ (440–4730).
3. Fruit and berry plantations 7500 id$ ha^{-1} $year^{-1}$ (6100–7580).
4. Olive groves 10,810 id$ ha^{-1} $year^{-1}$ (6540–10,910).

Price data were obtained from FAO database on Greek agricultural products for a 5-year period (FAO, 2014). Many of the areas are fertilized with inorganic fertilizer, and therefore, a proportion of the economic value can be attributed to the inorganic fertilizer (see Section 1.5.2). The following list is the soil ES contribution for the four main categories (provided by Technical University Crete):

1. Grassland, shrub land, and pasture 740 id$ ha^{-1} $year^{-1}$ or 100% of the biomass production is attributed to the soil ES.
2. Nonirrigated arable land 1115 id$ ha^{-1} $year^{-1}$ or 50% of the output is attributed to the soil ES.
3. Fruit and berry plantations 4500–5250 id$ ha^{-1} $year^{-1}$ or 60–70% can be attributed to soil ES.
4. Olive groves 6480–7560 id$ ha^{-1} $year^{-1}$ or 60–70% of the economic value can be attributed to soil ES.

Some of the areas on the biomass map (Fig. 6A) have no value data as no economic value could be found for the kind of crops allocated to these areas. It is then safe to assume that the values of the biomass production in the area are underrepresented. There is some uncertainty involved in allocating the different crops from the Greek national statistics on the agricultural output in Chania municipality to the correct CORINE land-use categories to obtain

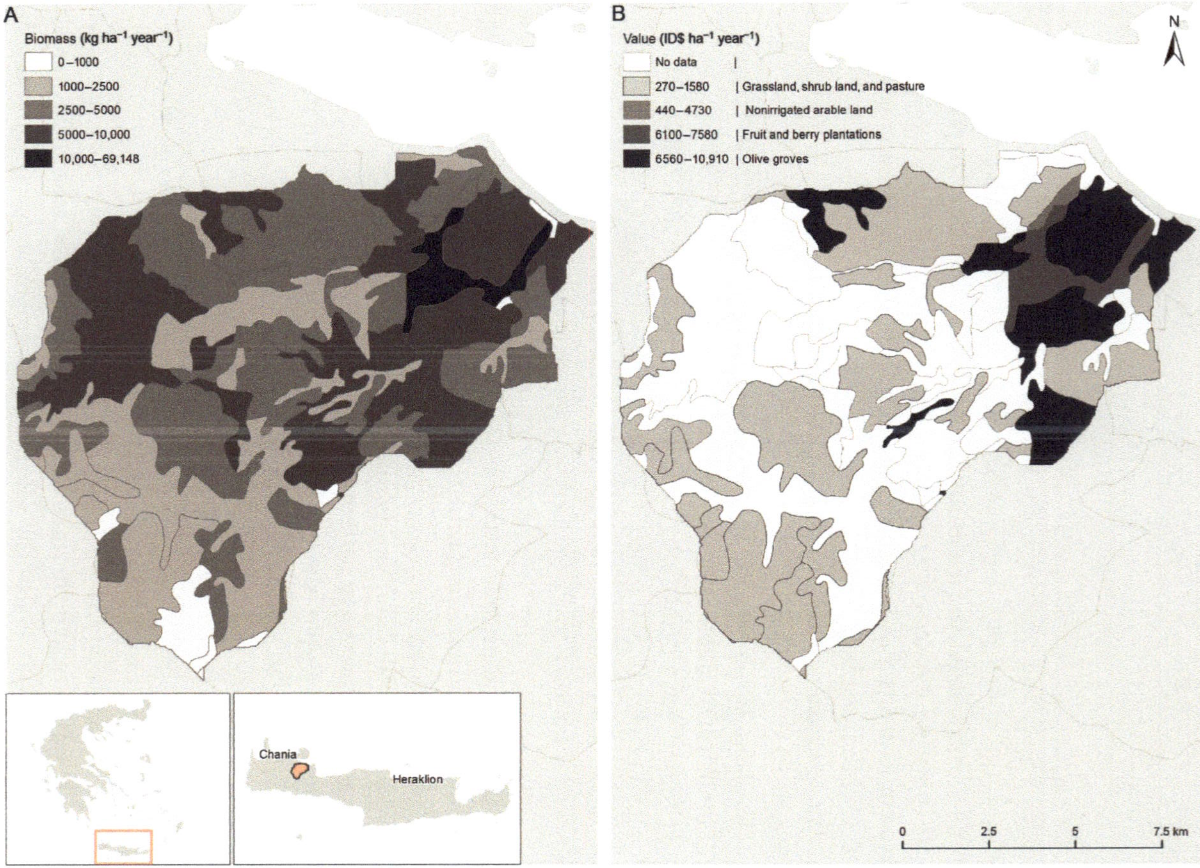

Fig. 6 Biomass provision from crop and livestock. (A) Biomass provision in kg ha^{-1} $year^{-1}$. (B) Value of biomass provision in id$ ha^{-1} $year^{-1}$.

the biomass values. The national statistics is not spatially allocated, thus expert judgment had to be used to allocate the crop categories, and this could potentially lead to misplacement. The allocation of the livestock biomass production faced similar problems as there was incomplete information available regarding the species of sheep and goats in the watershed. As different species produce different quantities of milk and meat, an average number for milk and meat production, for both goats and sheep, was assumed.

3.4 Summary

Owing to different scales of the data, it is difficult to evaluate trade-offs between services. The data on filtering of nutrient and contaminants are the most detailed, followed by the biomass provision and then climate regulation. Nevertheless, the following observation can be made: the value of the biomass production is highest in the lower part of the watershed, where the filtering service is low and the climate regulation service is high. The high values of the filtering service in the watershed are in the intensive grazing areas in the higher altitudes of the watershed, and also grazing in sloping areas. On flat land, as in the lower part of the watershed where most of the crop production takes places, there are low values for the filtering service. On the other hand, where there is intensive grazing on pasture and grassland in the higher altitudes and sloping areas of the watershed, there is a trade-off between the biomass production from livestock and the nitrate flow that results from the grazing activity. Grazing increases the livestock's biomass, but because of its intensity, it potentially decreases the capability of the soil system to retain nitrate as much as it could from the livestock manure. As we base the filtering service on the data that is available, it is only measuring the quantity of $(NO_3^-)\,ha^{-1}\,year^{-1}$ captured by the soil (the difference between the current output and the maximum output in the absence of soil ES). Therefore, it is hard to generalize about the relationship between inputs, retention, and outputs as it depends on many different soil parameters. For example, an area that receives high input (rain and fertilizer) might have high retention relative to other areas, measured in $kg(NO_3^-)\,ha^{-1}\,year^{-1}$, and still have high output of nitrate. More data are needed to establish the exact nature of this relationship.

In summary, the value for the three soil ES is:

- Filtering of nutrients and contaminants 0–278 id\$ ha^{-1} year^{-1}
- Climate regulation −2200 to −5610 id\$ ha^{-1} year^{-1}
- Biomass provision 740–7560 id\$ ha^{-1} year^{-1}

All of the services show a substantial range in value with the highest valued services being biomass provision. The regulation services, filtering of nutrients and contaminants, and climate regulation are assumed to be pure soil ES; that is, they are unaffected by the ecological subsidy that influences the biomass provision service.

3.5 Discussion

We have introduced a framework to assess the relevance of sustaining soil functions that links together the concepts of soil natural capital, soil biophysical support functions, and soil ES with beneficiaries and economic valuation. We have defined and categorized the different components of the framework, illustrating their functions within the framework and how these various parts interlink and create benefits for humans, which can be valued economically. As a proof of the concept, the soil framework was applied to a pilot study in Koiliaris watershed on the island of Crete, Greece. The results from the pilot study show that there are important and valuable soil ES in Koiliaris that are delivering benefits to humans. The relative economic values were demonstrated by biomass provisioning, climate regulation, and then filtering of nutrients and contaminants, in the order of highest to lowest value, respectively. When comparing the results to other studies (Jónsson and Davíðsdóttir, 2016), the economic values for biomass provision fall within the range that has previously been reported. However, the values for climate regulation and filtering of nutrients and contaminants are substantially different. Studies have reported values for biomass production ranging from 231 to 22,219 id\$ ha^{-1} $year^{-1}$ (Decaens et al., 2006; Dominati et al., 2014a,b; Haley, 2006; Porter et al., 2009; Sandhu et al., 2008) compared to 740 to 7560 id\$ ha^{-1} $year^{-1}$ for the biomass provision in Koiliaris. The reported values for filtering of nutrients and contaminants services range from 544 to 6402 id\$ ha^{-1} $year^{-1}$ (Dominati et al., 2014a,b), compared to 0 to 278 id\$ ha^{-1} $year^{-1}$ for Koiliaris. The climate regulation service from soil has been reported from 2 to 268 id\$ ha^{-1} $year^{-1}$ (Dominati et al., 2014b; Glenk and Colombo, 2011; Pretty et al., 2000; Rodriguez-Entrena et al., 2012; Sandhu et al., 2008; Xiao et al., 2005) compared to −2200 to −5610 id\$ ha^{-1} $year^{-1}$ at Koiliaris. The most noticeable difference can be seen in the climate regulation service from soils. This is because of the use of the annual mineralization rate of carbon, not the annual addition to its stock, along with the Greek climate mitigation cost estimates. By using annual mineralization rate of carbon as a proxy for CO_2 emission, we are looking at the potential value of the soil to prevent GHG emissions. As we do not know the annual carbon sequestration rate, we assume that it has to be at least equal to the emission rate to keep the system in balance. If GHG emissions from the soil profile would be totally unmitigated, it would result in cost that would have to be mitigated with other means. We thus report negative values for the carbon emitted from the different soil profiles. We use the mitigation cost numbers for climate regulation instead of market price of carbon, it better reflects the social cost of climate change as shown by the WTP for mitigation. The carbon stock exchanges have gone through turmoil during recent years, and there is substantial disparity between what the market is willing to pay for carbon and the social or mitigation cost. Thus, mitigation cost is opted for instead of market prices. If the market price of CO_2, based on the price of 6 id\$ per ton of CO_2 in the EU carbon stock exchange in December 2012 (World Bank, 2013), had been used, then the value of the climate regulation services for Koiliaris based on the carbon mineralization rate would have been in the range −63 to −127 id\$ ha^{-1} $year^{-1}$, depending on different soil types.

As in other valuation studies, there are study limitations as follows. Local economic data were difficult to find, and thus cost and price values had to be obtained from national databases and other sources. Although Koiliaris watershed has a plethora of biophysical measurements on the different aspects of the watershed, some of the biophysical data used in the pilot were unavailable as direct measurements or from modeling, and thus expert judgment was needed regarding the allocation and output of the services. Expert judgment was thus used regarding livestock concentration, livestock biomass yields, and crop biomass allocation within the watershed. With an improved land-use map of the watershed, the allocation of livestock and crop biomass can be improved.

With better data on land-use and carbon stocks and flow within the watershed, it would be interesting for future studies to look at the effects of different management decisions on soil ES and how climate change will affect the provisioning and economic value of services in the Koiliaris watershed. This could be done by further linking the soil framework proposed here to the modeling put forward in the SoilTrEC project (Banwart et al., 2012), thus tying further together the biophysical and economic aspect of soils, so both the physical and economic flows are visible.

Despite these limitations, the soil ES framework presented in this chapter can serve as a useful promoter of the sustainable management of soils, as it highlights in a holistic manner the economic benefit of different soil ES, in addition to estimating the economic cost of soil degradation and the associated loss of soil ES. This kind of information should be useful for land managers and policy makers as it gives them a better understanding of the possible trade-offs in land-use management. There are important caveats though. When relying solely on the economic value of soil ES in the context of land management, there is the inherent danger of focusing solely on provisioning services that are usually the highest yielding service in economic value, with possible negative consequences in the long run for other soil functions and services, such as soil biodiversity. This could lead to the mismanagement of the soil resource, and significant economic costs in the long run. Furthermore, a young soil that contains underdeveloped soil functions and services is likely to receive a low value and therefore be deemed unimportant. Therefore, it must be stressed that the economic value of soil ES certainly conveys important information, but should be presented in context with other metrics necessary for sustainable land management (see Jónsson et al., 2016, for examples, of soil sustainability indicators).

3.6 Conclusion

This chapter illustrates the importance of soils and their economic value and introduces a framework for the economic valuation of soil ES. A holistic framework for defining and assessing the economic value of soil ES is an important contribution in the quest for creating usable tools for decision makers that can illustrate the importance of fertile soils, and the economic costs of soil degradation. Soil ecosystems provide multiple benefits for humans but are often neglected in land-use decision-making processes. Soils play a central role in Earth's Critical Zone, which supplies this most important life-sustaining resource, and one way of protecting resources such as soil is to value their economic contribution and include it in land-use decision-making process.

ACKNOWLEDGMENTS

This work is supported by the European Commission 7th Framework Programme as a Large Integrating Project "Soil Transformations in European Catchments" (SoilTrEC, http://www.soiltrec.eu, Grant Agreement No. 244118). The authors would like to thank the SoilTrEC partners at the Technical University Crete, Greece, for their assistance on data collection and analysis, David Cook for his assistance with the manuscript, and Kristín Vala Ragnarsdóttir for her help and useful comments along the way.

REFERENCES

Ashman, M.R., Puri, G., 2002. Essential Soil Science: A Clear and Concise Introduction to Soil Science. Blackwell Publishing, Malden, MA.

Bank of Greece, 2011. The Environmental, Economic and Social Impacts of Climate Change in Greece. Climate Change Impacts Study Committee, Athens.

Banwart, S., Menon, M., Bernasconi, S.M., Bloem, J., Blum, W.E.H., de Souza, D.M., Davidsdotir, B., Duffy, C., Lair, G.J., Kram, P., Lamacova, A., Lundin, L., Nikolaidis, N.P., Novak, M., Panagos, P., Ragnarsdottir, K.V., Reynolds, B., Robinson, D., Rousseva, S., de Ruiter, P., van Gaans, P., Weng, L.P., White, T., Zhang, B., 2012. Soil processes and functions across an international network of critical zone observatories: introduction to experimental methods and initial results. C. R. Geosci. 344, 758–772.

Bardi, U., El Asmar, T., Lavacchi, A., 2013. Turning electricity into food: the role of renewable energy in the future of agriculture. J. Clean. Prod. 53, 224–231.

Bellamy, P.H., Loveland, P.J., Bradley, R.I., Lark, R.M., Kirk, G.J.D., 2005. Carbon losses from all soils across England and Wales 1978-2003. Nature 437, 245–248.

Belyazid, S., 2006. Dynamic Modelling of Biogeochemical Processes in Forest Ecosystems. Department of Chemical Engineering, Lund University, Sweden.

Boyd, J., Banzhaf, S., 2007. What are ecosystem services? The need for standardized environmental accounting units. Ecol. Econ. 63, 616–626.

Brantley, S.L., Goldhaber, M.B., Ragnarsdottir, K.V., 2007. Crossing disciplines and scales to understand the critical zone. Elements 3, 8.

Brink, C., van Grinsven, H., 2011. Costs and benefits of nitrogen in the environment. In: Sutton, M.A., Howard, C.M., Erisman, J.W., Billen, G., Bleeker, A., Grennfelt, P., van Grinsven, H., Grizzetti, B. (Eds.), The European Nitrogen Assessment. Cambridge University Press, Cambridge, pp. 513–540.

Bristow, K.L., Marchant, S.M., Deurer, M., Clothier, B.E., 2010. Enhancing the ecological infrastructure of soils. In: World Congress of Soil Science, Soil Solutions for a Changing World, Brisbane, Australia.

Brussaard, L., de Ruiter, P.C., Brown, G.G., 2007. Soil biodiversity for agricultural sustainability. Agric. Ecosyst. Environ. 121, 233–244.

Cao, Z.P., Han, X.M., Hu, C., Chen, J., Zhang, D.P., Steinberger, Y., 2011. Changes in the abundance and structure of a soil mite (Acari) community under long-term organic and chemical fertilizer treatments. Appl. Soil Ecol. 49, 131–138.

Costanza, R., Daly, H.E., 1992. Natural capital and sustainable development. Conserv. Biol. 6, 37–46.

Daily, G.C., Matson, P.A., Vitousek, P.M., 1997. Ecosystem services supplied by soil. In: Daily, G.C. (Ed.), Nature's Services: Societal Dependence on Natural Ecosystems. Island Press, Washington, DC, pp. 113–132.

Dale, V.H., Polasky, S., 2007. Measures of the effects of agricultural practices on ecosystem services. Ecol. Econ. 64, 286–296.

D'Costa, V.M., McGrann, K.M., Hughes, D.W., Wright, G.D., 2006. Sampling the antibiotic resistome. Science 311, 374–377.

Decaens, T., Jimenez, J.J., Gioia, C., Measey, G.J., Lavelle, P., 2006. The values of soil animals for conservation biology. Eur. J. Soil Biol. 42, S23–S38.

DEFRA, 2009. Safeguarding Our Soils. A Strategy for England. Department for Environment, Food and Rural Affairs, London, United Kingdom.

de Groot, R.S., Wilson, M.A., Boumans, R.M.J., 2002. A typology for the classification, description and valuation of ecosystem functions, goods and services. Ecol. Econ. 41 (3), 393–408, ISSN: 0921-8009. http://dx.doi.org/10.1016/S0921-8009(02)00089-7 (http://www.sciencedirect.com/science/article/pii/S0921800902000897).

Dominati, E., Patterson, M., Mackay, A., 2010. A framework for classifying and quantifying the natural capital and ecosystem services of soils. Ecol. Econ. 1858–1868.
Dominati, E., Mackay, A., Green, S., Patterson, M., 2014a. A soil change-based methodology for the quantification and valuation of ecosystem services from agro-ecosystems: a case study of pastoral agriculture in New Zealand. Ecol. Econ. 100, 119–129.
Dominati, E.J., Mackay, A., Lynch, B., Heath, N., Millner, I., 2014b. An ecosystem services approach to the quantification of shallow mass movement erosion and the value of soil conservation practices. Ecosyst. Serv. 9, 204–215.
Dominati, E.J., Robinson, D.A., Marchant, S.C., Bristow, K.L., Mackay, A.D., 2014c. Natural capital, ecological infrastructure, and ecosystem services in agroecosystems. In: Van Alfen, N.K. (Ed.), Encyclopedia of Agriculture and Food Systems. Academic Press, Oxford, pp. 245–264, ISBN: 9780080931395. http://dx.doi.org/10.1016/B978-0-444-52512-3.00243-6.
EEA, 2007. Corine land cover 2000, seamless vector data—Version 9. In: Agency, E.E. (Ed.), European Environment Agency. Copenhagen, Denmark.
eftec, 2005. The Economic, Social and Ecological Value of Ecosystem Services: A Literature Review. Economics for the Environment Consultancy, London.
European Commission Directorate General for Environment, 2000. 'Nitrates' Directive (91/676/EEC): Status and Trends of Aquatic Environment and Agricultural Practice: Development Guide for Member States' Reports. Office for Official Publications of the European Communities, Luxembourg.
European Commission, 2006. Thematic Strategy for Soil Protection. Commission of the European Communities, Brussels.
European Soil Bureau Network, E.C., 2005. Soil Atlas of Europe. Office for Official Publications of the European Communities, Luxembourg.
FAO, 2002. International Code of Conduct on the Distribution and Use of Pesticides. Food and Agriculture Organization of the United Nations, Rome.
FAO, 2014. Producers Price—Annual. Food and Agricultural Organization of the United Nations, Rome, Italy.
Frossard, E., Blum, W.E.H., Warkentin, B.P., 2006. Function of Soils for Human Societies and the Environment. Geological Society, London.
Giannakis, G.V., Panakoulia, S.K., Nikolaidis, N.P., Paranychianakis, N.V., 2014. Simulating soil fertility restoration using the CAST model. Procedia Earth Planet. Sci. 10, 325–329.
Giannakis, G.V., et al., 2017. Integrated Critical Zone Model (1D-ICZ): a tool for dynamic simulation of soil functions and soil structure. Adv. Agron. 142, 277–314.
Glenk, K., Colombo, S., 2011. Designing policies to mitigate the agricultural contribution to climate change: an assessment of soil based carbon sequestration and its ancillary effects. Clim. Change 105, 43–66.
Haley, S., 2006. Sugar and Sweeteners Outlook/SSS-246/May 30 2006. Economic Research Service, USDA, Washington, DC.
Haygarth, P.M., Ritz, K., 2009. The future of soils and land use in the UK: soil systems for the provision of land-based ecosystem services. Land Use Policy 26, S187–S197.
The Hellenic Statistical Authority (ELSTAT), 2014. Statistical Database—Agriculture, Livestock. Athens, Greece.
IPCC, 2014a. Climate change 2013—the physical science basis. In: Working Group I Contribution to the Fifth Assessment Report of the Intergovernmental Panel on Climate Change. Cambridge University Press, Cambridge, United Kingdom and New York, NY.
IPCC, 2014b. Climate change 2014—synthesis report. Cambridge University Press.
Johnston, R.J., Russell, M., 2011. An operational structure for clarity in ecosystem service values. Ecol. Econ. 70, 2243–2249.
Jónsson, J.Ö.G., Davíðsdóttir, B., 2016. Classification and valuation of soil ecosystem services. Agric. Syst. 145, 24–38.

Jónsson, J.Ö.G., Davíðsdóttir, B., Jónsdóttir, E.M., Kristinsdóttir, S.M., Ragnarsdóttir, K.V., 2016. Soil indicators for sustainable development: a transdisciplinary approach for indicator development using expert stakeholders. Agric. Ecosyst. Environ. 232, 179–189.

Kourgialas, N.N., Karatzas, G.P., Nikolaidis, N.P., 2010. An integrated framework for the hydrologic simulation of a complex geomorphological river basin. J. Hydrol. 381, 308–321.

Kragt, M.E., Robertson, M.J., 2014. Quantifying ecosystem services trade-offs from agricultural practices. Ecol. Econ. 102, 147–157.

Kumar, P., 2010. The Economics of Ecosystems and Biodiversity: Ecological and Economic Foundations. Routledge Taylor & Francis Group, London and New York.

MEA, 2005. Millennium Ecosystem Assessment, Ecosystem and Human Well-Being: A Framework for Assessment. Island Press, Washington, DC.

Milne, E., Banwart, S.A., Noellemeyer, E., Abson, D.J., Ballabio, C., Bampa, F., Bationo, A., Batjes, N.H., Bernoux, M., Bhattacharyya, T., Black, H., Buschiazzo, D.E., Cai, Z., Cerri, C.E., Cheng, K., Compagnone, C., Conant, R., Coutinho, H.L.C., de Brogniez, D., Balieiro, F.D.C., Duffy, C., Feller, C., Fidalgo, E.C.C., da Silva, C.F., Funk, R., Gaudig, G., Gicheru, P.T., Goldhaber, M., Gottschalk, P., Goulet, F., Goverse, T., Grathwohl, P., Joosten, H., Kamoni, P.T., Kihara, J., Krawczynski, R., La Scala Jr., N., Lemanceau, P., Li, L., Li, Z., Lugato, E., Maron, P.-A., Martius, C., Melillo, J., Montanarella, L., Nikolaidis, N., Nziguheba, G., Pan, G., Pascual, U., Paustian, K., Piñeiro, G., Powlson, D., Quiroga, A., Richter, D., Sigwalt, A., Six, J., Smith, J., Smith, P., Stocking, M., Tanneberger, F., Termansen, M., van Noordwijk, M., van Wesemael, B., Vargas, R., Victoria, R.L., Waswa, B., Werner, D., Wichmann, S., Wichtmann, W., Zhang, X., Zhao, Y., Zheng, J., Zheng, J., 2015. Soil carbon, multiple benefits. Environ. Dev. 13, 33–38.

Minton, N.P., 2003. Clostridia in cancer therapy. Nat. Rev. Microbiol. 1, 237–242.

Moraetis, D., Efstathiou, D., Stamati, F., Tzoraki, O., Nikolaidis, N.P., Schnoor, J.L., Vozinakis, K., 2010. High-frequency monitoring for the identification of hydrological and bio-geochemical processes in a Mediterranean river basin. J. Hydrol. 389, 127–136.

Moraetis, D., Paranychianakis, N.V., Nikolaidis, N.P., Banwart, S.A., Rousseva, S., Kercheva, M., Nenov, M., Shishkov, T., de Ruiter, P., Bloem, J., Blum, W.E.H., Lair, G.J., van Gaans, P., Verheul, M., 2015. Sediment provenance, soil development, and carbon content in fluvial and manmade terraces at Koiliaris River Critical Zone Observatory. J. Soil. Sediment. 15, 347–364.

Mozumdera, P., Berrens, R.P., 2007. Inorganic fertilizer use and biodiversity risk: an empirical investigation. Ecol. Econ. 62, 538–543.

Nikolaidis, N.P., 2011. Human impacts on soils: tipping points and knowledge gaps. Appl. Geochem. 26 (Suppl.), S230–S233.

Nikolaidis, N.P., Bouraoui, F., Bidoglio, G., 2013. Hydrologic and geochemical modeling of a karstic Mediterranean watershed. J. Hydrol. 477, 129–138.

Pimentel, D., 2006. Soil erosion: a food and environmental threat. Environ. Dev. Sustain. 8, 119–137.

Porter, J., Costanza, R., Sandhu, H., Sigsgaard, L., Wratten, S., 2009. The value of producing food, energy, and ecosystem services within an agro-ecosystem. Ambio 38, 186–193.

Power, A.G., 2010. Ecosystem services and agriculture: tradeoffs and synergies. Philos. Trans. R. Soc. Lond. B Biol. Sci. 365, 2959–2971.

Pretty, J.N., Brett, C., Gee, D., Hine, R.E., Mason, C.F., Morison, J.I.L., Raven, H., Rayment, M.D., van der Bijl, G., 2000. An assessment of the total external costs of UK agriculture. Agric. Syst. 65, 113–136.

Regnier, P., O'Kane, J.P., Steefel, C.I., Vanderborght, J.P., 2002. Modeling complex multi-component reactive-transport systems: towards a simulation environment based on the concept of a knowledge base. App. Math. Model. 26, 913–927.

Robinson, D.A., Lebron, I., Vereecken, H., 2009. On the definition of the natural capital of soils: a framework for description, evaluation, and monitoring. Soil Sci. Soc. Am. J. 69, 738–747.

Rodriguez-Entrena, M., Barreiro-Hurle, J., Gomez-Limon, J.A., Espinosa-Goded, M., Castro-Rodriguez, J., 2012. Evaluating the demand for carbon sequestration in olive grove soils as a strategy toward mitigating climate change. J. Environ. Manage. 112, 368–376.

Sandhu, H.S., Wratten, S.D., Cullen, R., Case, B., 2008. The future of farming: the value of ecosystem services in conventional and organic arable land. An experimental approach. Ecol. Econ. 64, 835–848.

Sandhu, H.S., Wratten, S.D., Cullen, R., 2010. Organic agriculture and ecosystem services. Environ. Sci. Policy 13, 1–7.

Sarathchandra, S.U., Ghani, A., Yeates, G.W., Burch, G., Cox, N.R., 2001. Effect of nitrogen and phosphate fertilisers on microbial and nematode diversity in pasture soils. Soil Biol. Biochem. 33, 953–964.

Scharlemann, J.P.W., Tanner, E.V.J., Hiederer, R., Kapos, V., 2014. Global soil carbon: understanding and managing the largest terrestrial carbon pool. Carbon Manag. 5, 81–91.

Science, 2004. Soils—the final frontier. Science 304, 1613–1637.

Sibetheros, I.A., Nerantzaki, S., Efstathiou, D., Giannakis, G., Nikolaidis, N.P., 2013. Sediment transport in the Koiliaris River of Crete. Procedia Technol. 8, 315–323.

Šimůnek, J., Šejna, M., Saito, H., Sakai, M., van Genuchten, M.T., 2009. The HYDRUS-1D Software Package for Simulating the One-Dimensional Movement of Water, Heat, and Multiple Solutes in Variably-Saturated Media Version 4.08. Department of Environmental Sciences, University of California Riverside, Riverside, CA.

Stamati, F.E., Nikolaidis, N.P., Venieri, D., Psillakis, E., Kalogerakis, N., 2011. Dissolved organic nitrogen as an indicator of livestock impacts on soil biochemical quality. Appl. Geochem. 26, S340–S343.

Stamati, F.E., Nikolaidis, N.P., Banwart, S., Blum, W.E.H., 2013. A coupled carbon, aggregation, and structure turnover (CAST) model for topsoils. Geoderma 211–212, 51–64.

Stockdale, E.A., Watson, C.A., Black, H.I.J., Philipps, L., 2006. Do farm management practices alter below-ground biodiversity and ecosystem function? Implications for sustainable land management: JNCC report no. 364. JNCC, Peterborough.

Sutton, M.A., Howard, C.M., Erisman, J.W., Billen, G., Bleeker, A., Grennfelt, P., van Grinsven, H., Grizzetti, B., 2011. The European Nitrogen Assessment: Sources, Effects and Policy Perspectives. Cambridge University Press, Cambridge, UK.

Swinton, S.M., Lupi, F., Robertson, G.P., Hamilton, S.K., 2007. Ecosystem services and agriculture: cultivating agricultural ecosystems for diverse benefits. Ecol. Econ. 64, 245–252.

Tilman, D., Cassman, K.G., Matson, P.A., Naylor, R., Polasky, S., 2002. Agricultural sustainability and intensive production practices. Nature 418, 671–677.

Turbé, A., Toni, A., Benito, P., Lavelle, P., Lavelle, P., Ruiz, N., der Putten, Van, Labouze, E., Mudgal, S., 2010. Soil Biodiversity: Functions, Threats and Tools for Policy Makers. Bio Intelligence Service S.A.S., Paris, France

Vitousek, P.M., Aber, J.D., Howarth, R.W., Likens, G.E., Matson, P.A., Schindler, D.W., Schlesinger, W.H., Tilman, D., 1997. Human alteration of the global nitrogen cycle: sources and consequences. Ecol. Appl. 7, 737–750.

Weber, J.-L., 2007. Accounting for Soil in the SEEA: Session 8 on Assets Accounts. European Environment Agency, Rome.

World Bank, 2013. Mapping Carbon Pricing Initiatives. ECOFYS, Washington, DC.

Xiao, Y., Xie, G.D., Lu, C.X., Ding, X.Z., Lu, Y., 2005. The value of gas exchange as a service by rice paddies in suburban Shanghai, PR China. Agric. Ecosyst. Environ. 109, 273–283.

Zhang, W., Ricketts, T.H., Kremen, C., Carney, K., Swinton, S.M., 2007. Ecosystem services and disservices to agriculture. Ecol. Econ. 64, 253–260.

CHAPTER THIRTEEN

European Contribution Towards a Global Assessment of Agricultural Soil Organic Carbon Stocks

Y. Yigini, L. Montanarella, P. Panagos[1]
European Commission, Joint Research Centre, Sustainable Resources, Ispra (VA), Italy
[1]Corresponding author: e-mail address: panos.panagos@jrc.ec.europa.eu

Contents

Abstract

The chapter discusses a study that predicts the global organic carbon stocks for agricultural soils using European databases with geostatistical analysis and modeling. The overall statistical model consists of two submodels namely *donor* and *donee* modules. The donor module uses statistics to quantify the relationships between soil organic carbon (SOC) and environmental covariates. The covariates were selected based on their availability at global scale and their roles as major drivers that affect the carbon cycle in terrestrial ecosystems. Multiple linear regression was used in the donor module with the selected covariates and dense SOC measurements coming from LUCAS soil database (Toth et al., 2013a). The LUCAS soil database has more than 22,000 SOC measurements from European countries and a standardized sampling procedure was used routinely to collect samples of around 0.5 kg of topsoil (0–20 cm) each. The *donor* module reveals and quantifies the relationships between SOC mass concentration in soil and the predictors to be used in the *donee* model to extend the prediction at global scale

Advances in Agronomy, Volume 142
ISSN 0065-2113
http://dx.doi.org/10.1016/bs.agron.2016.10.012

using the same set of predictors. We used the WorldClim dataset (Hijmans et al., 2005), which is comprised of global climate data layers representing long-term conditions for the years from 1950 to 2000. The land cover data were extracted from the GlobCover 2009 (ESA and Universite' Catholique de Louvain, 2010) provided by the European Space Agency (ESA), the terrain parameters were derived from CGIAR-CSI SRTM 90 m Database (Jarvis et al., 2008), the soil layers obtained from Harmonized World Soil Database (FAO/IIASA/ISRIC/ISSCAS/JRC, 2012), and the Normalized Difference Vegetation Index (NDVI) data were obtained from Copernicus Global Land Service Data Portal (Copernicus Global Land Service, 2015). The study yielded promising results which are broadly consistent with similar efforts predicting global agricultural SOC stocks. Our model fits the SOC data well ($R^2 = 0.35$) and preliminary results suggest a global agricultural SOC estimate of 100.34 Pg (Petagrams) in the first 20 cm. The study predicts the global agricultural SOC stocks using a geostatistical approach and the results are consistent with previous studies that used process-based SOC models.

1. INTRODUCTION

Carbon stored in soils worldwide exceeds the amount of carbon stored in the above ground biomass and the atmosphere. Despite the large quantity of carbon stored as soil organic carbon (SOC), consensus is lacking on the size of global SOC stocks, their spatial distribution, and the carbon emissions from soils due to changes in land use and land cover (Scharlemann et al., 2014). Even if limited only to agriculture, the role of soils in the Kyoto protocol underlines the necessity to account for the changes in greenhouse gas (GHG) emissions by sources and removals by sinks of agricultural soils worldwide. In the post-Kyoto period, the EU has committed to cutting its emissions to 20% below 1990 levels by 2020, or to 30% if other major emitting countries commit to undertake their fair share of a global emissions reduction effort (EC, 2010). In the climate change policies, C sequestration in soils is included in the land use, land-use change and forestry (LULUCF) sector.

The European Commission funded the project "Soil Transformations in European Catchments (SoilTrEC)" (Banwart et al., 2011) and among the objective of this research project was to up-scale modeling results found at small scale (catchment) toward larger scales (regional, national, continental). As such, novel methodologies developed in Critical Zone Science (CZS) could serve to better understand soil processes and influence policy makers (Montanarella and Panagos, 2015). Under the European Union Thematic Strategy for Soil Protection, the European Commission

Directorate-General for the Environment and the European Environmental Agency (EEA) identified a decline in SOC and soil losses by erosion as priorities for the collection of policy relevant soil data at European scale. Moreover, the estimation of SOC content is of crucial importance for soil protection and for climate change mitigation strategies (Panagos et al., 2013a).

Knowledge of SOC and SOC dynamics has grown considerably in recent years, driven by a number of current debates including on the potential of soils to contribute to climate change mitigation, ecological approaches to agriculture, and propoor agricultural development, among many others (Govers et al., 2013). Some 45% of soils in Europe have low or very low organic matter content (0–2% of SOC) (Jones et al., 2012). This is particularly evident in the soils of many southern European countries, but it is also evident in parts of France, the United Kingdom, Germany, Norway, and Belgium. A key driver is the conversion of woodland and grassland to arable crops. The soils of 27 Member States of European Union (EU27) are estimated to store between 73 and 79 billion tons of carbon (Jones et al., 2005).

Several modeling approaches were conducted to simulate regional or global development of agricultural SOC stocks. Smith et al. (2005) assessed the future changes in cropland and grassland SOC stocks to date, using a dedicated process-based SOC model and state-of-the-art databases of soil, climate change, land-use change, and technology change. The researchers used the Rothamsted carbon model developed by Coleman and Jenkinson (1996), to calculate soil carbon change on a European $10 \times 10'$ grid using climate data from four global climate models implementing four GHG emissions scenarios (SRES) from the Intergovernmental Panel on Climate Change (IPCC) (Smith et al., 2005).

Jobbagy and Jackson (2000) pointed out that, globally, the relative distribution of SOC with depth had a slightly stronger association with vegetation than with climate, but the opposite was true for the absolute amount of SOC. Total SOC content increased with precipitation and clay content and decreased with temperature. The researchers reported that the global SOC storage in the top 3 m of soil was 2344 Pg C, or 56% more than the 1502 Pg estimated for the first meter. Global totals for the second and third meters were 491 and 351 Pg C, and the biomes with the most SOC at 1–3 m depth were tropical evergreen forests (158 Pg C) and tropical grasslands/savannas (146 Pg C).

It is estimated that global soils contain between 1400 and 1600 Pg of carbon (1 Pg$=10^{15}$ g) in the upper meter, and that the next meter of soil contains an additional 500–1000 Pg C. These estimates imply that the SOC pool is more than twice the size of the atmospheric carbon pool (c.800 Pg) and that it contains about three times the amount of carbon in vegetation (Govers et al., 2013).

Lugato et al. (2014a) established a platform at a pan-European scale (EU+Serbia, Bosnia and Herzegovina, Croatia, Montenegro, Albania, Former Yugoslav Republic of Macedonia, and Norway) using the agro-ecosystem SOC model CENTURY. The researchers computed 164,000 combinations of soil–climate–land use, including the main arable crops, orchards, and pasture. They reported model estimation of the current 0–30 cm SOC stock of agricultural soils is 17.63 Gt (1 Pg=1 Gt).

Panagos et al. (2013b) calculated SOC stocks in EIONET partner countries using the EIONET data. Regarding the European SOC stocks, the sum of the provided data for the six participating countries were: Bulgaria: 315 Tg, Denmark: 370 Tg, Netherlands: 299 Tg (77% coverage), Poland: 1753 Tg (70% coverage), Italy: 994 Tg (57% coverage), and Slovakia: 122 Tg (54% coverage). The researchers compared the SOC stocks with the amount of SOC in Europe based on the OCTOP dataset (EC, 2008; Jones et al., 2005). The results suggested that the current estimates of SOC stock in Europe in the topsoil could be much less than the 73–79 Pg, as reported previously (EC, 2008).

Köchy et al. (2015) reviewed current estimates of SOC stocks and mass (stock × area) in wetlands, permafrost, and tropical regions of the world in the upper 1 m of soil using the Harmonized World Soil Database (HWSD) v.1.2 (FAO/IIASA/ISRIC/ISSCAS/JRC, 2012). HWSD provides one of the most recent global datasets of SOC, giving a total mass of 2476 Pg when using the original values for bulk density (BD). Adjusting the HWSD's BD of soil high in organic carbon results in a mass of 1230 Pg, and additionally setting the BD of Histosols to 0.1 g cm^{-3}, results in a mass of 1062 Pg.

Scharlemann et al. (2014) reported that many estimates of global SOC stocks have been published during the past 7 decades, most recently to support the calculation of potential CO_2 emissions from the soil under land use and land cover change (LULCC) and climate change scenarios. Although most studies report a global SOC estimate of roughly 1500 Pg of carbon (1 Pg C equal to 10^{15} g or 1 billion tons of carbon), there is considerable variation (median across all estimates: 1460.5 Pg C, range 504–3000 Pg C, $n=27$ studies).

Yigini and Panagos (2016) developed a two-phased geostatistical platform to predict current and future SOC stocks in European Union 26 Member States (EU26). The first phase of the study predicts current SOC content by using regression-kriging and the second phase of the study projects the SOC to the near future (2050) by using the same set of the environmental predictors used in the first phase. The researchers found that the larger part of the present SOC stocks is held in Europe's forest and seminatural soils which are around 16.40 Pg. While the agricultural areas store around 12.79 Pg of SOC, the pastures and wetlands store 8.53 Pg SOC in the first 20 cm in 26 European Countries.

2. DATA AND METHODS

2.1 Spatial Data

2.1.1 Point Data: LUCAS Soil Database

The SOC data derived from LUCAS Topsoil Database which is a module of the LUCAS Survey Project (Toth et al., 2013a). LUCAS is an in situ survey, which means that the data are gathered through direct field observations. The aim of the LUCAS survey is to gather fully harmonized data on land use/cover and their changes over time in the 27 European Union member states (EU27). In the LUCAS (2009) survey, 265,000 georeferenced points were visited by more than 500 field surveyors. The survey points were selected from a standard 2 km × 2 km grid based on stratification information provided by Martino and Fritz (2008). Approximately 22,000 points were selected out of the main LUCAS grid for the collection of soil samples. The descriptive statistics can be seen in Table 1. A standardized sampling procedure was used to collect around 0.5 kg of topsoil (0–20 cm) for each sample. The samples were dispatched to a central laboratory for physical and chemical analyses (Toth et al., 2013b). The selection of the LUCAS soil sampling sites has an inherent bias toward agricultural land (predominantly under arable cultivation), followed by grasslands and woodlands. This bias means that results based exclusively on LUCAS soil samples may over-represent properties from the more heavily sampled conditions while under-representing other land cover types. Each soil sample was taken from the topsoil zone (20 cm) with a mass of 0.5 kg. The samples were analyzed in a single ISO-certified laboratory that used harmonized chemical and physical

Table 1 Basic Statistics of LUCAS Topsoil Database

Mean (g kg^{-1})	StDev (g kg^{-1})	Variance	Median (g kg^{-1})	Minimum (g kg^{-1})	Maximum (g kg^{-1})	*N*
47.69	8.86	78.48	2.02	0.1	58.68	22,284

analytical methods (ISO standards, or their equivalent) in order to obtain a coherent and harmonized dataset with pan-European coverage. The analysis results formed the LUCAS soil database, including, among other things, SOC in topsoil (0–20 cm) expressed in g kg^{-1} (Panagos et al., 2013a).

As mentioned earlier, LUCAS soil samples overrepresent the agricultural lands compared to other land cover types in Europe. The distribution of the samples by land cover types is shown in Table 2. While 55% of the samples were taken from agricultural areas (GlobCover classes: 11, 14, 20, and 30) only 26% of the samples were collected from forestlands (GlobCover classes: 40, 50, 60, 70, 90, and 100).

2.1.2 Land Cover

The GlobCover (2009) (ESA and Universite' Catholique de Louvain, 2010) dataset was used to extract global agricultural lands to be used in the study. GlobCover is a European Space Agency (ESA) initiative which began in 2005 in partnership with JRC, EEA, FAO, UNEP, GOFC-GOLD, and IGBP. The aim of the project was to develop a service capable of delivering global composites and land cover maps using as input observations from the 300 m MERIS sensor on board the ENVISAT satellite mission. ESA makes available the land cover maps, which cover two periods: December 2004—June 2006 and January—December 2009. The classification module of the GlobCover processing chain consists in transforming the MERIS-FR multispectral mosaics produced by the preprocessing modules into a meaningful global land cover map. The global land cover map has been produced in an automatic and global way and is associated with a legend defined and documented using the UN LCCS. The GlobCover 2009 land cover map is delivered as one global land cover map covering the entire Earth. Its legend, which counts 22 land cover classes, has been designed to be consistent at the global scale and therefore, it is determined by the level of information that is available and that makes sense at this scale (Bontemps et al., 2011). We used here the agricultural areas which are coded as 11 for postflooding or irrigated croplands, 14 for Rainfed croplands, 20 for mosaic croplands (50–70%), and 30 for croplands (20–50%).

In this study, we reclassified the GlobCover 2009 data to extract the global agricultural lands to be used in the model. For the definition of the term "agricultural land," we used the UN FAO Statistics Division's definition of "agricultural area" where it means the collection of arable land,

Table 2 LUCAS Sampling Locations and GlobCover Land Cover Classes

GlobCover 2009 Land Cover Classes	GlobCover Codes	LUCAS Points	Share %
Postflooding or irrigated croplands (or aquatic)	11	234	1.05
Rainfed croplands	14	6006	26.95
Mosaic cropland (50–70%)/vegetation (grassland/shrubland/forest) (20–50%)	20	4021	18.04
Mosaic vegetation (grassland/shrubland/forest) (50–70%)/cropland (20–50%)	30	1833	8.23
Closed to open (>15%) broadleaved evergreen or semideciduous forest (>5 m)	40	0	0.00
Closed (>40%) broadleaved deciduous forest (>5 m)	50	2358	10.58
Open (15–40%) broadleaved deciduous forest/woodland (>5 m)	60	0	0.00
Closed (>40%) needleleaved evergreen forest (>5 m)	70	425	1.91
Open (15–40%) needleleaved deciduous or evergreen forest (>5 m)	90	1764	7.92
Closed to open (>15%) mixed broadleaved and needleleaved forest (>5 m)	100	1314	5.90
Mosaic forest or shrubland (50–70%)/grassland (20–50%)	110	553	2.48
Mosaic grassland (50–70%)/forest or shrubland (20–50%)	120	1283	5.76
Closed to open (>15%) (broadleaved or needleleaved, evergreen or deciduous) shrubland (<5 m)	130	229	1.03
Closed to open (>15%) herbaceous vegetation (grassland, savannas, or lichens/mosses)	140	903	4.05
Sparse (<15%) vegetation	150	997	4.47
Closed to open (>15%) broadleaved forest regularly flooded (semipermanently or temporarily)—fresh or brackish water	160	0	0.00

Continued

Table 2 LUCAS Sampling Locations and GlobCover Land Cover Classes—cont'd

GlobCover 2009 Land Cover Classes	GlobCover Codes	LUCAS Points	Share %
Closed (>40%) broadleaved forest or shrubland permanently flooded—saline or brackish water	170	0	0.00
Closed to open (>15%) grassland or woody vegetation on regularly flooded or waterlogged soil—fresh, brackish, or saline water	180	179	0.80
Artificial surfaces and associated areas (urban areas >50%)	190	121	0.54
Bare areas	200	33	0.15
Water bodies	210	31	0.14
Permanent snow and ice	220	0	0.00
No data (burnt areas, clouds, …)	230	0	0.00
Total		22,284	100.00

permanent cropland, natural or artificial grasslands, and shrubland able to be used for grazing livestock. We extracted the following classes from the GlobCover 2009 data:

- postflooding or irrigated croplands (or aquatic);
- rainfed croplands;
- mosaic cropland (50–70%)/vegetation (grassland/shrubland/forest) (20–50%); and
- mosaic vegetation (grassland/shrubland/forest) (50–70%)/cropland (20–50%).

Fig. 1 shows the global agricultural areas extracted from the GlobCover 2009 dataset.

2.1.3 Climate Data

The data representing the current conditions were downloaded from WorldClim Data Portal (Hijmans et al., 2005). The WorldClim data layers were generated by interpolation of average monthly climate data from weather stations on a 30 arc-second resolution grid. Variables included are monthly total precipitation, and monthly mean, minimum and maximum temperature, and 19 derived bioclimatic variables. The WorldClim precipitation data were obtained from a network of 47,554 stations, mean

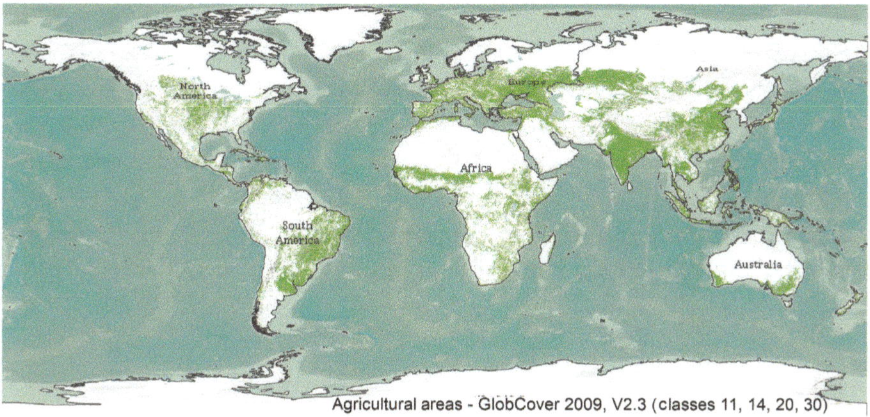

Fig. 1 Global agricultural areas (GlobCover classes: 11, 14, 20, 30).

temperature from 24,542 stations, and minimum and maximum temperatures from 14,835 stations (Hijmans et al., 2005).

In the modeling platform, we included monthly precipitation, monthly mean temperature, and 19 bioclimatic derivatives. The bioclimatic parameters are: annual mean temperature, mean diurnal range, isothermality, temperature seasonality, max temperature of the warmest month, minimum temperature of the coldest month, temperature annual range, mean temperature of the wettest quarter, mean temperature of the driest quarter, mean temperature of the warmest quarter, mean temperature of the coldest quarter, annual precipitation, precipitation of the wettest month, precipitation of the driest month, precipitation seasonality (coefficient of variation), precipitation of the wettest quarter, precipitation of the driest quarter, precipitation of the warmest quarter, and precipitation of the coldest quarter. Bioclimatic variables are the derivatives and were included in the base model as well as in the projection phase.

2.1.4 Soil Data

Percent SOC, depth (20 cm), soil BD, and topsoil gravel content (GC) were used to calculate SOC in tons/ha stored in the soils of the global agricultural lands. The soil BD and the topsoil GC data layers were extracted from the Harmonized World Soil Database v. 1.21 (HWSD). The HWSD is a 30 arc-second raster database with over 16,000 different soil mapping units that combine existing regional and national updates of soil information worldwide (SOTER, ESD, Soil Map of China, WISE) with the information contained within the 1:5 m scale FAO-UNESCO Soil Map of the World. The resulting raster database consists of 21,600 rows and 43,200 columns, which are linked to harmonized soil property data. The use of a standardized structure allows for the linkage of the attribute data with the raster map to display or query the composition in terms of soil units and the characterization of selected soil parameters (organic carbon, pH, water storage capacity, soil depth, cation exchange capacity of the soil and the clay fraction, total exchangeable nutrients, lime and gypsum contents, sodium exchange percentage, salinity, textural class, and granulometry) (FAO/IIASA/ISRIC/ISSCAS/JRC, 2012).

2.1.5 Vegetation Data (NDVI)

The vegetation index maps (10-daily, 2013) were obtained from the Copernicus data portal. The product was generated by the land service of Copernicus, the Earth Observation program of the European Commission. The research leading to the current version of the product has received funding from various European Commission Research and Technical Development programs. The product is based on vegetation density coverage data. The data were converted from 10-daily format into monthly (January–December) and included in the modeling platform.

The Normalized Difference Vegetation Index (NDVI) is an indicator of the greenness of the biomes. As such, it is closely linked to the Fraction of Absorbed Photosynthetically Active Radiation (FAPAR). Even though it is not a physical property of the vegetation cover, it's very simple formulation makes it widely used for ecosystems monitoring (Copernicus Global Land Service, 2015) (Eq. 1).

$$NDVI = (REF\_nir - REF\_red)/(REF\_nir + REF\_red) \quad (1)$$

where REF_nir and REF_red are the spectral reflectances measured in the near infrared and red wavebands, respectively.

2.2 Methods

We tested in this study a two-step geostatistical approach to predict agricultural global organic carbon stocks using a European database. The overall statistical model consists of two submodels namely *donor* and *donee* modules (Fig. 2). The donor module uses statistics to quantify the relationships between SOC and 68 environmental covariates (land cover, terrain, climate, and soil). The covariates were selected based on their availability at global scale and their role as the major drivers that affect the carbon cycle in

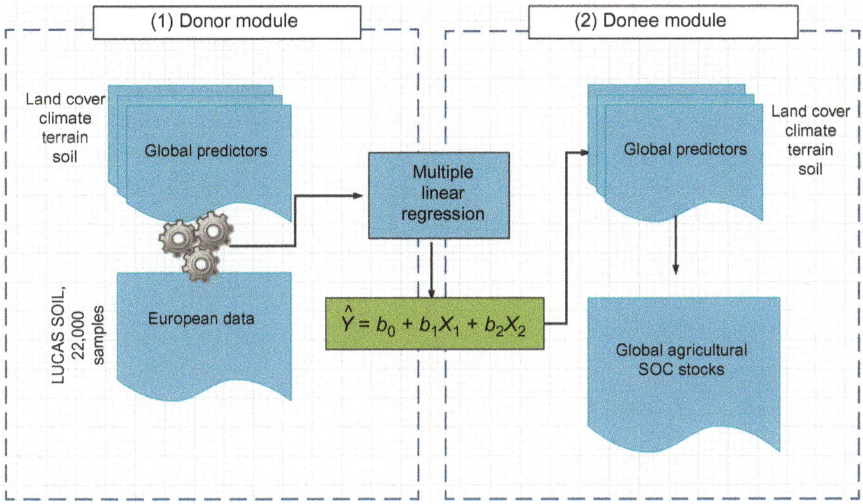

Fig. 2 Flowchart diagram of the calculation of the global agricultural soil organic carbon stocks.

terrestrial ecosystems. Multiple linear regression (MLR) was used in the donor module with the selected covariates and dense SOC measurements coming from the LUCAS soil database. The donor module reveals and quantifies the relationships between SOC and the predictors to be used in the donee model, in order to extend the prediction at global scale using the same set of predictors (Fig. 2).

The donor module of the two-step model uses MLR and generates the regression equation with coefficients for each of the predictors to be transferred to the donee module. In practice, the regression equation provides incisive information regarding SOC–covariates interactions. In other words, the regression coefficients reflect the knowledge of different processes affecting SOC at European level. In the second phase, the donee module receives the mathematical relations (regression coefficients) from the donor module and extends the SOC prediction to global level. We used MLR to identify the relationships between SOC and auxiliary data layers (terrain attributes, climate data, land cover, and vegetation index). In the donor module, we included climate data as 21 layers (19 bioclimatic parameters, annual mean temperature, and annual precipitation), land cover data as 3 layers (agriculture, forest, and wetlands), terrain data as 3 layers (aspect, elevation, and slope), and NDVI as 12 layers (January–December, 2013).

We combined multiple regression and principal component analysis to reduce the number of the data layers in the raster calculation and to avoid multicollinearity among the bioclimatic, terrain, and NDVI data layers. Principal component analysis is a procedure for identifying a smaller number of uncorrelated variables, called "principal components," from a large set of data. The goal of principal component analysis is to explain the maximum amount of variance with the fewest number of principal components. Principal component analysis is commonly used in the social sciences, market research, and other industries that use large datasets. Principal component analysis is commonly used as one step in a series of analyses. Principal component analysis can be used to reduce the number of variables and to avoid multicollinearity when there are too many predictors relative to the number of observations (Abdul-Wahab et al., 2005). We applied principal component analysis and created seven components to be used in the MLR. MLR is a widely used and simple approach to develop predictive models from soil C observations and to extrapolate predictions to locations without observations (Kunkel et al., 2011). The LUCAS data are clearly right skewed as shown in Fig. 3. The data were transformed by taking the Box–Cox

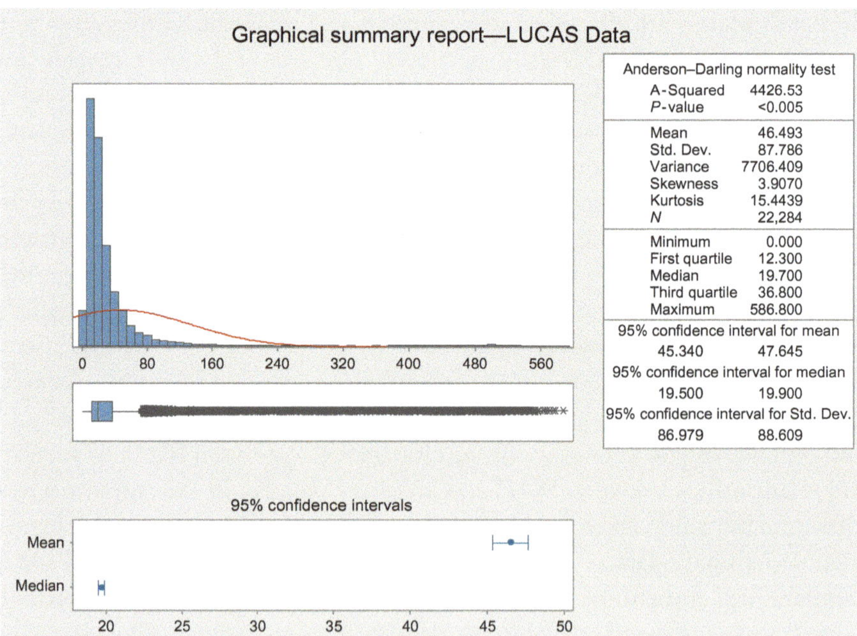

Fig. 3 The graphical summary of the LUCAS soil data.

transformation ($\lambda = 0$) to achieve an approximately normal distribution to stabilize the variance, thereby satisfying the conditions for a valid linear model.

We fitted a MLR model to reveal the relationships between SOC and the predictors and transferred the regression equation to the donee module to calculate the global SOC stocks using the same set of the predictors used in the donor module. In the end of the calculation session, a raster file was created as a raster map of the global SOC content in units of mass percent and it was used to calculate the global agricultural SOC stocks (Eq. 2).

2.2.1 Calculation of SOC Stocks

Calculation of SOC stocks (tons ha^{-1}) requires data on SOC content (%), soil BD, sampling depth, and the proportion of GC. SOC stocks were calculated by Eq. (2):

$$SOC_{stock} = SOC\% \times \rho b \times d \times \left(1 - \frac{GC}{100}\right) \tag{2}$$

where SOC_{stock} represents SOC stock in tons ha^{-1}, SOC is the SOC content (%), ρb is the BD (g cm^{-3}), d is the sampling depth which is 20 cm for the LUCAS sampling campaign, and GC is the gravel content (%). In case of no data on the GC, the GC multiplier becomes 1.

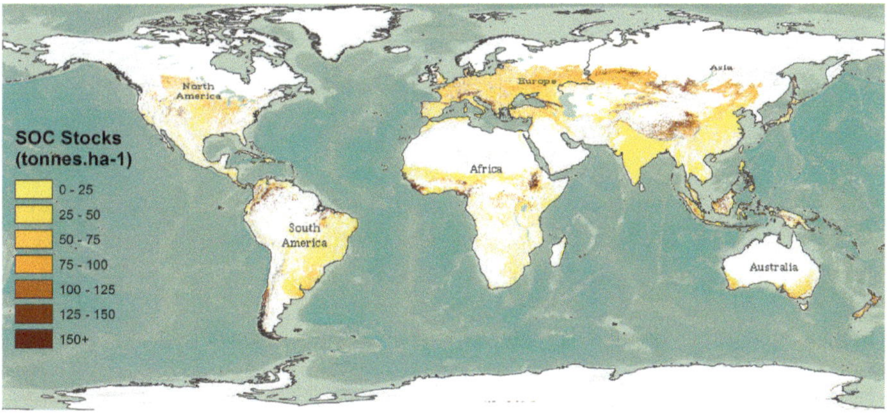

Fig. 4 Global agricultural soil organic carbon map. *Base Map: World Terrain Base (Source: ESRI, DeLorme, USGS, NPS, and NOAA).*

3. RESULTS AND DISCUSSION

The results suggest that present SOC stocks which are held in the World's agricultural soils are around 100.34 Pg (GlobCover classes: 11, 14, 20, 30). The global SOC stocks map (Fig. 4) shows the global agricultural SOC stocks in the first 20 cm.

3.1 Model Performance

For a global scale with huge heterogeneity of covariates, the model performance has achieved $R^2 = 0.35$. At this stage and with the limited point data availability (covering only Europe), the model performance is considered satisfactory. The model summary table of the MLR output is shown in Table 3. However, future introduction of SOC point data measurements from other continents can further improve the model prediction and accuracy.

A typical result of geostatistical interpolation is a map of predictions and prediction error, which is an estimate of prediction uncertainty (Ballabio et al., 2016; Hengl et al., 2004). The prediction uncertainty map for the donor module is given in Fig. 5.

3.2 Continental Distribution of Agricultural SOC Stocks

According to the results, Asia's agricultural soils hold 50 Pg SOC, followed by Europe which is around 21 Pg SOC. The SOC stock figures for each of the continents are shown in Table 4. The continental distribution of agricultural SOC stocks is obviously affected by the extent of the agricultural areas.

3.3 Countries and Agricultural SOC Stocks

The agricultural SOC stocks for the first 50 countries (Table 5) holds 92% of the total agricultural SOC stocks in the world. The country with the highest estimated agricultural SOC stock is India (17 Pg) as the main land cover is agricultural land, followed by Russia (15.3 Pg), and China with 9.5 Pg. The complete list of the countries is given in Appendix.

Table 3 The Model Summary of the Multiple Linear Regression Analysis

S	R^2	R^2 (Adjusted)	R^2 (Prediction)
0.80	35.39%	35.31%	35.23%

S: Standard error of the regression.

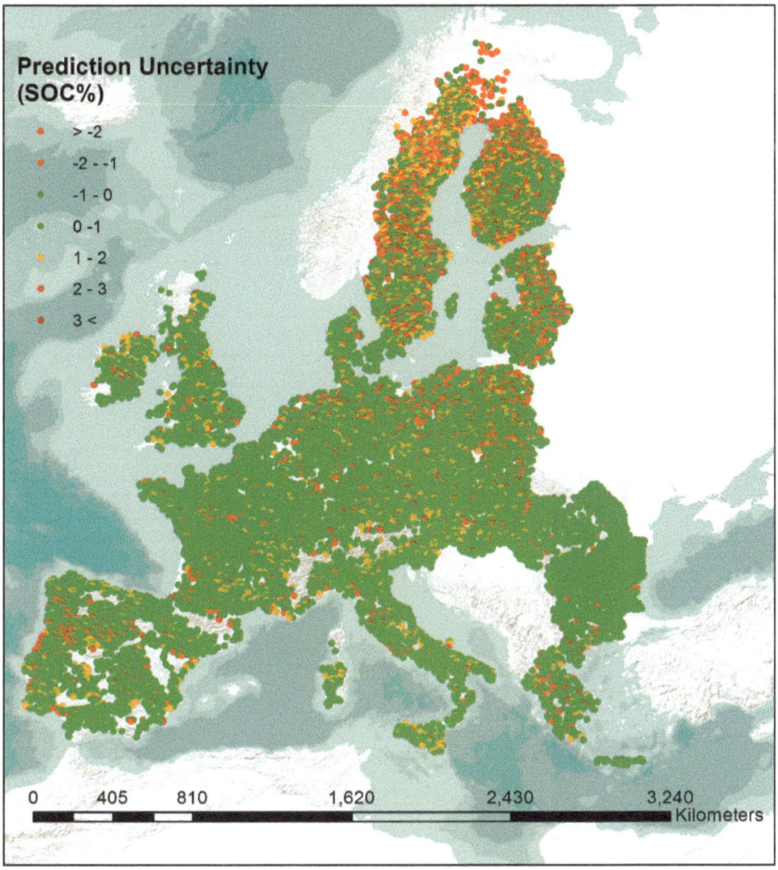

Fig. 5 Prediction uncertainty map. *Base layer: ESRI, USGS, and NOAA.*

Table 4 Agricultural Soil Organic Carbon Stocks by Continent

Continent	SOC Stock (Pg)
Africa	9.83
Asia	50.38
Australia	1.92
Europe	20.94
North America	7.74
Oceania	1.50
South America	7.82

For the definition of "Continents," we used the ESRI (2002).

Table 5 Agricultural Soil Organic Carbon Stocks of the First 50 Countries (by SOC Stocks, Pg)

Country	SOC Stock (Pg)	Global Share (%)	Country	SOC Stock (Pg)	Global Share (%)
India	17.062	17.01	Philippines	0.741	0.74
Russia	15.357	15.31	Italy	0.724	0.72
China	9.483	9.45	Malaysia	0.670	0.67
United States	5.302	5.29	Spain	0.658	0.66
Kazakhstan	5.047	5.03	Liberia	0.637	0.63
Brazil	2.958	2.95	Romania	0.601	0.60
Mongolia	2.396	2.39	Belarus	0.538	0.54
Ukraine	2.229	2.22	Japan	0.521	0.52
Indonesia	2.162	2.16	Venezuela	0.499	0.50
Australia	1.921	1.91	DR Congo	0.475	0.47
France	1.811	1.80	United Kingdom	0.469	0.47
Canada	1.686	1.68	Iran	0.463	0.46
Sierra Leone	1.553	1.55	Kyrgyzstan	0.398	0.40
Turkey	1.551	1.55	Uruguay	0.385	0.38
Ethiopia	1.547	1.54	Cameroon	0.349	0.35
Guinea	1.478	1.47	South Africa	0.308	0.31
Colombia	1.330	1.33	Austria	0.303	0.30
Chile	1.172	1.17	Mexico	0.284	0.28
Papua New Guinea	1.150	1.15	Hungary	0.281	0.28
Germany	1.067	1.06	Lithuania	0.261	0.26
Argentina	1.040	1.04	Thailand	0.257	0.26
Poland	1.013	1.01	Latvia	0.234	0.23
New Zealand	0.918	0.91	Georgia	0.231	0.23
Nigeria	0.807	0.80	Pakistan	0.222	0.22

3.4 Comparison of SOC Stocks in Agricultural Lands With Regional and National Assessments

The available data and analyses presented earlier suggest that global agricultural soils hold 100.34 Pg C in their first 20 cm. The distribution of the SOC stocks is consistent with the previous modeling efforts using process based and the other modeling approaches.

According to our results, the United States' agricultural soils hold 5.3 Pg SOC in the first 20 cm which is consistent with an earlier research study (Kern, 1994) that reported the US agricultural SOC stocks as 7 billion metric tons (7 Pg).

In a pan-European assessment, Lugato et al. (2014a) used the process-based CENTURY model for predicting SOC stocks. According to Lugato et al. (2014a), CENTURY estimated the current 0–30 cm SOC stock of agricultural soils as 17.63 Gt (Pg). According to results for the individual countries; France's agricultural soils hold 2.2 Pg, Poland 1.1 Pg, and Italy 0.8 Pg SOC in the first 30 cm. The comparison for European Countries between two studies (Lugato et al., 2014a and this study) is given in Table 6.

The current study corresponds well (±10% difference) with the results of CENTURY model in Austria, Bulgaria, Cyprus, Czech Republic, Estonia, Spain, Greece, Hungary, Italy, Latvia, Lithuania, Poland, Portugal, Romania, and Slovakia. The low values in our study in Finland and Sweden are due to the non-existence of agricultural areas in GlobCover dataset. We also predicted lower SOC stocks in Ireland, United Kingdom, Denmark, Netherlands, France, and Germany because the grasslands are underrepresented in the included GlobCover classes.

In Europe, recent research studies have estimated the carbon stocks in agricultural lands close to our results. Khalil et al. (2013) reported that, in Ireland the SOC stocks for arable lands are 13.5 (0.013 Pg) Tg for the 0–10 cm, 36.7 Tg (0.036 Pg) for the 0–30 cm, 50.2 Tg (0.050 Pg) for the 0–50 cm and 67.0 Tg (0.067 Pg) for the 0–100 cm soil depth. We calculated the total agricultural SOC stocks as 0.033 Pg in Ireland (Appendix). Gregory et al. (2014) estimated that the cultivated lands and permanent managed grasslands of England and Wales contain 0.419 and 0.660 Pg (0–30 cm) SOC, respectively. Our results show that the UK's (England, Wales, Scotland, and Northern Ireland) agricultural soils contain 0.46 Pg SOC in the first 20 cm. Barančíková et al. (2013) reported that Slovakia's agricultural lands contain 60 million tons (0.06 Pg) of SOC in the first 20 cm. This study suggests 0.12 Pg SOC is contained in the agricultural soils of Slovakia.

Table 6 Comparison Between Lugato et al. (2014b) and This Study

Countries	CENTURY Model (0–30 cm) Agricultural SOC Stocks (Pg) (Lugato et al., 2014a, 2014b)	Multiple Linear Regression (0–20 cm) Agricultural SOC Stocks (Pg) (This Study)
Austria	0.340	0.303
Belgium	0.135	0.105
Bulgaria	0.217	0.201
Cyprus	0.008	0.005
Czech Republic	0.237	0.209
Germany	1.306	1.067
Denmark	0.263	0.209
Estonia	0.162	0.158
Spain	0.731	0.658
Finland	0.430	<0.001
France	2.272	1.811
Greece	0.201	0.193
Croatia	0.205	0.141
Hungary	0.300	0.281
Ireland	0.106	0.034
Italy	0.763	0.724
Latvia	0.270	0.234
Luxembourg	0.012	0.007
Lithuania	0.240	0.261
Malta	>0.005	<0.001
Netherlands	0.133	0.100
Poland	1.074	1.013
Portugal	0.143	0.120
Romania	0.668	0.601
Sweden	0.640	0.065
Slovenia	0.055	0.034
Slovakia	0.149	0.123
United Kingdom	0.798	0.469

The current effort to estimate the agricultural SOC stocks solely relies on a statistical approach and does not use data representing land use practices which have influence on agricultural SOC stocks. The introduction of such practices (reduced tillage, conversion to grassland, cover crops, plant residues, etc.) in our model could even improve the carbon stock estimates. However, such data exist only at European level (Lugato et al., 2014b). The output data could not be validated with measured data but are comparable with the results reported for previous studies in the literature.

4. CONCLUSIONS

This study represents an extrapolation study to predict SOC stocks at global scale based on a European dataset and various environmental predictors used in this study. The results provide coherent and comparable results. The approach presented here assesses the present global agricultural SOC stocks. Given the difficulty of measuring changes in soil carbon stocks, modeling is a useful tool and it has been used as an effective methodology for analyzing and predicting the effect of land management practices on SOC stocks.

In this study, we hypothesized that SOC is driven by land cover, soil, climate, and terrain parameters. The challenge we addressed in this study was to examine how, and to what extent, the SOC predictions can be extended to larger scale and how the information is transferable from densely sampled small areas to nonsampled larger areas using a geostatistical approach. The predicted agricultural SOC stock figures are generally in agreement with values reported by previous studies in the literature. Better estimates of agricultural SOC stocks are needed for an improved understanding of the carbon balance and the influence of land use practices. Existing SOC estimations could be improved by collection of more data on soil profiles, and by including sampling to greater depths. Using standardized sampling methods and making data publicly available will provide a possibility to combine data from different sources to make better global and regional estimations. The methods used to calculate SOC stocks also need to be standardized to consistently take appropriate account of the spatial variation in depth, BD, and GC to allow comparisons among studies.

ACKNOWLEDGMENTS

This study was made possible by:

- LUCAS Top Soil Database which is distributed by European Soil Data Centre (ESDAC).
- WorldClim (V1) which was developed by Robert J. Hijmans, Susan Cameron, and Juan Parra, at the Museum of Vertebrate Zoology, University of California, Berkeley, in collaboration with Peter Jones and Andrew Jarvis (CIAT), and with Karen Richardson (Rainforest CRC).
- GlobCover V2.3 Land Cover Data which was processed by (European Space Agency) ESA and by the Université Catholique de Louvain. They are made available to the public by ESA.

APPENDIX. AGRICULTURAL SOIL ORGANIC CARBON STOCKS, COUNTRIES

Country	SOC Stock (Pg)	Global Share (%)	Country	SOC Stock (Pg)	Global Share (%)
Afghanistan	0.200	0.20	Cyprus	0.005	0.01
Akrotiri and Dhekelia	<0.001	0.00	Czech Republic	0.209	0.21
Albania	0.117	0.12	DR Congo	0.475	0.47
Algeria	0.186	0.19	Denmark	0.209	0.21
American Samoa	<0.001	0.00	Dominican Republic	0.027	0.03
Andorra	0.003	0.00	Dominica	<0.001	0.00
Angola	0.048	0.05	Ecuador	0.126	0.13
Antigua and Barbuda	<0.001	0.00	Egypt	0.028	0.03
Argentina	1.040	1.04	El Salvador	0.016	0.02
Armenia	0.095	0.09	Equatorial Guinea	0.005	0.00
Aruba	<0.001	0.00	Eritrea	0.011	0.01
Australia	1.921	1.91	Estonia	0.158	0.16
Austria	0.303	0.30	Ethiopia	1.547	1.54

Country	SOC Stock (Pg)	Global Share (%)	Country	SOC Stock (Pg)	Global Share (%)
Azerbaijan	0.121	0.12	Falkland Islands	0.023	0.02
Bahamas	0.001	0.00	Faroe Islands	0.001	0.00
Bahrain	<0.001	0.00	Fiji	0.007	0.01
Bangladesh	0.027	0.03	Finland	<0.001	0.00
Barbados	<0.001	0.00	France	1.811	1.80
Belarus	0.538	0.54	French Guiana	0.007	0.01
Belgium	0.105	0.10	Gabon	0.012	0.01
Belize	0.002	0.00	Gambia	0.004	0.00
Benin	0.051	0.05	Georgia	0.231	0.23
Bhutan	0.009	0.01	Germany	1.067	1.06
Bolivia	0.079	0.08	Ghana	0.111	0.11
Bonaire, Sint Eustatius, and Saba	<0.001	0.00	Greece	0.193	0.19
Bosnia and Herzegovina	0.148	0.15	Grenada	<0.001	0.00
Botswana	0.046	0.05	Guadeloupe	<0.001	0.00
Brazil	2.958	2.95	Guatemala	0.014	0.01
Brunei	0.003	0.00	Guernsey	<0.001	0.00
Bulgaria	0.201	0.20	Guinea-Bissau	0.008	0.01
Burkina Faso	0.173	0.17	Guinea	1.478	1.47
Burundi	0.015	0.01	Guyana	0.011	0.01
Cambodia	0.038	0.04	Haiti	0.018	0.02
Cameroon	0.349	0.35	Honduras	0.045	0.04
Canada	1.686	1.68	Hong Kong	<0.001	0.00
Cape Verde	<0.001	0.00	Hungary	0.281	0.28
Caspian Sea	0.001	0.00	India	17.062	17.01
Cayman Islands	<0.001	0.00	Indonesia	2.162	2.16

Continued

Country	SOC Stock (Pg)	Global Share (%)	Country	SOC Stock (Pg)	Global Share (%)
Central African Republic	0.046	0.05	Iran	0.463	0.46
Chad	0.122	0.12	Iraq	0.094	0.09
Chile	1.172	1.17	Ireland	0.034	0.03
China	9.483	9.45	Isle of Man	0.001	<0.001
Colombia	1.330	1.33	Israel	0.016	0.02
Comoros	<0.001	<0.001	Italy	0.724	0.72
Costa Rica	0.117	0.12	Jamaica	0.005	0.01
Côte d'Ivoire	0.199	0.20	Japan	0.521	0.52
Croatia	0.141	0.14	Jordan	0.005	0.01
Cuba	0.022	0.02	Kazakhstan	5.047	5.03
Curaçao	<0.001	<0.001	Kenya	0.161	0.16
Kosovo	0.032	0.03	Portugal	0.120	0.12
Kuwait	<0.001	<0.001	Puerto Rico	0.003	0.00
Kyrgyzstan	0.398	0.40	Republic of Congo	0.020	0.02
Laos	0.011	0.01	Reunion	<0.001	<0.001
Latvia	0.234	0.23	Romania	0.601	0.60
Lebanon	0.036	0.04	Russia	15.357	15.31
Lesotho	0.015	0.01	Rwanda	0.023	0.02
Liberia	0.637	0.63	Saint Kitts and Nevis	<0.001	<0.001
Libya	0.011	0.01	Saint Lucia	<0.001	<0.001
Liechtenstein	<0.001	<0.001	Saint Vincent and the Grenadines	<0.001	<0.001
Lithuania	0.261	0.26	Samoa	0.001	<0.001
Luxembourg	0.007	0.01	San Marino	<0.001	<0.001
Macedonia	0.048	0.05	Sao Tome and Principe	<0.001	<0.001

Country	SOC Stock (Pg)	Global Share (%)	Country	SOC Stock (Pg)	Global Share (%)
Madagascar	0.045	0.05	Saudi Arabia	0.005	<0.001
Malawi	0.015	0.01	Senegal	0.055	0.05
Malaysia	0.670	0.67	Serbia	0.212	0.21
Mali	0.130	0.13	Sierra Leone	1.553	1.55
Malta	<0.001	<0.001	Singapore	<0.001	<0.001
Martinique	<0.001	<0.001	Sint Maarten	<0.001	<0.001
Mauritania	0.013	0.01	Slovakia	0.123	0.12
Mauritius	<0.001	<0.001	Slovenia	0.034	0.03
Mayotte	<0.001	<0.001	Solomon Islands	0.029	0.03
Mexico	0.284	0.28	Somalia	0.060	0.06
Moldova	0.123	0.12	South Africa	0.308	0.31
Mongolia	2.396	2.39	South Korea	0.004	0.00
Montenegro	0.070	0.07	South Sudan	0.183	0.18
Morocco	0.197	0.20	Spain	0.658	0.66
Mozambique	0.056	0.06	Sri Lanka	0.013	0.01
Myanmar	0.105	0.10	Sudan	0.158	0.16
Namibia	0.014	0.01	Suriname	0.005	0.00
Nepal	0.061	0.06	Swaziland	0.003	0.00
Netherlands	0.100	0.10	Sweden	0.065	0.06
New Caledonia	0.005	0.01	Switzerland	0.210	0.21
New Zealand	0.918	0.91	Syria	0.079	0.08
Nicaragua	0.082	0.08	Taiwan	0.006	0.01
Niger	0.011	0.01	Tajikistan	0.130	0.13
Nigeria	0.807	0.80	Tanzania	0.115	0.11
North Korea	0.002	0.00	Thailand	0.257	0.26
Northern Cyprus	0.002	0.00	Timor-Leste	0.007	0.01

Continued

Country	SOC Stock (Pg)	Global Share (%)	Country	SOC Stock (Pg)	Global Share (%)
Oman	0.001	0.00	Togo	0.024	0.02
Pakistan	0.222	0.22	Tonga	<0.001	<0.001
Palestina	0.006	0.01	Trinidad and Tobago	0.001	0.00
Panama	0.124	0.12	Tunisia	0.043	0.04
Papua New Guinea	1.150	1.15	Turkey	1.551	1.55
Paraguay	0.115	0.11	Turkmenistan	0.099	0.10
Peru	0.109	0.11	Uganda	0.182	0.18
Philippines	0.741	0.74	Ukraine	2.229	2.22
Poland	1.013	1.01	United Arab Emirates	<0.001	<0.001
United Kingdom	0.469	0.47	Vietnam	0.062	0.06
United States	5.302	5.29	Virgin Islands, United States	<0.001	<0.001
Uruguay	0.385	0.38	Yemen	0.009	0.01
Uzbekistan	0.174	0.17	Zambia	0.027	0.03
Vanuatu	0.005	0.01	Zimbabwe	0.047	0.05
Venezuela	0.499	0.50			

REFERENCES

Abdul-Wahab, S.A., Bakheit, C.S., Al-Alawi, S.M., 2005. Principal component and multiple regression analysis in modelling of ground-level ozone and factors affecting its concentrations. Environ. Modell. Softw. 20 (10), 1263–1271.

Ballabio, C., Panagos, P., Montanarella, L., 2016. Mapping topsoil physical properties at European scale using the LUCAS database. Geoderma 261, 110–123.

Banwart, S., Bernasconi, S.M., Bloem, J., Blum, W., Brandao, M., Brantley, S., Chabaux, F., Duffy, C., Kram, P., Lair, G., Lundin, L., Nikolaidis, N., Novak, M., Panagos, P., Ragnarsdottir, K.V., Reynolds, B., Rousseva, S., de Ruiter, P., van Gaans, P., van Riemsdijk, W., White, T., Zhang, B., 2011. Soil processes and func tions in critical zone observatories: hypotheses and experimental design. Vadose Zone J. 10 (3), 974–987.

Barančíková, G., Makovníková, J., Skalský, R., et al., 2013. Changes in organic carbon pool in agricultural soils and its different development in individual agro-climatic regions of Slovakia. Agriculture 59 (1), 9–20. http://dx.doi.org/10.2478/agri-2013-0002. Retrieved 19 Sep. 2016.

Bontemps, S., Defourny, P., Van Bogaert, E., Arino, O., Kalogirou, V., Perez, J.R., 2011. GLOBCOVER 2009—product description and validation report.

Coleman, K., Jenkinson, D.S., 1996. Rothc-26.3—a model for the turnover of carbon in soil. In: Powlson, D.S., Smith, P., Smith, J.U. (Eds.), Evaluation of Soil Organic Matter Models Using Existing, Long-Term Datasets. In: NATO ASI Series I vol. 38. Springer-Verlag, Heidelberg, Germany, pp. 237–246.

Copernicus Global Land Service, 2015. NDVI, Global 10-Daily Spot-Vegetation Normalized Difference Vegetation Index. European Earth Observation Programme. http://land.copernicus.vgt.vito.be/(accessed 04/2016).

EC, 2008. Review of existing information on the interrelations between soil and climate change (CLIMSOIL). Final Report. Contract number 70307/2007/486157/SER/B1:208, In: Schils, R.E. (Ed.), European Commission, Page, Brussels, Belgium.

EC, 2010. Analysis of Options to Move Beyond 20% Greenhouse Gas Emission Reductions and Assessing the Risk of Carbon Leakage. European Commission, Brussels. COM (2010) 265 final.

Environmental Systems Research Inc (ESRI), 2002. World Continents: ESRI Data & Maps 2002. Environmental Systems Research Institute, Inc. (ESRI), Redlands, CA.

ESA, Universite' Catholique de Louvain, 2010. GlobCover 2009 (Global Land Cover Map). European Spatial Agency (ESA). http://due.esrin.esa.int/page_globcover.php. accessed 06/2016.

FAO/IIASA/ISRIC/ISSCAS/JRC, 2012. Harmonized World Soil Database (version 1.2). FAO and IIASA, Rome, Italy and Laxenburg, Austria.

Govers, G., Merckx, R., Van Oost, K., van Wesemael, B., 2013. Managing Soil Organic Carbon for Global Benefits: A STAP Technical Report. Global Environment Facility, Washington, DC.

Gregory, A.S., Kirk, G.J.D., Keay, C.A., Rawlins, B.G., Wallace, P., Whitmore, A.P., 2014. An assessment of subsoil organic carbon stocks in England and Wales. Soil Use Manage. 30 (1), 10–22.

Hengl, T., Heuvelink, G.B.M., Stein, A., 2004. A generic framework for spatial prediction of soil variables based on regression-kriging. Geoderma 120, 75–93.

Hijmans, R.J., Cameron, S.E., Parra, J.L., Jones, P.G., Jarvis, A., 2005. Very high resolution interpolated climate surfaces for global land areas. Int. J. Climatol. 25, 1965–1978.

Jarvis, A., Reuter, H.I., Nelson, A., Guevara, E., 2008. Hole-Filled SRTM for the Globe Version 4, Available From the CGIAR-CSI SRTM 90 m Database. http://srtm.csi.cgiar.org.

Jobbagy, E.G., Jackson, R.B., 2000. The vertical distribution of soil organic carbon and its relation to climate and vegetation. Ecol. Appl. 10, 423–436.

Jones, R.J.A., Hiederer, R., Rusco, E., Montanarella, L., 2005. Estimating organic carbon in the soils of Europe for policy support. Eur. J. Soil Sci. 56 (5), 655–671.

Jones, A., Panagos, P., Barcelo, S., Bouraoui, F., Bosco, C., Dewitte, O., Gardi, C., Hervás, J., Hiederer, R., Jeffery, S., Montanarella, L., Penizek, V., Tóth, G., Van Den Eeckhaut, M., Van Liedekerke, M., Verheijen, F., Yigini, Y., 2012. The State of Soil in Europe—a contribution of the JRC to the European Environment Agency's Environment State and Outlook Report—SOER 2010. No. EUR 25186 EN in EUR—Scientific and Technical Research Series, Publications Office of the European Union, Luxembourg.

Kern, J.S., 1994. Spatial patterns of soil organic carbon in the contiguous United States. Soil Sci. Am. J. 58 (1994), 439–455.

Khalil, M.I., Kiely, G., O'Brien, P., Müller, C., 2013. Organic carbon stocks in agricultural soils in Ireland using combined empirical and GIS approaches. Geoderma 0016-7061. 193–194, 222–235. http://dx.doi.org/10.1016/j.geoderma.2012.10.005.

Köchy, M., Hiederer, R., Freibauer, A., 2015. Global distribution of soil organic carbon—part 1: masses and frequency distributions of SOC stocks for the tropics, permafrost regions, wetlands, and the world. SOIL 1, 351–365. http://dx.doi.org/10.5194/soil-1-351-2015.

Kunkel, M.L., Flores, A.N., Smith, T.J., McNamara, J.P., Benner, S.G., 2011. A simplified approach for estimating soil carbon and nitrogen stocks in semi-arid complex terrain. Geoderma 0016-7061165 (1), 1–11. http://dx.doi.org/10.1016/j.geoderma.2011.06.011.

Lugato, E., Panagos, P., Bampa, F., Jones, A., Montanarella, L., 2014a. A new baseline of organic carbon stock in European agricultural soils using a modelling approach. Glob. Chang. Biol. 20 (1), 313–326. http://dx.doi.org/10.1111/gcb.12292.

Lugato, E., Bampa, F., Panagos, P., Montanarella, L., Jones, A., 2014b. Potential carbon sequestration of European arable soils estimated by modelling a comprehensive set of management practices. Glob. Chang. Biol. 20 (11), 3557–3567.

Martino, L., Fritz, M., 2008. New Insight Into Land Cover and Land Use in Europe, Statistics in Focus, 33. Eurostat, Luxembourg.

Montanarella, L., Panagos, P., 2015. Policy relevance of critical zone science. Land Use Policy 49, 86–91.

Panagos, P., Ballabio, C., Yigini, Y., Dunbar, M., 2013a. Estimating the soil organic carbon content for European NUTS2 regions based on LUCAS data collection. Sci. Total Environ. 442, 235–246.

Panagos, P., Hiederer, R., Van Liedekerke, M., Bampa, F., 2013b. Estimating soil organic carbon in Europe based on data collected through an European network. Ecol. Indic. 24, 439–450.

Scharlemann, P.W.J., Tanner, V.J.E., Hiederer, R., Kapos, V., 2014. Global soil carbon: understanding and managing the largest terrestrial carbon pool. Carbon Manag. 5 (1), 81–91. http://dx.doi.org/10.4155/cmt.13.77.

Smith, J., Smith, P., Wattenbach, M., Zaehle, S., Hiederer, R., Jones, R.J.A., Montanarella, L., Rounsevell, M.D.A., Reginster, I., Ewert, F., 2005. Projected changes in mineral soil carbon of European croplands and grasslands, 1990–2080. Glob. Chang. Biol. 11, 2141–2152. http://dx.doi.org/10.1111/j.1365-2486.2005.001075.x.

Toth, G., Jones, A., Montanarella, L., 2013a. The LUCAS topsoil database and derived information on the regional variability of cropland topsoil properties in the European Union. Environ. Monit. Assess. 185 (9), 7409–7425.

Toth, G., Jones, A., Montanarella, L., Alewell, C., Ballabio, C., Carre, F., De Brogniez, D., Guicharnaud, R.A., Gardi, C., Hermann, T., Meusburger, K., Nocita, M., Panagos, P., Rusco, E., Stevens, A., Van Liedekerke, M., Van Wesemael, B., Weynants, M.M.A., Yigini, Y., 2013b. LUCAS Topsoil Survey—methodology, data and results: JRC Technical Reports. ISBN: 978-92-79-32542-7.

Yigini, Y., Panagos, P., 2016. Assessment of soil organic carbon stocks under future climate and land cover changes in Europe. Sci. Total Environ. 557–558, 838–850.

INDEX

Note: Page numbers followed by "*f*" indicate figures, and "*t*" indicate tables.

L

M

P

Q

R

FSC
www.fsc.org
MIX
Papier | Fördert
gute Waldnutzung
FSC® C083411

Kolibri 360 GmbH
Ferdinand-Jühlke-Straße 7
99095 Erfurt, Deutschland
produktsicherheit@kolibri360.de